# INVERTEBRATE ZOOLOGY



**FATIK BARAN MANDAL**

Associate Professor
Department of Zoology
Bankura Christian College
Bankura, West Bengal

**PHI Learning Private Limited**

New Delhi-110001
2012

₹ 475.00

**INVERTEBRATE ZOOLOGY**
Fatik Baran Mandal



**ISBN-978-81-203-4615-4**



Published by Asoke K. Ghosh, PHI Learning Private Limited, M-97, Connaught Circus, New Delhi-110001 and Printed by Rajkamal Electric Press, Plot No. 2, Phase IV, HSIDC, Kundli-131028, Sonepat, Haryana.

***To***

My Teachers

# Contents


*Preface* .......... *xiii*

*Acknowledgements* .......... *xv*

**Chapter 1 INTRODUCTION .......... 1–10**

1.1 General Characteristics of Animals .......... 1
1.2 The Living World .......... 2
1.3 Invertebrates .......... 3
1.4 Invertebrate's Habitat .......... 4
1.5 Tissues .......... 5
1.6 Organ Systems .......... 5
1.7 Digestive System .......... 5
1.8 Circulatory System .......... 6
1.9 Respiratory System .......... 6
1.10 Excretory System .......... 6
1.11 Locomotor System .......... 7
1.12 Nervous and Sensory System .......... 7
1.13 Reproductive System .......... 8
1.14 Origins of Major Groups of Metazoa .......... 8
1.15 Origin of the Coelomic Condition .......... 9
*Review Questions* .......... *10*

**Chapter 2 EVOLUTION AND TAXONOMY .......... 11–21**

2.1 Brief History of Life .......... 11
2.2 Geological Time Scale .......... 12
2.3 Speciation .......... 12
2.4 Extinction .......... 14
2.5 Taxonomy .......... 15
2.6 Species Concept .......... 15
2.7 Various Approaches of Taxonomy .......... 17
2.8 Isolating Mechanisms .......... 18
2.9 Zoological Nomenclature .......... 18
2.10 International Code of Zoological Nomenclature .......... 19
2.11 Classification .......... 20
2.12 Taxonomic Hierarchy .......... 20
*Review Questions* .......... *21*

Chapter 3 **SUBKINGDOM PROTOZOA** ........ **22–61**

3.1 Protozoan Polymorphism ........ 23
3.2 Features at a Glance ........ 23
3.3 Metazoan Tendencies ........ 24
3.4 Habitats ........ 24
3.5 Food Habit ........ 24
3.6 Support and Locomotion ........ 24
3.7 Nutrition ........ 25
3.8 Reproduction ........ 25
3.9 Classification with Characters ........ 26
3.10 Classification of Phylum Sarcomastigophora ........ 27
3.11 Classification of Phylum Ciliophora ........ 29
3.12 Type Study: *Plasmodium vivax* ........ 31
3.13 *Trypanosoma cruzi* ........ 35
3.14 *Trypanosoma brucei rhodesiense and T. b. gambiense* ........ 37
3.15 *Leishmania* ........ 39
3.16 *Entamoeba histolytica* ........ 42
3.17 Geologic History and Uses of Foraminiferan ........ 45
3.18 Locomotion in Protozoa ........ 46
3.19 Flagella ........ 49
3.20 Locomotion in *Paramoecium* ........ 51
3.21 Ciliary Locomotion ........ 52
3.22 Feeding and Digestion in *Amoeba* ........ 54
3.23 Excretory System in *Paramoecium* ........ 55
3.24 Reproduction in *Paramoecium* ........ 57
*Review Questions* ........ *59*

Chapter 4 **PHYLUM PORIFERA** ........ **62–81**

4.1 General Characteristics ........ 62
4.2 Evolution and Fossil Records ........ 63
4.3 Size, Structure and Colour ........ 63
4.4 Distribution ........ 64
4.5 Classification with Characters ........ 64
4.6 Body Structure and Cell Types ........ 66
4.7 Support ........ 68
4.8 Skeleton ........ 68
4.9 Spicules ........ 68
4.10 Nutrition, Excretion and Gas Exchange ........ 71
4.11 Morphological Types and Water-Current System ........ 72
4.12 Key Points of Canal System in Sponges ........ 73
4.13 Reproduction ........ 75
4.14 *Sycon* sp. ........ 77
4.15 Economic Importance ........ 81
*Review Questions* ........ *81*

Chapter 5 **PHYLUM CNIDARIA** ........ **82–110**

5.1 General Characteristics ........ 82
5.2 Habitat ........ 83
5.3 Tissues and Organs ........ 83
5.4 Classification with Characters ........ 84

5.5 The Body Wall ........ 86
5.6 The Polypoid Form ........ 87
5.7 The Medusoid Form ........ 90
5.8 Support ........ 90
5.9 Movement ........ 91
5.10 Cnidae ........ 92
5.11 Type Study ........ 94
5.12 Anthozoa ........ 97
5.13 *Aurelia* ........ 97
5.14 Polymorphism in Siphonophora ........ 101
5.14.1 Origin of Polymorphism ........ 102
5.15 Coral Reefs ........ 104
5.15.1 Kinds of Coral Reefs ........ 105
*Review Questions* ........ *110*

Chapter 6 **PHYLUM CTENOPHORA ........ 111–117**

6.1 Support and Locomotion ........ 111
6.2 Feeding and Digestion ........ 112
6.3 Circulation, Excretion, Gas Exchange and Osmoregulation ........ 112
6.4 Sense Organ ........ 112
6.5 Reproduction ........ 112
6.6 Classification with Characters ........ 113
6.7 Oceanic Ctenophores ........ 114
6.8 Benthic Ctenophores ........ 114
6.9 Bioluminescence ........ 114
6.10 Ctenophores: Role in the Food Web ........ 114
6.11 Ctenophores and Medusae ........ 115
6.12 *Pleurobrachia pileus* ........ 115
*Review Questions* ........ *117*

Chapter 7 **PHYLUM ROTIFERA ........ 118–127**

7.1 Characters of Rotifers ........ 118
7.2 Habitat ........ 119
7.3 Shape, Size and Colouration ........ 119
7.4 External Anatomy ........ 119
7.5 Body Wall, Body Cavity, Support and Locomotion ........ 121
7.6 Corona ........ 121
7.7 Body Wall and Pseudocoel ........ 122
7.8 Digestive System ........ 122
7.9 Feeding and Digestion ........ 124
7.10 Circulation, Gas Exchange and Excretion ........ 125
7.11 Nervous System and Sense Organs ........ 125
7.12 Reproductive System ........ 126
*Review Questions* ........ *127*

Chapter 8 **PHYLUM PHORONIDA ........ 128–133**

8.1 Habit and Habitat ........ 128
8.2 Distinctive Characters ........ 128
8.3 External Anatomy ........ 129
8.4 Internal Structures ........ 130
8.5 Development ........ 131

8.6 Actinotrocha Larva ........ 131
8.7 Affinities ........ 132
*Review Questions* ........ *133*

**Chapter 9 PHYLUM ECTOPROCTA ........ 134–142**

9.1 Classification with Characters ........ 134
9.2 Habitat ........ 135
9.3 Shape, Size and Colouration (Colony Forms) ........ 135
9.4 External Structure ........ 135
9.5 Body Wall, Coelom, Muscles and Movement ........ 137
9.6 Lophophore, Feeding and Digestion ........ 138
9.7 Circulation, Gas Exchange and Excretion ........ 139
9.8 Nervous System and Sense Organs ........ 139
9.9 Reproduction and Development ........ 139
9.10 Affinities ........ 141
*Review Questions* ........ *142*

**Chapter 10 PHYLUM PLATYHELMINTHES ........ 143–161**

10.1 General Characteristics ........ 143
10.2 Classification with Characters ........ 144
10.3 *Fasciola* ........ 145
10.4 Tapeworms ........ 151
10.4.1 *Taenia solium* ........ 152
10.4.2 *Taenia saginata* ........ 153
10.4.3 *Diphyllobothrium latum* ........ 154
10.4.4 *Hymenolepis nana* ........ 154
10.4.5 *H. diminuta* ........ 155
10.4.6 *Diphylidium caninum* ........ 155
10.4.7 *Echinococcus granulosus* ........ 156
10.4.8 *Clonorchis sinensis* ........ 157
10.4.9 *Opisthorchis spp* ........ 157
10.4.10 *Spirometra mansonoides* ........ 158
10.4.11 *Paragonimus westermanni* ........ 159
10.4.12 Schistosomes ........ 159
10.5 General Notes ........ 160
*Review Questions* ........ *161*

**Chapter 11 PHYLUM ASCHELMINTHES ........ 162–189**

11.1 General Characters ........ 162
11.2 Class Kinorhyncha ........ 162
11.3 Class Gastrotricha ........ 163
11.4 Class Nematomorpha ........ 163
11.5 Class Nematoda ........ 163
11.6 Characters of the Class Nematoda ........ 163
11.7 Morphology ........ 164
11.8 Muscles ........ 164
11.9 Digestive System ........ 164
11.10 Nervous System ........ 165
11.11 Excretory System ........ 165
11.12 Reproductive System ........ 165

11.13 Life Cycle .......... 167
11.14 Intestinal Nematodes .......... 167
11.14.1 *Trichuris trichiura* .......... 167
11.14.2 *Trichinella spiralis* .......... 168
11.14.3 *Strongyloides stercoralis* .......... 170
11.14.4 Hookworms .......... 170
11.14.5 *Ancylostoma duodenale* .......... 171
11.14.6 *Necator americanus* .......... 174
11.14.7 *Angiostrongylus cantonensis* .......... 175
11.14.8 *Angiostrongylus costaricensis* .......... 175
11.14.9 Anisakiasis .......... 175
11.14.10 *Gnathostoma spinigerum* .......... 176
11.14.11 *Ascaris lumbricoides* .......... 177
11.14.12 *Enterobius vermicularis* .......... 179
11.14.13 Filariasis .......... 180
11.14.14 *Wuchereria bancrofti* .......... 181
11.14.15 *Brugia malayi* .......... 183
11.14.16 *Brugia timori* .......... 185
11.14.17 *Loa loa* .......... 185
11.14.18 *Onchocerca volvulus* .......... 186
11.14.19 *Mansonella* .......... 186
11.14.20 *Mansonella perstans* .......... 187
11.14.21 *M. ozzardi* .......... 187
11.14.22 *Dracunculus medinensis* .......... 187
*Review Questions* .......... *188*

Chapter 12 **PHYLUM ANNELIDA .......... 190–214**

12.1 General Characters .......... 191
12.2 Classification of the Annelida .......... 191
12.3 Earthworm .......... 193
12.4 Locomotion in *Nereis* .......... 210
12.5 Nephridia in Annelids .......... 213
*Review Questions* .......... *214*

Chapter 13 **PHYLUM ARTHROPODA .......... 215–266**

13.1 General Characteristics .......... 215
13.2 Body Wall .......... 216
13.3 Support and Locomotion .......... 216
13.4 Circulation and Gas Exchange .......... 217
13.5 Excretion and Osmoregulation .......... 217
13.6 Nervous System and Sense Organs .......... 217
13.7 Classification with Characters .......... 218
13.8 Class Crustacea .......... 221
13.9 *Limulus polyphemus* .......... 224
13.10 Insecta .......... 227
13.11 *Periplaneta americana* .......... 233
13.12 *Peripatus* .......... 243
13.13 Affinities .......... 248
13.14 Excretory System of Prawn .......... 249
13.14.1 Antennary Gland, or Green Gland .......... 249

13.14.2 Lateral Excretory Ducts ... 249
13.14.3 Renal Sac ... 250
13.15 Mosquitoes ... 250
13.16 *Aedes* ... 253
13.16.1 *Aedes aegypti* ... 253
13.17 *Culex quinquefasciatus* ... 254
13.18 *Anopheles* ... 255
13.19 Respiration in *Palaemon* ... 257
13.20 Respiration in Scorpion ... 259
13.21 Larval Forms of Crustaceans ... 259
13.22 Ommatidia in Prawn and Cockroach ... 262
*Review Questions* ... *265*

Chapter 14 **PHYLUM MOLLUSCA ... 267–291**
14.1 General Characters ... 267
14.2 General Features ... 268
14.3 Classification with Characters ... 271
14.4 *Pila* ... 274
14.5 Respiratory system in *Unio* ... 285
14.6 Torsion in Gastropoda ... 286
14.7 Foot in Mollusca ... 287
*Review Questions* ... *290*

Chapter 15 **PHYLUM ECHINODERMATA ... 292–315**
15.1 Aboral Surface ... 292
15.2 Oral Surface ... 292
15.3 Endoskeleton ... 293
15.4 Section of an Arm ... 293
15.5 Body Wall and Coelom ... 294
15.6 Support and Locomotion ... 295
15.7 General Characters ... 295
15.8 *Asterias* ... 295
15.9 Echinoderm Larvae ... 307
15.10 Classification with Characters ... 310
*Review Questions* ... *315*

Chapter 16 **PHYLUM HEMICHORDATA ... 316–327**
16.1 Characters of Hemichordata ... 316
16.2 Characters of Enteropneusta ... 317
16.3 Characters of Pterobranchia ... 317
16.4 Body Wall, Support and Locomotion ... 317
16.5 Feeding and Digestion ... 318
16.6 Circulation, Gas Exchange and Excretion ... 319
16.7 Nervous System and Sense Organs ... 319
16.8 Reproduction and Development ... 319
16.9 *Balanoglossus* ... 320
16.10 Affinities ... 326
*Review Questions* ... *327*

Chapter 17 **NEED FOR STUDYING AND CONSERVING INVERTEBRATES** ........ **328–339**

17.1 Protozoa ........ 329
17.2 Porifera ........ 329
17.3 Cnidaria ........ 330
17.4 Platyhelminthes ........ 331
17.5 Annelida ........ 332
17.6 Mollusca ........ 332
17.7 Arthropoda ........ 335
17.8 Echinodermata ........ 338
*Review Questions* ........ *339*

***Glossary*** ........ ***341–356***

***Suggested Reading*** ........ ***357***

***Index*** ........ ***359–368***

# Preface

A wealth of information on invertebrate zoology is available. However, information providers are mostly the foreign authors and foreign publishers. There is no doubt that such information are source for the furtherance of the subject. However, in our country, the mother tongue of the people is not English. So, English as a language is not easily accessible to the undergraduate students of our country. The field of invertebrate zoology is vast and the information are mostly scattered.

Textbooks in the above mentioned field for students of India appear to be a few. Thus, a consolidated account of invertebrate zoology would have significance for the students. Each book has its own strength and weakness. However, there is a gap between the information required by the students and the information readily available to them.

Invertebrate diversity is vast. It is difficult to acquire knowledge from such vast diversity without systematic study of the invertebrates. However, knowledge of invertebrates is fundamental to zoology. This book aims to present the classical information about the invertebrates in a lucid manner covering the syllabus of undergraduate and postgraduate zoology courses prescribed in the universities of our country. While writing this book, student-friendly approach has been emphasized and care has been taken to the scientific accuracy.

I welcome constructive suggestion for further improvement of the book.

**Fatik Baran Mandal**

# Acknowledgements

Invertebrate zoology covers a vast area of the subject and is very fundamental in starting the study of animal life, their diversity in morphology, physiology, anatomy and in many other aspects. It also provides a picture of attainment of gradual complexity of life forms. Writing a book of such nature is a difficult task. I would like to express my gratitude to those who encouraged me in this endeavour. I am particularly thankful to my teacher, Mr. Kartick Sinha and my student, Mr. Sourav Baral for their assistance.

I would like to express my appreciation to my family members for their support without which this work would not have been completed. Finally, I would like to extend my sincere thanks to the Publisher, PHI Learning, New Delhi, for extending his every support in finalizing the book. Thanks are also due to the editorial staff members, especially Ms. Babita Misra, Editorial Co-ordinator, for her suggestions. I would also like to convey my sincere thanks to Mr. Jyoti Kumar (Copy Editor) and Mr. Mayur Joseph for their tolerance and hard work without which this book would not have been published timely.

**Fatik Baran Mandal**

CHAPTER 1

# Introduction

Zoology, a branch of life science with its roots in the prehistoric time, deals with all the biological aspects of the animals. Early Egyptians had the significant knowledge about the animals. The Greek philosophers of 5th and 6th centuries speculated on the origin of animals. One earliest animal classification, recorded in a Greek medical book, classifies animals primarily based on whether they are edible or inedible. *Historia animalium*, written by Aristotle (384–322 B.C. Figure 1.1), deals with the animals of Greece and nearby regions of the Asia Minor. Zoology mainly deals with the animal diversity, their structure, behaviour, reproduction, development, origin and distribution, interrelationship with environment, along with their significance in nature and to human life. Zoology includes both a tremendous body of facts and theories about animals and means for learning more.

The ultimate goal of zoology is to reduce the apparent complexity of the animal life into the simple fundamental ideas, to discover facts and relationship among them. Zoology is basic in many ways to the fields of human and veterinary medicine, public health, agriculture, conservation biology, and to some other social sciences. Zoology including the Invertebrate and Vertebrate Zoology has become a dynamic and increasingly important subject with the fascinating and important new insights into the various aspects of animal's life.

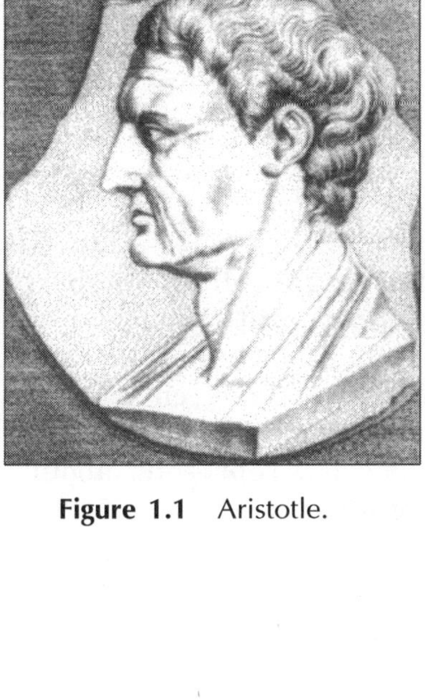

**Figure 1.1** Aristotle.

Zoology is taught on a wide range of animals from the microscopic protozoa, through the crustaceans and arthropods to fish, amphibians, reptiles, birds and mammals. Diversity in animals is vast. Acquiring scientific knowledge about this diversity is a difficult task. On the other hand, similarity among animals is also common, although animals are not united into groups at random but following a system. In the present endeavour, attempt has been made to provide information on animal life in a very simple way. Recently, *Balanoglossus* has been shown to possess characters, which indicate the relationships with the chordata, even though they lack a vertebral column. One of these characters is the presence of a notochord. Technically, the terms *chordate* and nonchordate have the sound basis for distinguishing the higher group of animals from all the lower groups, but the widespread and popular acceptance of the terms *vertebrate* and *invertebrate* is still continuing.

## 1.1 GENERAL CHARACTERISTICS OF ANIMALS

1. Generally multicellular, except the subkingdom protozoa, with tissues arranged into organs and organs into systems

2. Heterotrophic
3. Require oxygen for the aerobic respiration
4. Reproduce sexually or asexually, or both
5. Mostly motile in at least during part of their life cycle
6. Life cycle involves a period of embryonic development
7. Germ tissue layers produce the adult organs through differentiation
8. Animals which possess backbones are known as *vertebrates* and those lack a backbone are known as *invertebrates*
9. Presence of radial, biradial or bilateral symmetry
10. Food is digested in the gut
11. Complete digestive tracts have two openings—the mouth and the anus
12. Coelom, derived from the mesoderm, lies between the gut and the body wall
13. Flatworms do not have a coelom. Their body is packed solidly with the tissue between the gut and the body wall
14. Pseudocoelm not lined with the peritoneum is found in the flatworms
15. Segmented animals are composed of repeating body units or metameres
16. Segments may be grouped, and modified for performing the specialized tasks

## 1.2 THE LIVING WORLD

The living world is broadly divided into two categories—the prokaryotes and the eukaryotes. In the prokaryotes membrane-enclosed organelles, nucleus and linear chromosomes are not found. The eukaryotes possess membrane-bound organelles, nucleus and linear chromosomes. Prokaryotes are divided into two distinct groups—Eubacteria and Archaea (Archaebacteria). Eubacteria corresponds to traditional bacteria. Archaea resembles Eubacteria and is unique in their genetic and metabolic characteristics. Archaea differs from the Eubacteria and eukaryotes in their ribosome, cell wall, and lipid profile of the plasma membrane. Some Eubacteria can perform chlorophyll-based photosynthesis while the Archaea cannot. The genetic differences among prokaryotes are vast than those found in eukaryotes. Prokaryotes appear to rule on earth for at least 2 billion years before the appearance of modern eukaryotic cell. Some 4,000 species of prokaryotes are described with an estimated 1 to 3 million species that are yet to be described.

### The Prokaryotes

In the true bacteria including Cyanobacteria and spirochetes membrane-enclosed organelles or nuclei, or a cytoskeleton are absent. Some of them use chlorophyll for performing photosynthesis and are provided with peptidoglycan in cell wall and a single known *RNA polymerase*.

#### *Kingdom: Archaea (Archaebacteria)*

This kingdom includes anaerobic or aerobic, mostly methane-producing microorganisms and lacks membrane-enclosed organelles or nuclei, or a cytoskeleton. None uses chlorophyll for photosynthesis and lack peptidoglycan in cell wall although they are provided with several RNA polymerases.

### The Eukaryotes

Eukaryote cells are provided with a variety of membrane-enclosed organelles and with a membrane-enclosed nucleus, with structural support from a cytoskeleton.

#### *Kingdom: Fungi*

Fungi constitute probably a monophyletic group including molds, mushrooms, yeasts, which are heterotrophic and multicellular. The earliest fossil record of fungi goes back to the Middle Ordovician. The 72,000 described species represent 5–10% of the species diversity.

#### *Kingdom: Planate*

Photosynthetic, autotrophic, multicellular plants including some groups of algae, bryophytes and vascular plants constitute this kingdom. The described species represent about 50% of the earth's plant diversity.

#### *Kingdom: Protista*

Eukaryotic, single-celled microorganisms and certain algae constitute this polyphyletic group of perhaps 18 phyla, including green algae, diatoms and some brown algae, ciliates, dinoflagellates, foraminiferans, amoebae and others. The 80,000 described species represent about 10% of the protist diversity.

***Kingdom: Animalia***

This is a monophyletic taxon containing multicellular animals of heterotrophic, multicellular organisms. About 1.3 million living species is described. The number of undescribed species range between 10–30 million to 100–200 million. Three main divisions of life forms viz., Eubacteria, Archaea and Eukaryota are recognized at new taxonomic level called the *domains*. The eukaryote is a natural group as defined by their unique trait of a nucleus and linear chromosomes, whereas Eubacteria and Archaea may not be the natural groups. Two eukaryotic kingdoms are the Animalia (Metazoa) and certain 'animal-like' protist phyla loosely termed *protozoa*. The members of Animalia or Metazoa possess an acetylcholine/cholinesterase-based nervous system, special types of cell-cell junctions, and a unique family of connective tissue proteins called *collagens*. Over a million species of living animals are described. Among the metazoans, those possess a backbone are known as *vertebrata* and those do not possess a backbone are known as *invertebrata*. Thus, the division of animals into invertebrates and vertebrates is based mostly on tradition and convenience. About 10,000 to 13,000 new species are named and described each year mostly from invertebrates.

## 1.3 INVERTEBRATES

The present diversity of invertebrates is the result of billions of years of evolution. The presence of prokaryotic organisms in some oldest sediment on earth suggests that life first appeared in the seas when the planet cooled and was suitable for the existence of life. The first eukaryotic cells estimated to be appeared 2.7 billion years ago. Molecular clock data suggest that the last common ancestor of plants and animals existed on earth about 1.6 billion years ago. The ancestors of both plants and animals were almost definitely the protists. Representatives of nearly all the extant animal phyla were present early in the Paleozoic era. Life on land did not appear until fairly recently and terrestrial radiations began only about 470 million years ago Figure 1.2(a)–(e).

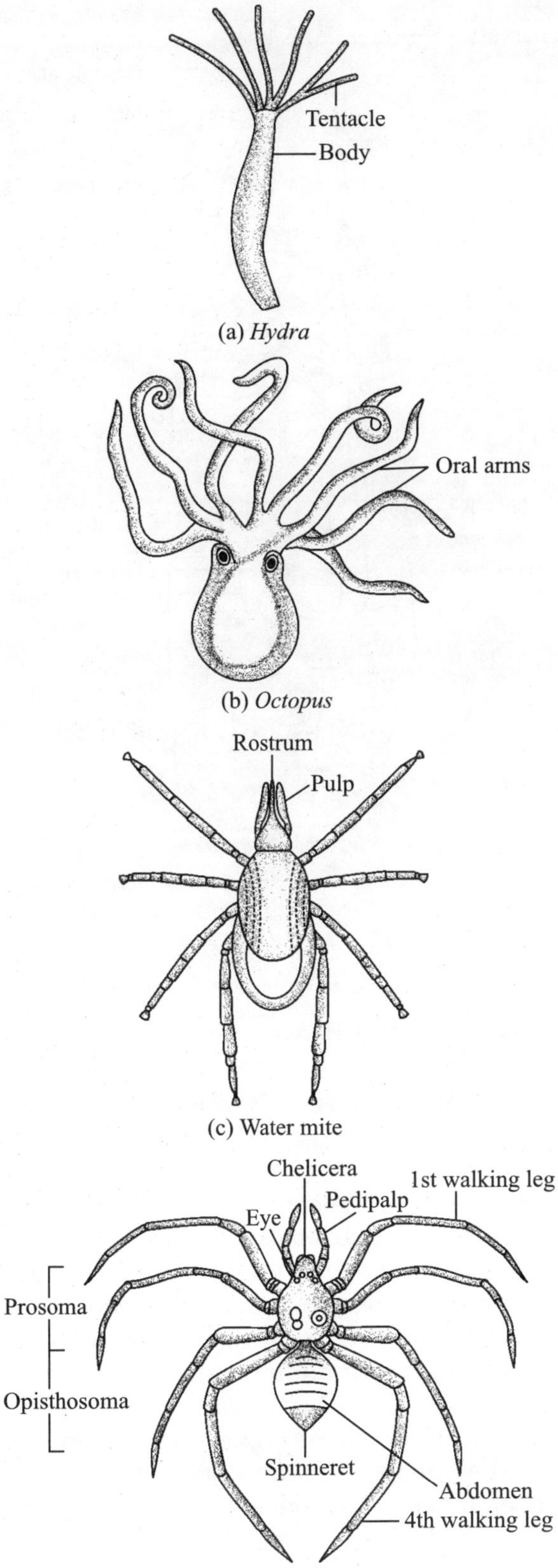


(a) *Hydra*

(b) *Octopus*

(c) Water mite

(d) Spider

**Figure 1.2** (*Contd.*)

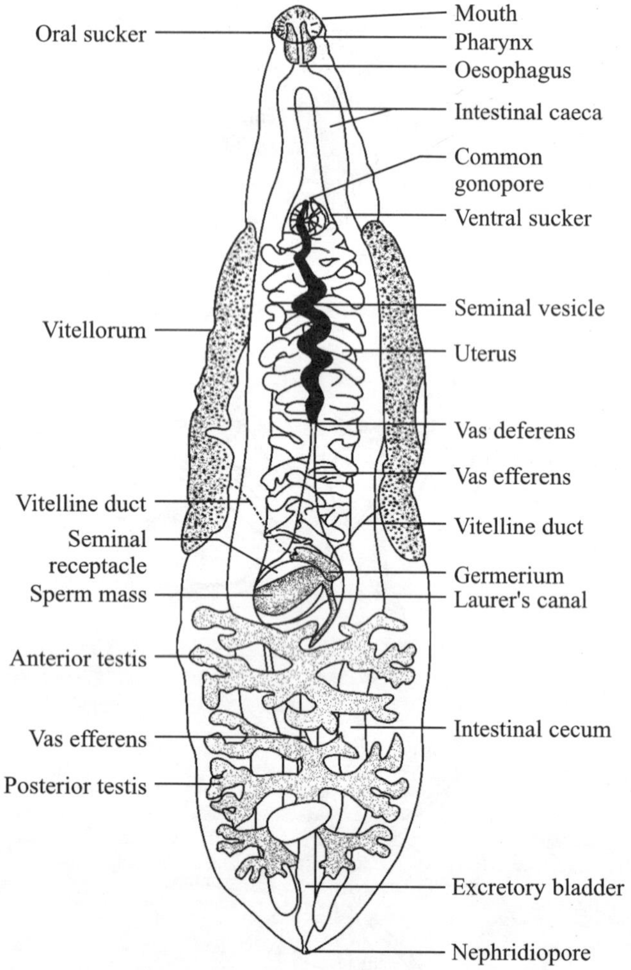


(e) Ventral view of *Opisthorchis sinensis*

**Figure 1.2** Glimpses of diversity in invertebrates.

## 1.4 INVERTEBRATE'S HABITAT

### *Marine habitat*

Salt water covers 71% of earth's surface. The seas provide 99% habitable space for living organisms. The diversification of invertebrates started in late Proterozoic and early Cambrian shallow seas. Marine environment provides protection to the living organisms by minimizing the physical and the chemical stresses on organisms and harbour the largest diversity of higher taxa. Some Phyla such as Echinodermata, Sipuncula, Chaetognatha and Ctenophora are exclusively marine. Besides, the high productivity of the oceans contributes to the high diversity of marine animal largely. Water is a very efficient thermal buffer. It is slow to heat up or cool down. Of about 1,300,000 estimated described species of *Animalia* excluding protists, about 1,250,000 (96%) are invertebrates.

Oceans absorb and lose vast amounts of heat with slight change in water temperature. Oceanic temperatures are stable in comparison to the freshwater and terrestrial environments. Short-term temperature extremes often occur in intertidal habitats and estuarine habitats. Invertebrates living there have to adapt temperature changes. The salinity of sea water averages about 3.5% giving it a high density and buoyancy, thereby minimizing energy expenditures for floatation. The total ionic concentration in sea water is more or less similar to the body fluids of most animals. The pH of sea water is mostly stable in most of the ocean. The naturally occurring carbonate compounds buffer sea water at about pH 7.5–8.5. In shallow and near shore water, carbon dioxide, various nutrients and sunlight are available in sufficient quantities to allow photosynthesis at high level. Benthic animals live on the surface of the substratum or burrow within soft substrata.

### *Estuaries and coastal marshlands*

Animals inhabiting these areas suffer from lower oxygen level, generally migrate to regions of higher oxygen levels, store oxygen in certain body fluid pigments, or sometimes switch temporarily to metabolic processes that do not require oxygen-based respiration. The inhabitants face stresses which results from the human activity. A large number of invertebrates adapted life in these dynamic environments.

### *Freshwater habitat*

Changes in temperature and other conditions in freshwater habitats such as ponds, streams and lakes occur quickly. Seasonal changes are the extreme, and may include complete freezing during winter and complete drying in summer. Ponds that hold water for only a few weeks during, and after rainy seasons are called *ephemeral pools*.

### *Terrestrial habitat*

Life on land is more rigorous than life in fresh water. In terrestrial environment, temperature extremes are encountered almost on a daily basis. Water balance is a critical problem. Water provides a medium for support, for dispersing gametes, larvae, and adults, and is a source of dissolved materials needed by animals. The number of invertebrate species occupying the above-said habitats are given in Table 1.1.

**Table 1.1** Approximate number of invertebrate species (modified from Science 1987)

| Phylum/ Suphylum/ Class/ Order/ | Approximate number of species |
|---|---|
| Protozoa | 260,000 |
| Porifera | 10,000 |
| Cnidaria | 10,000 |
| Echinodermata | 6,000 |
| Nematodes | 1,000,000 |
| Annelida | 15,000 |
| Brachiopoda | 350 |
| Bryozoa | 4,000 |
| Mollusca | 100,000 |
| Arthropods | |
| Crustacea | 39,000 |
| Chelicerata | |
| Arachnids | 63,000 |
| Uniramia | |
| Insecta | 1,000,000 |
| Coleoptera | 300,000 |
| Diptera | 85,000 |
| Lepidoptera | 110,000 |
| Hymenoptera | 110,000 |
| Hemiptera | 40,000 |

## 1.5 TISSUES

Four chief classes of tissues are epithelial tissue, connective tissue, muscle tissue and nervous tissue. Other kinds are recognizable but often these are modifications of one or more general classes. The nettling cells of a hydra or the pigment cells of a squid may be considered to be *special cells*. These special cells are referred as *epithelium* or a *connective tissue*. Both form and structure of an organism depend on the grouping of the component tissues. Thus, both morphology and physiology of the tissues and frequently the form and structure of the individual component cells are required to be studied.

## 1.6 ORGAN SYSTEMS

The individual cells in the many-celled animals become dependent one on another through the histological differentiation and individual tissue become interdependent. Tissues combined in group to form the organ, which are recognizable morphological units of the metazoan body. The tissues forming an organ cooperate in the performance of particular function. Not all tissues in an organ function to the same extent, but all kinds of tissues become the essential tissues which helps in carrying out functions of other tissues of the same organ.

In a digestive organ, the essential process of digestion is performed through the glandular epithelia, which form substances necessary for the digestion. The cooperative muscles, under whose action the content of the digestive tract is kept agitated, so that all food material meets the digestive fluids produced by the epithelia, aid the epithelia. When similar organs are grouped for the performance of the same or related functions, such a complex is termed *organ system*. All the activities, which characterize the living animals, are fundamentally interrelated and the numerous organs and tissues of the metazoan may be grouped under a relatively small number of systems. These systems have their interrelations and they function not as an independent unit but as a correlated whole of the individual.

## 1.7 DIGESTIVE SYSTEM

Transforming nonliving food substances into the protoplasm is the key of survival of the living organisms. The process of digestion of food is the first step in the transformation of food into the living matter. In some parasites, for example, tapeworms, food taking is restricted to the direct absorption of liquids. Organs of digestion are lacking in such instances, and the organs of the host perform the entire digestive process. They take the benefit from the host. In some protozoa and in a few metazoans, food substances are taken directly into the protoplasm. The ingested particles within the protoplasm remain as foreign substance to it until they have undergone changes, which render them capable of assimilation by the protoplasm. This type of digestion is termed *intracellular digestion*. Among the sponges and some flatworms, in addition to the protozoan, the entire digestion is intracellular. Ontogenetically, the archenteron of the gastrula stage represents the first organ specialized for digestion. In many animals, the larva on reaching the gastrula stage is an independent organism capable of self-maintenance. The food to be taken through the blastopore of gastrula is rendered liquid through the action of secretions formed by the endoderm cells. Thus, the gastrula cavity becomes a space within which intercellular digestion takes place and is termed an *archenteron* or *primitive digestive system*.

## 1.8 CIRCULATORY SYSTEM

Whether digestion is intercellular or intracellular, the products of digestion are directly available to only part of the cells. In the simple metazoa, the direct transfer of the materials accomplishes circulation from cell to cell. In the highly complicated organisms, such transfer is not possible. In such cases, a special system for the distribution of digested food has caused the appearance of a circulatory system. The gastrovascular system of cnidarians serves both the purpose of digestion and distribution. Trematodes and Turbellaria exhibit the condition of a combined gastrovascular system. Among the coelomate animals, digested food materials pass through the wall of digestive tract into coelomic cavities where it becomes recognizable as the body fluid. When the fluid remains continuously within the vessels, the system is termed closed type of circulatory system, but when in any part of its course, the fluid is emptied into the body cavity or into the sinuses, the system is designated as open type of circulatory system. The muscles become associated with at least part of the vessels, which then function as a pumping organ or the heart, the centre of function of the circulatory system. In many invertebrates, the conspicuous pumping organ is the vessel, which lies dorsal to the digestive tract, but hearts are often located in other regions of the body. The gill hearts of molluscs and the modifications of the circular vessels in the annelids are mentionable. In the molluscs, the heart becomes differentiated into various chambers. Those, which receive the blood, are termed *auricles* and those, which force the blood out into the vessels, are termed the *ventricles*. A three-chambered heart is found in the molluscs, while the fishes have a two-chambered heart.

## 1.9 RESPIRATORY SYSTEM

The energy manifested by a living organism results largely from the union of oxygen with the protoplasm and is contained in the food substances. The complicated organic compounds are broken down into the simpler waste substances and in the process transform the potential energy of the chemical compounds (food) into the kinetic energy of the living organism. Respiration is the process of carrying oxygen into living protoplasm and giving off carbon dioxide. This process is termed *aerobic respiration*. Some organisms obtain their energy without oxygen and are said to be *anaerobic*. Respiration is essential for the existence of every living cell. Since, the respiratory process involves an exchange of oxygen and carbon dioxide, any surface, which serves for this exchange, must be moist and delicate to permit a diffusion of the gases. In diploblastic animals such as cnidarians, conditions for respiration are not essentially different from those found in the protozoa, as every cell has a surface exposed to the water through which the gases diffuse. Even in some coelomate animals like earthworms, the body surface provides an area sufficient for the respiratory exchange. The body fluid plays an important part in respiration as it absorbs or loosely combines the gases within the body. In many animals, the body surface is unable to supply all the cells with oxygen because of dryness, which inhibits diffusion.

Special organs for respiration in invertebrates are gills, tracheae, book lungs and lung sacs. Two different kinds of gills are blood gills and tracheal gills. The gills occur on almost any part of the body, wherever their function may be carried out. Tracheae are invaginations of the body wall to form a system of air tubes, which carry atmospheric oxygen directly to all parts of the body, and are mostly found in insects. In the arachnids, an invagination of the body wall leads into the book lungs, the walls of which are composed of much-folded delicate membranes through which the oxygen is taken from an air-filled cavity of the book lung and carbon dioxide is released. In some molluscs, the cavity bearing the gills lack these structures and become modified as a lung sac into which atmospheric air is periodically admitted and from which the air bearing carbon dioxide is released.

## 1.10 EXCRETORY SYSTEM

Because of catabolic processes, wastes are

continuously formed within the living cell. Carbon dioxide, water, ammonia, urea, uric acid and other nitrogenous compounds accumulate within the living cell. Many wastes are soluble in water and diffuse through the cell walls. Contractile vacuoles facilitate the collection and elimination of these wastes in some protozoa. Among cnidarians and porifera, direct diffusion through the cell surfaces into the surrounding water eliminates waste products. The first specialization for waste elimination is found in the protonephridial system of flat worms, rotifers and trochophore larvae consisting of a series of tubules associated with the flame cells. Such a system is provided with an enlarged excretory vesicle within which the excretory matter accumulates before it is discharged through the excretory pore. Many coelomate animals possess a metanephridial system, which picks up excretory wastes from the coelomic cavities and the body fluid. The unit of structure in such a system is a nephrostome, which communicates with the exterior by a tubule. Nephridia are paired organs, segmentally arranged in annelids, while in the Crustacea they are modified in form and are limited to one or two pairs. A group of thin-walled Malpighian tubules communicates with the intestine of insects and releases wastes from the body fluid into the digestive canal.

## 1.11 LOCOMOTOR SYSTEM

In the single-celled protozoa, cilia, flagella and pseudopodia are characteristic partial differentiations of cells for movement. The first two of these especially are carried over into the metazoa as modifications of epithelial cells for locomotion. Many larval metazoa perform locomotion by cilia exclusively. Adult flatworms and rotifers perform locomotion by ciliary action. In the metazoa, movements and locomotion result mostly from the activity of the contractile fibers. Sometimes scattered contractile elements are found to be associated to form continuous sheets of muscles. The neuromuscular sac of flatworms is a definite stage in the development of a locomotor system. Alternate contraction and relaxation of the longitudinally and circularly directed fibers in such a sac produce successive shortenings and elongations of the body. When hard skeletal parts make their appearance, groups of muscle cells get attached to these hard parts and with fixed points of attachment perform movements that are more effective. Even the setae of an earthworm serve for muscle attachment and aid in locomotion. The exoskeleton of arthropods is the highly organized locomotor skeleton found in invertebrates.

## 1.12 NERVOUS AND SENSORY SYSTEM

The hydroid or polyp stage of the cnidarians is provided with a scattered arrangement of the nerve cells. In all metazoans above the cnidarians, the nerve cells are grouped to form definite tracts of nerve tissue, which occur in the cell bodies of the neurons and cytoplasmic fibres with their specializations and neuroglia cells. A tendency for the cell bodies of the neurons is to become massing in certain parts of the nerve tracts, called the *ganglia*. Nervous system of medusa shows tendency toward the centralization of the nervous system. In the hydromedusae, an undifferentiated nerve ring of ganglia parallels the margin of the bell. In the scyphomedusae, the nerve ring becomes specialized, and shows localization of the cell bodies to form ganglia. In metameric animals, tendency for ganglion formation in each somite is observed. This tendency is prominent in the anterior region of the body where the brain is located. In many worms, two or more lateral nerve cords pass through the body, while in some annelids and arthropods these are fused to form a single chain of ganglia ventral to the digestive tract.

The ganglionic part is the controlling centre for the activity of the organism and is designated as the central nervous system. From brain, bundles of fibres pass to the various body parts and constitute the peripheral nervous system. Through the latter, the sense organs are brought into relationship with the central nervous system. A sympathetic nervous system in higher arthropods controls the activity of internal organs. Tactile organs are commonly developed as bristles from epithelial cell surfaces. In arthropods, bristles may be formed as outgrowths of specialized cells. Organs of sight, hearing and balance are often highly organized. The association of pigment with sensory epithelial cells denotes the presence of a retina or some sort of structure for reception of light stimuli. Compound eyes are highly characteristic of arthropods. Numerous modifications for auditory and balancing functions are found as otocyst or statocyst, which depends on the movement of a free or suspended body, the otolith, against the walls of the vesicle for stimulating the central nervous system and causing the organism to right itself. In some insects a specialized auditory organ is found in which a tympanum receives sound waves and transmits the stimulus through underlying space to the nerve endings in the sensory epithelium.

## 1.13 REPRODUCTIVE SYSTEM

In some metazoa, the gonads are the only organs of reproduction. In the cnidarians, the germ cells are discharged directly from the body surface or into the gastrovascular cavity. The ducts, which lead from the gonads to the exterior, are present in most metazoa. Gonads may be single, paired or multiple. Often, a gonad becomes subdivided into follicles, each of which has the appearance of an independent organ. When each individual bears the organs of one sex, the species is called the *dioecious*. A single individual when produce both eggs and sperms is called the *monoecious*. Hermaphroditism results from the occurrence of a full set of gonads and accessory organs for each sex in the same individual. In some snails, both eggs and sperms are produced in the same gonad, called the *hermaphroditic gonad*. In most cases, the ducts, which carry the germ cells from the ovaries and testes, are derived as the modifications of the excretory ducts.

The vasa deferentia of male are usually directly continuous with gonads. The oviducts of female have communication with ovaries only through body cavity. A portion of the oviduct is often enlarged as an uterus within which the eggs are held for a brief time or through the full period of the embryonic development. Following copulation, sperm cells are stored within female in a receptaculum seminis, from which they escape to fertilize the eggs. The vitellaria are also accessory glands of the female, which furnish nutritive support and supply the substances of which the eggshell is formed. When there are no separate vitellaria, a portion of the ovarian cells become functional ova, while others act as nurse cells, which supply the ova with additional reserve food material. An intromittent organ known as the *cirrus* is often found in the male, which may be a modification of the terminal portion of vas deferens or, as is often the case, may be purely accessory structures such as spicules or appendages modified to aid in the transfer of sperm. The clasping organs are characteristics of the males of some arthropods. The ovipositors are frequent accessories of the female reproductive system in arthropods.

## 1.14 ORIGINS OF MAJOR GROUPS OF METAZOA

Life seems to be existed on earth for 4 billion years. We have been observing life for a few hundred years. A number of theories describe the sequences of hypothetical metazoan ancestors from the protozoans. The origin of the metazoa has received considerable attention and one of the most remarkable phenomenons in the fossil record is the sudden appearance and diversification of about all the metazoan phyla living today at the Precambrian–Cambrian transition. Metazoan evolves as a monophyletic group from protist ancestor perhaps 700 million years ago. The two brilliant hypotheses in this regard are the syncytial theory and the colonial theory. The syncytial theory suggests the metazoan ancestor as benthic multinucleate, bilaterally symmetrical, ciliated protist. The surface nuclei of the ciliate became partitioned off from one another by the formation of cell membranes, producing a cellular epidermis surrounding an inner syncytial mass. The cellular nature of the acoel inner mass is a relatively recent discovery. The support of the syncytial theory, rest on the similarities in size, shape, symmetry, mouth location and surface ciliation, between modern ciliates and acoel flatworms. Objections surrounds on the developmental matters and differences in levels of adult complexity. Like flatworms, acoels undergo a complex embryonic development but nothing of this sort occurs in ciliates. Acoels probably were not the most primitive flatworms. The syncytial theory suggests that the flatworms were the first ancestral metazoan.

The colonial theory suggests that a colonial flagellate gave rise to a planuloid metazoan ancestor. The ancestor was a hollow sphere of flagellated cells that developed anterior-posterior locomotor orientation, and specialization of cells into separate somatic and reproductive functions. Such conditions are common in living colonial protists. Haeckel called this hypothetical protometazoan ancestor the blastea. The first metazoa arose by invagination of blastea. The resulting animals had a double-layered, gastrula-like body (Haeckel's gastrea) with a blastopore-like opening similar to the gastrulae of modern animals. These ancestral creatures were recapitulated in the ontogeny of modern animals, and the gastrea was viewed as the metazoan precursor to the cnidarians. Otto Bütschli in 1883 suggested that the primitive metazoan was a bilaterally symmetrical creature of two layers of cells and he called this hypothetical animal, a *plakula*. Plakula crawled about ingesting food through its ventral cell layer and eventually became hollow by the separation of its dorsal and ventral cell layers which allowed an invagination of the nutritive cells. Thus formation of a gut chamber increased the digestive surface area and produced inner and outer cell layers, an arrangement approaching the metazoan grade of complexity.

The evidence that the metazoa had their origin in a flagella-bearing protist is strong. Choanoflagellata was the likely ancestral group of the metazoan. Proterospongia, Sphaeroeca, and colonial choanoflagellates are cited as typifying a potential metazoan. However, there is a common problem in the search for intermediates between the Protista and metazoa. Probably the actual precursor of the metazoan long ago became extinct. The presence of modern-day forms that somewhat combine protist and metazoan traits keeps debate alive. These organisms include multinucleate ciliates, the colonial *Volvox*, and some little multicellular animals of uncertain position. *Trichoplax adhaerens* was placed in the phylum Placozoa and was thought to be larval stage of some invertebrate, but now it is established that it is an adult of uncertain affinity at the mesozoan grade of construction. During feeding, *Trichoplax* hunches up to form temporary digestive chamber on its underside producing a form strikingly similar to Bütschli's hypothetical plakula. *Trichoplax* may represent the most primitive of all living metazoan. Frenzel in 1892 described a mesozoan animal *Salinella*. *Salinella* does not possess the layered construction of the metazoan, but display a higher level of functional organization than colonial protists. Two other so-called *mesozoanphyla*, *Rhomboza* and *Orthonectida* are structurally simple but display complicated life cycles. Some workers think that they are derived from another group of metazoa, perhaps from parasitic trematodes,

## 1.15 ORIGIN OF THE COELOMIC CONDITION

Most ideas concerning the evolutionary origin of the coelom advocate that the coelomic condition arose only once. There is a difficulty to relate existing coelomate animals to a single common coelomate ancestor. Perhaps it is reasonable to suggest that the coelomic condition arose twice. The coelom may have originated by the pinching off and isolation of embryonic gut diverticula as occurs in the development of many extant enterocoelous animals. Many authors have supported this so-called *enterocoel theory*. Some authors also cite examples of noncoelomate animals in which gut diverticula exist in arrangements that resemble possible ancestral patterns.

The gonocoel theory suggests that the origin of first coelomic spaces by way of mesoderm ally derived gonadal cavities that persisted subsequent to the release of gametes. The serial arrangement of gonads in flatworms and nemerteans could have resulted in serially arranged coelomic spaces and linings as occurs in annelids. Argument against this hypothesis is the fact that in no modern-day coelomate do gonads develops before coelomic spaces. Heterochrony can account for such turn about. Another idea on coelom origin is called the *nephrocoel theory*, which states that the protonephridia of flatworms expanded to coelomic cavities, arguing that the coelom first arose from ectoderm ally derived structures. Another view is that coelomic spaces arose as cavities within the mesoderm and served as storage areas for waste products. Schizocoely could have evolved by the formation of spaces within the solid mesoderm of acoelomate animals and then have been retained in response to the positive selection for the resulting hydrostatic skeleton.

The basic developmental differences between the two clades of coelomate animals suggest that the coelom may have arisen separately in these two lineages. Similarities between the coelomate protostomes and acoelomate flatworms and nemerteans suggest that the protostome clade arose from triploblastic acoelomate ancestor.

Schizocoely and the origin of mesoderm in the ancestral protostome may have arisen simultaneously, and the acoelomate condition may be secondarily derived from coelomate precursors. Enterocoely may have accompanied a tendency toward benthic life, giving the animal a peristaltic burrowing ability. The archenteron may have then opened anteriorly as a mouth, and the new coelomate creature adopted a deposit-feeding lifestyle. If such an idea began at the level of diploblastic metazoa the radial cleavage seen today in deuterostomes was also present in the ancestor to this group. Nielsen and Nørrevang have advocated in their hypothesis that a pelagic gastrea gave rise to the cnidarians and to a second lineage-another larval-like creature they call a *trochaea*. This ancestor was the precursor to the protostomes and the deuterostomes, but the coelom arose separately in each group. The origin of major metazoan groups from radially symmetrical gastrea are also suggested. From this planktonic ancestor there evolved two separate lines. One line led to the protostomes via the gastroneuron and the trochaea. The other line, to the deuterostomes, was by way of hypothetical notoneuron ancestor.

## REVIEW QUESTIONS

### Short Answer Questions

1. Define animal.
2. Mention the purpose of studying zoology.
3. Give the name of two prokaryotes.
4. Mention two differences between prokaryotes and eukaryotes.
5. Give the names of four types of tissues.
6. What do you understand by intracellular digestion?
7. Name an invertebrate phylum where three chambered heart is present.
8. Name the excretory organ of annelids.
9. Name the locomotor organelles of protozoa.
10. What do you understand by the term, *monoecious*?

### Long Answer Questions

1. Discuss the origin of major groups of metazoan.
2. Give an idea of classification of the living world.
3. Write a note on invertebrate habitats.
4. Write a note on the origin of coelomic condition.

CHAPTER 2

# Evolution and Taxonomy

Understanding the evolution is most important in the science. Evolution explains the diversity of life through speciation, help explains many phenomena of life and provides the foundation for the modern biology. The passing of biological traits from the parents to the offspring and the subsequent change during passing is the basis of evolution. When a new trait in an offspring works better in terms of fitness in its natural environment and produces more offspring, the trait becomes widespread over time. If the new trait makes the offspring less able to survive or less fit and leaves the offspring, the trait tends to be extinct. Some traits succeed and others fail in biological race in the environment by the process of natural selection.

Natural selection is the driving force for these changes. Based on the fossil records, it is evident that species evolve over time. Common structures and behaviours among many species show their evolution from the common ancestor. Comparing one fossil to another in structure and in various geological times shows the evolution of successor species from the ancestral one. The fossil records, DNA research, and the evidence that species have common ancestors, constitute the strong evidence for evolution by the process of natural selection. The fossils of ancient fish lived in shallow waters in ancient times and fossils of the four-limbed land dwellers appeared later. Evolutionary theory predicts that there would creatures with characteristics of both ancient fish and later land dwellers. According to Theodosius Dobzhansky, “nothing in biology makes sense except in the light of evolution.” Biological evolution refers to populations and the passing of changes in the next generation. Evolution is a process that results in heritable changes in a population over many generations. In other way, it means the detectable change in the frequency of genes in a population.

## 2.1 BRIEF HISTORY OF LIFE

Life evolved in the sea and is known to be stayed there for the most of the history of the earth. The first replicating molecule was supposed to be RNA. Some RNA sequences have the catalytic capabilities. Certain RNA sequences act as polymerases. Replication is believed to be the important step for the formation of life. This is explained beautifully in the RNA world hypothesis. The common ancestor of all life forms probably used RNA as genetic material and produced three major lineages of life namely the prokaryotes, archaebacteria and eukaryotes. The first cell was supposed to be anaerobic, probably thermophilic and fermentative.

The evolution of biosphere and organisms appears to be involved for origin of new species that results in the species diversity. During the beginning of life on earth, there was no free oxygen. In the absence of oxygen, no ozone layer protected the earth from the deadly ultraviolet radiation of the sun. This radiation is believed to trigger a chemical evolution leading to the formation of amino acids. However, life could develop only under the protective cover of water, since the atmospheric oxygen and ozone remained scarce. Photosynthetic bacteria, particularly the cyanobacteria release oxygen into the atmosphere about 3.5 to 2 billion years ago. The origin of photosynthetic bacteria, however, is still not clear. The gradual buildup of photosynthetically produced oxygen and its release into the stratosphere about 2 billion years ago brought about marked changes in the geochemistry of the earth which rapidly expanded the life and development of the eukaryotic cell leading to evolution of larger and more complex living organisms. Many minerals such as iron were precipitated from water and formed characteristic geological formations.

First nucleated cells appeared, when there was 0.6% oxygen in the atmosphere. Margulis advocated that eukaryotic cell originated as a mutalistic coming together of once independent microbes. During the Precambrian era, only the single-celled prokaryotes are known to be existed. The transition to eukaryotic cells appears to be occurred about 1.2 to 1.5 billion years ago. However, recent studies suggest that the eukaryotes diverged from the prokaryotes closer to 2 billion years ago. The fossil records do not yet show consistency with this date. The Kingdom Protista has been broken into many new kingdoms through new studies and techniques that attempted to explain the true phylogenetic sequence of life. The protists exhibit much variation in their life histories. They exhibit an alternation between the diploid generation and haploid generation that is similar to the alternation of generations found in the plants. The protist's life cycles vary from the diploid dominant to haploid dominant. The diversity of form, habitat, mode of nutrition and life history exhibited by the eukaryotes show their evolution several times from the prokaryotes in the history of life. This makes the Protista a polyphyletic group.

Eukaryotes are generally larger with several membrane-bound organelles, much complex internally than the prokaryotes, and with meiotic cell division that is considered to be a prelude to true sexual reproduction. The multicellular organisms such as sponges, corals, worms, shell and vertebrates appeared later. During the Paleozoic era, green mantle of terrestrial vegetation is believed to provide more oxygen and food for the subsequent evolution of large creatures such as dinosaurs, birds, mammals and finally human.

Present day life forms including humans are believed to be evolved during the last 570 million years consisting of the Paleozoic era, Mesozoic era, and Cenozoic era. The evolution results mainly from simplicity to complexity, and from imperfection to perfection which ultimately leads to the highly evolved and specialized species of birds and mammals. These advanced life forms are perfectly matched with the ecosystem where they live in. It seems likely that life on earth originated only once. Later it multiplied, diversified and colonized the already existing ecosystems. Life forms that could survive and flourish in various virgin ecosystems at that time would have largely been determined by abiotic components to begin with.

## 2.2 GEOLOGICAL TIME SCALE

Based on the study of rocks with its contained fossils, geologists have divided the earth's history into several eras, periods and epochs, and from which a convenient time scale has been worked out in order of their age which is known as geological time scale. Major divisions of the earth's history are called *eras*. Eras in their turn are divided into the periods, the names of which are largely geographic, or they may be of historic significance. Subdivisions of periods are called the *epochs*. As evolution has preceded, animals have come to occupy the environments differing even more widely from the sea in which the life probably arose. Life in these more difficult environments was possible by the development of special device, making the later animals more complex than the earlier, and in this sense higher. It remains uncertain what influences were responsible for producing the changes in the organic form.

Geological evidence shows that many changes have occurred in the climate and geography, some of them are quite fast but most proceeding at very slow rate in comparison with the rhythms of individual live. It is uncertain whether the evolutionary changes follow these slow geological changes are a result of competition and the instability imposed on living things by the climatic rhythms with shorter periods such as those of days and years (Table 2.1).

## 2.3 SPECIATION

Ernst Mayer recognized two factors for speciation.

**Isolation of population:** Two populations become sufficiently distinct when interbreeding becomes difficult or impossible between them. However, at that time there must be relatively

**Table 2.1** Geologic Time Scale

| Time Units of the Geologic Time Scale | | | | | Development of Plant and Animals |
|---|---|---|---|---|---|
| **Eon** | **Era** | **Period** | **Epoch** | | |
| Phanerozoic | Cenozoic | Quaternary | Holocene | | |
| | | | | 0.01 | |
| | | | Pleistocene | | Earliest *Homo sapiens* |
| | | | | 1.6 | |
| | | Tertiary | Pliocene | | |
| | | | | 5.3 | Earliest hominids |
| | | | Miocene | | |
| | | | | 23.8 | |
| | | | Oligocene | | Age of Mammals |
| | | | | 33.7 | |
| | | | Eocene | | |
| | | | | 55 | |
| | | | Palaeocene | | |
| | | | | 65 | |
| | Mesozoic | Cretaceous | Age of Reptiles | | Extinction of dinosaurs and many other species |
| | | | | 145 | First flowering plant |
| | | Jurassic | | | First bird |
| | | | | 208 | Dinosaurs dominant |
| | | Triassic | | | First mammal |
| | | | | 248 | |
| | Palaeozoic | Permian | Age of Amphibians | | Extinction of trilobites and many other marine animals |
| | | | | 286 | First reptile |
| | | Carboniferous: Pennsylvanian | | | Large coal swamps |
| | | | | 320 | Amphibians abundant |
| | | Carboniferous: Mississippian | | | |
| | | | | 360 | |
| | | Devonian | Age of Fishes | | First amphibian |
| | | | | 410 | First insect fossil |
| | | Silurian | | | Fishes dominant |
| | | | | 438 | |
| | | Ordovician | Age of Invertebrates | | First land plant |
| | | | | 505 | First fish |
| | | Cambrian | | | Trilobites dominant |
| | | | | 545 | |
| | | Vendan | Soft-bodied faunas | | First organism with shell<br>Abundant Ediacaran Fauna |
| | | | | 650 | |
| Proterozoic | | Collectively called Precambrian comprises about 87% of the geological time scale | | | First multicelled organism |
| Archean | 2500 | | | | First one-celled organism |
| | 3800 | | | | Age of oldest rocks |
| Hadean | 4600 Ma | | | | Origin of the earth |

*Source:* Sci. Waikato.ac.nz/evolution/zoological.shtml
The University of Waikato, Evolution for Teaching, School of Science and Engineering.

little gene flow between them. Genetic changes in one population soon become widespread in other as well when much gene flow occurs.

**Genetic divergence:** Two populations become separate species, if during the period of isolation, they evolve sufficiently large genetic differences, so they can no longer interbreed or produce vigorous fertile offspring if they are reunited. In both small and large populations, different environmental pressures in separate environment may favour evolution of large genetic differences.

The origin of new species falls into the following categories.

### *Allopatric speciation*

In Greek, allopatric means 'having a different fatherland'. Allopatric speciation occurs when two populations become geographically isolated from one another. When population becomes physically separated, little or no migration can occur between them. The physical separation occurs when some members of a population drift, swarm or flow to remote island and when a part of a population remains in a patch of suitable habitat that became isolated by a change in climate, or by geographical process such as volcanism or continental drift. After such separation, pressures of natural selection differ in two locations, or if the populations are small enough for genetic drift to occur, then population may accumulate large genetic differences and become separate species. Allopatric speciation, otherwise known as *geographical speciation*, is a slow process. Among Mexican platy fish, *Xiphophorus conchianus* inhabiting northern most part is restricted to a single river system and is sterile in crosses with other species. *X. variatus* with three distinct populations are found in three different rivers, but they can produce fertile hybrids with each other. *X. maculatus* is found in many streams and hybridize with those from other populations. It is assumed that sometime in the past, an ancestral platy fish colonized all these streams. Subpopulations isolated in different streams have been diverging since then. Some have attained full reproductive isolation whereas others have not.

Large areas are required for the geographical speciation, so that different population can be isolated from one another for a time sufficient to evolve reproductive isolation. The smallest islands, from where a single species is believed to have caused two species, are known for several organisms. Birds and mammals require larger area than fish, amphibians and reptiles. The isolation of a subpopulation from remainder of ancestral species is the first prerequisite for allopatric speciation. Secondly, divergent selection pressure establishes new barriers of reproductive isolation based on many genetic differences.

### Sympatric speciation

Sympatric means, 'having the same fatherland'. Sympatric speciation occurs within a single geographical area as occurred in Hawaiian land snails. This speciation requires limited gene flow. Two mechanisms, which reduce the gene flow between members of a single population in a given area, are ecological isolation, and chromosomal aberrations. In chromosomal aberration, gametes and somatic cells containing multiplies of haploid number of chromosomes are called the *polyploids*. Polyploidy is much more common in plants than in animals. Most animals cannot self fertilize, or reproduce asexually. Therefore, if an animal produced a tetraploid offspring, offspring would have to mate with a member of diploid parental species, and would produce all triploid offspring. Triploid offspring would almost certainly be sterile. Polyploidy is known from the asexually reproducing groups of arthropods such as isopods, bagworm moth, weevils and flies. This may be a result of duplication of chromosome of a single species (autopolyploidy), or by the combination of chromosomes from two different species (allopolyploidy).

### Parapatric speciation

This is a variation of sympatric speciation and does not require geographic isolation. When structural rearrangements of chromosomes or other genetic changes take place in a small number of individuals, reproductive isolation may develop. Individuals carrying new rearrangements become isolated from other members of population and become new species. Such speciation is reported in *Spalax ehrenbergi*. Its two northern subspecies have 52 and 54 chromosomes, central Israeli subspecies has 58 and southern subspecies has 60 chromosomes. Hybrids of these subspecies rarely occur, and when occurs, fertility is reduced. Fossil evidences show that all subspecies are descended from an ancestral species *S. mimtus* which lived in the said area about 500,000 years ago. Speciation, the key to understanding evolution, is the process for splitting of a single species into two or more species.

## 2.4 EXTINCTION

A species can be competitively excluded by a closely related species by the disappearance of suitable habitat, or due to many other reasons. Some species are long-lived while others are short-lived. Species are programmed to be extinct. Mass extinctions shape the macroevolution. The history of life includes several mass extinction, which wiped off many organisms. Mass extinctions were followed by the periods of adaptive radiation when new species were evolved to occupy the empty niches. Surviving a mass extinction is mostly regarded as luck. Contingency plays a major role in the macroevolution. The largest mass extinction is known to occur at the end of the Permian, about 250 million years ago coinciding with the formation of Pangaea II. During that time, all the continents were brought together by the plate tectonics. A worldwide drop in sea level occurred at that time.

The well-known extinction occurred at the boundary between the Cretaceous and Tertiary Periods, commonly called as the *K/T Boundary* dated about 65 million years ago. This extinction eradicated the dinosaurs probably by causing environmental disruption by a large impact of an asteroid with the earth. The mammalian radiation occurred following that extinction. Mammals coexisted for a long time with the dinosaurs, but were confined mainly to the niches of nocturnal insectivore. With the eradication of the dinosaurs, the mammals radiated to occupy the vacant niches. Presently, human alteration of the ecosphere is continuing to cause the global mass extinction.

## 2.5 TAXONOMY

Taxonomy provides a picture of existing biodiversity on earth along with the information needed for entire biology. It provides the most information needed for the reconstruction of phylogeny of life, ordering systems or classification, reveals numerous interesting evolutionary phenomena, and makes them available for the casual study. The word taxonomy is derived from two words—*taxis* meaning arrangement and *nomos* meaning law. Thus, taxonomy is the science of classification of organisms following some definite laws or rules. This is divided into microtaxonomy and macrotaxonomy. The discrimination of species is called the *microtaxonomy*. The classification of organisms above the species level is called the *macrotaxonomy*. There are three levels of taxonomy. *Alpha taxonomy* deals with characterization and naming of the species. *Beta taxonomy* deals with arranging of species into a natural system of lower and upper taxa. *Gamma taxonomy* deals with various biological aspects of taxa, ranging from the study of intraspecific population to studies of speciation, and of evolutionary rates and trends.

Several species concepts are in use by the biologists. These include folk-, morphological-, genetic-, paleontological-, evolutionary-, biological and phylogenetic taxonomy. The term *taxonomy* was first coined by A.P. de Candolle in 1813. The history of taxonomy is divided into three periods. In the first period, Greek scholars notably Hippocrates and Democritus included animals in their study. Linnaeus in his tenth edition of *Systema Naturae* introduced binomial system of nomenclature. He presented the first character-based classification, which serves for arranging of specimens from the vast biological data, for which he is rightly called the father of taxonomy. In the second period, Charles Darwin (Figure 2.1) conceived the theory of evolution. Darwin's *'On the origin of species by natural selection'* resulted in the stimulation of biological thought and work. Ernst Hackel introduced the method of representing phylogeny by trees or branching diagrams. The study characterized the third period of evolution within the species. During this period, the foundation for establishment of population genetics was proposed.

**Figure 2.1** Charles Darwin.

## 2.6 SPECIES CONCEPT

Species is the most important taxonomic category for the taxonomists as well as for the general biologists. The word species means originally kind. The species concept of biologists goes back to John Ray (1686), who in his book *Historia plantarum* used the term *species*. Typological species concept originated in philosophies of Plato and Aristotle and subsequently accepted by Linnaeus and his followers. According to this concept, the observed diversity in nature may be generalized into several types. Individuals were regarded as merely expression of the same type. The variations observed in individuals are imperfect expressions of the basic type.

According to nominalistic species concept, only individuals have a real existence. Species are conceptual or man made abstractions with no existence in nature. Both the concepts described above have been discarded and are now stand as historic events. Biological species concept includes the concept of Dobzhansky and Mayer. Dobzhansky noted, "Species are most inclusive Mendelian populations, which share the common gene pool". Mayer stated, "Species are groups of actually or potentially interbreeding natural population which is reproductively isolated from other such groups".

### *Folk concept of species*

Naturalists observed that the individual organism can be mentally grouped into several taxa in which the individuals are similar. In societies that are close to nature, each taxon is given a name. Such folk taxonomies have two common characters. One is the idea of reproductive compatability and continuity within a species. The second is the discontinuity of variation between the species.

### *Biological species concept*

Over the last few decades, the theoretically preeminent species definition has been transformed into the biological species concept. This concept defines a species as a reproductive community. Du Rietz defined species as "... the smallest natural populations permanently separated from each other by a distinct discontinuity in the series of biotypes." Dobzhansky defined species as "... that stage of evolutionary progress at which the once actually or potentially interbreeding array of forms becomes segregated into two or more separate arrays which are physiologically incapable of interbreeding. Mayr (1942) defined species as "... groups of actually or potentially interbreeding natural populations which are reproductively isolated from other such groups. Mayr later amended this definition to include an ecological component. In this form of the definition a species is a reproductive community of populations (reproductively isolated from others) that occupies a specific niche in nature.

### *Drawbacks of biological species concept*

There has been considerable criticism of the theoretical validity and practical utility of this concept. This definition does not apply to taxa, which are obligatory asexual including bdelloid rotifers, euglenoid flagellates, some Oocystaceae, chloromonad flagellates, and diatoms. Asexual forms within the normally sexual organisms are reported. Obligatory asexual populations of *Daphnia* are recorded in the arctic lakes. Genes can be exchanged among the bacteria. Sexuality as defined in the eukaryotes is unknown in the prokaryotes. The concept suggests breeding experiments to validate the species membership, which is rarely made. Consistent criterion is lacking for determining speciation event that has occurred in such cases, where the biological species concept does not apply. This is particularly applicable for obligatory asexual organisms. Thus, the drawbacks can be summarized in the following points:

1. This concept is not applicable to species living either in geographically isolated areas, or at different times as it does not tell whether populations isolated from each other in space or in time could actually interbreed or not.
2. The concept is not workable to museum taxonomists and paleontologists, who work with dead specimens and fossils as they have no means to determine whether the organisms under consideration are reproductively isolated or not.
3. The concept is also limited to asexually reproducing species. Asexual and obligatory parthenogenic organisms can never be tested by this criterion.

### *Significance of biological species concept*

During the early period, there was no criterion for defining a species. A group of taxonomists regarded every polymorphic variant of a population as a species. The second group termed every geographically isolated population, a species, creating confusion. Biological definition of species helps in solving the riddle of taxonomy for such abnormal cases in which the distinct species bear great morphological similarities. For example, *Drosophila pseudoobscura* and *D. persimilis* exhibit great similarities but are reproductively isolated.

### *Evolutionary species concept*

Simpson defined evolutionary species as a lineage evolving separately from the others and with its own unitary evolutionary role and tendencies. A species is a single lineage of populations or organisms that maintains its identity from other such lineages and which has its own evolutionary tendencies and histories.

### *Taxonomic problems*

Sibling species poses a challenge to the taxonomists. Morphologically identical, sympatric or allopatric populations, which are reproductively isolated, are called the sibling species. Sibling species are common in insects. A well-studied example of sibling species is found in *Drosophila* in which most of the species complexes contain the groups of sibling species. *D. pseudoobscura* and *D. persimilis* are established as sibling species. *Anopheles maculpennis, A. atroparvas, A. labranchiae, A. sacharovi, A. messae and A. melanoon* are described as *A. maculipennis* group of sibling species. Second problem is posed to the taxonomists by asexual reproducing organisms because criterion of interbreeding cannot be used in their cases. For example, some taxonomists recognize about 200 species of blackberry while others recognize only 20 species. Asexual reproduction in unicellular organisms makes their classification difficult. We are primarily concerned with large or conspicuous organisms of temperate regions although biodiversity is reported to be the greatest among, small, inconspicuous organisms inhabiting the tropics. Taxonomists have overlooked several species on land and in shallow water, and presently our knowledge about deep-sea biodiversity at depths of 3250 feet is insignificant.

### *Phenetic (morphological) species concept*

Cronquist (1988) proposed an alternative concept in place of the biological species called a 'renewed practical species definition'. According to him species are "... the smallest groups that are consistently and persistently distinct and distinguishable by ordinary means." Three comments about this definition are as follows:

First, 'ordinary means' includes widely available, cheap and relatively easy applying techniques. These show difference among various groups of organisms. Second, the requirement that species be persistently distinct implies a certain degree of reproductive continuity due to the phenetic discontinuity between the groups that cannot persist without a barrier to interbreeding. Third, this definition emphasizes more on morphological characters and recognizes phenetic characters such as chromosome number and morphology, ultra structure of the cell, secondary metabolites, habitats and other characters.

### *Phylogenetic species concept*

Two types of phylogenetic species concepts were described by Baum. A species is the smallest unit of organisms that possesses at least one diagnostic character which is either morphological, biochemical or molecular and must be fixed in reproductively cohesive units. Such reproductive continuity is not used in the way as for biological species concept. Phylogenetic species may be the reproductive communities. Reproductively compatible individuals need not have the diagnostic character of a species; the individuals need not to be conspecific.

A species must be monophyletic comprises all the descendants of a common ancestor or may be a group of closely related organisms than to any other organisms and share one or more derived character. Recently, phylogenetic species concepts and the biological species concepts are suggested to be highly incompatible. 'Parallel speciation' is the repeated independent evolution of the same reproductive isolating mechanism. This is exemplified when a species colonizes several newly isolated but environmentally similar areas. Similar selective pressures in such environments cause parallel evolution among the traits which are responsible for the reproductive isolation.

## 2.7 VARIOUS APPROACHES OF TAXONOMY

**Morphological approach:** This approach is based on morphological characters such as the genitalia, antennae, wings and use of scanning electron microscope. Morphology differs between the races and populations, as well, of course between the species. Most taxonomic study is based on the morphological approach.

**Ecological approach:** This is based on ecological differences. *Anopheles maculipennis* complex was broken into six species based on ecological differences. Darwin's finches are well known. The Hawaiian honeycreepers are even more extraordinary. From finch like ancestors, they have produced the nectarivorous, insectivorous, frugivorous, as well as the seed eating forms. Ecological differences are found across the clines that are under extrinsic selection across an environmental gradient. Once again there is no clear dividing line between races and species in the degree of ecological differentiation.

**Cytological approach:** This approach includes DNA hybridization and karyological studies.

**Embryological approach:** This is based on studies of embryos.

**Behavioural approach:** Sexually selected colours, tail length in birds, pheromones in moths and other insects, and even mammals are all involved in species recognition. In many crickets and grasshoppers as well as frogs species specific sounds are recognized. In fireflies, species recognize each other by the coded flashes. These kinds of differences are easily derived from mate choice differences within species, perhaps caused by sexual selection or for ecological reasons of efficiency. Various species of *Apicotermes* may be separated on nest structure. Praying mantis can be identified on forms of their egg cases.

### *Primary and secondary subdivisions*

Each phylum is made up of subphylum, super classes, orders, suborders, families, genera, and species in descending sequence. Each primary division may be divided into subgroups and supergroups. The entire system of classification, based on interpretation and judgment, aims to express the degrees of relationship between the various organisms. Phylogenetic relationships found in the structure and development of the individual and through fossil remains serve as the connecting links between the modern and ancient groups. The simplicity of organization is not always safe in interpreting relationships, because the simplicity in form and structure may be primitive or secondarily derived.

## 2.8 ISOLATING MECHANISMS

Mechanisms which produce the reproductive isolation fall into two broad categories—premating mechanisms and postmating mechanisms. Premating isolating mechanisms keep species separate before the mating occurs and act to prevent the mating altogether. Premating mechanisms are the ecological-, temporal-, behavioural- and mechanical-mechanisms. Ecological isolation occurs when the species occupy or breed in different habitats. Similarly, temporal isolation occurs when species breed at different periods of the year or different times of the day. Behavioural isolation relies on organism's choice of whether to mate and a choice of who to mate with. Differences in courtship behavior may be sufficient to prevent mating. Behavioural isolation should result in somewhat positive assortative mating. Mechanical isolating mechanisms occur when the morphological or physiological differences prevent the normal mating. Postmating Isolation prevent hybrid offspring from developing or breeding when the mating occur. Several examples of postmating mechanisms are recorded. Mechanical postmating isolating mechanisms occur where the mating is possible, but the gametes are unable to reach each other. Mortality acts as an isolating mechanism when the hybrid dies before attaining maturity. Sterility of hybrids also serves as an isolating mechanism. A reduction in the fitness of the hybrid offspring may isolate two populations. This occurs when the F1 hybrid is fertile, but the F2 hybrid has lower fitness than either of the parental species.

## 2.9 ZOOLOGICAL NOMENCLATURE

Nomenclature (*nomen* meaning name; *calore meaning* to call) means allocation of names to the taxa. The scientific names should be unambiguous, unique, universal and stable, as they serve the only means of reference, thereby avoiding the need of continuous usage of cumbersome descriptive phrases. An animal name is the key to its literature. Proper naming of an animal is thus the fundamental principle of nomenclature. The formation and the use of scientific names of animals follow the International Code of Zoological Nomenclature (ICZN).

### *Origin of the code*

Linnaeus was the first to mention a set of rules of nomenclature in *Critica Botanica* and *Philosophica Botanica*. Fabricius, improved on the Linnaeus, rules of nomenclature in *Philosophica Entomologica* for naming the insects. In 1801, Rudolphi, a parasitologist, suggested a set of rules which were so much confusing that no one followed them at that time. During the 19th century, more and more new and local sets of rules originated in different countries. In June, 1843 *Stricklandian Code* was formed which was later amended, republished and continued to be followed by American- and British- zoologists for several years. Another code, the *Dall's code*' was made in 1877. Later more and more codes were added. By the end of 19th century, it became apparent that more and more problems had been created, and it became essential to formulate International code. Therefore, Professor Blanchard prepared a set of rules, which were presented at the First International Congress of Zoology held in Paris in 1889, but accepted only in the Second Congress in 1892. Later at the Third Congress in Leyden in 1895, Blanchard's code was labelled as a fresh product and a commission was appointed to draft a set of rules agreeable to all the zoologists. Later, several discussions were held in subsequent congress to revise and re-revise the rules and finally a draft was prepared and circulated to the International Commission of Zoological Nomenclature on 11th January 1961. It was then approved by the commission and formally accepted by voting as "The International Code of Zoology Nomenclature" adapted by the XV International Congress of Zoology, London in July 1958 in English and French, and published by International Trust for Zoological Nomenclature, London in 1961.

## 2.10 INTERNATIONAL CODE OF ZOOLOGICAL NOMENCLATURE

This code consists of three main parts—the code proper, appendices and glossary. The International Code of Zoological Nomenclature is the system of rules and recommendations authorized by the International Congress of Zoology. The object of the code is to promote the stability and universality in the scientific names of animals, and to ensure that each names is unique and distinct. When stability of nomenclature is threatened in an individual case, the strict application of the code may under specified conditions be suspended by the International Commission of Zoological nomenclature. International Commission on Zoological Nomenclature formulates a code of rules governing the problems of naming animals. One basic principle of this code is the law of priority. According to the law of priority, the valid name of a genus or a species can only be the scientific name under which it was first described, and the name is binomial and has not been used before for some other animal. The same specific namesmay be used for any number of various animals belong with various genera. When a generic name has been used previously for other genus of animals, only the earliest use of the name stands as the valid. The name used for the second time is treated as a homonym. The name of a family is formed by adding the suffix *idae* to the stem of the name of the type genus. Family names are also governed by the same law which governs the use of generic names.

Thus, classification intends to express a linear relationship of all forms that each group has arisen from or has evolved from the one preceding. However, the modern idea of phylogenetic relationship of the animal groups is best expressed as a branching dendrite structure in which the phylum and various groups are represented by the branches. Closely related groups would be represented by a forking of a common ancestral branch.

### *Uni-, bi- and tri- nominalism*

The scientific names of animals from subgenera and above are uninominal. These are plural nouns or adjectives used as nouns for names above genus and singular nouns for genus and subgenus. They are written with a capital letter. The zoological code stipulates standardized endings for the names of super family (-oidae), family (-idae), subfamily (-inae), tribe (-ini) and rarely subtribe (idi). The names of species are binominal and those of subspecies are trinominal. For example,

| Genus | Subgenus | Species | Subspecies | Author |
|---|---|---|---|---|
| Dacus | (*Afrodacus*) | *aberrans* | *nigritus* | Hardy, 1955. |

The first (genus) and the third (species) words formed the binominal nomenclature. The second word in parenthesis is the subgenus and the fourth indicates the subspecies, which forms the trinominal nomenclature. The presence of subgenus does not affect the status of nomenclature. Hardy is the original author of this taxon published in 1955. The zoological code does not recognize the names of taxa of infra subspecific rank. The basic language of the scientific names is Latin and names from other languages are Latinized or treated as such.

## 2.11 CLASSIFICATION

There are two type of classification—artificial and natural. An artificial classification is based on one or a few easily observed characteristics, and is usually designated for a practical purpose with an emphasis on convenience and simplicity. The ancient Chinese system is an artificial classification. Linnaeus included all worm-like organisms in a single group—the *vermes.* This included a wide range of animals from the simple nematode worms, and earthworms to snakes. This was an artificial classification because it did not consider the important natural relationships like the fact that snakes have backbones and earthworms do not. Snakes have more in common with other vertebrates than with worms. An example of an artificial classification of fish could be to group them as freshwater fish, brackish water fish, and marine fish because of their environment which would be convenient to the purpose of investigating their mechanisms of osmoregulation. Similarly, all microscopic organisms are known as *microorganisms*, a convenient group for study, but they are not natural groups.

A natural classification tries to use the natural relationships between the organisms and considers more evidence than artificial classifications, including the internal and external features. Similarities of embryology, morphology and anatomy, physiology, biochemistry, cell structure and behaviours are all relevant. Most classifications in use today is natural and phylogenetic which is based on evolutionary relationships. In such a system, organisms belong with the same groups are believed to have a common ancestor. The phylogeny of a group can be shown by a family tree. Another way to classify organism is to use a phenetic classification—an attempt to avoid the problem of establishing evolutionary relationships, which is very difficult and controversial, especially if there is little or no fossil evidence. The word *Phenetic* comes from the Greek word *Phainomenon meaning*, 'that which is seen'. This classification is based solely on the observable characteristics and all characters used are considered of equal importance. All features of an organism can be considered, the better and they do not necessarily have the evolutionary significance. Masses of data are collected and the computer calculates the degrees of similarities between the different organisms usually. The use of computers in taxonomy is known as *numerical taxonomy*.

## 2.12 TAXONOMIC HIERARCHY

The actual method of establishing a classification consists in defining groups or categories on a hierarchic scale. Each of those categories includes one or more groups from the next lower level, which is the next lower category. The results are that all the animals can be classified in a taxonomic hierarchy consisting of a series of categories of ascending rank from the species to the kingdom. Each successive category embraces one or more of the next lower category. Taxonomic categories reduce the diversity of nature to a comprehensible system. Groups may be understood and remembered more readily than the countless units of which they are composed. Every classification involves the two steps following:

1. The arranging of lower units into groups.
2. To join these groups in an ascending hierarchy.

Both steps involve numerous practical and scientific questions. Linnaeus, who founded the hierarchy of taxonomic categories, recognized within the animal kingdom only five classis, ordo, genus, species and varietas. Now it became necessary to make the finer divisions, of which two are universally accepted, the family and the phylum. The varietas was an optional category under which were placed various types of variation. The remaining categories from the basic taxonomic hierarchy of animals are kingdom, class, order, genus and species. The respective position of animals in the zoological system may be expressed by this hierarchy with fair accuracy. In most groups of animals, the need has arisen for an even more precise definition of the taxonomic position of a species. This has been accomplished historically by a splitting of the original categories and by inserting additional categories between the seven basic ones. Most of these are formed by combining the original names with the prefixes super or sub. Thus, there are superorders, suborders, super classes, and subclasses, etc. Many other names have been proposed for the higher categories, but none of them is in general use, except perhaps the term tribe between genus and family. Authors like Simpson used a category cohort between the order and class. The generally accepted categories are the followings.

Kingdom
Phylum
Subphylum
Super class
Class
Subclass
Cohort
Superorder
Order
Suborder
Superfamily
Family
Subfamily
Tribe
Genus
Subgenus
Species
Subspecies

A small group of relatively simple, many-celled animals which are composed of a layer of ectoderm cells covering a single or several endoderm cells has been standing intermediate between the protozoa and the metazoa and is termed the mesozoa. Probably their simplicity of organization is an accompaniment of degeneracy traceable to the adaptation to the parasitic existence. All metazoans differ with the protozoa, in that the cells of the former become specialized for restricted functions. Certain groups of cells become similarly specialized for carrying out specific function effectively. Such a group of similarly specialized cells is termed a *tissue* which is made up of units of cytoplasm, their contained nuclei and frequently the substances formed by the cytoplasm.

In the protozoa, any specialization of somatic cells affects all equally, so there is no differentiation. The metazoa is known to be originated from protozoa. Embryology often offers a clue to and outlines the steps in phylogeny. During development, all metazoa typically pass through a single-celled stage—the fertilized egg, which furnishes the basis for an analogy of metazoan origin from the single-celled organisms. During development of the fertilized egg, large numbers of fundamentally similar cells are produced through the partition of the single cell to form the adult. These cells through histological differentiation assume different forms and perform the diverse functions.

## REVIEW QUESTIONS

### Short Answer Questions

1. Define evolution.
2. What is natural selection?
3. Define life.
4. What was the first replicating molecule?
5. What do you understand by the alternation of generations?
6. Define geological time scale.
7. What do you understand by the term speciation?
8. Define allopatric speciation.
9. Define sympatric speciation.
10. Define parapatric speciation.
11. Define extinction.
12. What do you understand by mass extinction?
13. Define taxonomy.
14. What is alpha taxonomy?
15. What is beta taxonomy?
16. What is gamma taxonomy?
17. Who wrote 'systema naturae'?
18. Who wrote 'on the origin of species by means of natural selection'?
19. What do you understand by the folk concept of species?
20. Define biological species.
21. What are the key criteria for designating a species?
22. What do you understand by isolating mechanisms?
23. Define zoological nomenclature.
24. Define zoological classification.

### Long Answer Questions

1. Describe geological time scale. Why we read geological time scale from bottom instead of the top.
2. Give a comparative account of sympatric and allopatric speciation.
3. Write an essay on the biological species concept.
4. Discuss the various approaches of taxonomy.
5. Critically describe the International code of Zoological nomenclature with examples.
6. Discuss taxonomic hierarchy.

CHAPTER 3

# Subkingdom Protozoa

The protozoans are the single-celled animals and the cell aggregates without histological differentiation. They are microscopic and unicellular. Study on protozoa started with the introduction of the microscope. As the single cell performs all the functions of an individual, the protozoans are best designated as acellular. Protozoa exhibit wide range of structural complexity and adaptations for various niches. Two important components of the protozoan cell are the nucleus and the cytoplasm. Nucleus may be compact or vesicular. Except ciliates, the nucleus is vesicular with scattered chromatin. Vesicular nucleus sometimes contains endosome or karyosome. DNA is absent in endosome of *Entamoeba* and *Trypanosoma*. In Apicomplexa, vesicular nucleus possesses one or more DNA containing nucleoli. Ciliates possess both micro- and macro- nucleus.

The cytoplasm is generally differentiated into ectoplasm and endoplasm. Various structures of the single-celled protozoa are cell organs or organelles. Larger alveoli in the protozoa are called *vacuoles* which are used for storage, digestion, drainage and hydrostatic skeleton. Protozoan shell may be nitrogenous or calcareous. Mineral shells contain groundwork of organic material. The simplest organization of the protoplasm is found in the protozoa. Though the single cells constitute the individual, the cytoplasm of these cells exhibits various types of differentiation and specialization. These cells are not simple from the functional adaptation point of view. In protozoan, a single mass performed functions of the protoplasm. Specialization for the particular function in the single cell is due to specialization of a cell part. In the metazoa, cells depend on each other, cells unite to form tissue; tissues form the organ, which in turn constitutes the system and finally the whole individual. Thus, there is a major difference in cellular organization between protozoa and metazoa. The concept of the protozoan as simple organism creates misunderstanding about the nature of these animals. Considering the protozoa as the noncellular means the body of a protozoan as a mass of protoplasm which modify for various functions, though the protoplasmic mass never divides into the cellular units. About 31,000 living protozoan species and about 10,000 parasitic protozoan are known.

The term *protozoon* was coined as far back as an equivalent for the German word Urthiere, meaning animals of primitive nature. von Siebold in the year 1853 pointed out for the first time the protozoa as a single cell. Anton van Leeuwenhoek observed protozoa for the first time using microscopes constructed with simple lenses. Human-inhabiting protozoans are unicellular eukaryotes, generally less than 50 μm in size. About 80,000 species of protozoa have been recorded recently, out of which 20,000 are found in fossil forms and about 10,000 species are parasitic. With a few exceptions, all protozoans are microscopic. Parasitic protozoan infect vertebrates, invertebrates, plants and other parasite. Protozoa exhibits the holozoic, saprozoic and holophytic mode of nutrition. They perform locomotion by pseudopodia, flagella or cilia. Pseudopodia, the temporary structure, are formed when required and retracted when not needed. Flagella arise from a basal granule or kinetoplast in the cytoplasm. Cilia are fine, short structures remaining embedded in the pellicle or ectoplasm. Flagellum and cilium show 9 + 2 stranded organization. Both sexual and asexual reproduction occurs in protozoa. Binary fission, multiple fission, and budding are the asexual mode of reproduction. Conjugation, syngamy and sporogony are the sexual mode of reproduction. Contractile vacuoles are found in fresh water forms but absent in parasitic forms. Excretion occurs directly through the body wall or by contractile vacuoles.

Parasitic forms are characterized by short generation times, high reproduction rate and a tendency to reinfect hosts, as host exhibits premunition, or sterile immunity against protozoan infection. Structurally, a protozoan is equivalent to a single eukaryotic cell. Their characteristic features are mega- and micro- nucleus of ciliates, kinetoplast, undulating membrane and even shell. Kinetoplast, a DNA containing structure is found in the mitochondrion of kinetoplastid flagellates. Typically plastids are associated with the photosynthetic organisms, but interestingly members of the Phylum Apicomplexa contain an organelle which appears to be a vestigial plastid.

Some protozoans possess a cytostome or cell mouth for ingestion. Contractile vacuole occurs in *Balantidium* for osmoregulation. In the apicomplexa, subpellicular microtubules provide a means for slow movement. A distinct undulating membrane is present in *Trypanosoma* between the body wall and flagella. Golgi apparatus, mitochondria, lysosomes, food vacuoles and conoids are present in the Apicomplexa. Extracellular chromatin granules, the so called *chromidia* may be found scattered throughout the whole or some part of the protoplasmic body in some species. Chromidia may be present besides the principal nucleus, or may be produced during certain phases of the life cycle. A protozoan generally passes through several stages in course of its life cycle. Trophozoite, the active or feeding stage is usually associated with pathogenesis. Some species form cysts containing one or more infective forms. Trace of nervous system is very seldom found but protozoa exhibit responses to the physical and chemical cues for their survival. Chemical signalling in protozoan involves molecules such as serotonin, acetylcholine and cyclic AMP.

## 3.1 PROTOZOAN POLYMORPHISM

This is considered under three heads

***Polymorphism in relation to life conditions***

Some protozoa produce a protective covering to form cyst or sporocyst to protect themselves in unfavourable conditions. During cyst stage, the vital body functions may be wholly or partly suspended, body becomes rounded off, and protoplasm becomes denser. In favourable conditions, the cyst envelope is dissolved and the contained organisms are set free to start the strenuous life. In sporocyst, multiplication of sporoplasm occurs to produce sprozoites.

***Polymorphism in relation to growth and development***

There is hardly any observable difference among different individuals during their active phases except in size. The individuals about to multiply by fission are slightly above the normal in dimensions. The resulting form produced from recent fission is smaller than the normal. Such differences are more marked, when the fission is unequal binary type.

***Polymorphism in relation to sex***

Commonly a double series of generation—asexual generation or schizogony, and sexual generation or sporogony occurs. Coccidia multiply by schizogony, which begins as a preparation for infecting new hosts when nutrient is abundant. In sporogony, the micro- and macro- gametocytes are well known. Microgametocyte is distinguished by more clear protoplasm free from coarse granules and a large nucleus richer in chromatin and produces microgamete. Macrogametocytes characterized by coarsely granular cytoplasm are rich in reserve food stuffs and a relatively small nucleus and produce microgamete.

## 3.2 FEATURES AT A GLANCE

1. Single celled eukaryotic organism
2. Single cell performs all the functions of an individual
3. They lack cell walls
4. Most protozoans are motile and microscopic
5. Terrestrial as well as aquatic
6. Locomotion occurs by the cilia, flagella, or by pseudopodia
7. Free-living or parasitic
8. Absorb the dissolved organic nutrients through the cell membrane, or ingest the solid particles or liquid droplets
9. Ingested food remains enclosed in the food vacuole

10. Some protozoa exist in a single form while some exhibit trophic stages and cyst stages
11. Parasitic forms complete their life cycle through the various stages often involving more than one host
12. Reproduction is either asexual or sexual
13. Asexual reproduction occurs through the budding, binary fission or multiple fission
14. Sexual reproduction occurs mainly through syngamy and conjugation

## 3.3 METAZOAN TENDENCIES

There is no broad gap for separating the simple, single-celled protozoa from the metazoa. Many protozoans, especially the members of the Mastigophora and Ciliata, show tendencies toward specialization of cells. Often the cells, produced from the division of a single cell, remain attached, forming a group, which functions as a unit and is termed a *colony*. Within such a group, the part of the cells engages in reproduction while the others carry on other functions. This mark an early separation of two types of cells, the germ cells and somatic cells. Separate germ cells occur in many protozoans.

## 3.4 HABITATS

Protozoans are found in diverse habitats. Except the Sporozoa, all classes have numerous free-living members, which occur in fresh water and salt water. Soil-inhabiting species are also common. Such broad distribution is perhaps associated with the cyst stage of the Protozoa. During the cyst stage, a protective wall surrounds the body wall. This encysted condition may be the part of a life cycle or may help to overcome the unfavourable conditions. When the cysts come into the favourable conditions, the cyst wall is dissolved and the protozoa start its active life. Many species are almost cosmopolitan in distribution. Protozoans inhabiting the open ocean and the deep sea are restricted in their distribution. Different regions in the same ocean may be inhabited by different protozoan fauna. Certain species inhabit the surface of the sea water, while some are found everywhere except the extreme depths of the seas. Many forms inhabit as cysts in the soil. Rhizopods and flagellates remain active in the ordinary garden soil and feed on bacteria of the soil. Active protozoans feed on important soil bacteria, although their effect on the soil remains unknown. The parasitic habit is exclusive among the Sporozoa, but in other classes the same habit is also found. Some species lead either independent or parasitic life, as opportunity is presented to the animal. Several diseases of man and domestic animals, some of which are serious, result from the protozoan parasites. Malaria, sleeping sickness, dysentery, Rocky Mountain spotted fever, and Texas fever are some of the diseases for which the causative agents are the protozoans.

## 3.5 FOOD HABIT

Many chlorophyll-bearing Mastigophora synthesize the food substances in a plant-like manner. The living forms in which the food is built up from the simple compounds by the photosynthesis are called the *autotrophic*. Most protozoans are holozoic and require complicated organic compounds as foods, which may be ingested as the solid particles through the action of pseudopodia as in *Amoeba,* or through a cytostome as in *Paramecium* and other ciliates and flagellates. In the endoplasm, solid food particles undergo digestion in the food vacuoles before the assimilation. Some species exhibit predaceous habit. *Didinium* lives largely on paramecia, the bodies of which ingests through a specialized cytostome. Suctoria through their hollow tentacles suck out the protoplasm from other organisms as the food. Organic matter is absorbed through the body surface in many species, especially which lack a cytostome. These organisms, dependent on the absorption of elaborated foodstuffs, may be either parasitic, or may utilize decomposing organic matter. In the latter case, they are called *saprophytic*.

## 3.6 SUPPORT AND LOCOMOTION

The cell surface exchanges materials with the environment, provides protection and structural integrity to the cell, and serves as a mechanical and chemical boundary. Many species maintain a spherical, radial or bilaterally symmetrical shape by thickening the cell membrane to form a pellicle, or by secreting scales, or a shell-like covering called the *test*. Locomotor capabilities are also due to the interactions between the cell surface and the surrounding medium. Pseudopodia, cilia and flagella provide the means by which protozoans perform the locomotion.

## 3.7 NUTRITION

Protozoans may be either autotrophic or heterotrophic, and some may be both. Photosynthetic members possess chloroplasts, although all do not use the same chloroplast pigments. Heterotrophic members acquire food from the environment. They may be saprozoic, which takes dissolved organics by diffusion, active transport, or pinocytosis. They may be holozoic by taking in the whole prey by the phagocytosis. Many are symbiotic on or within the other organisms. Protozoans that engage in pinocytosis or phagocytosis rely on the formation of food vacuoles, which may form at nearly any site on the cell surface, or at particular site associated with some sort of cell mouth or cytostome. Once a food vacuole is formed and moved into the cytoplasm, it swells as various enzymes and other chemicals are secreted into it. The vacuole first becomes acidic, and the vacuolar membrane develops numerous inwardly directed microvilli. During digestion, the vacuolar fluid becomes increasingly alkaline. The cytoplasm inside the vacuolar membrane becomes distinctive in appearance from the products of digestion. The vacuolar membrane forms tiny vesicles that pinch off and carry these products into the cytoplasm. The result is numerous, tiny, nutrient-carrying vesicles offering a greatly increased surface area for absorption of the digested products into the cell cytoplasm. The original vacuole gradually shrinks and undigested materials finally are expelled from the cell. In some species, the spent vacuole may be discharged anywhere on the cell surface. In ciliates and others in which a relatively impermeable covering exists around the cell, the covering bears a permanent pore called the cytoproct through which the vacuole releases material to the outside. Mitochondria produce ATP. The mitochondria have two membranes, but the inner membrane, or cristae has different forms—tubular, discoidal and lamellar.

## 3.8 REPRODUCTION

Many complex cycles seen in certain species involving alternation between the sexual- and asexual- processes, with repeated asexual divisions between brief sexual phases. Protozoans undergo several asexual reproductive processes including binary fission, multiple fission and budding. Binary fission involves a single mitotic division, which results in two daughter cells. During multiple fission, the nucleus undergoes several multiple divisions before cytokinesis and finally results in many daughter cells. The most frequent asexual reproduction is binary fission in which the nucleus undergoes an indirect division followed by the constriction and final division of the cytoplasm to produce two new individuals. Fission does not occur in an inactive state. Particular type of multiple fission is characteristic of particular group. In multiple fission, the nucleus undergoes repeated divisions. Each of the newly formed nuclear mass becomes surrounded by a cytoplasmic layer after the rupture of the mother cell wall, which releases numerous small cells called the *spores*. The process differs from mitosis of the metazoan cells. When the centrosome occurs within the nuclear wall, the division process may be accomplished without the disappearance of nuclear membrane. Chromatin is found as one or more large bodies and is called the *karyosome*. This constricts and divides without chromosome formation. In some flagellates and rhizopods, the chromatin is not confined to a nucleus rather spread uniformly in the cytoplasm. Such distributed chromatin granules form secondary nuclei (*Actinosphaerium*) which become the nucleus of the conjugating gametes. Some engage in plasmotomy, the process which is considered by some workers as a form of budding. Some members of the *Apicomplexa* undergo internal budding called the *endopolyogeny*, during which daughter cells form within the cytoplasm of the mother cell. Reduction division occurs, and haploid cells or nuclei of one kind or another are produced and then fuse to restore the diploid condition. Cells producing the gametes are usually called the *gamonts*. Syngamy may involve gametes those are similar in size and shape (isogamy), or the more familiar condition of gametes of two distinct types (anisogamy). Many species have a definite life cycle involving two distinct types of adult individuals—sporonts which produce gametes, and schizonts which produce asexual individuals. There may, thus, be an alternation of generations involving two distinct developmental cycles—a cycle of sporogony, and a cycle of schizogony, during which no sexual process occurs. These alternating cycles are correlated with the environmental conditions in the free-living forms. In the parasitic species, they are often correlated with change of the host. In the malarial parasites, schizogony continues during the development of the organism in the vertebrate host, and sporogony is restricted in the mosquito body. Budding differs from fission in the relative sizes of the resulting parts. In fission, two or more about equivalent parts result from the partition of one individual. Through budding, the identity of a parent organism is retained, as the buds arise as smaller outgrowths from the body of the daughter individual. Conjugation involves fusion of two or more individuals of the same species.

#### *Endomixis*

An intricate process of the nuclear reconstruction without the fusion of the individuals occurs in some species which is termed *endomixis* and is observed especially in *Paramecium*, where the macronucleus gives off budlike fragments at regular intervals. These fragments are absorbed in the cytoplasm. Macronucleus is being broken up. Micronucleus undergoes repeated divisions and result in the formation of eight products. In *P. caudatum* which has a single micronucleus, three divisions produce the eight micronuclear bodies. In *P. aurelia* which has two micronuclei, only two divisions of the micronuclei occur. Part of the eight micronuclear bodies disintegrates, some among the remainder form micronuclei, and others produce macronuclei. Thus, wholly new nuclear material is formed from a portion of a micronucleus in a manner strikingly similar to the nuclear reorganization following conjugation. In *P. caudatum*, endomixis takes place at regular intervals of about 60 days, while in *P. aurelia,* the period is approximately 30 days. In some other ciliates, endomixis is observed, but there it occurs during inactive period.

## 3.9 CLASSIFICATION WITH CHARACTERS

1. Single-celled, eukaryotic organisms.
2. Size varies between 1 μm to 50 μm or more.
3. Free-living or parasitic.
4. Solitary or colonial.
5. With a single vesicular nucleus or multinucleate.
6. Holozoic and saprozoic, a few are holophytic.
7. Locomotion occurs by pseudopodia, flagella or cilia.
8. Reproduction is sexual or asexual.
9. Encystment occurs.
10. Contractile vacuoles are found in the fresh-water forms.

According to Levine et al., subkingdom protozoa includes seven phyla. The scheme of classification followed here is given by Levine et al.

#### *Phylum Sarcomastigophora*

1. Single type of nucleus is present.
2. Sexuality occurs by the syngamy.
3. Flagella, pseudopodia or both are present.

**Example:** *Noctiluca, Euglena, Trypanosoma, Amoeba, Entamoeba, Potystomella.*

#### *Phylum Labyrinthomorpha*

1. Spindle shaped or spherical.
2. Nonamoeboid cells are found in the ecto-plasmic network.
3. Senogenetosome is associated with the ecto-plasmic network.
4. Heterokont zoospores are found.
5. Saprozoic and parasitic.

**Example:** *Labyrinthula, Thraustochytrium.*

#### *Phylum Apicomplexa*

1. Apical complex is present.
2. Generally, microspore is present.
3. Cilia are absent.
4. Sexuality occurs by the syngamy.
5. All species are parasitic.
6. The shape of cell is maintained by the pellicle consisting of alveolar vesicles.
7. Some species produce flagellated- or amoeboid-gametes
8. Mitochondria with vesicular cristae is found.
9. Single vesicular nucleus is present.
10. Asexual reproduction occurs by binary fission, multiple fission, or endopolygeny.
11. Sexual reproduction is gametic. Gametes are either isogametous, or anisogametous.
12. Meiosis involves single division after the formation of zygote.

13. Plastids are absent.
14. Carbohydrates are stored as paraglycogen.

**Example:** *Monocystis, Plasmodium, Eimeria, Babesia.*

***Phylum Microspora***

1. Obligatory, intracellular parasites.
2. Mitochondria is absent.
3. Often dimorphic in the sporulation sequence.
4. Unicellular spores with imperforate wall are found.
5. Polar cap and polar tubes are present.

**Example:** *Nosema, Thelohania, Pleistophora.*

***Phylum Ascetospora***

1. All are parasitic.
2. Spores are multicellular.
3. Spores are provided with one or more sporoplasms.
4. Polar capsules are absent.
5. Pores may be present, or absent in the spore wall.

**Example:** *Haplosporidium, Urosporidium.*

***Phylum Myxozoa***

1. Spores of multicellular origin are found.
2. Spores are provided with one or more polar capsules and sporoplasm.
3. All species are parasitic.
4. Spore is composed of one to three valves.
5. Polar capsule is provided with coiled polar filaments.

**Example:** *Myxidium, Myxobolus.*

***Phylum Ciliophora***

1. Simple cilia and compound ciliary organelles are present.
2. Usually two types of nuclei-macronucleus and micronuceus are found.
3. Binary fission is transverse.
4. Budding and multiple fission also occur.
5. Sexuality involves conjugation, autogamy, and cytogamy.
6. Nutrition is heterotrophic.
7. Most species are free living.
8. Presence of cytostome and cytopyge.
9. Anterior part of the body contains the buccal ciliature.
10. Shape of the cell is maintained by the pellicle consisting of alveolar vesicles and fibrous layer.
11. There are two microtubular roots and one fibrous root in association with the basal bodies of cilia. These roots plus basal bodies are collectively known as infraciliature.
12. Mitochondria with tubular cristae is present.
13. Macronucleus divides amitotically by simple constriction.
14. Plastids are absent.
15. Carbohydrates are stored as glycogen.

**Example:** *Paramoecium, Batantidium, Vorticella, Stentor.*

## 3.10 CLASSIFICATION OF PHYLUM SARCOMASTIGOPHORA

1. Single type of nucleus is present, except in Foraminiferida.
2. Sexuality, when present, by syngamy.
3. Flagella, pseudopodia or both types of locomotor organelles are present.

***Subphylum Mastigophora***

1. One or more flagella, typically, are present.
2. Asexual reproduction takes place by the binary fission.
3. Sexual reproduction is known to occur in some groups.

***Class Phytomastigophorea***

1. Chloroplast is present
2. Mostly free living.

**Example:** *Euglena, Ceratium, Volvox.*

***Class Zoomastigophorea***

1. Chloroplast is absent.
2. One or many flagella are present.
3. Amoeboid forms are provided with or without flagella.
4. Sexual reproduction occurs in few groups.

**Example:** *Trypanosoma, Leishmania, Giardia.*

***Subphylum Opalinata***

1. Numerous cilia occur in oblique rows over the entire body surface.
2. Cystostome is absent.
3. Life cycles involve syngamy with anisogamous flagellated gametes.
4. All are parasitic.

*Class Opalinatea*

Characters are same as that of the subphylum.

**Example:** *Opalina*

*Subphylum Sarcodina*

1. Pseudopodia are present or locomotive protoplasmic flow may occur without distinct pseudopodia.
2. Flagella are found in the developmental, or other temporary stages.
3. Body is naked, or provided with test or skeleton.
4. Asexual reproduction occurs by the fission.
5. Sexuality is associated with the flagellated gametes, or more rarely with amoeboid gametes.
6. Most species are free living.

*Superclass Rhizopoda*

Locomotion occurs by the lobopodia, filopodia, or reticulopodia, or by protoplasmic flow without distinct pseudopodia.

*Class Lobosea*

1. Pseudopodia are lobose, filiform and are produced from the broader hyaline lobe.
2. Usually uninucleate.
3. Multinucleate forms are not flattened.
4. No sporangia or similar fruiting bodies are found.

**Example:** *Amoeba, Entamoeba, Naegleria.*

*Class Acarpomyxea*

1. Small plasmodia or much expanded similar uninucleate forms, usually branching, sometimes forming reticulum of coarse branches.
2. Advancing tips are lobose.
3. Test is absent.
4. Spores of fruiting bodies are absent.
5. No regular reversal of streaming.

**Example:** *Leptomyxa, Corallomyxa.*

*Class Acrasea*

1. Uninucleate amoebae with eruptive, lobose pseudopodia.
2. Amoebae aggregate to form the pseudoplasmodium that produces fruiting bodies without the stalk tube.
3. Flagellate cells are known in only one species.
4. Sexuality is unknown.

**Example:** *Acrasis.*

*Class Eumycetozoea*

1. Myxamoebae with filiform subpseudopodia.
2. Flagella sometimes are present.
3. Aerial fruiting bodies are found.
4. Stalk tubes are typically present.

**Example:** *Licea, Trichium.*

*Class Plasmodiophorea*

1. Obligate intracellular parasites with plasmodia.
2. Zoospores are produced in the zoosporangia and bears anterior pair of unequal flagella.
3. Resting spores are formed in the compact sori or loose clusters within the host cells.
4. Sexually is reported in some species.

**Example:** *Plasmodiophora, Sorosphaera.*

*Class Filosea*

1. Filiform pseudopodia, often branching, and sometimes anastomosing.
2. Spores or flagellate stages are absent.

**Example:** *Vampyrella, Euglypha.*

*Class Granuloreticulosea*

Delicate finely granular reticulopodia or rarely finely pointed, granular, but nonanastomosing pseudopodia.

**Example:** *Elphidium, Rosalina.*

*Class Xenophyophorea*

1. Multinucleate plasmodium, remain enclosed in the branched tube system which is composed of the organic substance.
2. Barium sulfate crystals are found in the cytoplasm.
3. Faecal pellets remain outside the organic tube system.
4. Test of foreign matter is present surrounding the tube system and faecal pellet masses.
5. Marine.

**Example:** *Psammina, Stannophyllum.*

*Superclass Actinopoda*

1. Often spherical, usually planktonic.
2. Axopodia is present.
3. Skeleton is composed of organic matter and silica, or of strontium sulfate.

4. Reproduction is asexual and/or sexual.
5. Trophic cells are rarely flagellated.

### Class Acantharea

1. Skeleton is made by strontium sulfate, usually made of 20 radial or 10 diametrical spines, rarely 16 diametrical or 32 radial spines is present.
2. Sometimes many spines are randomly oriented.
3. Spines are about joined in cell centre.
4. Extracellular outer cortex and inner envelops usually are present.
5. Inner envelope often closely line the central cell mass.
6. Marine, usually planktonic.

**Example:** *Acanthocolla, Conacon.*

### Class Polycystinea

1. Siliceous skeleton is present in most species.
2. Mostly planktonic, one genus is benthic.
3. Capsular membrane is composed usually of grossly polygonal plates and conjoining more than three pores.
4. Marine, planktonic.

**Example:** *Plagonium, Octodendron.*

### Class Phaeodarea

1. Skeleton of mixed silica and organic matter consists of usually hollow spines and/or shells are present.
2. Capsular membrane is very thick.
3. Marine, planktonic.

**Example:** *Astracantha, Phaeodina.*

### Class Heliozoea

1. Central capsule is absent.
2. Skeletal structures, if present, are siliceous or organic.
3. Axopodia radiate on all the sides.
4. Most species inhabit fresh water, some are marine.

**Example:** *Ciliophrys, Clathrulina.*

## 3.11 CLASSIFICATION OF PHYLUM CILIOPHORA

1. Simple cilia, or compound ciliary organelles are typically present in at least one stage of life cycle.
2. Subpellicular cilia are present, even when surface cilia is absent.
3. Two types of nuclei are found with rare exception.
4. Binary fission is transverse.
5. Budding and multiple fissions occur.
6. Sexuality involves conjugation, autogamy and cytogamy.
7. Nutrition is heterotrophic.
8. Contractile vacuole is typically present.
9. Most species are free-living.
10. Many are commensal, and some are truly parasitic.
11. Large numbers are found as symphorionts on a number of hosts.

### Class Kinetofragminophorea

1. Oral cilia are slightly distinct from the somatic cilia.
2. Cystostome is often apical, or mid-ventral on the body surface, or at bottom of atrium, or vestibulum.
3. Cytopharyngeal apparatus is prominent.
4. Compound cilia are typically absent.

**Example:** *Balantidium, Heliochona, Stylochona.*

### Class Oligohymenophorea

1. Oral apparatus are well defined, although absent in one group.
2. Oral cilia are distinct from the somatic cilia.
3. Cytostome is usually ventral and/or near anterior end, present at bottom of the buccal, or infundibular cavity.
4. Cysts are common.
5. Various species are loricate.
6. Colony formation is common in some groups.

**Example:** *Paramoecium, Vorticella, Trichodina.*

### Class Polymenophorea

1. Somatic cilia are complete, or reduced.
2. Cystostome is found at the bottom of the buccal cavity.
3. Cytoproct is often absent.
4. Cysts, particularly the loracae, are common in some groups.
5. Large, free-living, free-swimming forms are found in several habitats.

**Example:** *Stentor, Nyctotherus, Lagotia* (Figures 3.1–3.10).

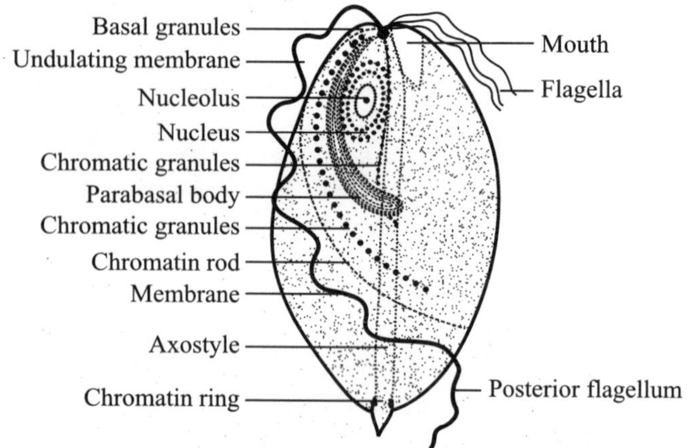


**Figure 3.1** *Trichomonas.*

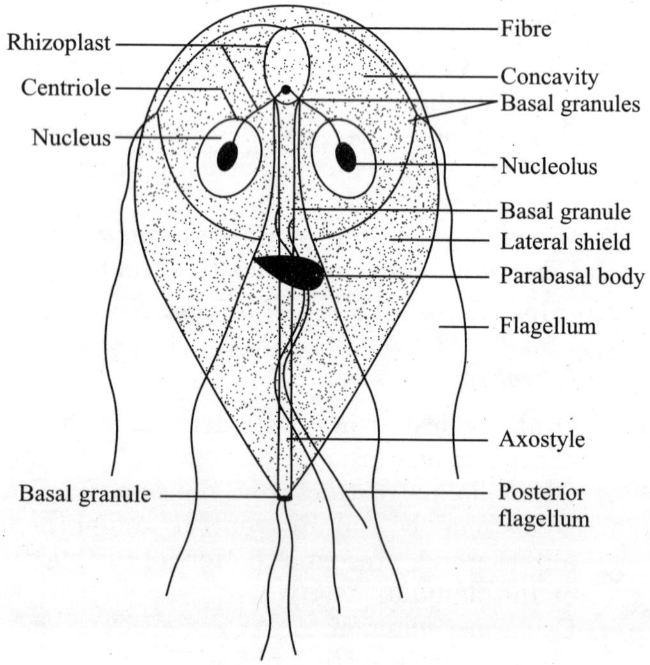


**Figure 3.2** *Giardia.*

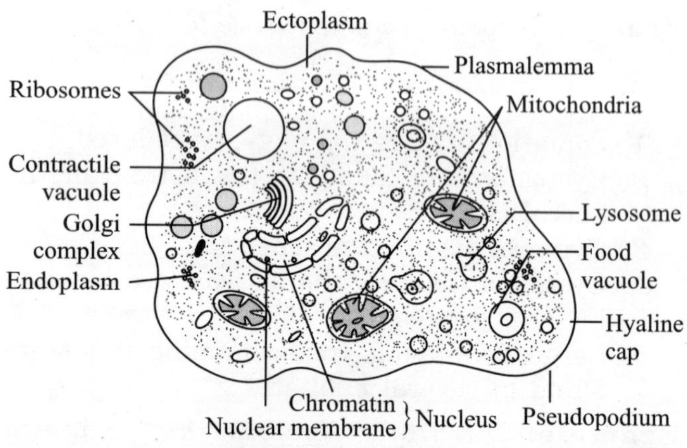


**Figure 3.3** *Amoeba.*

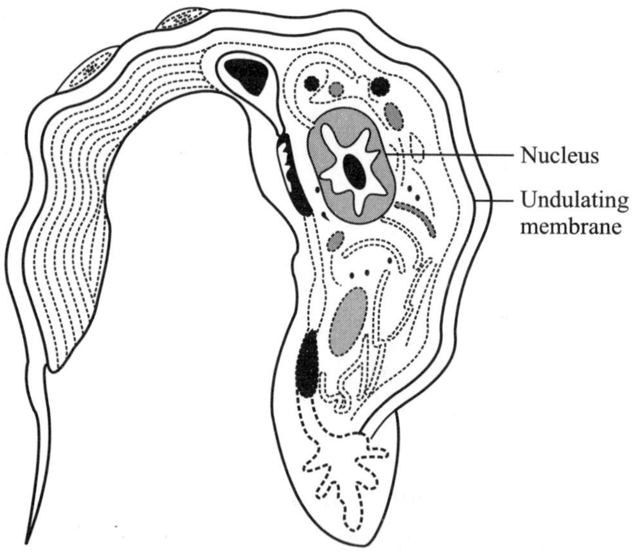


**Figure 3.4** *Trypanosoma.*

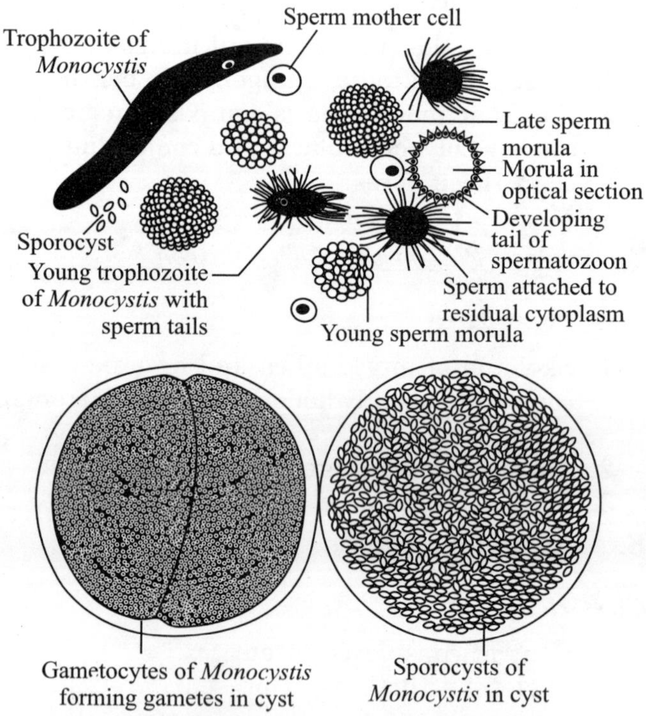


**Figure 3.5** *Monocystis.*

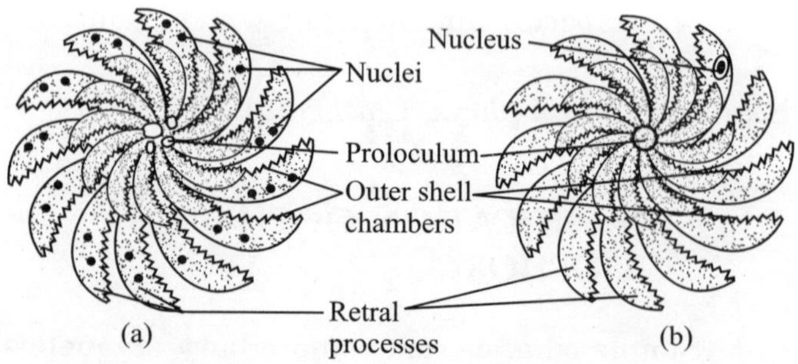


**Figure 3.6** *Elphidium* (*Polystomella*): (a) Dorsal view (b) Ventral view.

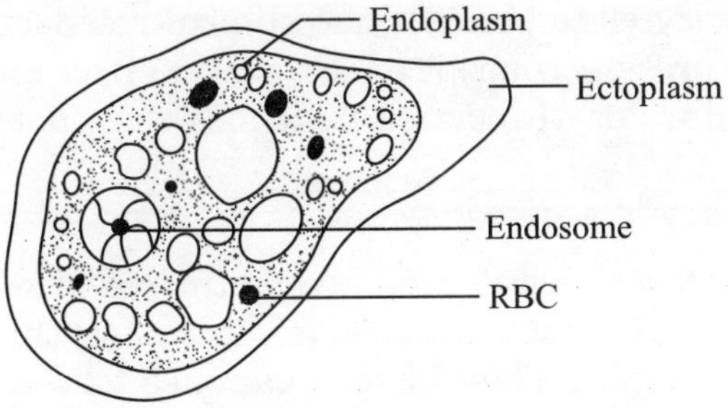


**Figure 3.7** *Entamoeba histolytica.*

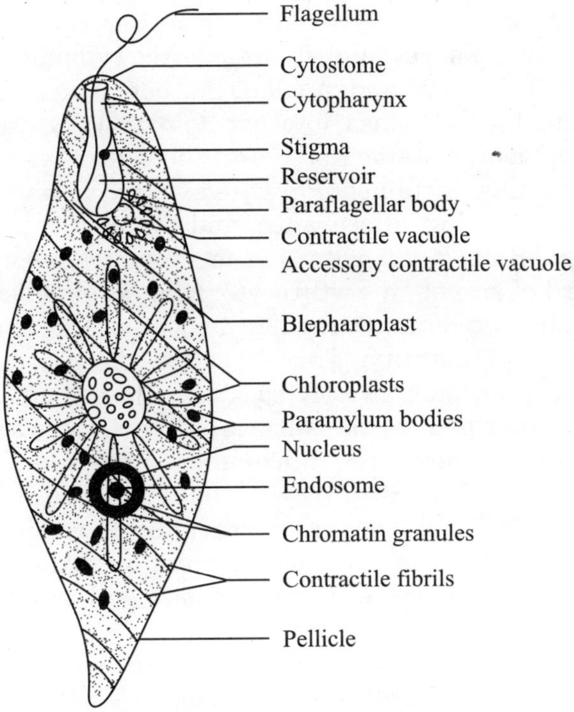


**Figure 3.8** *Euglena.*

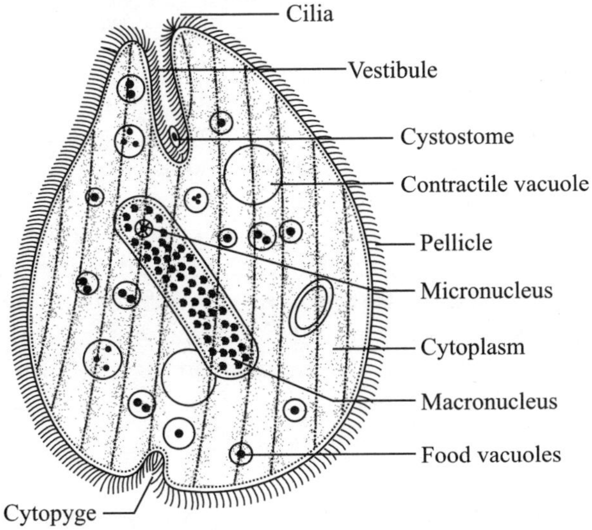


**Figure 3.9** *Balantidium.*

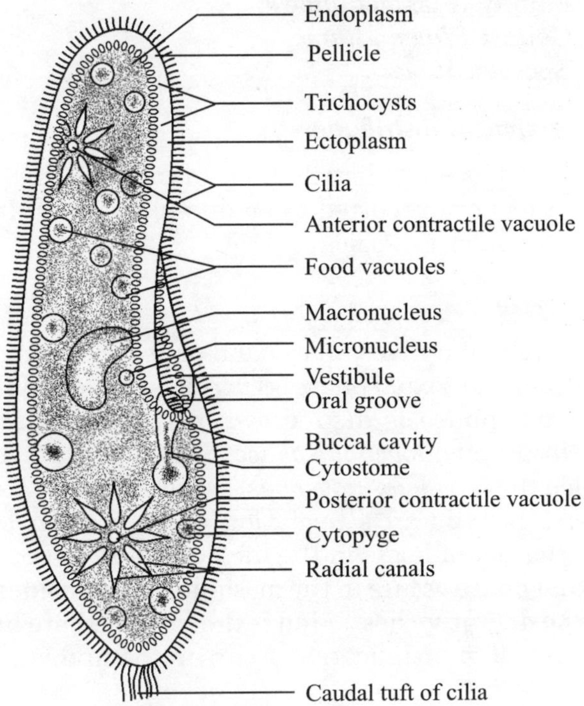


**Figure 3.10** *Paramoecium.*

## 3.12 TYPE STUDY: *Plasmodium vivax*

*Plasmodium vivax*, the parasitic protozoa, causes the disease malaria. Malaria remains one of world's great health problems. Laveran discovered the causative agent of malaria in red blood cells of a patient in 1880. In 1886, Golgi in Italy described asexual development of parasite in RBC, which is therefore called the Golgi cycle. In 1894, Manson hypothesized that mosquito transmitted the malaria. Romanowsky in Russia, in 1891, developed a method of staining the malaria parasites in blood films. In 1897, Ronald Ross, in Secunderabad (India), found malaria parasites (oocysts) on the stomach wall of the *Anopheles* mosquito. Ross completed the work on the life cycle of bird's malaria parasite in Calcutta in 1898.

### *Systematic position*

Subkingdom: Protozoa
Phylum: Apicomplexa
Subphylum: Sporozoea
Class: Telosporea
Subclass: Coccidia
Order: Eucoccidia

Suborder: Haemosporina
Family: Plasmodiidae
Genus: *Plasmodium*
Species: *Vivax*

### *Geographical distribution*

*P. vivax* has wide distribution. In India, 65 to 69% infections are reported to be due to *P. vivax*. It is less common in Africa.

### *Life cycle*

Life cycle comprises an asexual phase occurring in the man and sexual phase occurring in the mosquito. Asexual phase is also known as the *vertebrate*, *intrinsic*, or *endogenous phase*. Sexual phase is also called the *extrinsic*, *exogenous*, or *invertebrate phase*. There is thus an alternation of generation and alternation of hosts in the life cycle of *P. vivax*. As sexual phase occurs in the mosquito, it is considered as the definitive host. Man is the intermediate host as asexual multiplication occurs in the human.

### *Asexual phase*

Man gets infected through the bite of infected female *Anopheles*. Sporozoites, the infective forms of the parasite, are present in the salivary gland of the infected mosquito. Sporozoites are injected into the human blood capillaries, when mosquito feeds on the blood.

### *Exo-erythrocytic schizogony*

In blood stream, the phagocytes destroy many sporozoites. Some reach the liver and enter hepatocytes. On entering the liver cell, narrow and slightly curved sporozoites become rounded. They enlarge in size and undergo repeated nuclear division. Cytoplasm surrounds each daughter nucleus. This stage of parasite is called the *pre-erythrocytic* or *primary exo-erythrocytic schizont*. Pre-erythrocytic schizogony involves a very small portion of the liver cells, and causes no significant damage or clinical illness. Mean duration of the pre-erythrocytic schizogony is 8 days. Mean number of merozoites produced per tissue schizont is 10,000. Mature schizont bursts and liberates the merozoites. In 1980, it was discovered that some sporozoites, on entering hepatocytes, become rounded and remain dormant, a stage called the hypnozoites (*hypnos*, meaning sleep). Hypnozoites activate from time to time to become the schizonts. Schizonts release the merozoites, which infect erythrocytes to produce the clinical relapses. This is the present concept of relapse in vivax malaria. Secondary erythrocytic schizogony does not occur.

### *Erythrocytic schizogony*

Merozoites thus released invade the red blood cells. Merozoites attach themselves to the erythrocytes by their apex. They have organelles for secreting a substance that produce a pit on the erythrocyte membrane. Merozoite then enter erythrocyte by the endocytosis and red cell membrane seals itself. In erythrocyte, the merozoite appears as a rounded body with a vacuole in the centre, cytoplasm in the periphery, and nucleus at one pole. When stained with Giemsa, or other Romanowsky stains, cytoplasm is stained blue and the central vacuole remaining unstained. Parasite exhibits an annular or 'signet ring' appearance, and is called the *ring form*. Ring is 2.5 μ to 3 μ in diameter, about a third of size of an erythrocyte. Ring form enlarges in size, becoming irregular in shape and shows amoeboid motility. This is called the *amoeboid form*. Infected erythrocytes show enlargement and Schuffner dots on the surface. At particular stage of development, the nucleus of amoeboid form starts dividing. Erythrocyte inhabiting parasite till the time its nucleus starts dividing is called the *trophozoite* (from trophos, meaning growth). This form is called the early trophozoite and amoeboid form the late trophozoite.

Parasite with dividing nucleus within the erythrocyte is called the schizont. Nucleus first divides into several small nuclei, while cytoplasm remains undivided. These are called the early schizont. When each daughter nucleus becomes surrounded by the cytoplasm (merozoite), then it is called the *late schizont*. Mature schizont bursts and releases the merozoites into the circulation. Schizont occupies virtually whole of the enlarged red cell. There are about 12 to 24 (usually 16) merozoites per schizonts. Erythrocytic schizogony lasts for about 48 hours. Merozoites, after releasing from schizont, invade fresh erythrocytes in which they go through the same process of development. Interval between the entry of sporozoite into the host and earliest manifestation of the clinical illness is called the incubation period.

### *Gametogony*

After many cycles of erythrocytic schizony, some merozoites that infect red cells do not form the schizonts, but develop into the sexually differentiated gametocytes. Gametocytes appear usually within 3 days after the first appearance of trophozoites. Both male and female gametocytes are large, nearly filling the enlarged RBC. Macrogametocytes have dense, deep blue staining cytoplasm and a small compact nucleus. Microgametocyte has pale staining cytoplasm and the large diffuse nucleus. Pigment granules are prominent in the gametocytes (Figure 3.11).

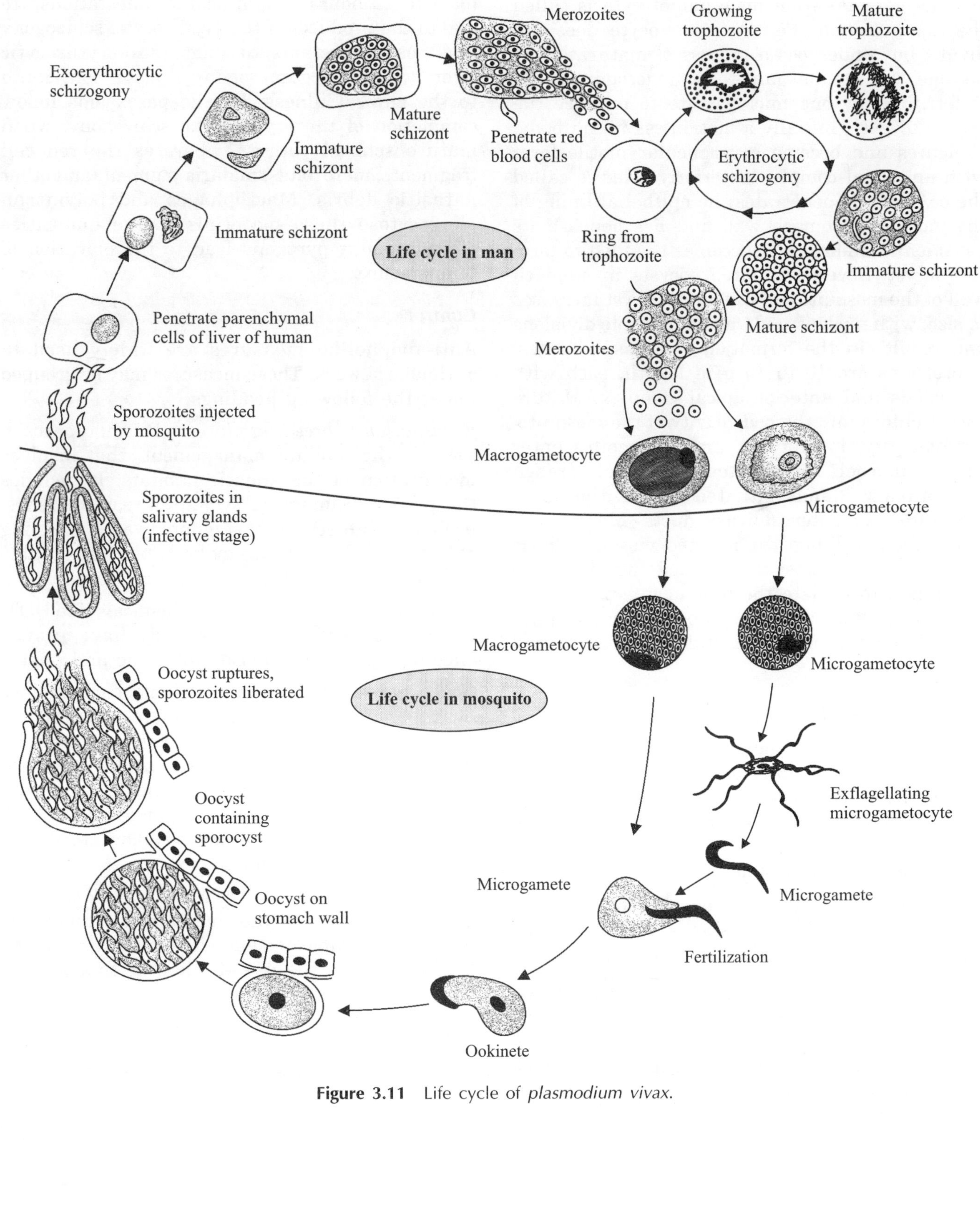


**Figure 3.11** Life cycle of *plasmodium vivax.*

### Sexual phase

Within the fifteen minutes of entering into mosquito stomach, male gametocyte divides into 4 to 8 nuclei, from each of which protrudes a long, actively motile, whip-like filament, or male gametes (microgametes). The process of formation of male gametes from male gametocyte is called the *exflagellation*. Female gametocyte does not divide, but undergoes a process of maturation to become the female gamete or macrogamete. It is fertilized by one microgamete to produce the zygote. Zygote, initially a motionless round body, elongates and become a vermicular motile form with an apical complex anteriorly. This is called the *ookinete*. It penetrates the epithelial lining of the mosquito stomach wall and lies beneath its basement membrane. It becomes rounded to form the oocyst. Several hundred oocysts in stomach wall of the mosquito can be found. Oocyst increases in size, with nucleus undergoing multiple divisions and results in the formation of the sporozoites. Sporozoites are 10 to 15 µ in length, each with a nucleus and anterior apical complex. Mature oocyst bulges into the body cavity of the mosquito. On maturity, it ruptures, and sporozoites enter the haemocoel. Sporozoites, ultimately, reach the salivary glands situated in the thorax of mosquito and enter salivary ducts. Mosquito is now infective. When the infected mosquito feeds on man, sporozoites are injected into the skin capillaries to initiate the human infection. Time taken for completion of sporogony in mosquito depends on the environmental temperature and species.

### Incubation period

Interval between the entry of sporozoite into the host and earliest manifestation of the clinical illness is the incubation period. In the case of *P.vivax* the period ranges from 10 to 17 days (average 15 days).

### Clinical features of malaria

Febrile paroxysm starts generally in the early afternoon, although it may start at any time. Each paroxysm shows a succession of three stages.

*Cold stage:* This stage lasts from 15 to 60 minutes. The patient feels intense cold and uncontrollable shivering.

*Hot stage:* This stage lasts for 2 to 6 hours. The patient feels intensely hot. Fever mounts to 102 degrees centigrade or higher. Severe headache, nausea and vomiting are common.

*Sweating stage:* This stage lasts for 2 to 3 hours. The patient is drenched in profuse sweat, and the temperature drops quickly. Patient falls into the deep sleep to wake up refreshed. Paroxysm lasts for 8 to 12 hours. All clinical manifestations are due to the products of the erythrocytic schizogony and host's reactions to them. Exo-erythrocytic liver cycle and gametogony do not contribute to the clinical illness. Febrile paroxysms follow completion of the erythrocytic schizogony, when mature schizont bursts, liberates the red cell fragments, merozoites, malaria pigment, and other parasitic debris. Macrophages and polymorph phagocytose these materials. Large quantities of endogenous pyrogens lead to the elevation of temperature.

### Control

Anti-anopheline measures are widely used in antimalaria work. These measures may be grouped under the following headings.

*Elimination of breeding places:* It is an approach towards the "habitat management" that involves modification of the aquatic habitats. It includes the proper drainage of swampy areas, marshes, pools and ditches with stagnant water, cleaning of bushes and shrubs, to make habitat unsuitable for the development of the mosquito larvae.

*Chemical larvicides:* Certain chemicals like DDT, DDD, and benzene hexachloride have proved effective against the mosquito larvae and pupae.

*Genetical control:* Attempts to control anophelines by the mass production of male sex and its sterilization by chemosterilants have been confined to the *A. albimanus*. An initial, small-scale field experiment in a 12 sq km area in the region of Lake Apastepeque in El Salvador was carried out in 1972 using ENT-61585. This produced 99.8% sterility in the males and 91.3% in females.

*Biological control:* Some fishes such as *Haplochilus* and *Gambusia* eat larvae, and in certain conditions constitute a simple and effective biological measure. It also includes the use of microbes that attacks mosquito larvae.

### Prophylaxis or prevention of infection

- Mosquito nets should be used, especially during the night.

- Exposed body parts may be protected by repellents, such as odomos, mustard oil, dimethyl phthalate or dimethyl carbamate.
- House should be built on the high grounds, having good drainage and away from the vegetation and marshy places.
- House should be made mosquito proof by the adequate screening of all the doors, windows and ventilators.

### Treatment

Antimalarial therapy should ideally destroy all the asexual forms of the parasite to cure the clinical illness, eliminate sporozoites, and exo-erythrocytic forms to prevent relapse, and kill gametocytes. No single drug satisfies all these objectives. Combinations of drugs are, therefore, necessary. Treatment strategy is to cure clinical disease with the blood schizonticidal drugs like chloroquine (600 mg stratum, 300 mg after 6 hours, 300 mg for next 2 days). Chloroquine does not destroy exo-erythrocytic parasites, and a tissue schizonticidal, drug like primaquine (15 mg daily for 5 days) should be administered. Primaquine is active against the gametocytes. For travellers, visiting endemic areas, chemoprophylaxis provides the effective protection. Prophylaxis should begin on the day of arrival and be continued for 4–6 weeks after the departure. Drugs recommended for chemoprophylaxis are chloroquine, amodiaquine and fansidar in weekly doses.

### Comment

As malaria has no extra human reservoir, its eradication is feasible by the elimination of the vector mosquito and treatment of the patients and carriers. In India, National Malaria Control Programme operated very successfully for 5 years, bringing down annual incidence of malaria, from 75 million in 1953 to 2 million in 1958. National Malaria Eradication Programme was introduced in 1958 with the objective of ultimate eradication of the disease.

## 3.13 *Trypanosoma cruzi*

Carlos Chagas accidentally found *T. cruzi* in the intestine of a tritomid bug and found the trypanosome as a causative organism for the disease, South American trypanosomiasis. Trypomastigotes of *T. cruzi* is about 20 μm in length, characteristically curved, generally C-shaped, 12–30 microns in length, with a narrow membrane and with the flagellum of medium length. Its posterior kinetoplast is larger than the kinetoplast of *T. brucei*. Nucleus is central in location. Metacyclic trypanosomes remains in the posterior position of vector's gut and are passed out in faeces. Infection is contaminative. Trypanosomes development occurs in the posterior region of the vector and therefore, they are called the *stercocarian*. Infective forms appear in the insect's faeces and hosts are infected through the contaminative route.

### Geographical distribution

In the southwestern USA, silvatic *T. cruzi* is mainly found in the wood rat populations. Only 3 cases of disease caused by the accidental contact with the silvatic cycle are known from the USA. About 7 million people suffered from the disease in Latin America in 1960 as per WHO estimation. As per WHO, some 16 million individuals were infected or at least serologically positive for *T. cruzi* and 65 million people were exposed to the risk of infection. Transmission via blood transfusion causes the disease in urban areas. The disease is endemic in 21 countries. Around 100 million people in Central and South America are reported to be exposed to reduviid bugs and the prevalence of Chaga's disease is estimated to be about 16–18 million. The migration of rural people to urban areas during 1970s and 1980s are known to change the traditional epidemiology of Chaga's disease and it became established as an urban disease, due to the unscreened blood transfusion. Between 1960 and 1989, the prevalence of infected blood in blood banks in selected cities of the South America is reported to range from 1.7% in the Sao Paulo, Brazil to 53% in the Santa Cruz, Bolivia, a percentage far higher than that of HIV infection.

### Life cycle

When an infected reduviid bug bites an individual, trypomastigotes enter their alimentary canal, ultimately reach the midgut where they are changed into the epimastigotes. Epimastigotes migrate to the hindgut, proliferate and subsequently develop into the metacyclic trypomastigotes, the infective forms, and are excreted along with insect faeces. Human infection occurs through rubbing the bite wound or entry through the mucosal surfaces, particularly the conjunctiva by the person's fingers. The parasite spreads through the lymphatic system and is transformed into the amastigote forms, which divide by the binary fission. To complete the life cycle, these procyclic forms make their way to the salivary glands. The exact mechanism by which the parasite migrates from tsetse gut to the salivary glands is unknown. Two proposed routes are:

1. the classical route in which the parasite backtracks through digestive system and migrates up the salivary duct,
2. the direct route in which the parasite penetrates the peritrophic membrane and gut epithelium of the vector to gain the access to the hemolymph.

Both routes present barriers and illustrate the complex interactions between the parasites and vector. The parasite converts to an epimastigote form after reaching the salivary gland, attaches to the epithelial cells of salivary glands via its flagellum and undergoes further replication. Differentiation into metacyclic trypomastigotes involves the appearance of surface coat and changes in mitochondria to accompany the migration of kinetoplast to the posterior end. These developmental changes are necessary preparations of the parasite for becoming infective to the host. Mature metacyclic trypomastigote detaches from the epithelial cells and remains in lumen of the salivary gland until being expelled during the tsetse feeding (Figure 3.12).

After passing through the promastigote and epimastigote forms, the parasites transform into the trypomastigote, which are released into the blood stream. No multiplication of trypomastigote occurs. Metacyclic transmission requires to be separated from the mechanical transmission. Metacyclic transmission requires a lapse of time to allow the trypanosomes to attain the infective

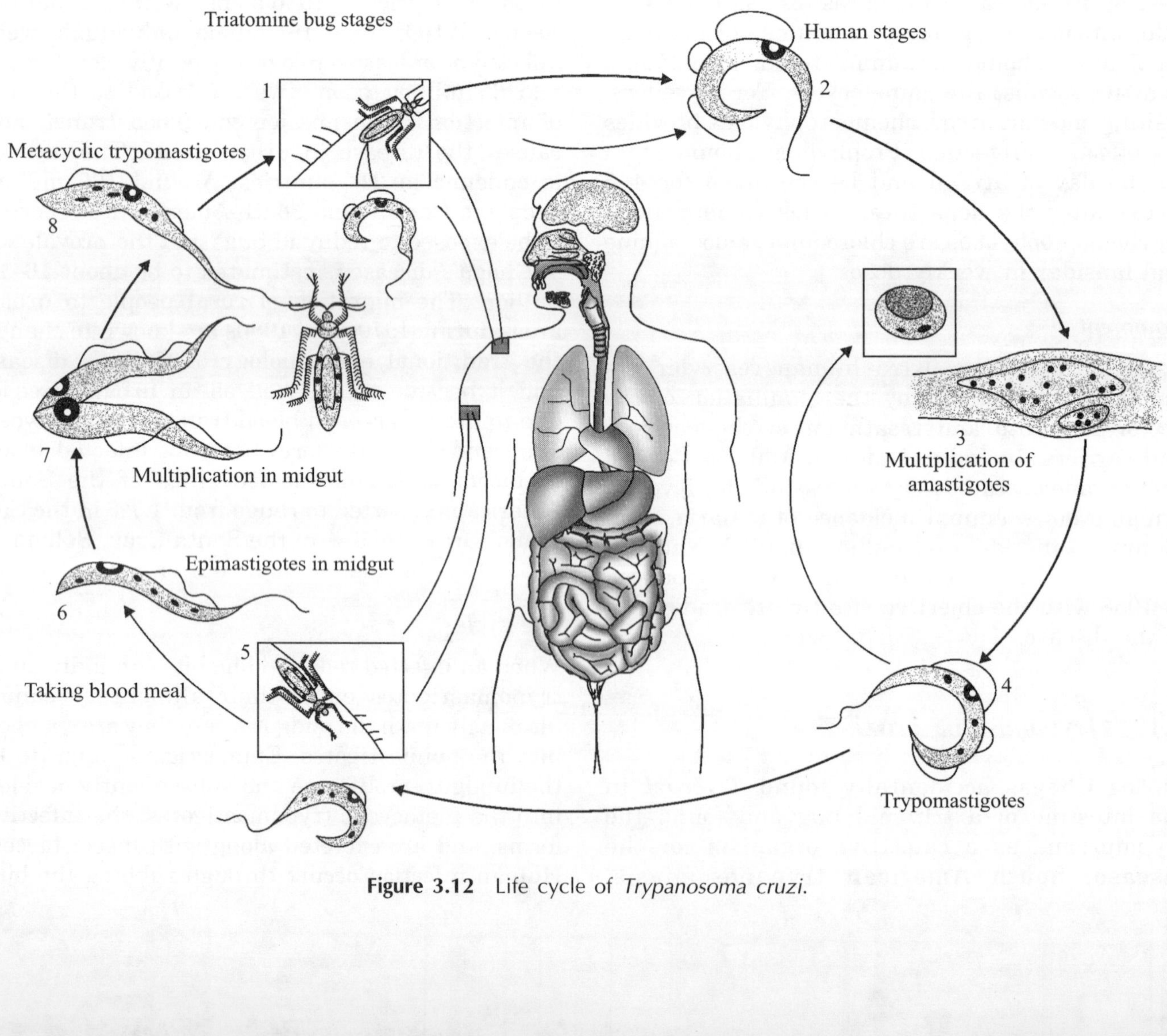


**Figure 3.12** Life cycle of *Trypanosoma cruzi*.

stage by a development sequence in the vector. Trypomastigotes are deposited in faeces on the skin, when the triatomine bug feeds. The bug generally bites round the edges of the mouth and eyes. The life cycle completes, when a triatomine bug vector ingests the circulating trypomastigotes. In the vector, the trypomastigotes develop into the epimastigotes, multiply by the binary fission in the gut of bug. Within 15 days, the metacyclic trypomastigotes are found in the hindgut of the bug (Figure 3.12).

#### Symptoms

Many people infected with the *T. cruzi* remain asymptomatic and free from Chagas disease, or experience only an acute infection without progressing to the chronic stage. The severe form of disease is commonly seen in the children younger than 5 years of age. Multiplication of *T. cruzi* at the infection site produce an inflamed swelling (chagoma) which persists for weeks. Trypomastigotes or amastigotes may be found in the aspirate of chagoma. Regional lymph nodes get infected involving one side of the face. Unilateral oedema of the upper and lower eyelid may occur. This is known as *Romana's sign*. In the acute stage, trypomastigotes are found in the blood.

Symptoms pass unnoticed but there may be fever, increased pulse rate, enlargement of lymph glands, liver and possibly spleen. Muscle aches and pains are also recorded at this stage and parasites may be found in the blood films. The acute form is often seen in the young children and causes serious damage to the heart and other problems leading to the death involving central nervous system. Chronic manifestations include the signs of cardiac muscle damage. Dilation of the digestive tract results in the megaoesophagus and megacolon may occur. About 10% of the infected persons develop chronic Chagas cardiopathy.

*T. (T.) b. gambiense* has a mammalian reservoir other than the man, which may be important for the persistence of foci. *T. (T.) b. rhodesiense* is fundamentally a parasite infecting ruminants, which are the principal reservoir hosts. Sleeping sickness occurs at a low level of transmission in most countries of the tropical Africa with occasional epidemic outbreaks and at least 50 million people are at risk. *T. cruzi* infection exists in many areas as a silvatic cycle between silvatic Triatominae, rodents, and marsupials. In many areas, there is no human contact with silvatic cycle.

Human infection occurs only through accidental contact with that cycle or by initiation of a domestic cycle of transmission, when the bugs invade and colonize houses. In the acute phase, a small red nodule (Chagoma) may form at the site of the bite. Romana's signs comprise fever, oedema and ophthalmia. Most acute phases are asymptomatic and resolve to an asymptomatic chronic stage within 3 months. Children may develop the meningoencephalitis. Congenital infection may lead to the abortion, or still birth.

#### Treatment and control

There is no satisfactory treatment. Extended therapy with the benznidazole, or nifurtimox only kills extra-cellular parasites. Optimum efficacy (60%) may be achieved during the acute phase. For control, improvements in housing, health education and insecticidal treatment of dwellings is suggested. Vaccines and chemoprophylaxis are ineffective. For control of blood-transmitted infections, screening all blood donors from the endemic countries for *T. cruzi* antibodies, and strengthening health service infrastructures are also suggested.

## 3.14 *Trypanosoma brucei rhodesiense* and *T. b. gambiense*

Metacyclic trypanosomes are found in the fly's proboscis. Infection is inoculative. It is a zoonotic species as it multiplies in the blood of animals, domestic and wild animals, as well as in man. *T. b. rhodesiense* is responsible for acute form of sleeping sickness in the East Africa, while *T. b. gambiense* causes the chronic sleeping sickness in the West Africa. Both parasites are spread by the tsetse flies. Symptoms are caused when the parasite invades the nervous system. These are called the *salivarian* as they complete their development in the salivary system of the insect vector.

#### African trypanosomiasis

Human African trypanosomiasis, sleeping sickness, African lethargy, or Congo trypanosomiasis is caused by *T. brucei* and transmitted by the tsetse fly. This disease is known in Africa since the 14th century. The parasite and vector were identified in 1902–1903 by Sir David Bruce. Differentiation between the subspecies of the causative protozoan was made in 1910. Pau developed Atoxyl, an arsenic-based effective drug. The drug was introduced in the 1910 with blindness as a side effect. Three severe epidemics in Africa, one between 1896 and 1906, mostly in the Uganda and the Congo Basin; one in 1920 in several African countries; and one that began in 1970 and is still in progress are recorded. The disease disappeared between the 1960 and 1965.

### *Geographic distribution*

*T. b. gambiense* is found in the central and western Africa. The parasite extends in a passive phase for months, or years before symptoms emerge. *T. b. rhodesiense*, found in the southern and eastern Africa, is the acute form of the disease with a very limited range. Its virulent infection emerges in a few weeks and develops faster. The disease is known to occur in 36 countries, all in the subSaharan Africa, and the disease is endemic in southeast Uganda, western Kenya and kills more than 40,000 Africans a year. The disease affects the rural areas.

### *Morphology*

The parasite is elongated with a single nucleus, which usually lies near the centre. Each cell bears a single flagellum, which arises from kinetoplast. Kinetoplast is a specialised part of mitochondria and contains the DNA. Length and position of flagellum is variable. In trypanosomes from host's blood, flagellum originates near the posterior end of the cell. Flagellum passes forward over the cell surface. Its sheath is expanded and forms the undulating membrane. The occurrence is characterized by the development of three types of blood forms (polymorphic). These are long and thin slender forms measuring about 29 μm long and with free flagellum; thick and short stumpy forms with average length 18 μm typically with no free flagellum, but a short one may be present, and intermediate forms are about 23 μm long with a moderately thick body and a free flagellum of medium length. The body is covered by a thin, elastic pellicle, which is made of fibrils. Such fibrils are called *microtubules*. Single flagellum is present.

### *Life cycle*

The blood-sucking insect, *Glossina,* carries out transmission from one vertebrate host to another. Metacyclic (infective) trypomastigotes are inoculated via the skin when a vector fly takes a blood meal. The parasites develop into the slender trypomastigotes which multiply at the inoculation site where ulceration occurs. Trypanosomes continue to develop and invade the lymphatic tissues, heart and various organs, and in later stage the central nervous system. The vector (male and female) takes up trypomastigotes during a blood meal. The parasites develop in the midgut of vector where they multiply. After 2–3 weeks, trypomastigotes move to the salivary glands, transforms from epimastigotes to the metacyclic (infective) trypomastigotes. The fly remains infective for about 3 months. During biting, the fly injests trypomastigotes and development cycle of the parasite takes place within vector (Figure 3.13).

### *Transmission*

Transmission occurs in the following two ways:

**Mechanical or direct transmission:** When the vector fly bites an infected person, some trypanosomes enter its proboscis, and remain there, and if such a fly bites another healthy individual within the 24 hours, trypanosomes attached in the fly's proboscis are transferred to the new host, and as the parasite does not undergo any change within the vector, such type of transmission is known as the *mechanical transmission*.

**Cyclical transmission:** When the infected blood remains within the vector's body for two days and undergo multiplication and subsequent transformation of the parasite, bite of such a vector fly causes the cyclical transmission.

### *Symptoms*

The early stages of the African trypanosomiasis may be asymptomatic with a low-level parasitiaemia. This period may last from several weeks to several months. The disease, if untreated at this stage, invades the lymph glands, which are generally accompanied by irregular fever with shivering, sweating and an increased pulse rate. The lymph glands near the site of *T. b. rhodesiense*, the glands under jaw become affected (Winterbottoms sign).

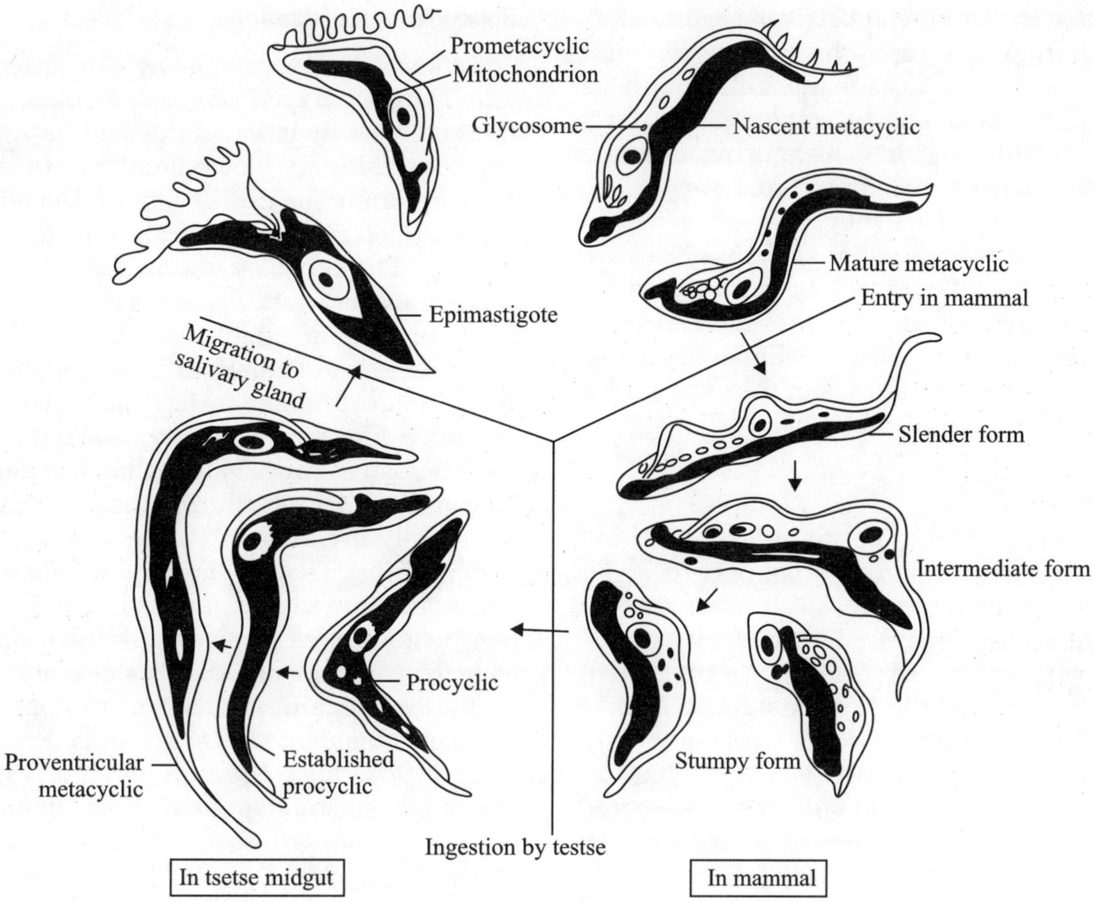


**Figure 3.13** *Trypanosoma brucei*: Development cycle.

As the disease progresses, oedema of the eyelids, face and sleeplessness along with increasing lethargy and listlessness occur. Trypanosomes may invade the central nervous system with symptoms of the meningoencephalitis, confusion, apathy, and excessive sleeping. At this stage, the cerebro-spinal fluid contains the mononuclear cells and trypanosomes can be detected. If untreated, mental deterioration and coma develops. This finally results in the death. Such signs are more common with *gambiense* than in *rhodesiense* in which patients generally die before these symptoms develop fully.

## 3.15 *Leishmania*

*Leishmania* species are parasites of man and other animals especially dogs and rodents. The diseases they cause collectively known as *Leishmaniasis*, which is endemic in many parts of Africa, Asia and South America. It is transmitted by the sandfly, *Phlebotomus* spp. Visceral Leishmaniasis (VL) is the serious and potentially fatal, if untreated. Cutaneous Leishmaniasis (CL) is also a common form causing a sore at the bite site, which heals in a few months to a year. This form can transform to any of the other three forms. Diffuse cutaneous leishmaniasis produces widespread skin lesions, resembles leprosy and is difficult to treat. Mucocutaneous Leishmaniasis (ML) commences with the skin ulcers, which spread and cause tissue damage to nose and mouth. In CL, the lesions are confined to skin, which though slow to heal are not generally life threatening. It occurs around the Mediterranean, Spain, France, Italy and Bulgaria. Leishmania is transmitted by the bite of two genera—*Lutzomyia* in the Americas and *Phlebotomus* elsewhere. ML destroys mucous membranes, the linings of nose and mouth and is horribly disfiguring. Death due to the complications is not uncommon. *L. braziliensis* complex is found in the Brazil, Eastern Peru, Paraguay, Columbia and Venezuela. Most forms of the disease are zoonotic, some can be spread between the humans. Human infection is caused by 30 species that infect the mammals. Cutaneous Leishmaniasis and Mucocutaneous Leishmaniasis are particularly dangerous, debilitating, and disfiguring diseases and occur throughout the old and new world. The parasites are unusual as they live solely within the cells of reticulo-endothelial systems. They are perfectly adapted as the proteolytic enzymes, which invade other foreign bodies in the blood stream, do not destroy them. In 1901, Leishman identified certain organisms in smears taken from the spleen of a patient who had died in 1903.

### *Epidemiology*

Leishmaniasis is found in many tropical and subtropical countries and in different parts of about 88 countries. Approximately 350 million people live in these areas. Geographic boundary extends from the rainforests in Central- and South-America to deserts in the West Asia and Middle East. It affects 12 million people worldwide with 1.5–2 million new cases per year. VL has an estimated incidence of 500,000 new cases and 60,000 deaths per year. More than 90% of the world's cases are found in India, Bangladesh, Nepal, Sudan and Brazil.

Leishmaniasis is found through most of Americas from the northern Argentina to the southern Texas, though not in Uruguay or Chile, and has been shown to be spreading to the North Texas. The disease is found through much of the Asia. In Afghanistan, Leishmaniasis occurs commonly in Kabul partly due to the bad sanitation and waste left uncollected in streets allowing sand flies a favourable environment. In Kabul, number of infected individuals are estimated to be at least 200,000 and in three town, viz. Herat, Kandahar and Mazar-i-Sharif, there may be about 70,000 more cases, according to WHO. Africa, in particular the East and North is home to the Leishmaniasis. The disease is spreading to the Southern Europe but is not found in Australia or Oceania. This is mostly a disease of developing countries and is rarely reported in the developed countries, mostly in the instances where troops are stationed away from their home countries. Leishmaniasis is reported by the U.S. troops stationed in the Saudi Arabia and the Iraq since the Gulf War of 1990 including the VL.

### Life cycle of *Leishmania*

All *Leishmania* species follow the same basic life cycle. They begin their trip from a human host when a female sand fly bites an infected individual. The fly during taking a blood meal ingests the human cells containing the tiny Leishman-Donovan bodies, or amastigotes, the vertebrate tissue stage of the parasite. These bodies are oval, tiny, about three to five one-thousandth of a millimeter long. Inside the vector fly, amastigotes reach the middle, or far end of its gut to transform into the promastigote, and attach to the gut wall. Promastigotes, with a long whip-like flagellum get attached at one end, are about three times longer than the amastigotes. They multiply by reproducing their nuclear material and simply dividing in two, a process called the *binary fission*. Though the sand fly ingests a small number of parasites, a large number are found in the fly's gut. In less than a week the promastigotes move forward to lodge the fly's oesophagus. When the fly feeds again, it clear out the promastigotes and expels them into the bite wound. Promastigotes enter into a host through fly bite invade host cells, often macrophages which are involved in the immunity. There, they transform into amastigotes and multiply by the binary fission. When human macrophages become filled with the amastigotes, they rupture and destroyed. The released parasites invade new cells and multiply again. Amastigotes—Leishman-Donovan bodies—inside the host cells is the typical feature of Leishmaniasis. Multiplying amastigotes spread to new cells and new tissues causing lesions and tissue destruction. Amastigotes present in the blood and a sand fly ingests tissues of an infected person and the life cycle repeats itself.

The sequenced genomes of *L. major, L. infantum* and *L. braziliensis* provide information of the parasite biology and a rich source of vaccine candidates. In *Leishmania,* the protein-coding genes are organized as the large polycistronic units in a head-to-head or tail-to-tail manner. RNA polymerase II transcribes long polycistronic messages in the absence of defined RNA pol II promoters. The parasite has unique features in regulation of the gene expression in response to the environmental changes. One study tested 100 randomly selected genes as the DNA vaccines against *L. major* infection in mice. Fourteen protective novel vaccine candidates were identified. A separate study attempted to identify the T-cell antigens. Six unique clones—glutamine synthetase, a transitional endoplasmic reticulum ATPase, elongation factor 1 gamma, kinesin K-39, repetitive protein A2 and a hypothetical conserved protein were identified. Twenty antigens known in these two studies are being further evaluated for vaccine development. *L. tropica* causes old world CL and in India, it is particularly prevalent in Rajasthan. Morphologically and life cycle pattern is similar to *L. donovani* except that its amastigote forms are found in large mononuclear cells of skin. *L. tropica* minor causes painless dry ulcerative lesions, while *L. tropica* major causes moist ulcer (Figure 3.14).

### *Visceral leishmaniasis*

Visceral leishmaniasis (VL) known as Kala-azar is caused by *L. donovani* complex; *L. donovani* and *L. donovani infantum* in the old world and *L. donovani chagasi* in the new world. Clinical features of kala-azar are similar. *L. donovani* occurs in the Asia and Africa and can affect people of all ages. *L. d. infantum* occurs in countries bordering the Mediterranean, affects young children and infants, although now being seen in the immunocompromised individuals. In the new world, VL is mainly a disease of the young children. *L. d. chagasi* being closely related to, but slightly different from the *L. donovani*. The geographical foci of VL are the Latin America. Small foci are found in the northern Argentina, Columbia and Venezuela. Sporadic cases are found in the Central American countries including the Mexico. Incubation period of the VL may be up to 18 months and its onset is usually insidious with fever, sweating and weight loss. The important symptom are fever, hepatosplenomegaly and anaemia. The affected site are the liver, spleen and bone marrow. Enlargement of the liver is due to the hyperplasia of kupffer cells, which are loaded with the amastigotes. Bone marrow is infiltrated with the parasitized macrophages. Kidneys show pathological changes. In advanced cases, the ascites and oedema may develop. VL may involve serious damage to the liver and spleen and causes immunity impairment. Untreated victims generally die within 3 years due to secondary bacterial infections like pneumonia, tuberculosis or dysentery. The disease is characterized by the dry, rough and dark pigmented skin, in addition to britle hairs and bleeding gums.

### *Morphology*

Various species are not distinguishable from one another by their appearence. When stained with Giemsa, the amastigotes appear as round bodies ranging from 2–3 μm in diameter with a well-

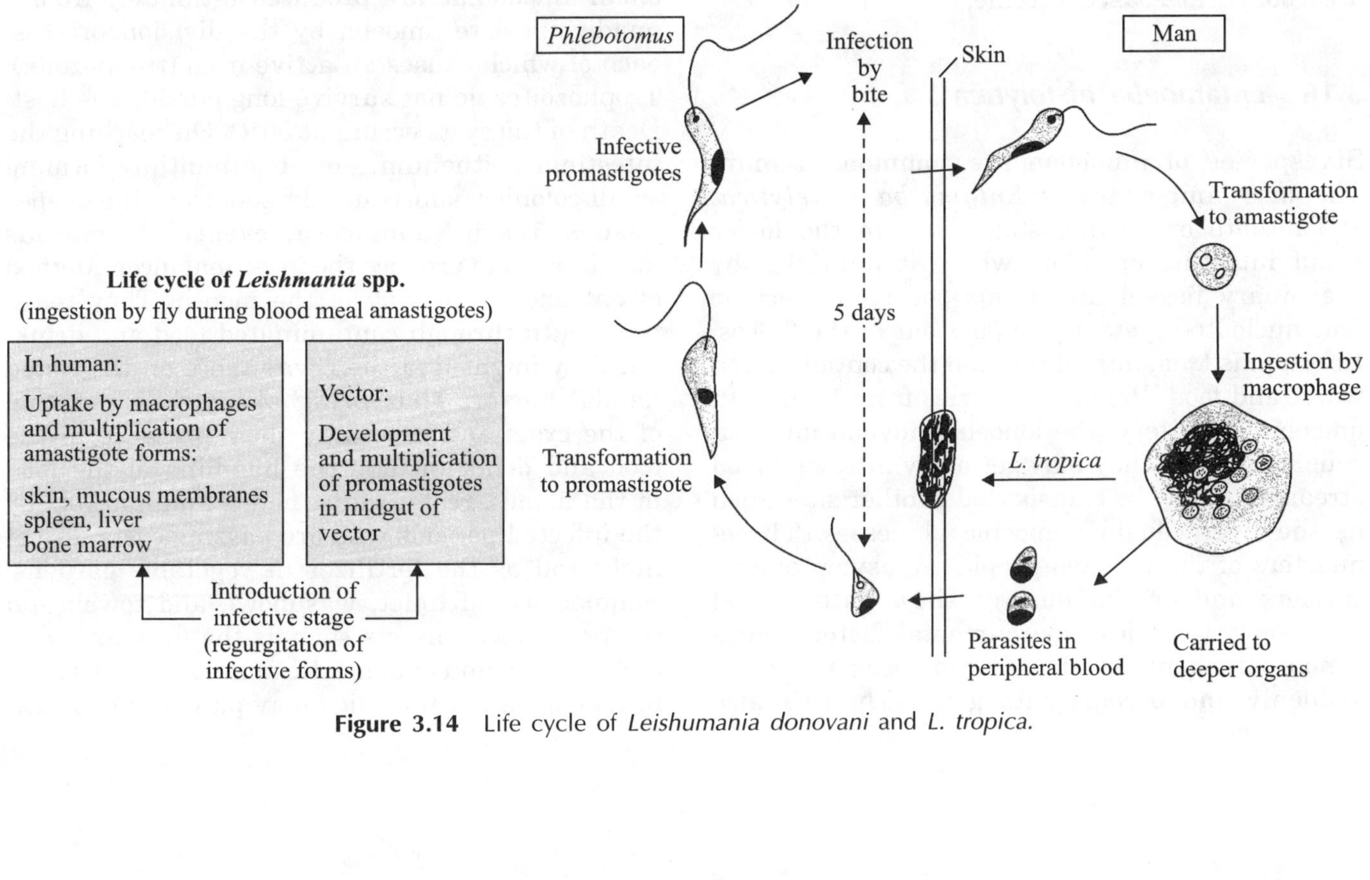


**Figure 3.14** Life cycle of *Leishumania donovani* and *L. tropica*.

defined nucleus and kinetoplast containing the extranuclear DNA. The flagellated promastigote is spindle shaped, measuring 10–20 μm in length excluding the length of flagellum. Like amastigote form, a nucleus and kinetoplast are clearly visible.

### Diagnosis

Bone marrow aspirate smear is prepared for VL. Leishmaniasis is diagnosed by the direct visualization of the amastigotes. Buffy-coat preparations of the peripheral blood, or aspirates from marrow, spleen, lymph nodes, or skin lesions should be spread on a slide for making a thin smear, and stained with Giemsa's stain (pH 7.2) for 20 minutes. Amastigotes are seen in the monocytes, or less commonly in the neutrophil and in macrophages in aspirates.

### Treatment

Paromomycin is an inexpensive and effective treatment.There are two common therapies containing antimony, meglumine antimoniate and sodium stibogluconate. Amphotericin is now the treatment of choice. Its failure in some cases to treat VL is reported in Sudan. The Indian Government approved paromomycin for sale in August 2006. A team at the laboratory for organic chemistry at the Swiss Federal Institute of Technology in Zürich has been trying to design a carbohydrate-based vaccine.

## 3.16 *Entamoeba histolytica*

Six species of amoebae are common in man. The most important is *Entamoeba histolytica*. Trophozoite or feeding stage lives in the lower small intestine or colon, where it multiples by the binary fission and forms the characteristic four nucleated cysts. Cysts pass out in the faeces. Infection is transmitted through the contaminated water and food. Prevalence varies from 1–40%. In amoebic dysentery, the amoebae invade mucosa-causing ulcers. They may get entry into the blood stream and may be transported to other sites such as the liver causing amoebic abscesses. Three-quarters of the infected people are asymptomatic carriers and only a quarter show pathological symptoms. Complex environmental factors cause a non-pathogenic strain of *Entamoeba* to switch suddenly and become pathogenic. There is also some evidence that infection of amoebae with a virus may influence virulence. There may be two strains of the parasite. *E. histolytica* is a serious protozoan pathogen. This is the third cause of parasitic death in the world. The parasite was discovered by Friedrich Losch in 1873 in Russia.

### Morphology and transmission

The cyst wall is composed of chitin and has a smooth refractile appearance. Cyst maturation involves two rounds of nuclear replication without cell division. Nuclear morphology of the cyst is similar to that of the trophozoite except that nucleus become progressively smaller following each division. The young cysts would have a glycogen vacuole, which appears, as a clear area in the stained specimens. Vacuole sometimes displaces and alters the morphology of nuclei. Chromatid bodies tend to disappear as cyst matures. Cysts are generally 12–12 mm in diameter, immediately infective on excretion with the faeces and remain viable for weeks to months depending on the environmental conditions.

*E. histolytica* exhibits a typical faecal oral life cycle. While taking contaminated water or food, pathogenic amoebae enter the human intestine in the cyst form. Cyst wall dissolves there in the presence of an enzyme and results in the emergence of quadrinucleate amoeba. Eight small amoebulae are produced ultimately from a quadirinucleate amoeba by the division process, each of which causes an active from (trophozoite). Trophozoites do not survive long outside the host. Death of the cysts occurs at 50°C. On reaching the intestinal epithelium, amoebae multiply forming small colonies, subsequently penetrate the deeper tissues, reach submucosa, eventually mucous membrane and produce the intestinal ulcer. Method of entrance is by way of the mouth. They reach the mouth through contaminated food and drink, but they might find their way there on the soiled hands. Factors that bring about dissemination of the cysts and especially their presence in the food and drink include the handling of the food in the homes, restaurants, hotels and markets by the infected persons who are passing cysts; use of night soil as the fertilizer in vegetable gardens; common use of toilet, washbowl, and towel; and the presence of insects such as the flies, ants and cockroaches, and of domestic animals such as rats, mice, dogs and cats. Flies may play an important role in the dissemination. Principal reservoirs are human carriers who are passing cysts. Association of such carriers with susceptible hosts results in the infections.

### *Life cycle*

Parasite is found in three forms, *viz*., trophozoite, precyst and cyst. Immature trophozoites, sometimes called the *amoebulae*, colonize the large intestine, especially the caecal and sigmoidal regions, where they feed on the bacteria and cellular debris, and undergo repeated round of binary fission. *Trophozoites* with an amorphous shape is generally 15–30 µm in diameter. They move by extending a finger like pseudopodium and pull the rest of the body forward. Such a movement is known as *amoeboid movement*. Pseudopodia and sometimes outer edge of trohozoite have a refractile appearance and is known as *endoplasm*. Rest of the cytoplasm with a granular appearance is the endoplasm. Occasionally, a glycogen vacuole is present. Nucleus is a finely granular ring of peripheral chromatin with centrally located karyosome. Some trophozoites undergo encystment in the intestinal lumen. Encystment does not occur in the tissues. Before encystment, the trophozoite extrudes its food vacuoles and becomes round or avoid, about 10 to 20 mm in size. This is the precystic stage. It secretes a highly refractile cyst wall around it and becomes the cyst (Figure 3.15).

### *Pathogenesis*

Craig states that he has observed several cases in which typical amoebic ulcerations containing

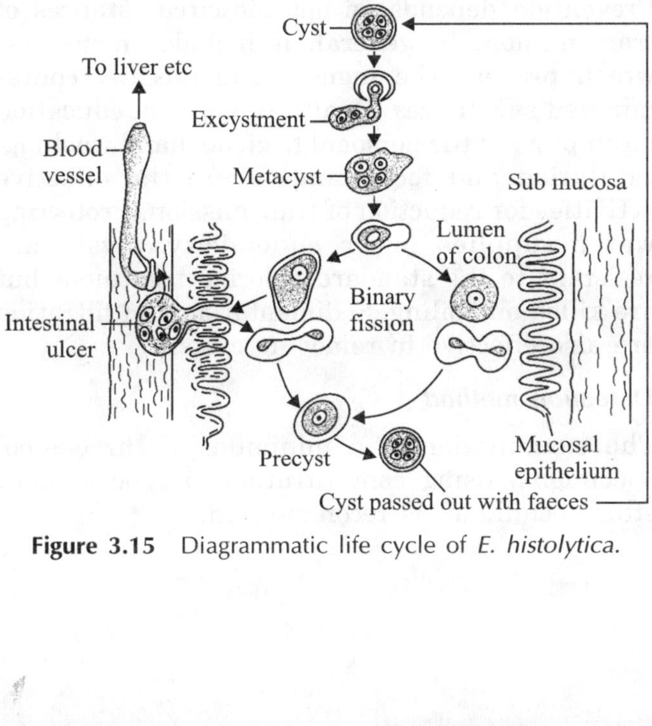


**Figure 3.15** Diagrammatic life cycle of *E. histolytica*.

the amoebae were present in lower portion of ileum in region just above the ileocaecal valve. Other secondary sites of the infection are noted besides the ileum. Many of these are parts of the body at a considerable distance from the large intestine which can be explained on assumption that amoebae are carried there through blood or lymph. In amoebic ulcer, walls of large intestine and blood vessels are opened. Amoebae may obviously be carried to any part of body and may start infections wherever suitable conditions are encountered.

Liver is the most frequent secondary site of infection. Amoebae gain access to this organ by way of portal vein and give rise there one or more abscesses. Such abscesses are often found in the right lobe but may be found in any part of the liver. Sometimes they are very large. They may rupture into the lung or other regions and thus may be responsible in spreading the infection. Investigators have reported the percentage of liver abscesses in intestinal amoebiasis. The total conditions caused by *E. histolytica* in human host are termed *amoebiasis*. Initial stage of infection is called the *acute amoebic dysentery*, which occurs in the caecum and ascending colon. Infection in the later stage is called the *chronic intestinal amoebiasis*. Abscesses of amoebic origin are known to occur in the lung, brain and spleen. Pulmonary abscesses occur due to the rupture of a liver abscess. Lower lobe of the right lung is the common location. Abscesses in lung may rupture into the air passages and pus containing amoebae may be expelled. Amoebae when carried into the brain may cause the cerebral abscesses. About fifty such cases are recorded, all ending fatally. Several cases of splenic abscess are reported. Amoebae are reported from the urinary tract, testis, skin, eye, bone marrow and lymph glands. Whether the causative organism described in all these cases were *E. histolytica* and whether they were responsible for various disease conditions are questionable.

Thus, primary or intestinal amoebiasis may cause the amoebic diarrhoea or amoebic dysentery, and secondary amoebiasis with various symptoms characteristic of amoebic liver abscess, or amoebic abscesses in the lungs, brain and spleen, and of lesions in other parts of the body. The symptoms that result from invasion of these organs are well known. Origin of the symptoms observed is practically unknown. The common symptom of intestinal amoebiasis is diarrhoea. This is abnormal frequency and liquidity of faecal discharges. Two factors are involved here (i) exudation into the intestinal lumen of serous fluid with a tendency to putrefy and, (ii) an increase in the peristaltic activity. Auerbach's plexus is recognized as the centre for intestinal peristalsis and its abnormal stimulation may bring about in various ways. In the case of amoebic diarrhoea, the probable stimulus is irritation of the intestinal wall. If mild, such irritation would result in an increased production of the mucus. If more severe, a more copious transudation might ensue. The exact *modus operandi* in this and other types of symptoms due to amoebic invasion is the unknown.

*E. histolytica* is a facultative pathogen that exhibits a wide range of virulence. Noninvasive disease is often asymptomatic but can cause diarrhoea. Noninvasive infection can progress to an invasive disease in which the trophozoites penetrate their intestinal mucosa and kill the epithelial cells. Early lesion is a small area of necrosis or ulcer. Amoeba spread laterally and downward in submucosa and kill the host cells as they progress. This results in classic flask-shaped ulcer with a small opening and a wide base. Ulcerative process may continue to expand laterally or downward. If large numbers of ulcers are present, they may coalesce which could lead to a localized sloughing of intestinal wall. Amoebic ulcers can also become secondarily infected with bacteria. Infection can occasionally lead to formation of amoebic granuloma, also called an *ameboma*. Haematogenous spread of trophozoites to other sites such as lungs or brain is rare, but does occur. Pulmonary infections generally result from a direct extension of hepatic lesion across the diaphragm and into pleura and lungs. Cutaneous lesions formed as a result of hepatic or intestinal fistula can also occur although rare. Other cutaneous lesions include perianal ulcers and involvement of genitalia including penis of homosexuals.

Probably infection once established persists throughout the life of the host. Chances of reinfection are so favourable that no definite conclusion can be drawn. Three principal points of view exist regarding the host-parasite relations during the carrier period. The first is that the amoebae live as harmless commensals in the intestinal lumen. This was considered impossible until recent cultivation experiments proved that tissue element are not necessary for the growth and reproduction of trophozoites. The second view is that the amoebae require access to the tissue of intestinal wall. They must have tissue elements for their growth and reproduction, and therefore always injure the host. These injuries, because of host resistance, are so slight that they are repaired by regeneration of tissue as rapidly as they are incurred, the result being a sort of equilibrium between the host and parasite. The third view is that a large proportion of carriers suffer from "chronic" amoebiasis.

Studying this phase for several year states that 'it has been possible for the last two years to say that there is a definite disease entity that can be recognized by the clinical means as chronic amoebiasis. This is so definite that fully 95% of cases can be diagnosed clinically before laboratory confirmation is obtained. Persons with amoebiasis commonly complain of fatigability, feel bed tired, and arise in the morning tired. They are tired whether they work or rest. Many patient are so tired that they are in anguish. Very commonly associated with fatigability is constipation or constipation interspersed with the evanescent diarrhoea. Occasionally, one will elicit the passing of considerable quantities of mucus, or blood. In amoebiasis, the patient often suffers no pain rather comfortable in the abdominal region. If an acute infection appears, the patient may recover with or without treatment. Such a recovery is often followed by relapses as in many other protozoan diseases.

### Prevention

Prevention depends on specific circumstances of transmission. In general, it includes measures, which prevent the ingestion of faecally contaminated substances. Health promotion, education for improving the personal hygiene, hand washing, sanitation and food handing are the effective activities for reduction of transmission. Protecting water supplies lower endemicity. Cysts are resistant to the standard chlorine treatment but are killed on boiling. Sedimentation and filtration are also effective in removing cysts.

### Detection method

The examination of a minimum of three stool specimens, using concentration and permanent stain techniques is recommended.

## GENERAL NOTES

## 3.17 GEOLOGIC HISTORY AND USES OF FORAMINIFERAN

Foraminiferans are large sized marine, or brackish water forms, excluding family Allogromiidae which inhabits fresh water. A very few species are found in the damp earth. Their presence is characterized by the reticulopodia, and shell of secreted organic material, or of cemented foreign mineral particles, or of secreted calcium carbonate. Many species are dimorphic having two distinct forms. Megalospheric phase has a large initial chamber. Microspheric phase has a smaller initial chamber. They are mostly benthic, but some are pelagic. Pelagic foraminiferans are important in the Cretaceous and Tertiary stratigraphy.

### *Geologic distribution*

Foraminiferans are reported to occur throughout the geologic periods from Cambrian to Recent. Tests of dead pelagic forams sink to sea bottom to form a layer of grey mud, known as the foraminiferan ooze. About 78 million sq km of present ocean bottom is covered with such ooze. Since Precambrian period, the oozes have undergone fossilization all over the world. Limestone ridges of the North America, great limestone beds of the Paris, and Cairo and White Chalk Cliff of England are almost exclusively of these fossils foraminiferan shells.

### *In Paleozoic era*

Foraminiferans first appeared in the Cambrian period were mainly vitreous, arenaceous, and structurally primitive forms, and there is no record of multiform forams. Mostly, they were agglutinated forms. Calcareous forms first appeared in the Ordovician. In Silurian, a diverse assemblage is observed. A most common form, *Ammodiscus* appeared first in the Silurian along with *Saccamina*. A few more genera of families Ammodiscidae and Astrorhizidae appeared for the first time in the Devonian. Genus *Semtextulana* is restricted to the Devonian. From Carboniferous onwards, foraminiferans became very significant in the diversity and quantity. Permo-Carboniferous time experienced luxuriant growth of these microorganisms. *Fusulina* enjoyed dominance throughout the world during this period. Important genera of Carboniferous period were *Reophax, Trochamminoides*, and *Textularia*. Microgranular calcareous forms were abundant in the Carboniferous. Indian Carboniferous forms include *Hyperammina, Glomospira, Lituotuba* and *Trochammina* from the Manendragarh (M.P.) and *Fusulina* from the Ladakh, Kashmir. Permian forms were complex, and many species of *Schwargerina* are found in Permian rocks of Ladakh.

### *In Mesozoic era*

During Triassic period, family Laginidae was dominant. Triassic assemblage differs from the Permian in disappearance of Endothyridea, and first appearance of septate miliolids, definite polymorphinids and rotalids. New forms of said group were added in the Jurassic along with the first appearance of earliest representative of family Buliminidae, i.e. *Bolivina*. Larger forams of family Lituolidae came into existence in the late Jurassic. Genus *Quinqueloculina* is recorded from the Jurassic rocks of Kutch. In the early Cretaceous, the planktonic oozes of *Globigerina* were observed. Family Orbitolinidae of this period is represented by the *Orbitolina bulgarica* in orbitolina limestone of the Kashmir. Calcareous perforate forams of this period are characterized by the great expansion in number of coiled genera. Subfamily Virgerininae (of family Buliminidae) and family Rotaliidae appeared during this period.

Genus *Globotruncana* is a marker foram of upper Cretaceous. In India, Cretaceous rocks of Trichinopoly, Pondichery, Kutch, Andaman and the Himalayas have yielded a diverse assemblage. In fact, forams have played a significant role in establishing the Cretaceous Paleocene boundary all over the world, because at the end of Cretaceous many important genera such as *Globotruncana* became extinct, and were replaced by the new forms for example *Globorotalia*.

### *In Cenozoic era*

In the Tertiary, large and well represented smaller forms were diversified assemblages. Large forms like *Assilina, Nummulites*, and *Discocyclina* appeared for the first time in the Paleocene, while smaller forms like *Globigerina* and *Globorotalia* were abundant. During this period, earliest *Hentkenina* was also observed along significant development of families like Rotalidae, Milionidae and Buliminidae. In Eocene, the larger forms occupied tropical regions and temperate regions was occupied by smaller forms. Characteristic forms of this period are *Hentkenina, Nummulites, and Lepidocyclina*. The Oligocene witnessed emergence of *Miogypsina* along additional planktonic genera like *Hestigerina, Condiena, Orbulina* and *Sphaeroidiella*. New genera like the *Pulleniatina* and *Pseudorotalia* were emerged during the Pliopleistocene times. Forams of this time are similar to recent forms. Almost all forms of the Cenozoic rocks of India are recorded from Andaman Nicobar Islands, Coromondal coast, Dwarka formation, Mihoite limestone of Kutch, Tertiary formations of Assam, Rajasthan and Gujarat (Figure 3.16).

### *Use of foraminifera*

1. Oceanic ooze converts into the sedimentary rock, which has been put to various commercial uses such as filtering agent, abrasives, chalk, building stones, etc. Most buildings of Paris are made of limestone, composed almost exclusively of shells of foraminiferan genus *Miliolina*. Similarly, the great pyramids of Egypt were carved form limestone deposits by tests of an early Tertian foraminiferan, *Nummulites*.
2. Fossil foraminiferans are extremely useful in determining approximate age of sediments encountered in drilling for oil, and in making stratigraphic correlations between oil wells in a single field, different fields or even different basin.

### *Various forms of foraminifera*

Foraminifera may be divided into three groups, characters of which are based on structure and composition of the shell, but this cannot be regarded also in some types which are usually calcareous. We occasionally meet with species in which shell consists largely of sandy material.

**Porcellaneous forms:** Shell calcareous, porcellaneous, not perforated by canals, but provided with one or two large apertures, through which pseudopodia pass out.

**Example:** *Miliola, Orbitolites, Alveolina*.

**Arenaceous forms:** Shell composed of grains of sand or other particles cemented together by the chitinous, calcareous or ferruginous material. Young stages sometimes are calcareous.

**Example:** *Saccammina, Lituola, Orbitolina, Endothyra, Textularia*.

**Vitreous forms:** Shell of calcite, vitreous, perforated by the numerous minute canals for the passage of pseudopodia.

**Example:** *Lagena, Nodosaria, Crystellaria, Globigerina, Orbulina, Rotalia, Calcarina, Fusulina, Amphistegina, Operculina, Lepidocyclina*.

## 3.18 LOCOMOTION IN PROTOZOA

Three major types of locomotor organelles occur among the protozoa. Protozoan classification considerably depends on the kinds and distribution of these locomotor parts. The amoeboid movement, movement with the help of undulipods (i.e. cilia and flagella) and gliding, affects locomotion in protozoa.

### *Locomotor organelles*

**Pseudopodia:** Pseudopodia is characteristic of Sarcodina, although occur in flagellates and other groups as well, are semi-permanent to transitory extensions of the protoplasm. Pseudopodia assume various shapes, but five types recognized among the Sarcodina are as follows:

In some cases, *Amoeba* flows without forming definite extensions of the body, and is called a limax form because of its sluggish shape. Broad, advancing margins at the tip of body are protoplasmic waves.

**Lobopodia:** Most large amoebae and many smaller ones move by the lobopodia. These are finger shaped, round tipped pseudopodia containing both the ectoplasm and endoplasm.

**Example:** *Vahlkamptia*.

**Filopodia:** Slender, ectoplasmic pseudopodia, which are pointed at the tips and sometimes branching. They never anastomose to form networks and are commonly met within *Euglypha*, many testaceans and chlamydepheons.

**Reticulopodia:** These are endoplasmic, branch extensively, and fuse together to form the meshworks that act as food traps. Reticulopodia of several individuals may unite to form a multicellular reticulum as are found in foraminifera such as *Elphidium, Chlamydophyrs, Microgromia*.

**Axopodia:** These are slender and composed of the granular ectoplasm. They contain an axial

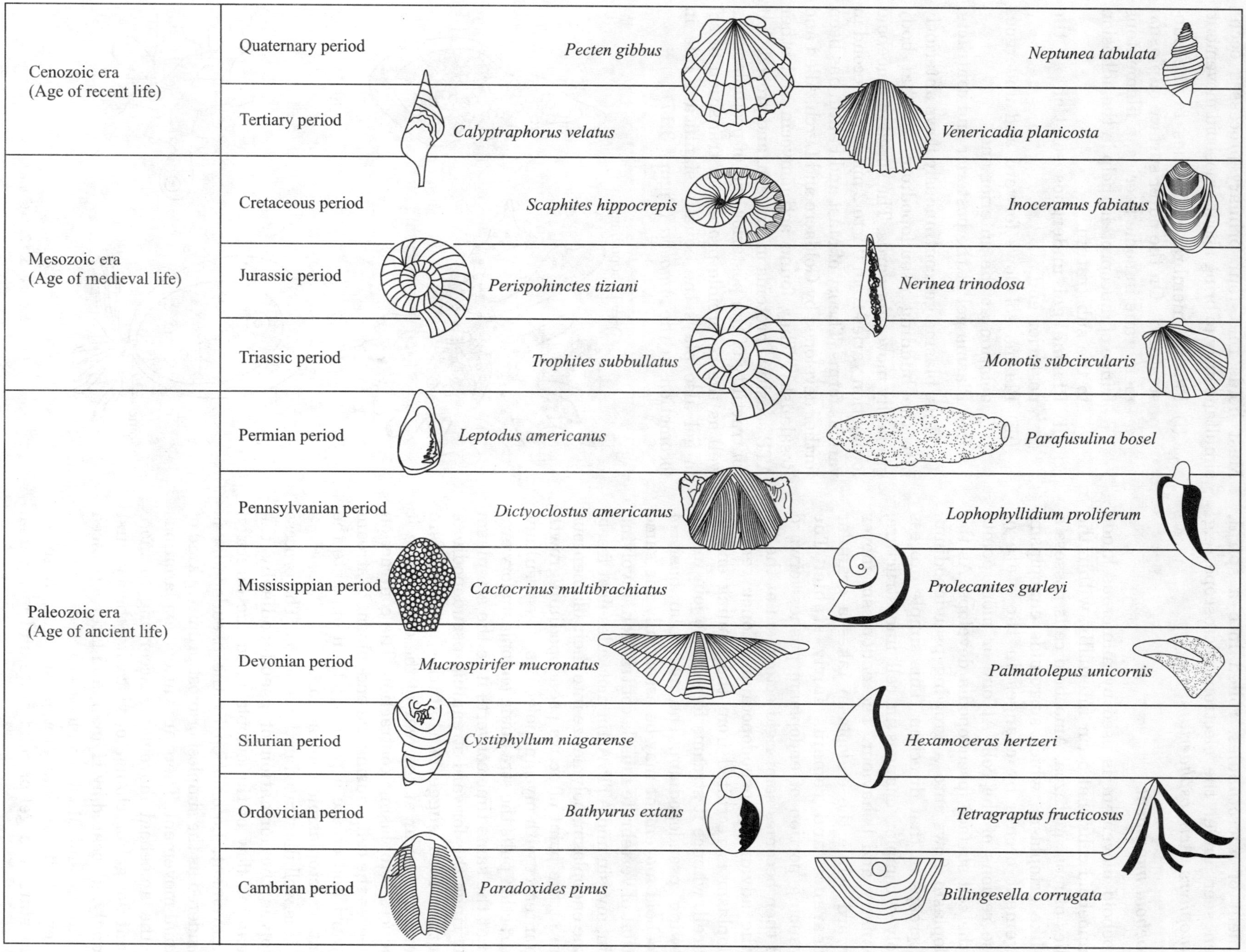


**Figure 3.16** Fossil foraminiferan. Redrawn after the http://mapslinomaha edu/maher/geott7/part2/india

filament of microtubules, rolled into a spiral, when seen under the electron microscope, e.g. *Actinophyrs, Actinosphaerium.*

### *Amoeboid movement*

Amoeboid movement is used by numerous kinds of amoeboid cells that occur internally, within the bodies of most metazoa. Amoeboid cells possess a gel-like ectoplasm, which surrounds endoplasm. Movement occurs by changes in the state of these regions of the cell. At one or more points on the cell surface, pseudopodia develops. As the endoplasm flows into a growing pseudopodium, cell creeps in that direction. This simple process involves complex changes in cell fine structure, chemistry and behaviour. The endoplasm moves 'forward' while the endoplasm takes a granular appearance and remains fairly stable. The advancing portion of endoplasm pushes forward and then becomes semi rigid ectoplasm at the tip of the advancing pseudopodium. Concurrently, endoplasm is recruited from the trailing end of the cell, whence it streams forward to join in the 'growing' pseudopodium. The molecular basis of amoeboid movement may be essentially the same as that of vertebrate muscle contraction, involving actin, myosin and ATP. Actin molecules float freely in the endoplasm, polymerize into their filamentous form at the point of active pseudopodium growth and interact with myosin molecules. The resultant contraction pulls the streaming endoplasm forward, and at the same time converts it to the ectoplasm that rings the forward-streaming pseudopodium. Another idea suggests that the actin-myosin interaction occur at the rear of the cell, where it produces ectoplasmic contraction. The contraction squeezes the cell, causing the endoplasm to stream forward and create a pseudopodium opposite the point of ectoplasmic contraction.

It is well illustrated in the *Amoeba.* The process involves the formation of pseudopodia by the streaming flow of the cytoplasm in the direction of movement. Locomotion by the pseudopodia is considered as the simplest or most primitive kind of animal movement. There are various explanations for the amoeboid movement. According to more recent sol gel or change of viscosity theory, first strongly advocated by Hyman, and later advanced by Pantin and Mast, changes effected amoeboid movement of colloidal protoplasm from a sol or fluid state to a gel or solid state and vice versa. Mast holds that four primary processes occur simultaneously to bring about amoeboid movement:

(i) Plasma membrane attaches to the substratum: On the outer surface of plasma membrane in some species, a filamentous fringe of mucoprotein, help in the adhesion to the substratum.
(ii) Plasma gel undergoes solation at the posterior end.
(iii) Plasma sol flows forward and undergoes gelation at the anterior end.
(iv) Plasma gel at the posterior end contracts.

Thus, plasma gel continuously flows anteriorly into the budding pseudopodium, so that body gradually moves forwards. The speed of amoeboid locomotion varies from 0.2 to 2 micra per second in various forms. Theory of Mast and Pantin has been further developed by Goldacre and Lorch, Allen and Rosalansky. They found actinomyocin and other ATP sensitive proteins in Amoeba, and proved that contractility of the posterior end of protoplasm as well as the reversible process of transition of sol to gel also plays the equally significant role in locomotion in the *Amoeba* (Figure 3.17).

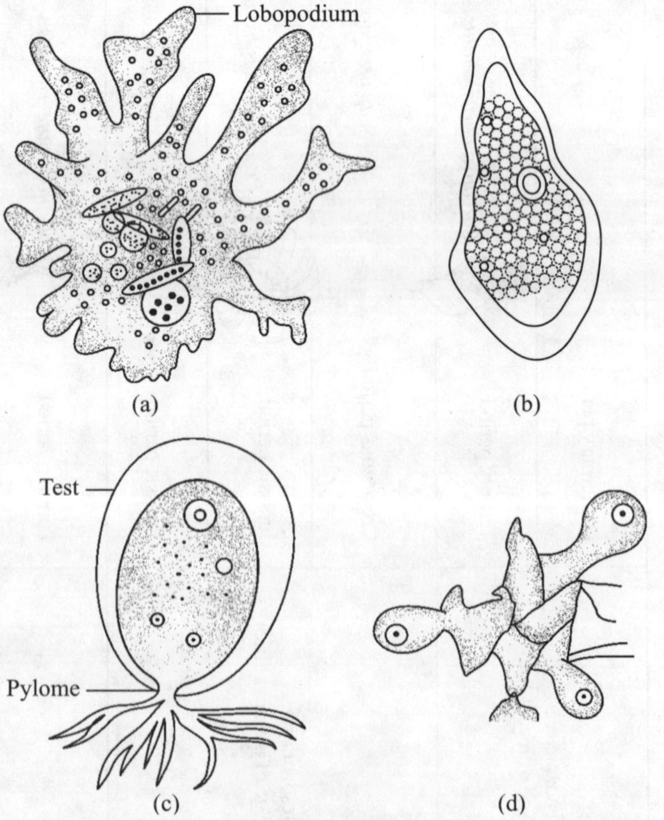


**Figure 3.17** *(Contd.)*

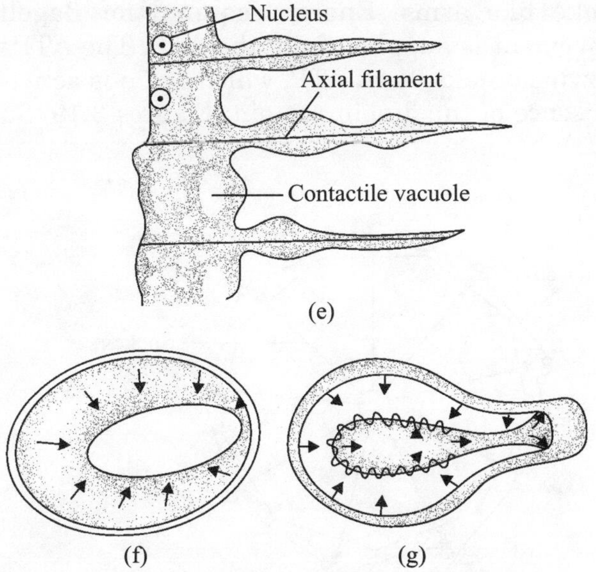


**Figure 3.17** Types of pseudopodia (a) *Amoeba* sp. with lobopadia (b) Protoplasmic waves in *Vahikampfia sp.* (c) Filopodia in *Chlamydophorus* (d) Anastomosing reticulopadia in *Microgromia sp.* (e) Axopadia in *Actinosphaerium.*

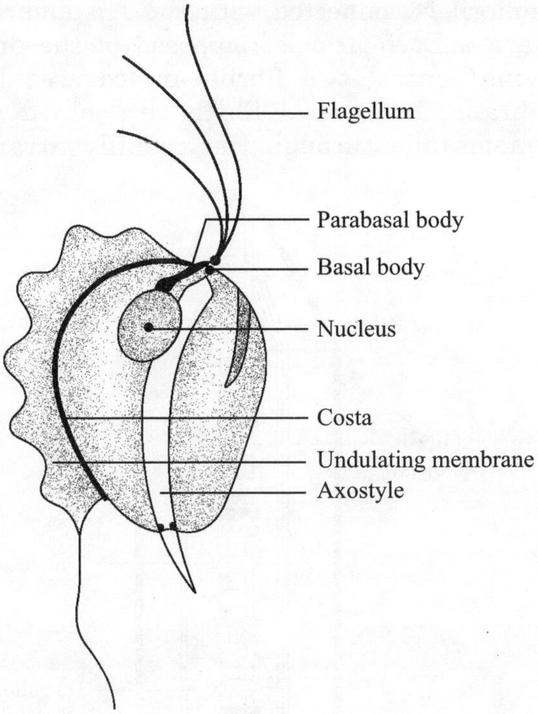


**Figure 3.18** Flagellar root system of *Trichomonas.*

## 3.19 FLAGELLA

Flagella are permanent or semi-permanent organelles, characteristics of the Mastigophora, although occur on the swarms or gametes of some Sarcodina and Sporozoa. Most flagellates bear flagella throughout the life. Some may reabsorb their flagella and move about with the amoeboid actions or become temporarily incapable of movement. Flagella are usually larger than the cilia, and occur singly or in small numbers. Cilia and flagella differ in nature of their root systems, and associated parts. Flagella are delicate, whip-like structures that beat to propel flagella through the water. Details structure of flagella is about same in regard to kind of cell or organisms they are part of. Outer covering is continual with the plasmalemma of the body surface. Main locomotory organs of *Euglena* are the flagellum. This is a whip-like organ which strikes water and results in free swimming locomotion. As *Euglena* moves with the flagellum, locomotion is known as the *flagellar locomotion* (Figure 3.18).

### *Light microscopic structure*

The flagellum emerges out of the cytostome through the cytopharynx. In *Euglena viridis* flagellum is as long as body of *Euglena.* Flagellum originates by two roots from the base of the reservoir, opposite to the contractile vacuole. Each root originates from a basal granule, which lies in the cytoplasm lining the reservoir. Flagellum consists of an outer contractile protoplasmic sheath and an inner axial filament, axoneme. A bundle of contractile fibrils known as the *mastigoneme* emerge from the terminal part of flagellum, which project along one side of the sheath, and therefore, flagellum is stichonematic type.

Warner and Satir described the electron microscopic structure of the flagellum. Various microtubules extending from the base to flagellum tip, and construct the basic framework of the flagellum. Out of 11 microtubules, 9 are situated at the periphery, and 2 at the centre, forming the 9 + 2 arrangements. All fibrils remain enclosed within a protoplasmic sheath continuous with the cell membrane. Nine peripheral fibrils and two central fibrils make the axoneme. Each peripheral tubule is made up of two subfibrils, viz., subfibril A and subfibril B. Subfibrils A and B are connected by the very short arms. All micro fibrils of axoneme are composed of the tubulin dimmer, which is formed of two monomers namely a and b monomers. Each subfibril of centre is composed of 13 protofilaments. Peripheral A and B subfibrils consist of 13 and 10 protofilaments, respectively.

B subfibril is connected with the A subfibril by two arms. Each arm is composed of the dinein protein. Central two fibrils are covered by a membrane. Peripheral fibrils are joined with the membrane through the radially arranged spoke like arms. Energy required for flagellar movement is supplied from the ATP. The ATPase enzyme breaks down ATP, which again is active in presence of the dinein protein (Figures 3.19, 3.20, 3.21).

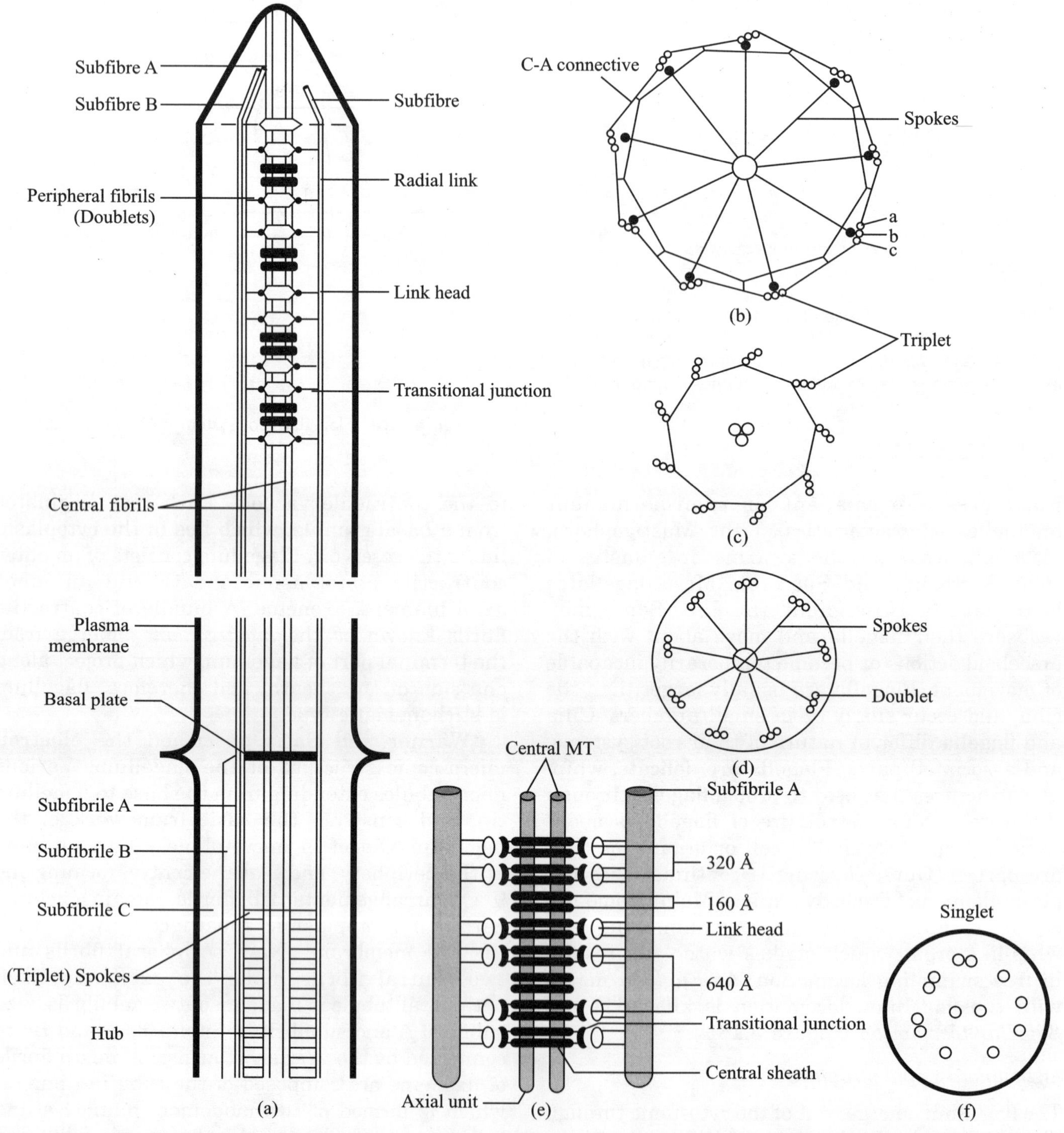


**Figure 3.19** Structure of flagellum under E.M. (a) The part below the basal plate is the basal body, the part above is the shaft (b) T.S. through basal body region (c) T.S. through the transitional region (d) T.S. through flagellar shaft (e) T.S. near the tip of flagellum (f) Longitudinal view of 9 + 2 flagellar axoneme.

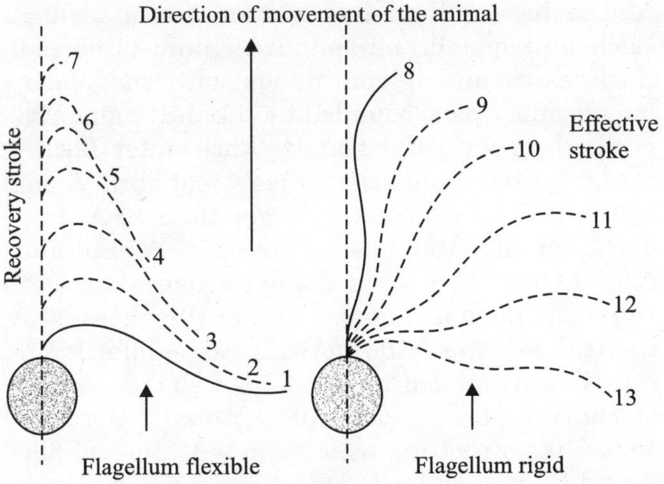


**Figure 3.20** Direction of the movement of the animal.

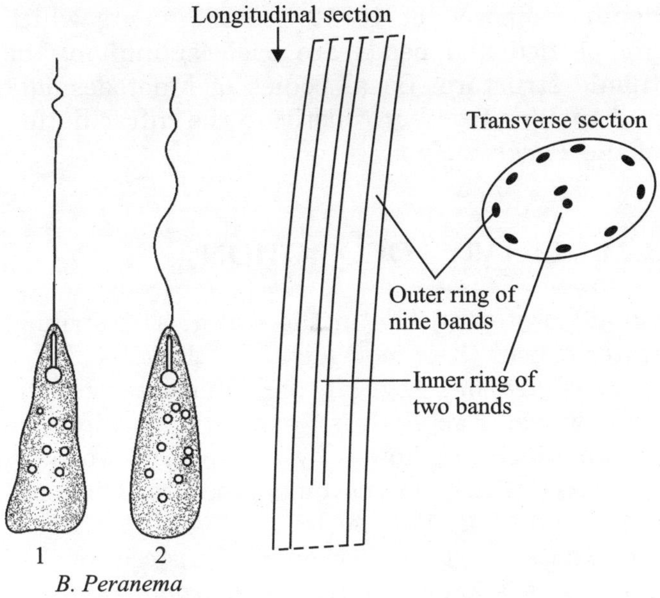


**Figure 3.21** Section of flagellum.

### *Flagellar locomotion*

To explain flagellar locomotion, several theories have been forwarded. They are:

**Butschli's screw-theory:** *Euglena* strikes the water surface laterally with its flagellum regularly. As a result, a pressure is created on the water at right angle to body axis, which creates two forces, one is distributed at right angle to body axis, and other is parallel to it. Parallel forces drive the animal forward. Force at right angle to axis helps the animal to rotate on its axis. As dual forces acting simultaneously, animal moves forward in the same way as a screw moves inside an object.

**Gray's (1928) wave action theory:** Several contraction waves, pass from the base to flagellum tip, creates two forces. One force acts in direction of movement, and other force acts in a circular way surrounding the main axis. Two forces act simultaneously. As a result, *Euglena* moves forward rotating on its own axis.

**Bowden's (1943) spiral wave action theory:** During locomotion, the flagellum remains posteriorly directed, and in such spiral, waves pass from the base to tip. Generally number of such waves are 12 per second. In rapid movements, number of waves per second increases. So following a spiral path, it moves forward rotating on its own axis.

### *Euglenoid locomotion*

Body rotates 180° on its own axis with verylittle displacement of the body. A peristaltic contraction wave starting from the cytopharynx area passes towards the posterior end. Cytopharynx are becoming thick and broad, middle portion and posterior portion becomes swollen. Due to peristaltic contraction, and relaxation, *Euglena* moves forward rotating on its axis. This special types of locomotion is known as *euglenoid movement*, which occurs due to contraction of micro fibrils of cytoplasm or myonemic fibrils present below the pellicle.

## 3.20 LOCOMOTION IN *PARAMOECIUM*

Whole body of a *Paramoecium* is covered with the cilia. It moves by cilia and movement is known as the ciliary movement.

### *Structure of cilia under light microscope*

Cilia are principal locomotory organs. Structurally cilia and flagella are similar. They differ in their arrangements as well as in number. Flagella are one, two or a few with localized arrangement, but cilia are found in rows. Flagella are long, but cilia are short. Cilia are arranged in longitudinal rows. Entire body is covered with the cilia, so called *holotrichus (hole*—entire, *trichus*—cilia). The cilia can be grouped into two categories. Cilia associated with the cytostome and the surrounding feeding area are the oral ciliature. Cilia of the general body surface are the somatic ciliature.

The cilia in both categories may be single or simple cilia or the kinetosomes may be grouped close together to form compound ciliature. Many workers treat cilia as short specialized flagella. However cilia are much more numerous and densely distributed than flagella, and the patterns of ciliation on the body are extremely varied. Each individual cilium undergoes an effective stroke as it beats. The cilium does not move on a single plane, but creates distorted cone as it beats. The counter clockwise movement is an intrinsic feature of cilia and occurs even when they are isolated. The beating of a ciliary field occurs in metachronal waves that pass over the body surface. The coordination of these waves is apparently due to hydrodynamic effects generated as each cilium moves. Micro disturbances created in the water by the action of one cilium stimulates movement in the neighbouring cilium, and so on over the cell surface. The effective stroke of the cilium, coupled with the direction of the metachronal waves, results in three main patterns of metachronal coordination.

#### *Under electron microscope*

Microfibrils forming the basic skeleton of a cilium collectively known as the *kinetodesmata*. Each kinetodesmata resides within the cytoplasm directly arises from a basal granule and cilia are arranged in a row. Each row is known as *kinety*. Each cilium exhibits complicated arrangement of basal granule and kinetodesmata, which appears to be a thread like structure, but under electron microscope, it is a complicated cluster of overlapping fibrils. Each fibril is connected with a basal granule. In a kinetodesmata, the number of striated fibrils may exceed 500 which form the frame work of the cilium.

There are contractile microfibrils below the pellicle, which may join the basal granules or motorium, a specialized fibrillar organ situated near the cytopharynx. These fibrils are not striated. Because of structure; each cilium is divided into two regions. They are

(i) kinetodesmal region, and
(ii) shaft region.

Kinetodesmal region shows the (9 + 2) fibrilar arrangement like a flagellum. There is 9 peripheral fibrils and 2 central fibrils. Each cilium take its origin from a double root, roots that join to the basal granules from which kinetodesmata arise. At the base, cilium has a diameter of 2000 Å. Each unit membrane bounded cilium is 90 Å in thicknesses and is continuous with the plasma membrane. Space bounded by the unit membrane is filled up with the matrix. Each outer fibril is made by two subfibrils namely subfibril A and subfibril B, of which A is larger than B. Average diameter of subfibrils is 180–250 Å. Subfibril A gives out two arms from its one side. Both subfibrils have a common wall of thickness 50 Å. Central two fibrils do not possess subfibrils and remain surrounded by a common sheath. Sheath of central fibrils gives out 9 radially arranged spoke like structure, which join with the subfibril A and this cause's cart wheel arrangement. Inner arm of subfibril A joins with subfibril B of next fibril in a clockwise direction. Each subfibril is again composed of 10–12 protofilaments which are beaded and beads are basic subunit of the tubule structure. Basal bodies of kinetodesmata and tubules together constitute the infraciliature of the *Paramoecium*.

## 3.21 CILIARY LOCOMOTION

Ciliate protozoan live in the water. The striking water causes their movements by the cilia. These strokes are not regular and characterized by only wave. Each cilium exhibits an effective stroke which is followed by a recovery stroke. In recovery stroke, no movement occurs. Cilia are many in number but shorter in length. So there is coordination in their movement. This coordinated movement is known as the *metachronal rhythmic movement*. During an effective stroke the protein molecules of central fibrils contract. As a result, cilia become straight, immediate peripheral fibrils contract and cilia bend to the strike water. This effective stroke creates the sufficient force to drive the animals forward. After completion of effective stroke, cilia tends to come to its original state, by applying minimum forces of the alternate events that helped in effective stroke called the *recovery stroke*. Effective strokes and recovery strokes are two main components of the ciliary locomotion. Some cilia in the kinety operate the effective stroke, whereas some perform the recovery stroke. In this way, a wave of metachronal rhythm pass from the anterior to posterior region and animal tends to move forward. As oral cilia are larger, and they beat more quickly animal rotate on its axis and moves forward while rotating. As position of kinetodesmata is on right side of cilia shaft, so its rotation is always dextric.

A cross section of a cilium has shown. Note that there is a circle of nine doublets, each of which have one complete (A tubule) and one incomplete (B tubule) microtubules. The core doublets are both complete. Extending from the doublets is set of arms that join the neighboring doublets. These are composed of the protein dynein. It is spaced at 24 nm intervals. Nexin links are spaced along the microtubules to hold them together. Projecting inward is radial spokes that connect with a sheath enclosing the doublets.

Figure 3.22 showed an electron micrograph of a cross section of a cilium. Note the dynein arms and the nexin links. The dynein arms have ATPase activity. In the presence of ATP, they can move from one tubulin to another and enable the tubules to slide along each other so the cilium can bend. The dynein bridges are regulated so that sliding leads to synchronize bending. Because of the nexin and radial spokes, the doublets are held in place, so sliding is limited lengthwise. If nexin and the radial spokes are subjected to enzyme digestion and exposed to ATP, the doublets will continue to slide and telescope up to 9X their length.

### *Control of metachronus rhythm*

Two theories have been proposed to explain controlling mechanisms of metachronal rhythm in the *Paramoecium* and other ciliates. They are:

**Neuroid fibre theory:** According to Taylor (1964), the metachronus rhythmic movements of cilia are controlled by the infraciliature of ciliates and function of the infraciliature can be compared with function of the metazoan nerves.

*Criticism:* After removing infraciliature system in *Paramoecium*, no change in the metachronus rhythm has been observed. Therefore, this theory needs further verifications.

**Electropotential theory:** According to Paralucz, Naitoh and Eckert, potential differences created by the polarization and depolarization during the effective and recovery stroke are responsible to maintain the metachronic rhythm.

*Criticism:* Present state of our knowledge tells that coordination between the infraciliature and microfibrils of kinetodesmata and their coherent activities are responsible for the metachronus ciliary movement in *Paramoecium* and other ciliates.

### *Molecular mechanism of ciliary movement*

Two theories have been proposed. They are:

**Localized contraction theory:** According to this theory, contractile units are found at regular intervals through the length of axoneme. Contraction of the units aids in bending of cilia. During ciliary movement, there is a change in length of subfibres of a doublet. During ciliary movements, 9–4 appear sharply defined, and 5–8 are blurred in appearance and vice versa. The sharpness of fibres is due to contraction caused by molecular level change, and in blurred fibres, they are in a relaxation state. Such molecular contraction and relaxation explain the effective as well as recovery stroke. In effective stroke, one set undergoes contraction, while other set stretches and during the recovery stroke, reverse would take place.

*Criticism:* Uncoiling of helically arranged protein molecules in walls of the tubules would also give same picture.

### *Sliding filament model*

This theory states that bending of cilium is started by the sliding of tubules of peripheral fibrils relative to each other. Two subfibres A and B of a doublet remain fixed in relation to each other, because they share a common wall, although termination distance of the subfibres remains constant during effective and recovery stroke. So, it is supposed that filament must be moving against adjacent filament. It is believed that subfibres A of one fibril slides against the subfibres B of an adjacent doublet, and thus, effects a ciliary movement. Flagella contains a matrix in which a ring of nine doublet microtubules surrounds two central microtubules. Such an arrangement is also seen in the cilia. The 9 + 2 pattern is enclosed in an outer sheath. Two central fibres confer the bilaterality. The plane of flagellar beat may be associated with the orientation of the central filaments. Flagellar roots are extremely diverse. In very simple flagellar substructure of trypanosomes, a small dark staining body occurs at the base of flagellum and a larger DNA containing body lies near it. Small body is called a *basal body* or *kinetostome*. Large body is termed a *parabasal body* or *kinetoplast*. Flagella is divided into the following types, because of presence and absence as well as distribution of the mastigonemes.

**Stichonematic:** With only one row of mastigonemes disposed in the same side, e.g. *Euglena*.

**Pentacronematic:** Near the tip naked, and provided with the mastigonemes in two rows, e.g. *E.S. Urcalns.*

**Acronematic:** A terminal filament mastigonemes are absent, e.g. *Polytoma*.

**Pantonematic:** With two or more rows of mastigonemes, e.g. *Paranema*.

**Anematic:** Without mastigonemes and terminal filaments, e.g. *Noctiluca* (Figures 3.22–3.25).

## 3.22 FEEDING AND DIGESTION IN *AMOEBA*

*Amoeba* is a microphagus animal. The term *microphagus* is derived from micros meaning microscopic and phagus meaning who eats. *Amoeba*, a microscopic single celled animal, feeds on the microscopic organisms in a complicated way. It could not procure food from the simple inorganic substances. *Amoeba* depends on the solid living materials for food and takes solid food as whole. Therefore, its nutrition is called the *zootrophic* or *holozoic*. The animal ingests algal cells, algal filaments, diatoms, bacteria, ciliate protozoans, rotifers, and small microscopic nematodes as their food, preferably the ciliates. *Amoeba* identifies organic and inorganic materials. When carbon

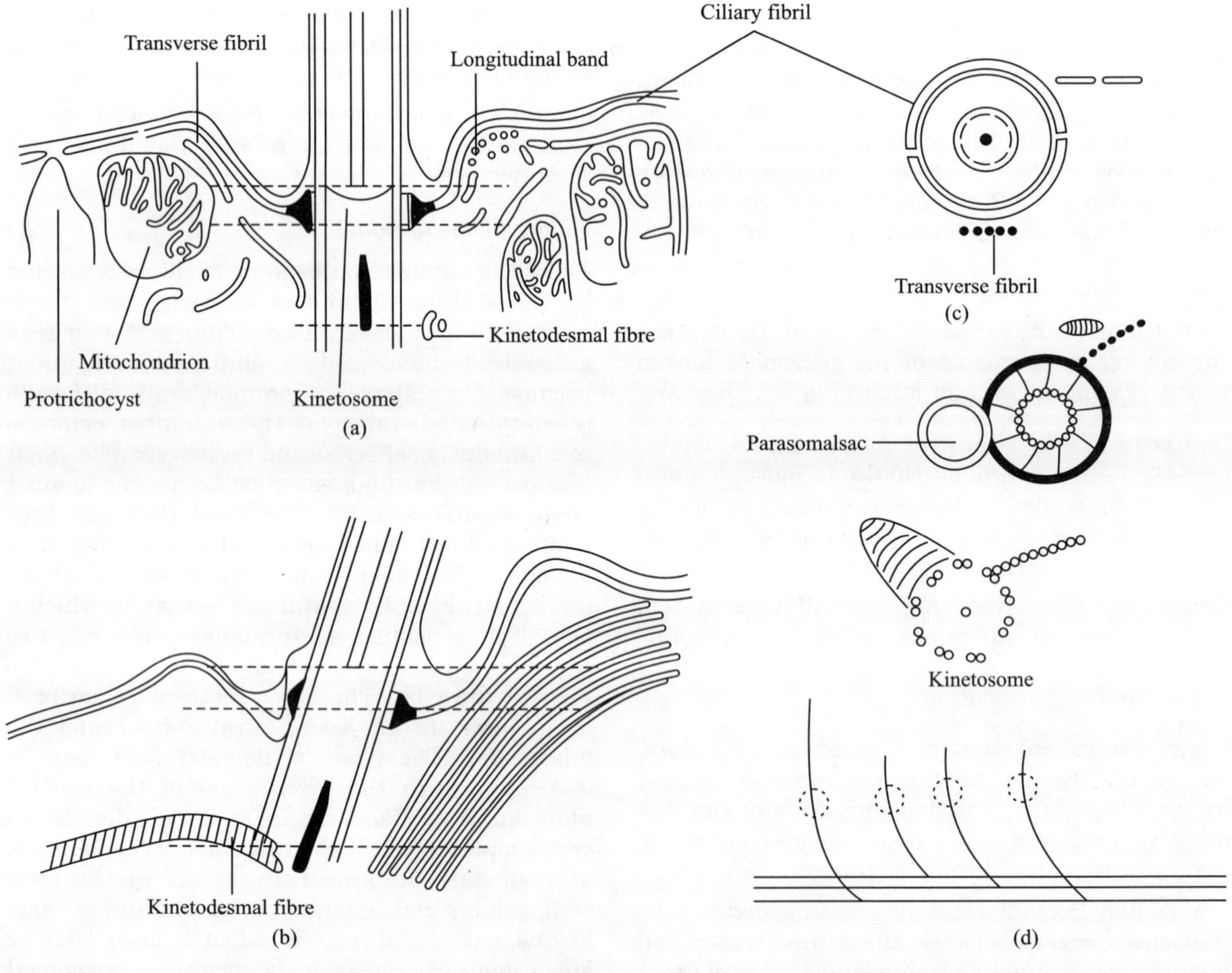


**Figure 3.22** E.M. of cortex of *colpoda* (a) Transverse section (b) Longitudinal section (c) Cross section of cilium (d) Kinetosome.

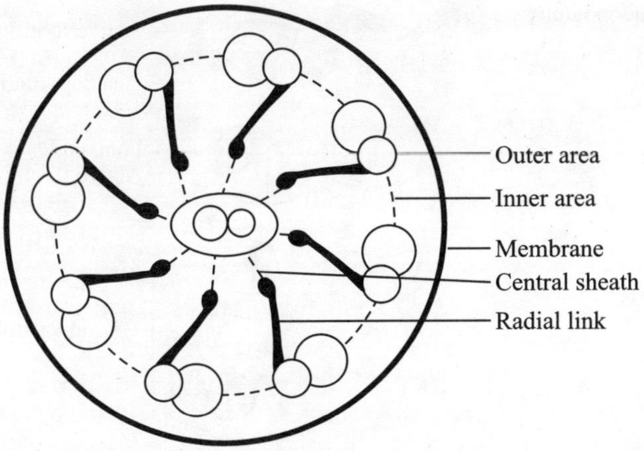


**Figure 3.23** Structure of cilium under E.M.

**Figure 3.24** Highly magnified centrioles of cilium.

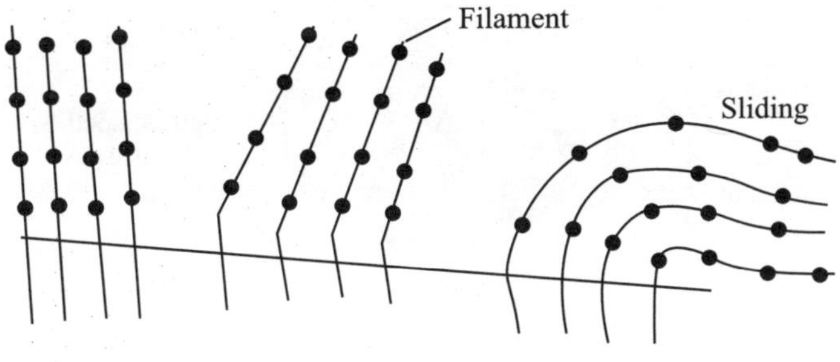


**Figure 3.25** Sliding filament model of ciliary movement.

particles are mixed up with its food, it ingests only the organic food avoiding carbon particles. Although it can take in food from any part of the body, but generally anterior pseudopodia help in capturing food. Ingestion takes place by the pseudopodia, and ingested food is kept in food cup in the cytoplasm. When it comes across a food particle, it encircles food particle by the pseudopodia without touching food and fix food on substratum.

**Circumvallation:** Food cup fuse to form a food vacuole with some water. This method of ingestion is applied in capturing the living prey.

**Circumfluence:** This method is employed when food is immobile. A pseudopodium comes in contact with food to form a food cup. Then food is pinned down to the substratum. Pseudopodium is extended below the food to complete food cup, and food along with the drop of water is enclosed in the food vacuole. By repeating this process, animal can ingest a long filament of algae.

**Import:** When *Amoeba* encounters the food, food sinks passively in the endoplasm and a food vacuole is formed within which food is placed. This passive method of taking food is known as the *import*.

**Invagination:** When *Amoeba* meets food, it adheres to the food particle. Ectoplasm along with food is invaginated as a tube into endoplasm, and food particle is sucked in. Plasmalemma disappears, and the ectoplasm changes into the endoplasm, a food vacuole is formed, and the tube disappears.

### *Digestion*

*Amoeba* engulf food particles by phagocytosis. The prey or food particles are enclosed in a vacuole into which digestive enzymes are secreted. As the vacuole follows a path through the cytoplasm, it shrinks as water and the products of digestion leave the vacuole. Finally, the vacuole reaches the cell boundary where any indigestible residue is expelled. The vacuole itself then disintegrates. This form of digestive arrangement, in which particles are first taken into a cell before being subjected to the action of enzymes, is termed *intracellular digestion*. This name is used even though digestion actually occurs inside the food vacuoles (Figures 3.26–3.28).

## 3.23 EXCRETORY SYSTEM IN *PARAMOECIUM*

Contractile vacuoles play dual role of excretion and osmoregulation. There are two contractile vacuole in a *Paramoecium*—one lying at the anterior end and other lying at the posterior part of body. $CO_2$ is produced in respiration and ammonia is

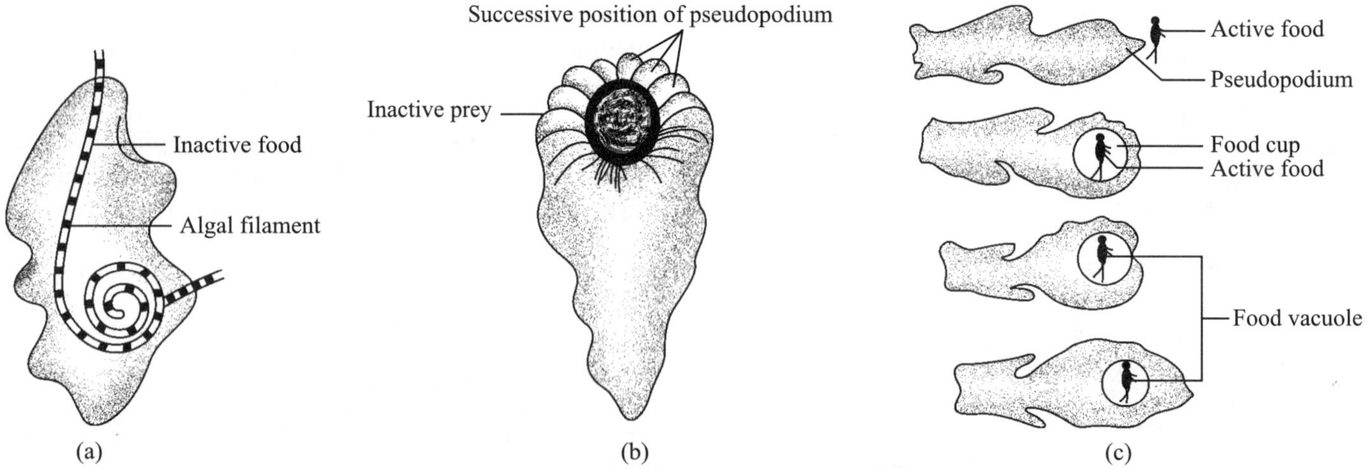


**Figure 3.26** (a) Food capture by import (b) Food capture by circumfluence (c) Food capture by circumvallance.

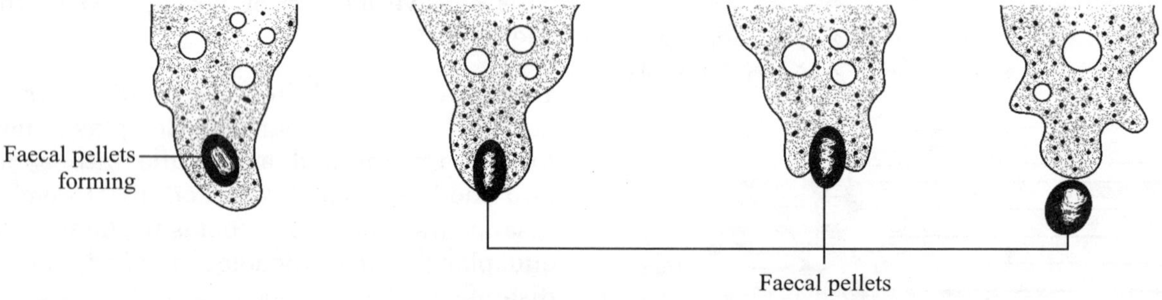


**Figure 3.27**

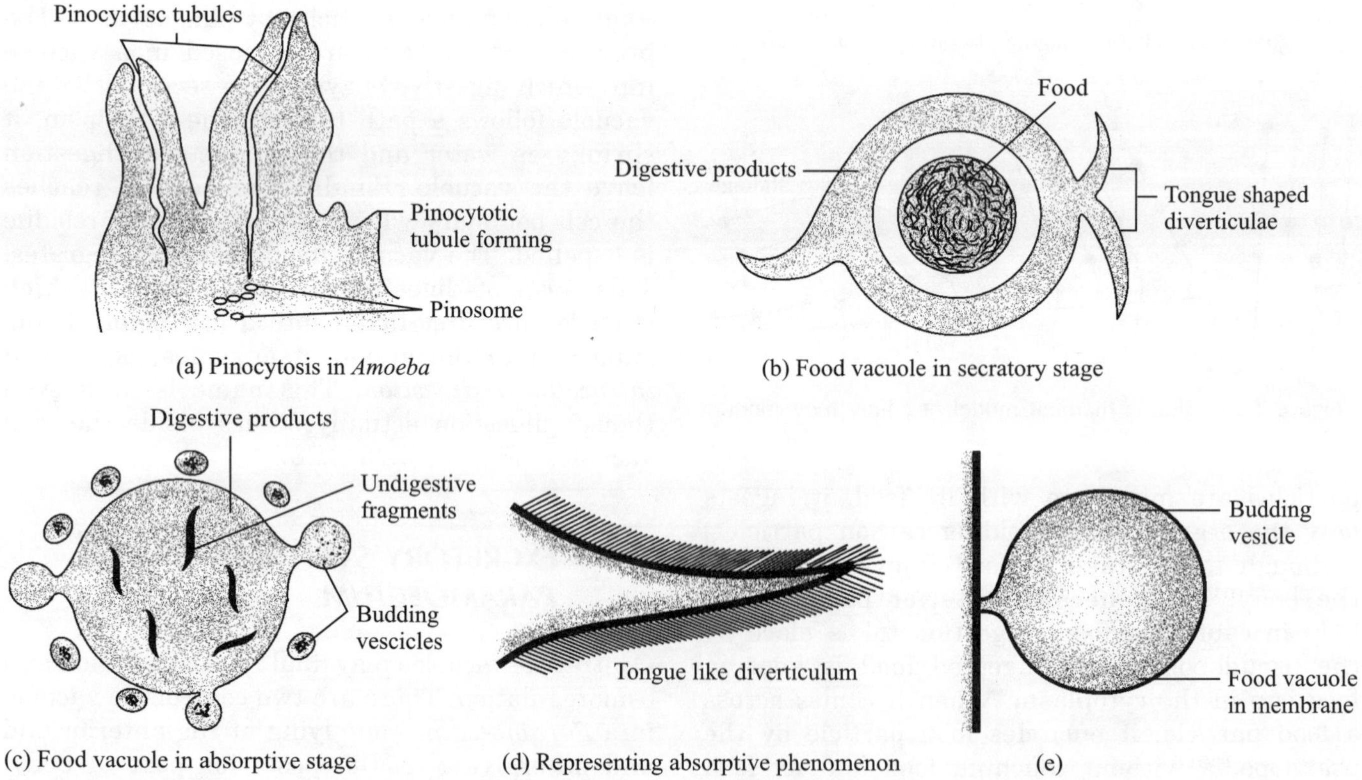


**Figure 3.28** Food vacuole formation and digestion in *Amoeba.*

produced as a metabolic process of nitrogenous material is expelled throughout the body surface by the diffusion. Contractile vacuoles play role in the excretion but its role as osmoregulator is very important.

### *Structure of contractile vacuole*

Contractile vacuoles form, become large and then discharge their content. A small vacuole appears first and to this vacuole 6–11 radially arranged terminally closed tubules open. These tubules are known as the *feeder canals*. Each feeder canal has its terminal end closed but opens by an opening into the vacuole. Each feeder canal has three part as terminal end is closed, middle portion is tubular and by absorbing excretory products, water and ions this portion swells to form an ampulla which is a temporary storage organ, and narrow canal runs from the ampulla to open into the vacuole proper. The formation of ampulla represents diastolic condition of the vacuole. After fully formation, the ampulla discharges its contents and disappears. This condition is known as the *systolic condition*. When contractile vacuole form the feeder canals, it becomes distinct and vacuole remains in systolic condition and when it is filled up with the excretory material and water reaches it the condition is diastolic. After diastole is completed, it suddenly contracts and discharges its contents in the pellicle. Of two contractile vacuoles, posterior one functions more rapidly than the anterior contractile vacuole as it is situated close to the cytopharynx. For this reason, water is collected very rapidly here. Each contractile vacuole works alternately and time between this alternative works is 10–20 seconds. *Paramoecium* lives in water. Water enters its body by the endosmosis and excess water is expelled through the contractile vacuole. Nitrogenous wastes like urea and ammonia area expelled through contractile vacuole (Figure 3.29).

## 3.24 REPRODUCTION IN *PARAMECIUM*

*Paramecium* reproduces asexually by the transverse binary fission and sexually by the conjugation, autogamy and cytogamy.

### *Transverse binary fission*

This is the most common type of reproduction found in *Paramecium*. In such reproduction plane of division occurs through the centre of the cell and in a plane at right angles to long axis of the body. *Paramecium* reproduces through such process during the favourable conditions. Micronucleus divides by mitosis into two daughter micronuclei, which move to the opposite ends of the cell. Macro nucleus elongates and divides transversely by the amitosis. Another cytopharynx is budded off and two new contractile vacuoles appear one near anterior end and another posteriorly. Meanwhile, a constriction furrow appears near the middle of body and deepens until cytoplasm is completely divided. Resulting two 'daughter' paramecia are of equal size, each containing a set of cell organelles. They grow to full size before another division occurs. The process of binary fission requires about two hours to complete and may occur one to four times per day, yielding 2 to 16 individuals. About 600 generation are produced in a year. Rate of multiplication depends on the external conditions of food, temperature, age of culture, population density, internal factors of heredity and physiology.

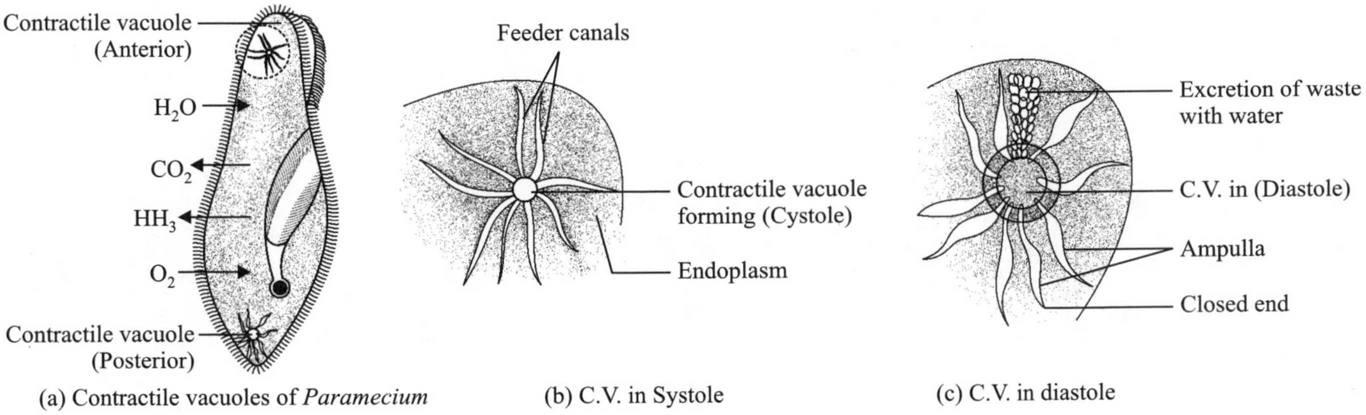


(a) Contractile vacuoles of *Paramecium* (b) C.V. in Systole (c) C.V. in diastole

**Figure 3.29** Excretion in *Paramoecium*.

*Conjugation:* This is a process which is unique among the protozoa, accomplishes exchange of genetic materials in *Paramecium*. After practising binary fission for a considerable number of generations, paramecia make a nuclear reorganization by the conjugation. The following account refers to *P. caudatum:*

1. Two individuals from two different mating types come in contact ventrally by their oral grooves and remain attached until the end of process. Pellicle and ectoplasm degenerate at the point of contact and a continuous protoplasmic bridge is formed between the two individuals, which are called conjugants. Conjugating pair continues to swim actively and a sequence of complicated nuclear changes takes place in the animal.
2. Vegetative macronucleus simply breaks up into the fragments, which are hydrolyzed by the enzyme and later absorbed by the cytoplasm.
3. Micronucleus of each conjugant first grows in size and then divides twice, one of them being a reduction division. Thus, four haploid daughter micronuclei are produced.
4. Three micronuclei degenerate and disappear in each conjugant, while remaining one divides again unequally forming two pronuclei or gametic nuclei, which are genetically identical.
5. Smaller one is active migratory or wandering and potential male pronucleus and bigger one is inert stationary and potential female pronucleus.
6. Migratory pronucleus of one conjugant then passes over through protoplasmic bridge into opposite individual.
7. There, it fuses with its stationary pronucleus forming a single diploid zygote nucleus, also known as *conjugation nucleus*, *synkaryon* or *amphinucleus*. Complete fusion of two nuclei form the two different individuals form a zygote nucleus, which is termed *amphimixis*.
8. Two paramecia now separate, after a union of about 12 to 48 hours and are called *ex-conjugants*.
9. In each conjugant the zygote nucleus divides three times in rapid succession producing eight nuclei.
10. Of these four enlarge to become the macronuclei and other four become micronuclei. Three micronuclei disintegrate and disappear.
11. Remaining micronucleus divides with binary fission of ex-conjugant. Thus, two daughter paramecia are obtained each containing two macronuclei and one micronucleus.
12. Micronucleus again divides with division of each daughter paramecia, forming two individuals each containing one macronucleus and one micronucleus. Thus, each conjugant produces four daughter individuals at the end of conjugation (Figure 3.30).

### *Conditions for conjugation*

1. Conjugation does not occur under the favourable living conditions.
2. Certain substances secreted by paramecia themselves probably initiate the conjugation under certain conditions.
3. A certain range of temperature is essential.
4. Probably light suppresses the process, while darkness favour it.
5. Individuals must have passed through a desirable number of asexual generations

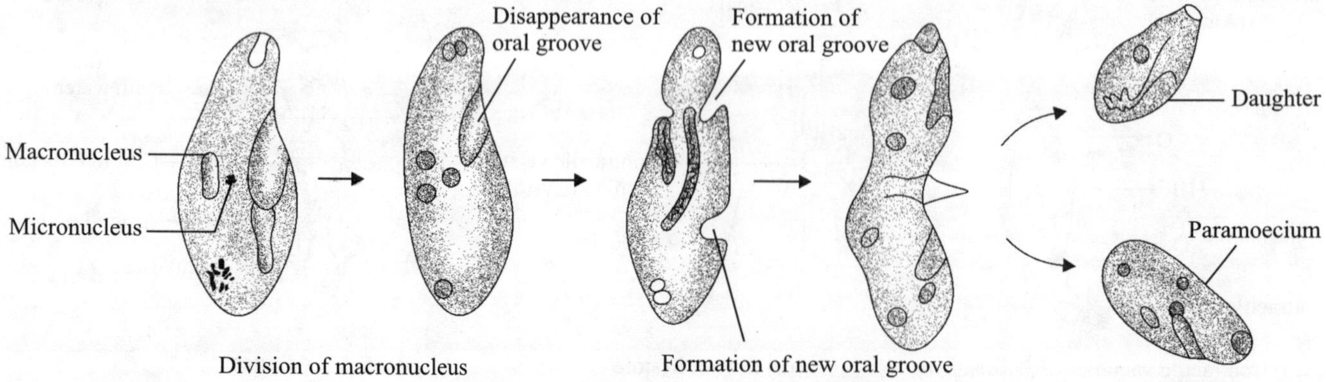


**Figure 3.30** Conjugation in *Paramoecium*.

and attained sexual maturity before they conjugate.

6. Ordinarily, conjugation occurs between the two different mating types belonging to the same karyon.

### *Significance*

Exact reasons for conjugation are not known, although following effect is attributed to this process.

**Rejuvenation:** The process seems to rejuvenate and revive the lost vigour for asexual reproduction. Woodruff and Jennings do not support the view that conjugation helps in the rejuvenescence.

**Nuclear reorganization:** During conjugation, the nuclear apparatus is reorganized and a readjustment occurs between it and the cytoplasm.

**Hereditary variation:** Periodic occurrence of conjugation ensures inherited variation. It brings about the blending of two lines of ancestry just as the bisexual reproduction does. Fusion of migratory pronucleus of one conjugant with the stationary pronucleus of other combines hereditary characters from two separate individuals belonging to the different strains or classes. This hybridization brings an increase in vigour and external diversity or variability of offspring which may be better suited than others to survive in conditions of life. This likelihood of hereditary variation may partially explain why the process evolved.

### *Autogamy*

In autogamy, two micronuclei of *P. aurelia* fuse together to form a synkaryon. During this process, macronucleus grows into an irregular skien like mass which breaks into pieces later to be absorbed in the cytoplasm. Two micronuclei divide twice, once meiotically to form 8 haploid nuclei, of which 7 degenerate and remaining one divides to form 2 sister gamete nuclei, homologous of stationary and migratory pronuclei. These enter a protoplasmic cone temporarily formed near the cytostome and then fuse together to form a synkaryon. This divides twice to yield four nuclei, two of which become macronuclei and two micronuclei. Cell body and micronuclei then divide to form two daughter individuals each with a new macronucleus and two micronuclei.

### *Cytogamy*

During this process, two individuals come together as in the conjugation. Nucleus undergoes changes as in conjugation but fails to make a reciprocal exchange. Individuals separate after the two pronuclei of each conjugant fuse among themselves.

## REVIEW QUESTIONS

### Short Answer Questions

1. Give two characters of Protozoa.
2. Name two phyla of Protozoa.
3. State two characters of the Phylum Apicomplexa.
4. State two characters of the Phylum Ciliophora.
5. State two characters of the Phylum Labyrinthomorpha.
6. Name the Phylum to which *Plasmodium* belongs.
7. Name the Class to which *Entamoeba* belongs.
8. Name the Phylum to which *Entamoeba* belongs.
9. Name two locomotor organelles in Protozoa.
10. Why protozoans are also called acellular.
11. What is karyosome?
12. What is kinetoplast?
13. What is blepheroplast?
14. What is cytostome?
15. What is trophozoite?
16. What do you mean by schizogony?
17. What do you understand by sporogony?
18. What is gametocytogenesis?
19. What is exflagellation?
20. Who got the Nobel Prize for his work with malaria parasite?
21. What the term 'malaria' signifies?
22. Name two malaria parasites.
23. Name most dangerous human malaria parasite.

**24.** What is hypnozoite?
**25.** What is sporozoite?
**26.** What is merozoite?
**27.** What is schizont?
**28.** Comment on pre-erythrocytic schizogony of malaria parasite.
**29.** Comment on erythrocytic schizogony of *Plasmodium*.
**30.** Comment on gametogony of *Plasmodium*.
**31.** What is malarial oocyst?
**32.** What is malarial ookinete?
**33.** What is febrile paroxysm?
**34.** What is schizogonic periodicity?
**35.** What is malarial microgamete?
**36.** What is malarial macrogamete?
**37.** Differentiate between microgametocyte and macrogametocyte of *P. vivax*.
**38.** What is pernicious malaria?
**39.** What is quotidian periodicity in malaria?
**40.** Name one schizonticidal drug for *P. vivax*.
**41.** What is tachysporozoite?
**42.** What is bradysporozoite?
**43.** What do you understand by the incubation period of a parasite?
**44.** What are haemozoin granules?
**45.** What do you understand by the term splenomegaly?
**46.** Who discovered *Trypanosoma cruzi?*
**47.** What are trypomastigotes?
**48.** What do you understand by metacyclic transmission?
**49.** Name the causative agent of African trypanosomiasis
**50.** What do you mean by prophylaxis?
**51.** Comment on biological control of malaria.
**52.** What is trypanosomiasis?
**53.** Comment on stercorian trypanosome.
**54.** Comment on salivarian trypanosome.
**55.** Name the causative organism of visceral leishmaniasis.
**56.** What do you understand by the term epidemiology?
**57.** What is metacyclic transmission?
**58.** Comment on Chagoma.
**59.** Give scientific name of vector of *Trypanosoma cruzi.*
**60.** What is amastigote?
**61.** What is promastigote?
**62.** What is epimastigote?
**63.** Comment on cutaneous leishmanisis.
**64.** Comment on mucocutaneous leishmaniasis.
**65.** Comment on visceral leishmaniasis.
**66.** What is Leishman–Donovan body?
**67.** What is promastigote?
**68.** What do you understand by pinocytosis?
**69.** What is the function of contractile vacuole?
**70.** What is circumvallation?
**71.** Who proposed the sol-gel theory?
**72.** What is undulating membrane and what function does it serve.
**73.** Define conjugation.
**74.** Distinguish between synchronus and asynchronus beating of flagella
**75.** What is endodyogeny?
**76.** What is pseudocyst?
**77.** What is the causative organism of amoebiasis?
**78.** How many paramecia are produced after conjugation?
**79.** Name a protozoa with a shell.
**80.** What is vector?
**81.** What is infraciliary system?
**82.** What is cytopharynx?
**83.** Comment on polar filament.
**84.** What are exconjugants?
**85.** Comment on merogony.
**86.** Comment on sporoblast.
**87.** What is axostyle?
**88.** In which group the locamotory organelles are absent.
**89.** What is the function of Trichocyst in *Paramoceium*?
**90.** What is the function of oral groove in *Paramoecium?*
**91.** By which process *Paramoecium* moves about.
**92.** What is the common name of *Paramoceium?*
**93.** Name a protozoa where filter feeding is found.
**94.** Mention the number of contractile vacuoles in *Paramaecium*.
**95.** By which the body of *Paramaecium* is enveloped.

**96.** What is kinety?
**97.** What is the function of cytopyge?
**98.** Name two phases of ciliary beat.
**99.** How pseudopodia are formed in *Amoeba*.
**100.** Name a protozoon where lopopodia is found.
**101.** Name a protozoon with filopodia.
**102.** Name a protozoon with reticulopodia.
**103.** Name various types of flagella.
**104.** What are stichonematic flagella?
**105.** What is pentacronumatic flagella and where it is found?
**106.** What is aeronematic flgella and where it is found?
**107.** What is pantonematic flagella and where it is found?
**108.** What is anematic flagella and where it is found?
**109.** Distinguish between cilia and flagella.
**110.** Name the reproductive nucleus in *Paramecium*.
**111.** Name the protozoa which are considered as connecting link between plants and animals.
**112.** Name the flagellate found in the gut of termites. Name two chemicals which are used in malaria control.
**113.** State two characters of foraminifera.
**114.** State two examples of vitreous forms of foraminiferans.
**115.** Comment on the use of Foraminifera.
**116.** What is reticulopoida?
**117.** What is axopodia?
**118.** What do you understand by 9 + 2 arrangement?
**119.** What is kinety?
**120.** What do you understand by effective stroke?
**121.** What are circumfluences?
**122.** Define autogamy.
**123.** Define cytogamy.

## Long Answer Questions

**1.** Describe the asexual phase of *Plasmodium vivax*.
**2.** Describe the sexual phase of *Plasmodium vivax*.
**3.** Write a note on control of malaria.
**4.** Describe the life cycle and geographical distribution of *Trypanosoma cruzi*.
**5.** Give an account of African trypanosomiasis.
**6.** Write an account on the life cycle, pathogenicity and control of *Leishmania donovani*.
**7.** Describe the process of conjugation in *Paramoecium*
**8.** Describe the locomotory organs of protozoans.
**9.** Describe the mode of locomotion in protozoa with suitable diagrams. Classify Protozoa up to Phylum with examples.
**10.** Classify Sarcomastigophora with character and examples.
**11.** Classify Ciliphora with characters and examples.
**12.** Write a note on the control of malaria.
**13.** Describe the geologic distribution of foraminifera through various eras.
**14.** Describe the structures of a flagellum with suitable diagram.
**15.** Describe the structure of cilia along with a note on ciliary locomotion.
**16.** Describe the feeding and digestion in *Amoeba* with diagram. Write a note on contractile vacuole in *Paramoecium*.

CHAPTER 4

# Phylum Porifera

Sessile nature and asymmetrical growth of sponges are the features to the early naturalists to treat them as plants. Lamarck, Linnaeus and Cuvier classified the sponges under zoophytes. They considered them as allied to anthozoan cnidarians. Grant coined the name Porifera. The classes of Porifera are designated based on the nature of internal skeletons. Sponges, belong with the Phylum Porifera, are unusual animals as definite organs are absent in them to carry out their various functions. The important structure found in the sponges is the canal system and chambers, called the *water-current system*, through which water circulates for bringing the food and oxygen to the sponge. In addition, the system also helps to disperse gametes, larvae, and in removing wastes. Sponges are distinct from other animals in structure, function and development. They vary in shape and get attached to substrata with no power of free locomotion. Numerous pores are found in their body. In fact, the term Porifera has been derived from the words pori and ferrae which together means 'pore bearing'. Thus sponges are the pore bearing animals. Nervous tissue and specialized organs are absent in them. Usually a horny, siliceous, or calcareous internal skeleton occur in large, sessile sponge colonies. Some 5,000 species are known to be permanently affixed. They are mostly marine, solitary or colonial. They inhabit both the shallow- and deep-waters.

Simple sponges are provided with a large opening called the *osculum* at the top through which water and wastes are expelled. A perforated outer thin epidermal layer covers the porous skeleton which is composed of interlocking spicules of calcium carbonate, silica or spongin. The sponge body may be finger-like, tree-like, or may be a shapeless mass. Flagellated cells play important role in moving water into the central cavity, the spongocoel, through the perforations. The individual cells digest food, excrete waste and absorb oxygen. Sponges reproduce either asexually or sexually. Larval forms are free-swimming, while adults are sessile. Sponges are used for holding water, bathing and scrubbing.

## 4.1 GENERAL CHARACTERISTICS

1. Mostly marine with varying size and shape.
2. Sometime inhabit fresh water.
3. Metazoa at the cellular grade of construction without true tissues.
4. Adults are asymmetrical, or superficially radially symmetrical.
5. Unique flagellated cells, the choanoytes, are present.
6. Flagellated choanoytes, or collar cells, line the spongocoel.
7. Choanocytes move water by beating flagella, traps suspended food particles, and transfer the food to cells.
8. Suspension feeders.
9. Larval stages are motile.
10. Outer and inner cell layers lack a basement membrane.
11. Middle layer called the mesohyl is variable, but includes motile cells, and skeletal materials.
12. Amoebocytes are present.
13. Skeletal elements are composed of calcium carbonate or silicon dioxide.
14. Range in height from about a few mm to 2 m.

15. Asymmetric body without true tissues and organs.
16. Flattened cells cover the exterior.
17. Most sponges are sequential hermaphrodites. Each individual produces both sperms and eggs in sequence.
18. Asexual reproduction occurs through fragmentation and gemmules.
19. Gametes arise either from the choanocytes or amoebocytes.
20. Sperm is carried out through the osculum by the water current.
21. Sperm is drawn into neighbouring individuals and fertilize eggs in the mesohyl.
22. Zygotes develop into the flagellated, swimming larvae.
23. Larva after finding the suitable substratum develops into a sessile adult.

## 4.2 EVOLUTION AND FOSSIL RECORDS

Sponges are believed to be evolved from the flagellated protozoan. Choanocytes of sponges resemble the choanoflagellate protozoans. Choanoflagellate protozoan colonies, however, do not develop by the embryological stages as found in the sponges. Porifera exhibits affinities with some protozoan colonies, in absence of integration of body parts. However, both have a skeletal element in which single elements are produced by a single cell or by a group of cells. Porifera appeared in the early Cambrian Period. The classes known in the Middle Cambrian Period are Hexactinellida, Heteractinellida and Demospongiae. Hexactinellida are now extinct. Calcarea appeared in the Carboniferous Period. Despite the mineral skeleton, the well represented fossils are pharetronids, lithistids and *Dictyonina*. The oldest sponge fossils from the Lower Cambrian are *Protospongia* and *Eiffelia*. They resembled a pouch and their spicules are arranged in a single layer with ascon type canal system. Various genera of the order Lyssacina are known from the Paleozoic era. In the Mesozoic era, the Hyalospongiae produced *Dictyonina* with the compact skeleton. Hyalospongiae recorded in the beginning of the late Cretaceous Period exhibit affinities with the recent ones. The calcareous sponges are believed to be appeared in the Carboniferous Period.

The first members of the order Pharetronida from the Permian period had the compact calcareous skeletons and spicules. Demospongiae would have followed a line of evolution which originated in the Tetractinellida and advanced after reduction of the tetraxon spicules into the monoaxons. The polyaxon microscleres evolved into the specialized forms. Spongin cemented the spicules, and ultimately a skeleton of horny fibres was evolved. Tetractinomorpha have the four-rayed megascleres, asters, but without spongin. Ceractinomorpha possess monaxon megascleres without asters and spongin. Lithistida possess a siliceous skeleton of netlike desmas and are believed to be polyphyletic. Lithistid modification of skeleton appeared independently in different lines in Demospongiae. Demospongiae without a compact skeleton are poorly represented in sediments. Phylogenetic relationships among the various classes of Porifera are difficult to draw. However, common characters may be found among the Calcarea, Hexactinellida and tetractinellid Demospongiae.

## 4.3 SIZE, STRUCTURE, AND COLOUR

Most sponges are a few centimetres in size. Some are urn-shaped, or shapeless. Some shape likes vases, tubes or branches. Some may attain two metres in tallness. Broad rounded masses of Porifera may be up to two metres in diameter. Size within a species varies with the age, surrounding environmental conditions and food supply. Some are bushy or tree-like with finger-like projections. Members belonging to Demospongiae are shapeless. A few species like *Tethya aurain* (Demospongiae) possess spherical shapes. Others may be cup- or fan-shaped. *Scypha (=Sycon)* are tubular, sac-like, with an opening at the tip. Members of the Hexactinellida are either erect, or cylindrical with a stalk like base. Deep-water sponges exhibit generally a neutral colour, drab, or brownish. Shallow-water sponges are often bright coloured, which range from red, yellow, and orange to violet and occasionally black. Calcareous sponges are generally white. Some are greenish as green algae and live in a symbiotic relationship with them. Some are violet or pinkish as they harbour symbiotic blue-green algae. The sponge surface may be smooth, velvety or rough with spicules.

## 4.4 DISTRIBUTION

Few sponges are found in the fresh water. Majority inhabit the marine habitats in benthic communities. They are filter feeders, feed on the bacteria-size particles, exhibit multicellularity and comprising relatively independent cells. They live in the conditions from the polar region to the tropics in both salt- and fresh- water. Some are found in the shorelines, even flourishing between the tidemarks, while others live only in the highest depths of the ocean. Spongillidae is adapted to fresh water and its cosmopolitan distribution is facilitated by the reproductive buds. They are also found in the tidal zone in the deepest regions at all latitudes and are particularly abundant in Antarctic waters. Calcarea and Demospongiae are found mainly on the rocky bottoms of the continental shelf. Members of the Hexactinellida occur in the deepest muddy bottoms of the oceans and the seas. Sponges are the dominating organisms covering wide areas, particularly the rocky overhangs and the caves of the littoral zone. A few species prefer brackish waters.

## 4.5 CLASSIFICATION WITH CHARACTERS

1. Mostly marine, a few inhabit fresh water.
2. Multicellular.
3. Sessile, solitary or colonial.
4. Asymmetrical or radially symmetrical.
5. Body is cylindrical, tubular or vase like.
6. Diploblastic with mesenchyme in between ectoderm and endoderm.
7. Body surface bears ostia.
8. Peculiar canal system is present.
9. Internal skeleton in the form of spongin fibres is present.
10. Digestion is intracellular.
11. Nervous and sensory cells are absent.
12. Mostly bisexual, asexual forms also exist.
13. Asexual reproduction occurs by the budding, fission or gemmule formation.

The grouping of sponge is based primarily on the nature of spicules. Such divisions are not unanimous particularly in case of horny and siliceous sponges. Scheme of classification adopted here is based on classificatory plan outlined by L.H. Hyman.

### *Class Calcarea*

1. Smaller in size.
2. Solitary and colonial.
3. Cylindrical and vase like.
4. Osculum is narrow, terminal and fringed with the bristles.
5. Skeleton consists of separate calcareous spicules.
6. Spicules may be monaxon, triradiate and tetraradiate.
7. Exhibit asconoid, syconoid or leuconoid structure.
8. Exclusively marine.
9. Occur in the shallow costal water.

#### *Order Homocoela*

1. Body wall is thin and unfolded.
2. Spongocoel is lined by the choanocyte cells.

**Example:** *Leucsolenia, Clathrina.*

#### *Order Heterocoela*

1. Body wall is thick and folded.
2. Choanocytes line only the radial canals.
3. Spongocoel is lined by the endoderm cells.

**Example:** *Sycon, Grantia.*

### *Class Hexatinellida, or Trioxinida, or Hyalospongiae*

1. Size moderate.
2. Usually solitary.
3. Osculum is wide and is covered by a siliceous sieve plate.
4. Triaxon or hexactinal siliceous spicules are present.
5. Choanocyte cells are restricted to the finger-like flagellated chambers.
6. Exclusively marine.

#### *Order Hexasterophora*

1. Spicules are hexasters.
2. Flagellated chambers are radially and regularly arranged.

**Example:** *Euplectella.*

#### *Order Amphidiscophora*

1. Spicules are amphidiscs.
2. Flagellated chambers are irregularly arranged.

**Example:** *Hyalonema, Pheronema.*

***Class Demospongiae***

1. Skeleton is composed of siliceous spicules or spongin fibres, or a combination of both or none.
2. Siliceous spicules are monaxon or tetraxon.
3. Choanocyte cells are restricted to the flagellated chambers.
4. Body is generally compact and brightly coloured.
5. Mostly marine.
6. A few inhabit the fresh water.
7. Occur in shallow as well as in the deep water.

***Subclass Tetractinellida***

1. Siliceous spicules, when present, are tetraxon.
2. Spongin fibres are absent.

***Order Myxospongida***

1. Body surface is velvet-like and structure is simple.
2. Spicules are lacking.

**Example:** *Oscarella.*

***Order Carnosa***

1. Structure is homogeneous.
2. Spicules are equal-sized.

**Example:** *Picking.*

***Order Choristida***

1. Body form is elaborate.
2. Small and large sized spicules are present.

**Example:** *Geodia, Ancorina.*

***Subclass Monaxonida***

1. Skeleton consists of monaxon spicules.
2. Spongin fibres are present, or absent.

***Order Hadromerina***

1. Megascleres are mostly tylostyles.
2. Star like microscleres are absent.
3. Spongin fibres are absent.

**Example:** *Cliona, Teyhya, Halichondria.*

***Order Poeciloselerina***

1. Megascleres are of two or more kinds.
2. Microscleres may be curved, bow-shaped or C-shaped.

**Example:** *Myxilla.*

***Order Haplosclerina***

1. Megascleres are diactinal and pointed at both ends.
2. Microscleres are present, or absent.
3. Spongin fibres are usually present.

**Example:** *Chaline, Spongilla.*

***Subclass Keratosa***

1. Body form is rounded
2. Massive in size
3. Occur in warm shallow water.
4. Skeleton is made up of spongin fibres.
5. Spicules are absent.
6. Outer surface of evocates lies at the point of contact invaginates.
7. Capsule containing sperm head penetrates into ovocyte.
8. Thus fertilization is completed and zygote is formed.

**Example:** *Euspongia, Hippospongia* (Figures 4.1–4.5).

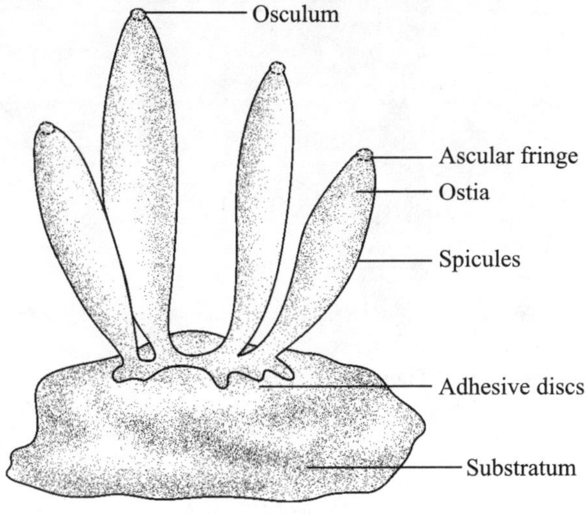


**Figure 4.1** *Sycon.*

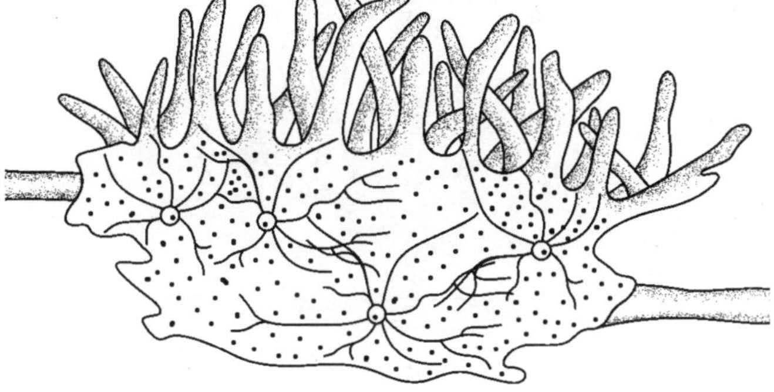

**Figure 4.2** *Spongilla.*

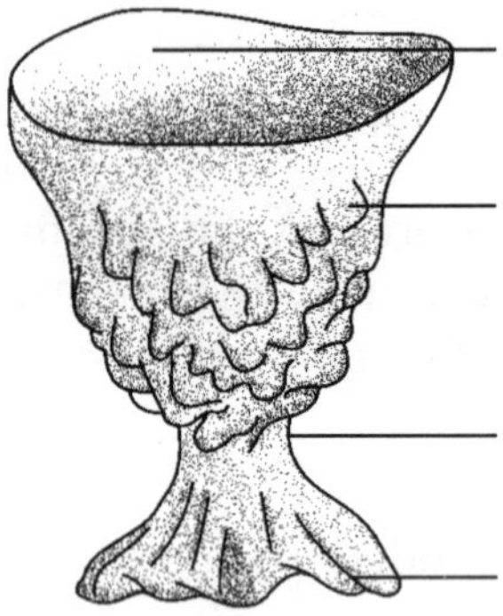


**Figure 4.3** Neptune's cup (Poterion).

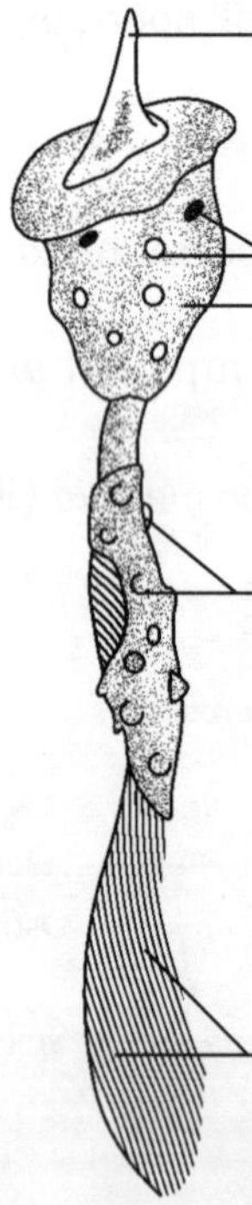


**Figure 4.4** *Hyalonema.*

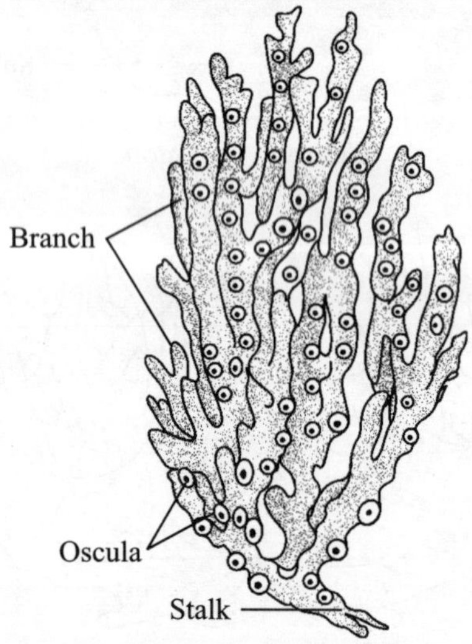


**Figure 4.5** *Haliclona.*

## 4.6 BODY STRUCTURE AND CELL TYPES

In the outer surface of a sponge, there is pinacoderm which is made by pinacocytes. Dermal pores perforate the pinacoderm. The inner surfaces mostly comprise the choanoderm and flagellated cells called *choanocytes*. Both of these layers are single cell thick. Between these two thin cellular sheets, the mesohyl exists. Water is pulled through these openings and is driven across the choanoderm by the beating of the flagella of choanocyte. The choanocytes pump large volumes of water through the body at low pressures and constitutes the water current system. Cuticle may cover the pinacoderm in some species. The main bulk of the body consists of the mesoderm cells. The cells covering the body surface and those linings the internal cavities have different history from the ectoderm and entoderm cells of other metazoans. Some chambers inside the sponge are lined with the flagellated cells. These cells because of their location and function in capturing and digesting food have considered to be the *entodermal cells*. The location of the flagellated cells are believed to arise from those cells of the embryo which in normal development cause the ectoderm and is marked by the hair like parallel lines.

Genus *Proterospongia* belonging to protozoa is thought to be the ancestor bridging the gap between the flagellate protozoa and sponge. Sponges lack a well-defined organization of tissues. Single layers of cells line the outer surface and the internal cavities. Other cells and fibres occur in a mesohyl which is gelatinous in nature. Homologies between various types of sponge cells and those of higher animals are difficult to draw. Each type of sponge cell performs the particular functions. The cells may gather in certain areas or form layers and membranes. They are easily modified in both form and function during larval- and adult- life. They migrate and transform from one cell type to another. Three principal types of cells—choanocytes, archaeocytes and pinacocytes may be distinguished. Choanocytes are provided with the flagellum, which is surrounded by a collar composed of cytoplasm. The main function of flagellum is to produce the water current for capturing food particles. Archaeocytes remain scattered in the mesohyl for transform into various other cell types particularly in Demospongiae. Some persist and reproduce without specializing, thus forming an embryonic reserve from which other cellular types may be derived.

Others become specialized to carry out the particular functions. Archaeocytes, often called the amoebocytes, are amoeboid cells. Their cytoplasm contains large quantities of RNA, and their large nuclei contain the nucleoli. Amoebocytes function in the regeneration and transportation of food particles acquired at the choanocytes. Amoebocytes laden with carotenoids and melanin confer various colours to the sponge. Archaeocytes may be important in sexual reproduction, as it is postulated that male and female reproductive cells are derived from them. Pinacocytes form the pinacoderm, a single cell layer found on body surface and lining the canals.

There are several types of pinacocytes. Basipinacocytes remain in contact with the surface to which the sponge is attached. Exopinacocytes are found on the sponge surface. Endopinacocytes line the canals. Pinacocytes are flattened cells containing many granules. They are capable of contracting and may cause a reduction in the volume of the sponge if it is disturbed. In the Calcarea, the outer surface also contains flattened granular cells called the *porocytes*. Porocytes contain the pores to allow watering into the sponge. The porocytes can contract and close the pores during the unfavourable environmental conditions. Collencytes found in the mesohyl secrete the fibres and often form a net in the cytoplasm.

The mesohyl contains cells like lophocytes and sclerocytes. Myocytes are believed to be derived from the archaeocytes. Lophocytes, similar to but larger than collencytes, have long cytoplasmic processes at one end, giving them the appearance of a comet. They apparently secrete the fibres that form the skeletal material. Sclerocytes which also produce skeletal material are classified according to the chemical nature of the spicules. Calcoblasts secrete calcareous spicules. Silicoblasts secrete siliceous spicules. Myocytes are elongated, contractile cells, particularly abundant near the oscula, where they control their expansion and contraction (Figures 4.6).

### *Cells forming the skeleton*

Several types of amoeboid cells in the mesohyl secrete the various elements of sponge skeletons. The entire supportive matrix is built on a framework of fibrillar collagen. The cells that secrete this

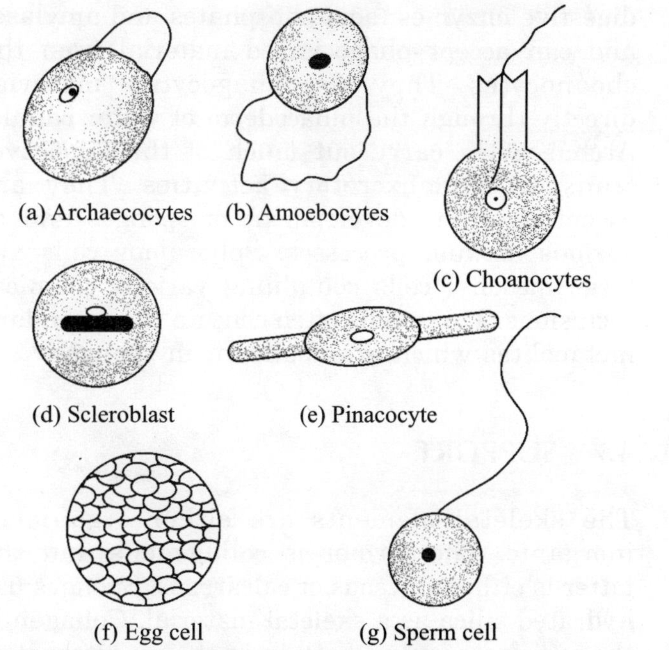


**Figure 4.6** *Haliclona.*

material are called the *collencytes*, *lophocytes*, and *spongocytes*. Collencytes are morphologically indistinguishable from the pinacocytes. Lophocytes are large, motile cells and can be recognized by a collagen tail. Both the cell types secrete the dispersed fibrillar collagen. Spongocytes produce the spongin. Spongocytes operate in groups. Sclerocytes produce calcareous and siliceous sponge spicules. They are active cells with abundant mitochondria, cytoplasmic microfilaments and small vacuoles. Numerous types of sclerocytes are described. These cells disintegrate after completion of spicule secretion.

### *Contractile cells*

Contractile cells called *myocytes* are found in the mesohyl and are usually fusiform and are grouped concentrically around oscula and major canals. Myocytes have the great numbers of microtubules and microfilaments. Because of the nature of their filament arrangement, it has been suggested that myocytes are homologous with the smooth muscle cells of higher invertebrates.

### *Other cell types*

Archaeocytes are ameboid cells give rise to virtually any other cell type. Archaeocytes are large, highly motile cells and play major role in the digestion and food transport. These cells possess a variety of digestive enzymes (acid phosphates and amylase) and can accept phagocytized material from the choanocytes. They also phagocytize material directly through the pinacoderm of water canals. Archaeocytes carry out much of the digestive, transport and excretory activities. They are essential to the development of sponges and to various asexual processes. Spherulous cells are large mesohyl cells containing various chemical inclusions. These cells often contain the secondary metabolites which are abundant in sponges.

## 4.7 SUPPORT

The skeletal elements are either organic or inorganic. The former is collagenous and the latter is either siliceous or calcareous. Sponges use hydrated silica as a skeletal material. Collagen is the major structural protein in the invertebrates. In sponges, it is either dispersed as thin fibrils in the intercellular matrix, or remain organized as spongin in the mesohyl. True spongin is found only in members of Demospongiae. Dispersed collagen fibrils are found in all sponges. The organic skeleton of the sponge is termed *spongin*. Spongin often cements siliceous spicules together at their points of intersection. The encysting coat of the asexual gemmules of fresh water and some marine sponges is also composed largely of spongin. Mineral skeletons of silica or calcium are found in almost all sponges. Several demospongiae lack both spongin and a spicule skeleton.

## 4.8 SKELETON

The skeletal structures are spicules which are formed by the mesoderm cells called the *scleroblasts*. Each of the large spicules starts as a microscopic crystal within the cytoplasm of a scleroblast. As the spicule grows, it extends the limits of the mother cell and breaks out from its walls. Spicules may be formed by the action of a group of scleroblasts. Spicules may be formed of either the calcareous or the siliceous material in a number of patterns. The spongin is another type of supporting structure formed by the spongioblasts. Spongin which is closely allied to silk in its chemical composition may appear either alone, or along with the siliceous spicules. Skeleton is of great taxonomic significance. It may be mineral in nature or may be composed of protein and other components. Mineral skeleton is formed for the most part by units called the spicules, either scattered throughout the sponge or united to form fibres. Spicules are classified as megascleres, which function in support, and microscleres which function in protection and aid in support.

### Mineral skeletons

Calcareous spicules, characteristic of the Calcarea, are composed chiefly of calcium carbonate in crystalline forms. Most calcareous spicules have one axis (monoaxon), which is usually pointed at both ends. These spicules are called the *oxeas*. Triaxons with three rays are called the *triacts*. Tetraxons with four rays are called *tetracts*. Siliceous spicules found in Demospongiae and in Hexactinellida, are made of silicic acid, some water, a small quantity of other sodium containing compounds, potassium, iron, chlorine and a small quantity of organic matter, called *spiculin* which forms an axial fibre. Spicules in Hexactinellida are variable in form and often have remarkable dimensions. Characteristic spicules of Hexactinellida are triaxon forms with six rays. Spicules are connected in a continuous network, and after the death of sponge and the loss of its soft parts, skeleton that remains has a delicate glass texture as found in *Euplectella*. Bundles of the large spicules form stalks that allow members of the Hexactinellida to attach to the muddy bottoms of the deep ocean in which they generally live. In *Monoraphis,* the stalk is one enormous spicule that may attain a length of two or three metres and a thickness of approximately one centimetre.

## 4.9 SPICULES

Few deep sea sponges possess a pseudoskeleton entirely of the foreign materials. Most sponges possess a true skeleton known as the *autoskeleton* composed of materials secreted by the sponge itself. Such autoskeleton is either spicules or spongin, or a combination of both. Spicules also known as the *sclerites* are definite small bodies, crystalline in appearance and consisting of simple spines, or spines radiating from a point. Spicules of spongin fibres take part in the formation of skeleton. Spicules possess an axis of organic substance surrounded by inorganic substance like $CaCo_3$, or clear hydrate silica. Variations in shape of spicules help in the identification of the sponges. Spicules may be large spicules, or megascleres, and small spicules, or microscleres.

### *Megascleres*

**Monaxon spicules:** These are formed by the growth in one or both directions along an axis. Growth in one direction results in formation of the monactional monaxon spicule (style). Styles are typically rounded at one end and pointed at other end. It may be.

**Tylostyle:** In which broad end is knobbed.

**Acanthostyle:** In which there is a covering of thorny process. The growth of monaxon spicule in both directions from a central part forms the diactinal.

**Monaxon** (rbabds or diactines). They may be:

*Oxeas:* Provided with pointed both ends.

*Tornotes:* Provided with lance-headed each end.

*Styongyles:* Rounded each end.

*Tylotes:* Pin head like at each end.

**Triaxon spicules:** This is also called *hexactional* and occur only in Hexactinellaida. Basically, it consists of 3 axes crossing at right angles resulting in the production of six rays, which extend from a central point at right angles to each others. This may result in much variation by loss or reduction of rays.

**Tetraxon spicules:** Also known as tetractines and quadriradiates. Basically each possess 4 rays, but not in the same plane. They may be:

*Calthrops:* Provided with 4 more or less equal rays.

*Triacnes:* When one ray is elongated (rhabdome) and bears 3 smaller rays (cladome). When one smaller ray is lost, it becomes diaene.

*Amphidisc:* With disc are present at both ends of a rhabdame. A triradiate spicule is most common in the calcareous sponges and results by the loss of rhadbome of a triaene.

**Platoon spicules:** These are provided with several equal rays radiating from a central point.

**Spheres:** These are almost round spicules in which growth occurs around a centre.

**Desma:** This is a special type of megasclere consisting of minute monaxon, triradiate or tetroxon spicule called the crepis on which silica is deposited. According to the shape of crepis, desmas are named as the monocrepid, tricrepid, and tetracrepid. They are generally united into a network resulting in the formation of lithistid skeleton.

**Microscleres:** These are minute spicules and are found scattered in the mesenchytma. They do not form the supporting framework and are of two types:

*Spires:* These are curved in one plane or spirally twisted. C-shaped forms, called the *sigmas* are common, bow-shaped forms called the *taxas* and forms with recurved hooks, plates or flukes at each end are called the *chelas*. Chelas with two similar ends are called the *isochelas*, when dissimilar, aniosochelas. Spirally twisted sigmas are termed sigmaspire.

*Asters:* Microscleres with small centres and long rays and large centres and small rays fall into this type. Small centred forms with pointed rays are called the *crylaster*, with rounded ends are called the *strongylaster*, with knobbed rays are called the *tylasters*. Large centred forms with definite rays are called *spherasters*, with rays reduced to small projections are called *sterrasters*. Short spiny microscleric monaxons are called *streptasters*. Microscleric forms of diactines may be *microrhabds*, microxeas and *microstrongyles*.

### *Development of spicules*

- **Monoxon spicules**
  - (i) In a binucleate scleroblast, a small deposition of $CaCO_3$ appears.
  - (ii) With development, the two nuclei are separated wide apart and the scleroblast divides into two cells.
  - (iii) They are lying towards the inner side, called the *founder* which gives shape and length to the spicule. Other cell lying outwards helps to thicken the spicule.
  - (iv) When the spicule is fully formed. Both the cells migrate in the mesenchyma.
- **Triradiate spicules**
  - (i) To form a triradiate spicule, 3 scleroblasts take part.

(ii) Each divides and in each pair, a small spicule is secreted.
(iii) Ultimately, these three spicules unite to form.

The siliceous spicules consisting of both the megascleres and microscleres of the Demospongiae have several forms. Megascleres may be monaxons with both the ends pointed (oxeas), with one end pointed and the other rounded, or with both the ends rounded. If one end is swollen, the styles are called *tylostyles* and *strongyles tylostrongyles*. Spicules with both ends swollen are called the *tylotes*. If the surface of spicules is spiny instead of smooth, the spicules are called, *acanthoxeas*, *acanthostyles* and *acanthostrongyles*, respectively. Megascleres also include triacts which are provided with three rays and tetracts, called the *calthrops* are provided with four rays. Tetracts, with one axis (rhabdome) longer than the other three, are called the *protriaenes*, *plagiotriaenes*, *anatriaenes* or *dichotriaenes*, depending on the way the rays of the cladome are directed.

All microscleres apparently are derived from a spherical type with many axons (polyaxon). The result is a series of star-shaped spicules or asters with various numbers of rays. Spicules with rays missing, or reduced often form protective armour around the sponge. Specialized types of microscleres include the sigmas, toxas, chelas and anchoras. The last two have plates or teeth at each end and may be distinguished as isochelas and isanchoras, or anisochelas and anisanchoras, depending on whether the ends are equal or unequal.

A few members of the Demospongiae lack skeletons. Ceractinomorpha has a type of spongin, which, in certain orders, cements the spicules in bundles, or meshes, thereby increasing the elastic nature of the skeleton. In Keratosa, spongin fibres constitute the entire skeleton. Spongin fibres may be branched, netlike (order Dictyoceratida), without inclusions or with inclusions. In *Ircinia*, thin spongin filaments that fill the mesohyl accompany the fibres. In some rare species of the Calcarea and Demospongiae, specialized types of skeletons in two groups have great paleontological importance. Calcarean sponges of order Pharetronida have skeletons formed by an amorphous mass of calcium carbonate, with which few spicules are associated. Those in the Demospongiae (Lithistida) form a heterogeneous group in which desmas form a compact skeleton. Some Demospongiae possess a compact calcareous skeleton, which incorporates both siliceous spicules and organic fibres (Figures 4.7–4.9).

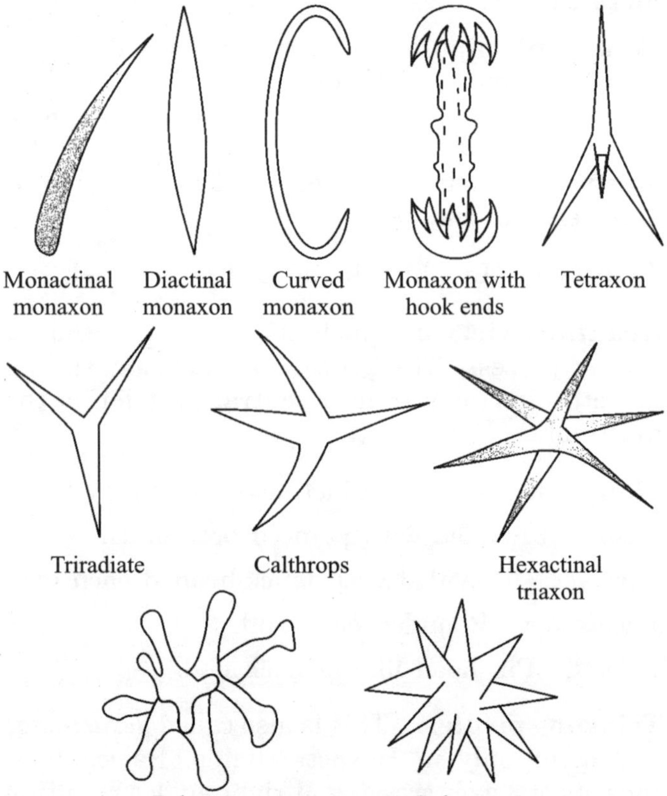


**Figure 4.7** Spicules and spongin.

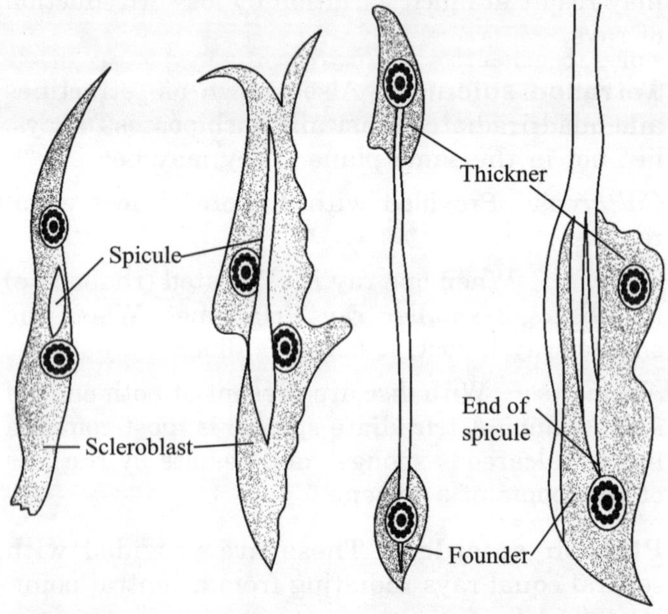


**Figure 4.8** Secretion of a monaxon spicule.

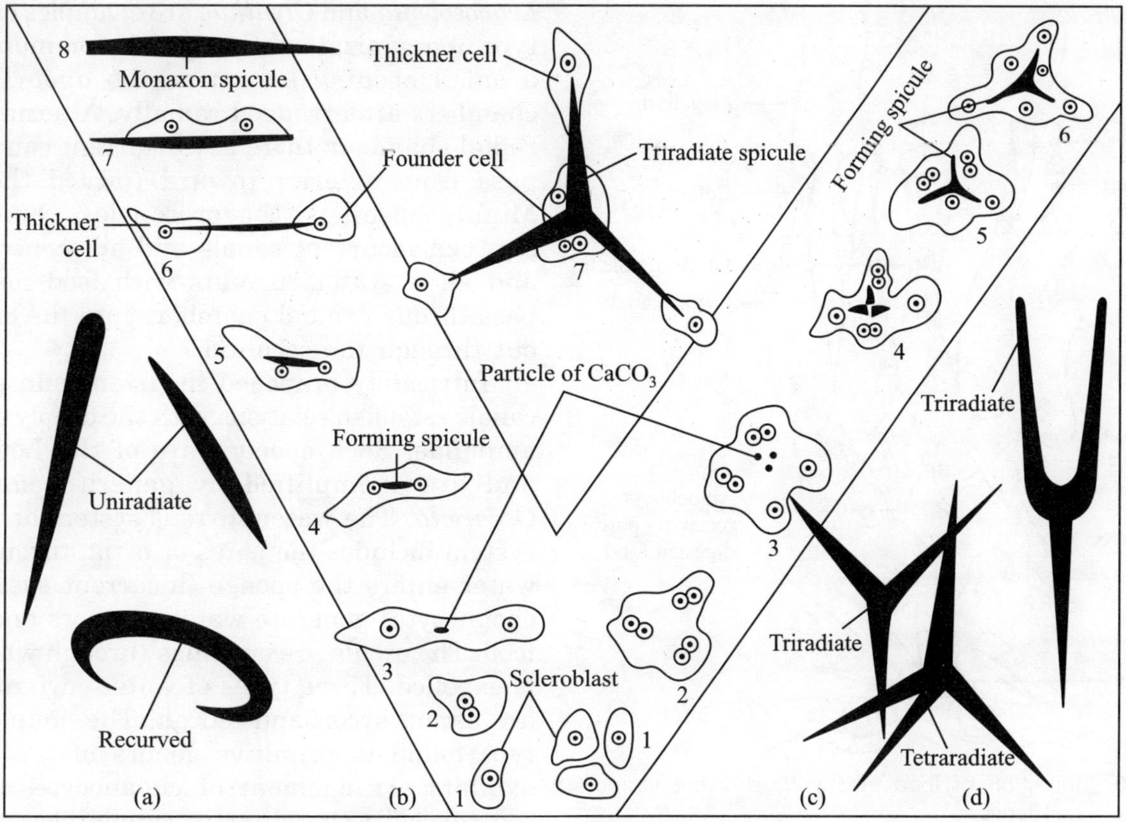


**Figure 4.9** Structure and development of spicules in *Scypha* (a) Kinds of monaxon spicule, (b) Development of monoxon spicule, (c) Development of triaxon spicule, (d) Kinds of triradiate spicule.

Sponge spicules are produced by special mesohyl cells called the sclerocytes, which accumulate calcium, or silicate. Sometimes, one sclerocyte produces one spicule; in others, several sclerocytes work together to produce a single spicule, often two cells per spicule ray. Unlike siliceous spicules, calcium carbonate spicules do not have an organic axial structure. Calcareous spicules are produced extracellularly in intercellular spaces bounded by several sclerocytes. Each spicule is a single crystal of calcite, or aragonite. According to their morphology, spicules are termed either microscleres or megascleres. The former is small to minute reinforcing spicules; the latter are large structural spicules. The demosponges and hexactinellids have both types. Calcareous sponges often have only megascleres.

## 4.10 NUTRITION, EXCRETION AND GAS EXCHANGE

The Porifera are primarily filter feeders, utilizing the food particles suspended in the water. Food particles consist of bacteria, and particles of organic debris. Sponges probably absorb dissolved organic substances. Cladorhizid sponges feed as carnivores by capturing the prey with numerous hairs that cover the body. A sponge absorbs a large quantity of food, and it can use several litres of water a day. Water-current system is efficient despite the lack of a nervous coordinating centre. The water flows is maintained by continuous movement of flagella of choanocytes. Flagellar movement pushes water into the choanocyte chambers; create pressure that extends to excurrent canals, as water passing into the pores is drawn along the incurrent canals (Figure 4.10).

Water flow, regulated by contraction and expansion of the pores and oscula, is slow at the pores and faster at the oscula. As a result, water expelled with force through the oscula cannot contaminate water entering the pores. Respiratory organs are absent in sponges. Oxygen is supplied by a direct exchange between tissues and surrounding water. Excretion occurs through oscula and the body surface. Special amoebocytes disintegrate in the mesohyl, and their granules are expelled

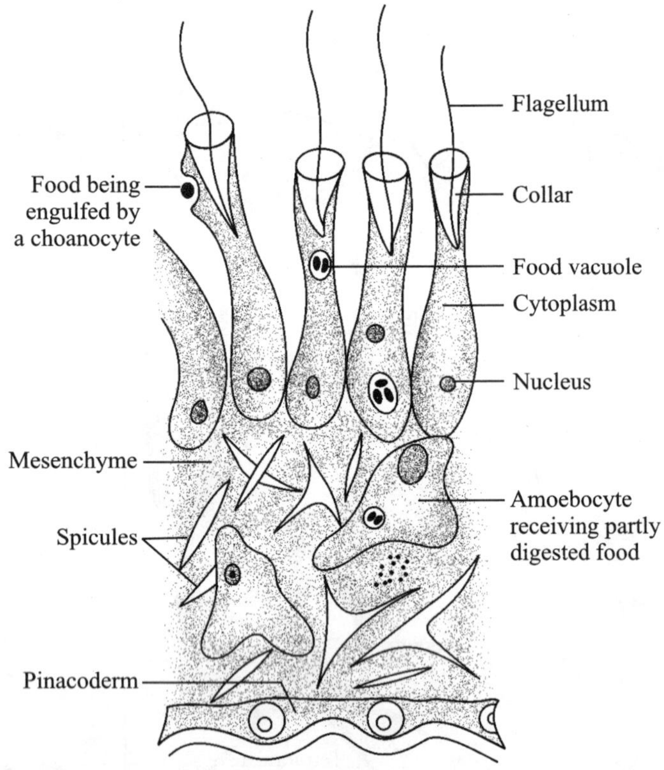


**Figure 4.10** Ingestion of food in *Scypha* showing various cell types.

through the canals. Excretory products—ammonia and other nitrogen-containing substances—account for their characteristic unpleasant odour. Many sponges *(Tedania ignis)* exude large quantities of mucus. Some species produce toxic substances. Excretion (primarily ammonia) and gas exchange is mediated by simple diffusion, much of which occurs across the choanoderm.

## 4.11 MORPHOLOGICAL TYPES AND WATER-CURRENT SYSTEM

Three types of sponge structure are found. In the asconoid type, the body is a thin-walled sac with a single large opening. The osculum lies at one extremity and numerous minute pores on the body wall communicates with a central spongocoel, which is lined by flagellate cells. Since digestion is intracellular, the cavity is not a true digestive cavity. When the pores remain open, the flagellar action causes the water currents bearing food material to pass through flagellated chamber and out through osculum. During this passage, the food is removed from the water and ingested by the cells. *Leucosolenia* and *Olynthus* are examples of simplest type of organization. Spongocoel communicates by a small opening known as the *apopyle*. Radial chambers are arranged radially. Alternating with radial chambers there are incurrent canals which pass from exterior inward toward the cloaca. Minute openings, the prosopyles, communicate between incurrent canals and adjacent ampullae and allow water currents with food material to pass through radial chambers into the cloaca and out through the osculum.

Intricately branched incurrent and excurrent canals establish relations with the deeply embedded ampullae. Such sponges are of the Leucon type and are exemplified by genera *Leucilla* and *Oscarella*. The water-current system or the canal system includes the pores or ostia, through which water enters the sponge (incurrent system). The choanocytes generate water currents and capture food. The oscula are openings through which water is expelled. Three types of water-current systems are ascon, sycon and leucon. The simplest ascon type found in primitive genera of the Calcarea, exhibits arrangement of choanocytes around a central cavity that directly communicates with the osculum. The sponge wall is thin, lack canals and bears pores which actually are porocytes.

The syconoid canal system, found also in calcareous sponges, is characterized by choanocytes that surround finger-like projections of the sponge wall. Water enters the projections directly through pores, makes its way into the spongocoel and leaves through osculum. In *Scypha*, the radial canals are bordered by incurrent canals through which passes the water entering the pores. Other openings called *prosopyles* allow water into the choanocytes, from which it passes into the internal cavity and out of it through osculum.

In the leucon type, the radial canals are replaced by several small flagellated chambers in which the choanocytes are localized. The chambers, scattered throughout the body, have pores through which water passes into an incurrent canals, then into spongocoel by way of excurrent canals. Water enters very small pores which are found among the pinacocytes, which line the outer surface of sponge. After passing through a system of incurrent canals and cavities, the water reaches the flagellated chambers, enters them through prosopyles, and leaves through apopyles. The water is expelled through the osculum after passing through a system of excurrent canals and cavities lined with pinacocytes.

The rhagon type is characterized by reduced excurrent canals and by a large central cavity. In some Demospongiae, the body is organized into an external ectosome without choanocytes and an internal choanosome with choanocytes. The cladorhizids, a group of deep-water and cave-dwelling demospongae lack a water-current system. They are carnivores, capture small prey with numerous long, thin filaments that cover the body. Flagellar movements in the flagellated chambers produce water currents through the canal system. Food particles entering the cells are ingested by the flagellated chambers. Digestion is intracellular. The sponge derives its nourishment and oxygen for respiration from the incurrent streams of water, while the water stream as it leaves the body through osculum carries the metabolic wastes.

## 4.12 KEY POINTS OF CANAL SYSTEM IN SPONGES

Canal system furnishes a peculiar kind of pump, driving in and out the current of water. Structural complexities in sponges are primarily due to possession of canal system, which constitutes the most vital system. Cell types work on background of this system. Four basic type of canal system in sponges are described below.

### *Asconoid type*

This type is the simplest of all type and is present in those sponges found only in sponges like *Olynthus, Leucosolenia* whose body is vase like and radically symmetrical. Walls are extremely thin. It encloses a large spongocoel lined by choanocytes, which opens at summit by a narrow osculum. Spongocoel wall is pierced by numerous microscopic apertures, incurrent pores, or ostia which extend from external surface to spongocoel. Each pore is intracellularly disposed in a porocyte. Important features of asconoid structure is simple wall and complete continuous lining epithelium of choanocytes interrupted only by inner ends of porocytes. The course of water current is as follows:

Dermal pores → Internal ostia → Spongocoel → Osculum → Exterior

### *Syconoid type*

This type, the first stage above asconoid type, is formed by out pushing of wall of an asconoid sponge at regular intervals into finger like projections, called the *radial canal*. In simple syconoid, syconoid stage-1, radial canals are free projections, which are completely surrounded by water without definite incurrent canals. In most syconoid sponges, walls of radial canal fuse at distal ends and a mesenchymal cortex is formed (syconoid-ll), which is pierced by dermal ostia and opens internally into an incurrent canal which is lined by pinacocytes. Short canals, prosopyles, in walls of incurrent canals communicate with the radial canals. These bring outpushings of spongocoel and are lined by choanoytes. Spongocoel is lined by endopinacocytes, and choanoytes are restricted to radial canal. Syconoid (stage-l) type is found in heterocoelous calcareous sponges as in *Scypha*.

Syconoid (stage-ll) type canal system is found in calcareous sponges like *Grantia, Heteropia*. Syconoid sponges differ from asconoid type (i) in thick folded walls contaning alternating incurrent and radial canals, and (ii) in breaking of choanocyte layer, which on longer lines whole interior but is limited to certain definite chambers (radial canal).

### *Leuconoid type*

As a result of further out folding of the choanocyte layer and thickening of body wall, leuconoid type develops. Choanocyte layer of radial canal of syconoid stage evaginates into many small chambers which repeats same process to produce a cluster of small flagellated chambers. In many cases, dermal pores open into sub dermal spaces. Sub dermal spaces and incurrent canals lead into small rounded flagellated chambers open by apopyles into excurrent canals to form large tubes. The largest one leads to osculum. Choanocytes are limited to small chambers. Great development of the mesenchyme and complexity of incurrent and excurrent canals are important features. This canal system exhibits numerous variations, but presents three stages of evolution, viz., eurypylous, aphodal and diplodal. The course of water current is as follows:

Dermal pores or ostia → Sub dermal spaces → Prosodus, if present → Prosopyle → Flagellated chambers → Apopyles → Aphodus, if present → Excurrent canals → Large canals → Osculum → Exterior.

**Eurypylous:** In this type, flagellated chambers are wide and thimble shaped, each opening directly into excurrent canal by a wide aperture called apopyle and receives water supply direct form incurrent canal through prosopyle. Current of water takes the following route. This type of canal system is found in *Leucilla*. The course of water current takes the following route:

Dermal pores → Sub dermal space → Incurrent canal → Prosopyle → Flagellated chambers—Apopyles → Excurrent canal → Spongocoel → Oscula → Exterior.

**Aphodal:** In this type, flagellated chambers are small and rounded. Opening of each chamber into the excurrent canal is drawn out into a narrow tube usually not of great length, termed aphodus. Relation of the flagellated chambers to incurrent canals remains as before. Route of water current is as follows:

Ostia → Sub dermal space → Incurrent canals → Prosopyles → Flagellated chambers → Aphodus → Excurrent canals → Spongocoel → Oscula → Exterior

**Diplodal:** In some cases there is also a narrow current tube, prosodus between incurrent canal and flagellated chambers. Such a condition is called *diplodal*. This system is found in *Oscarella. Spongilla*, etc. Current of water takes the following route:

Dermal pores or ostia → Sub dermal spaces → Incurrent canals → Prosodus → Flagellated chambers → Aphodus → Excurrent canals → Spongocoel → Oscula → Exterior.

### *Rhagon type*

Rhagon type of sponge has a broad base, conical in shape with a single osculum at summit. Base wall is termed the hypophare which is devoid flagellated chamber. Upper wall bearing a row of small oval flagellated chambers is the spongophare. Spongocoel is bordered by the oral flagellated chambers opening into it by the wide apopyles. Between chambers and epidermis lies a considerable thickness of mesenchyme traversed by incurrent canals and sub dermal spaces. Dermal pores or ostia open into the sub dermal spaces which extend below the entire surface of body. Branching incurrent canals lead from sub dermal spaces into the small flagellated chambers which are formed by breaking down of the radial canals. Flagellated chambers alone are lined by the choanocytes and from which excurrent canals lead into a spongocoel. Incurrent and excurrent canals may be complex and branched. Spongocoel opens by a single osculum. The course of water current is as follows:

Ostia → Sub dermal spaces → Incurrent canals- → Porsopyles → Flagellated chambers → Apopyles → Excurrent canal → Spongocoel → Osculum → Exterior.

Differences between four basic types of canals system in sponges are given in Table 4.1.

### *Functions of canals system*

Canal system helps in nutrition, respiration excretion and reproduction. Current of water which flows through the canal system brings the foods and oxgen and takes away carbon dioxides,

**Table 4.1** Types of canals systems in sponges

| Point | Asconoid type | Syconoid type | Leuconoid type | Rhagon type |
|---|---|---|---|---|
| Wall | Simple | Evaginated to produce radial and incurrent canals. | Irregular | Simple |
| Mesenchyme | Simple and thin completely traversed by porocytes. | Thickened and not completely traversed by porocytes. | Highly elaborated, traversed by current canals. | Considerable thickened traversed by current canals and sub dermal spaces. |
| Choanocytes | Lined the spongocoel. | Limited to the radial canals. | Limited to the flagellated chambers. Flagellated chambers open by narrow apopyles. | Limited to the flagellated chambers which open by wide apopyles. |
| Spongocoel | Spacious | Spacious | Usually obliterated. | Spacious |

nitrogenous wastes and feaces. It carries sperms from one sponge to the other for fertilization of ova and increase surface area of sponges in contact with the water and enables sponges to increase their volume as the surface volume ratio remains fixed (Figures 4.11–4.16).

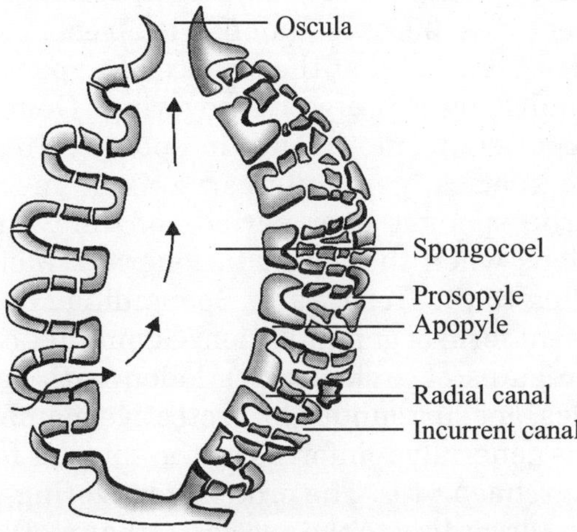


**Figure 4.11** Syconoid type of canal system.

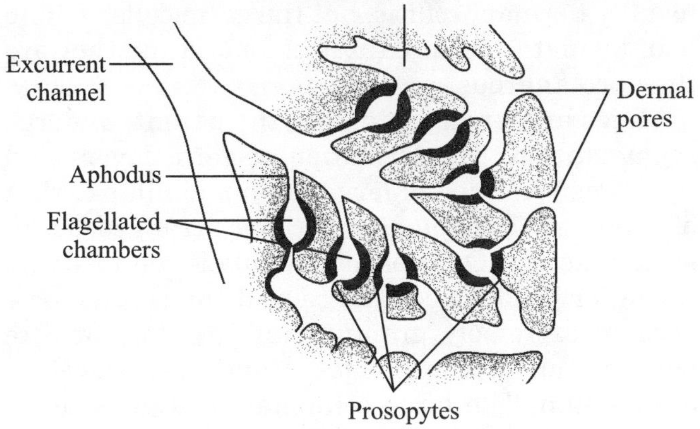


**Figure 4.12** Leuconoid trype of canal system with aphodal chambers.

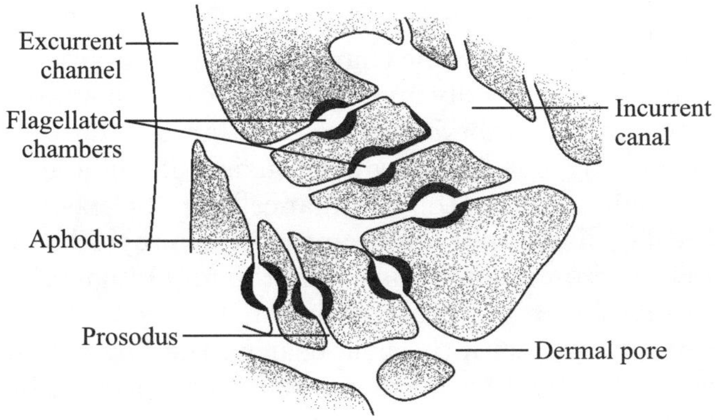


**Figure 4.13** Leuconoid trype of canal system with diplodal chambers.

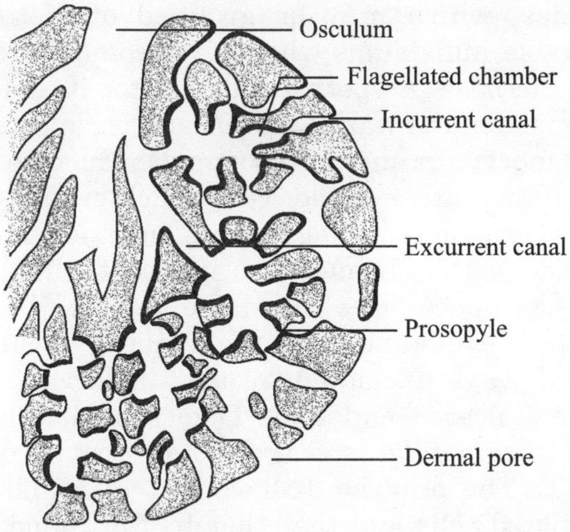


**Figure 4.14** Leuconoid type of canal system with eurypylous chambers.

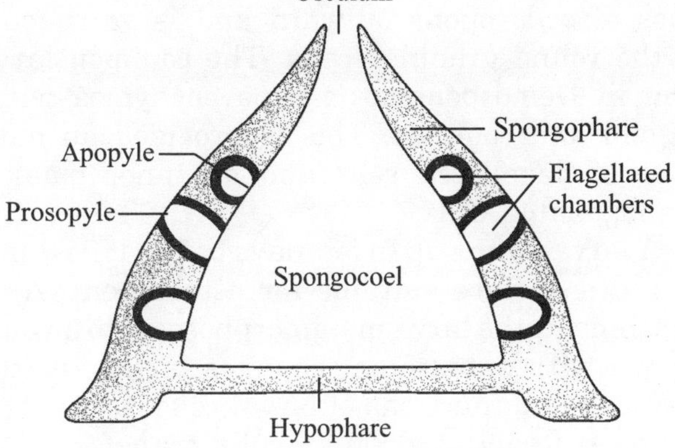


**Figure 4.15** Rhagon type of canal system.

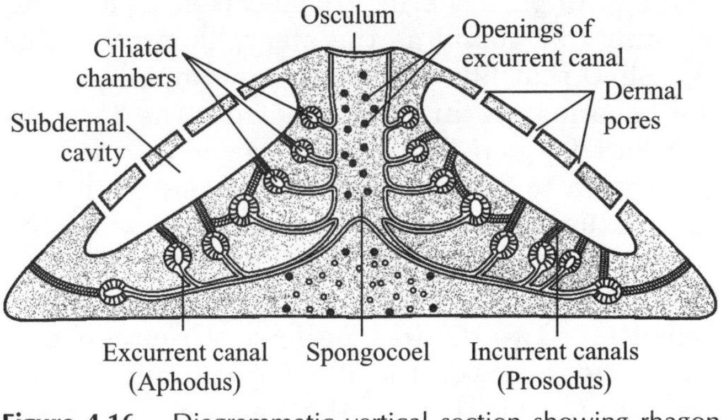


**Figure 4.16** Diagrammatic vertical section showing rhagon type of canal system.

## 4.13 REPRODUCTION

### *Sexual reproduction*

A sexual reproduction through budding is common for the production of masses of individuals forming colonies, which may be grouped of distinctly separable individuals with only a common region near the base. A spermatozoan, after its release, is carried by the water current and is captured by a choanocyte in another sponge. Choanocyte then transforms into a carrier cell, which gives up the spermatozoan to an egg, lying near a chamber formed by the choanocytes and containing long lash like appendages called the *flagella*.

The development of the embryo occurs in several ways ultimately results in larva. Larva of the Calcarea and some Demospongiae, called an *amphiblastula*, is oval and has a cavity in the middle. The anterior half consist of cylindrical, flagellated cells and the other half of round cells without flagella. Larva swims with the flagellated portion forward. Amphiblastula is preceded by a stomoblastula in which the central cavity of a hollow mass of cells opens outward and is surrounded by the round granular cells. The common larval form in Demospongiae is a parenchymea which is solid and compact. This is provided an outer layer of flagellated cells and an inner mass of nonflagellated cells.

A larva swims up to few days before it descends to find a surface suitable for attachment. After attachment, the larva metamorphoses into a young sponge with subsequent changes in the relative positions and functions of larval cells. In one type of larva, flagellated outer cells transform into choanocytes of interior of adult sponge. Interior cells give rise to pinacoderm, archaeocytes, collencytes in adult that found as the amorphous substance (mesohyl) that fills the sponge. In amphiblastula, the choanocytes are derived from the forward flagellated region. The other cells and the mesohyl are known to be derived from the posterior half. The sexual maturation is connected with the water temperature. In the temperate regions, the maturation occurs from spring to autumn. Sometimes two distinct periods of reproduction occur, one in spring, the other in autumn. *Scypha* mature at any time of the year. Sponges are viviparous; larvae are released through the canals of excurrent (outgoing) water system and an opening also involved in that system. Sponges such as *Cliona* and *Tethya* lay eggs. Life-span of sponges is not well known. Small encrusting forms probably live about a year, disappears during a season unfavourable for their survival. Small fragments of an individual may persist and reproduce new individuals in the next season. The large species have a much longer life-span. Both sponges attain a commercially desirable size after 7 years and may live up to 20 years.

### Asexual reproduction

The best known method of asexual reproduction is called the *gemmulation*. Gemmulation begins when the aggregates of cells, mostly archaeocytes. Archaeocytes become laden with reserve food granules get solated at the surface of a sponge and surrounded by a protective covering. Gemmules are expelled from the adult sponge. In some marine species, gemmule serve as a means to carry the sponges over periods of unfavourable conditions when the adults degenerate. Members belonging to the fresh water Spongillidae undergo a different form of gemmulation. Gemmules consist of aggregates of archaeocytes laden with reserve granules, get surrounded by protective membranes which is generally reinforced by the spicules formed by the archaeocytes. The protective covering varies in shape according to the species. Other methods of asexual reproduction are the formation of stolons and fragmentation of the individuals. Sponges are usually hermaphroditic, but the germ cells of male usually mature before those of female and they are thus protandrous.

Eggs are fertilized, within the parent, undergo segmentation. The formation of germ layers and the transformation of larva are not similar to that of Metazoa. Blastomeres become arranged as an ovoid blastula. One pole of blastula is rounded and the other is flattened. The cells at the pointed pole become extremely granular and at the opposite pole become much elongated and each develops a flagellum. The large columnar cells grow down partly enclosing the mass of the granular cells. During this stage, the larva breaks through the embryonic chamber and by way of the osculum and using the flagella swim freely in the water. Upon leaving the embryonic chamber, the granular cells which have partially invaginated into the blastocoel come on the embryo surface. The larva is now an ovoid body, one pole of which consists of flagellated cells and the other the granular cells. In the sponges the flagellated cells later come to lie inside the body. Free-swimming larva is called the amphiblastula. After a day or more, free-swimming amphiblastula settles down with the pole bearing the flagellated cells in contact with the suitable substratum. The flagellated cells invaginates to form a cavity within the granular cells which develop into the collared flagellated cells. The system of pores, canals and the osculum later make their appearance, laying down the general form and organization of sponge. During larval development some mesenchyme cells from the interior of the larva break through the layer of granular cells and spreadover the surface. Such apparently mesodermal cells form the body covering.

#### Regeneration

The extraordinary capacity of sponges to regenerate is manifested through restoration lost parts and complete regeneration of an adult from fragments. Sponge cells when get separated by the mechanical methods or by the chemical methods, the dissociated cells settle, migrate, and form active aggregates. The cells generally become attached to a surface, flatten and develop pinacocytes. This is called the *dimorph stage*. Reconstitution of the choanocyte chambers and the canal system follows and results in functional young sponge capable to grow. The reconstitution process is not comparable with the embryonic development, as many types of dissociated cells participate in formation of the new sponge by rearranging themselves, rather than through differentiation from primitive cell types. Sponge regeneration is important in studying the cell-to-cell recognition, adhesion, cell movement, and cell properties.

## 4.14 Sycon sp.

#### Habit and habitat

A small, marine sponge, found permanently attached to submerge rocks and other solid substrata, in the shallow water near coasts, just below the 10 m tidemark. It is especially abundant in the places where wave action provides with an adequate food supply and with well-oxygenated water. It is a branching, colonial sponge, though solitary individuals are also found.

#### External features

Usually occurs as a cluster of branching cylinders or individuals, connected basally to a creeping stoln over the substratum. Individual varies from 15 to 30 mm in height and 5 to 6 mm in diameter, each resembles a slender vase-like cylinder slightly bulging in the middle. Asexually reproducing animals may give out one or two basal or lateral tubular outgrowths of the body wall, called *buds*, thus looking like a branching tree.

#### Colour

Sponge shows different shades of grey or light brown, according to the habitats and species.

*Osculum:* Distal or free end of each cylinder communicates with the outside by a large terminal opening, exhalant or excurrent pore or osculum. Large, straight and needle like calcareous structures, the monaxon spicules, surround osculum on all sides in a circlet or funnel-like collar, which appears like a delicate fringe. It prevents the entry of other animals into sponge.

*Collar:* Below the osculum, distal end of a cylinder may be somewhat narrower forming a collar region.

#### Body surface

Body is covered externally by the ectoderm. Under a lens it appears bristly due to the minute, needle like spicules protruding beyond the body surface. Entire surface presents innumerable elevations of the polygonal shape separated from one another by the depressed lines or furrows.

#### Dermal ostia

The groups of microscopic intercellular apertures, the inhalant or incurrent pores or dermal ostia, pierce furrows.

#### Canal system

The innumerable intercommunicating cavities or canals of several types, which together form the canal system, traverse body. Each cylinder includes a large spongocoel, while its thick wall contains several small and radially arranged cavities of three types—flagellated canals, excurrent canals and incurrent canals. Canal system is known as *syconoid* (stage 1) type. Spongocoel is out pushed at regular intervals as finger like projections which are called *radial canals*. The wall of radial canal is lined with choanocytes. In between two successive radial canals, a tubular space, the incurrent canal is present. Thus, radial canals and incurrent canal are rearranged alternately and latter opens to exterior through ostia. The wall between incurrent and radial canals is pierced by numerous minute pores, the prosopyles. Openings of radial canals into spongocoel are called *internal ostia* or *apopyles*. Finally, spongocoel opens to exterior by osculum.

*Skeleton:* Soft and gelatinous body wall is supported and protected by an endoskeleton of several minute, crystalline bodies of clear glassy appearance, called *spicules*. These are embedded in mesenchyme and have a definite arrangement in respect to its canal system and belong to several types. Monaxon spicules are arranged in and around osculum forming oscular fringe. Simple spear like or club like monoaxon oxeo spicules project in dense masses and from dermal cortex opposite the outer ends of flagellated canals giving sponge a bristly appearance. Triradiate spicules are present along flagellated canals with one ray pointed towards closed distal ends of canal. Tetraradiate spicules present along with triradiate spicules in thick gastral cortex surrounding spongocoel.

*Nutrition, respiration and excretion:* Sponges feed on microorganisms which enter into body along with water current. Choanoytes engulf them and pass to amoeboid cells situated below choanoytes. Digestion takes place inside amoeboid cells. Assimilated products are conveyed to various parts of body. Thus, nutrition is holozoic and digestion is intracellular, a process comparable to that of protozoan. Some amoeboid cells, often contains chlorophyll or green pigments carries out autotrophic nutrition such as green plants. The presence of algae within sponge body also helps in nutrition. Respiration and excretion take place by diffusion. Contractile vacuoles are recorded in amoeboid cells of fresh water sponges which probably play important role in osmoregulation and excretion.

### *Reproduction and development*

Reproduction and development takes place by regeneration and external budding. Regeneration is the ability to replace the parts lost by injury. In budding, a bud arises basally near point of attachment. An osculum opens at its free distal end. Bud remains attached, grows in size and increases bulk of sponge or else. It finally breaks and leads separate existence.

### *Sexual reproduction*

Special sex organs are lacking. Ova and sperm, develop from undifferentiated amoebocytes called *archaeocytes*, are found in jelly-like middle layer of body wall or mesoglea. They do not mature at same time. So, the cross fertilization occurs. Here, the condition is protogyny, that is, ova become mature first.

**Oogenesis:** Egg mother cell or oocyte is derived from a large amoebocyte with a distinct nucleolus and is nourished by special nurse cells or trophocytes (modified choanocytes). Oocyte obtains food by engulfing or fusing with a trophocyte. When full grown, oocyte undergoes two maturation divisions to form ovum. A ripe ovum is a large oval or spherical cell containing a large nucleus with a distinct nucleolus. Ripe ovum lies in wall of a radial canal, ready to be fertilized by a sperm from another sponge.

**Spermatogenesis:** Sperm mother cell or spermatogonium is an enlarged amoebocyte, which remains surrounded by one, or more flattened cover cells to form spermatocyte. Cover cells are derived either by divisions of mother cells or from amoebocytes. Spermatogonium undergoes two or three divisions to form spermatocyte which give rise to spermatozoa. A mature sperm consists of a rounded nucleated head and vibratile tail. By lashing movements of tail, it moves through water to reach other sponges.

**Fertilization:** Fertilization is internal. Ova are fertilized *in situ*. Spermatozoa enter other sponges with water current. It first enters a nurse cell or a choanocyte, which lies adjacent to a ripe ovocyte. Spermatozoa then lose its tail and its swollen head becomes surrounded by a capsule. Chanocytes also loses its collar, flagellum and becomes amoeboid. Outer surface of ovocyte at the point of contact invaginates, so that sperm carrying choanocyte is received in conical depressions. Capsule containing sperm head penetrates into ovoclyte. Thus fertilization is completed and zygote is formed.

### *Early embryonic development*

Fertilized egg undergoes maturation and holoblastic cleavage which still *in situ*. First 3 cleavages are vertical forming a pyramidal plate of 8 cells or blastomeres. A horizontal cleavage then divides blastomeres unequally yielding two tiers of 8 cells each. Large blastomeres of long tier in contact with parent choanodermare the macromere. They are destined to form future pinacoderm. Small blastomeres of upper tier are micromeres, which give rise to future choanoderm. At 16-cell stage, embryo resembles a flattened disc and lies just beneath maternal choanocyte layer. Blastomeres become arranged around a central cavity or blastocoel. Small cells multiply rapidly, elongate and each acquires a flagellum on its inner end facing sphere. Large cells remain undivided for sometime, become rounded and granular and develop in the middle and large opening that functions as a mouth to ingest adjacent maternal cells. This stage of embryo is termed stomoblastula.

### *Larval development*

Stomoblastula undergoes inversion during which it turns inside out through its mouth. As a result, flagellar ends of micromeres are brought to outside. Mouth of stomoblastula disappears. After inversion, embryo is known as an *amphiblastula*, which is enclosed in a trophic membrane, or brood pouch formed of maternal cloaneytes, to supply food to the growing embryo. When fully formed, amphiblastula is set free into adjacent radial canal and finally escapes to exterior through osculum with outgoing water current. Animal pole of amphiblastula is kept in front while swimming. After swimming for a few hours, an embolic gastrulaion takes place. Macromere multiply more rapidly than micromeres and flagellated half of larva is invaginated into or overgrown by granular nonflagellated half. It now becomes a typical double-walled gastrula with a blastopore at invaginated side. Outer layer of gastrula is made of non-flagellated granular cells while its inner layer of flagellated cells. Inversion that took place at stomoblastula stage is reversed after gastrulation.

### *Metamorphosis*

Soon after gastrula fixes to some substratum by its blastoporal end and lengthens into a cylinder at free distal end. Blastopore closes by growth of cells towards substratum. Several small perforations inside cylinder are made and nonflagellated granular cells give rise to dermal epithelium or pinacoderm and to scleroblasts and porocytes. Inner flagellated cells develop collars and produce functional choanoderm, archaeocytes and other amoebocytes. Mesenchyme cells are thus derived from both embryonic layers. Young sycon reaches Olynthus stage 4 which closely resembles a simple ascon sponge. Its inner cavity or future sponogocal is lined by flagellated choanocytes. Water enters spongocoel through dermal ostia and goes out through osculum (Figures 4.20–4.25). Adult stage is derived by budding of radial canals first at middle of young stage, which further grows vertically. Flagellated choanocytes shift into radial canals and body wall further increases in thickness by growth of middle layer, with spicules and canal system. Colony develops by further branching and differentiation (Figures 4.17–4.20).

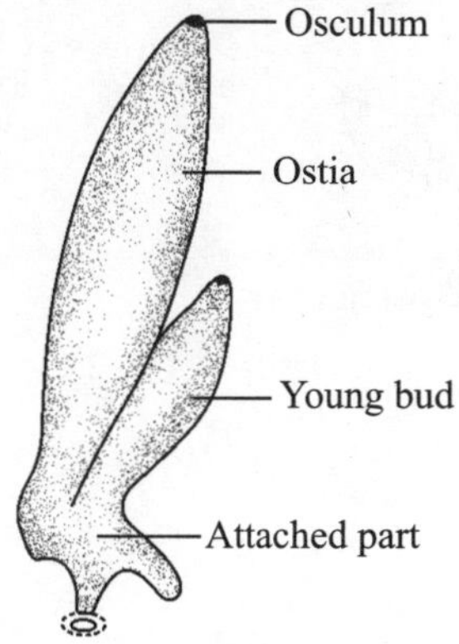


**Figure 4.17** Budding in *Sycon*.

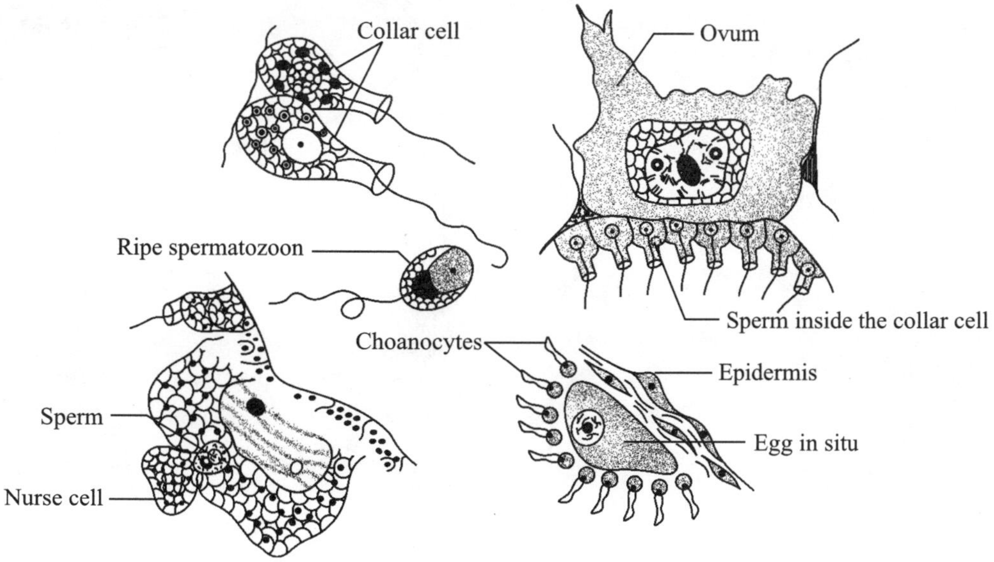


**Figure 4.18** Fertilization in *Grantia*.

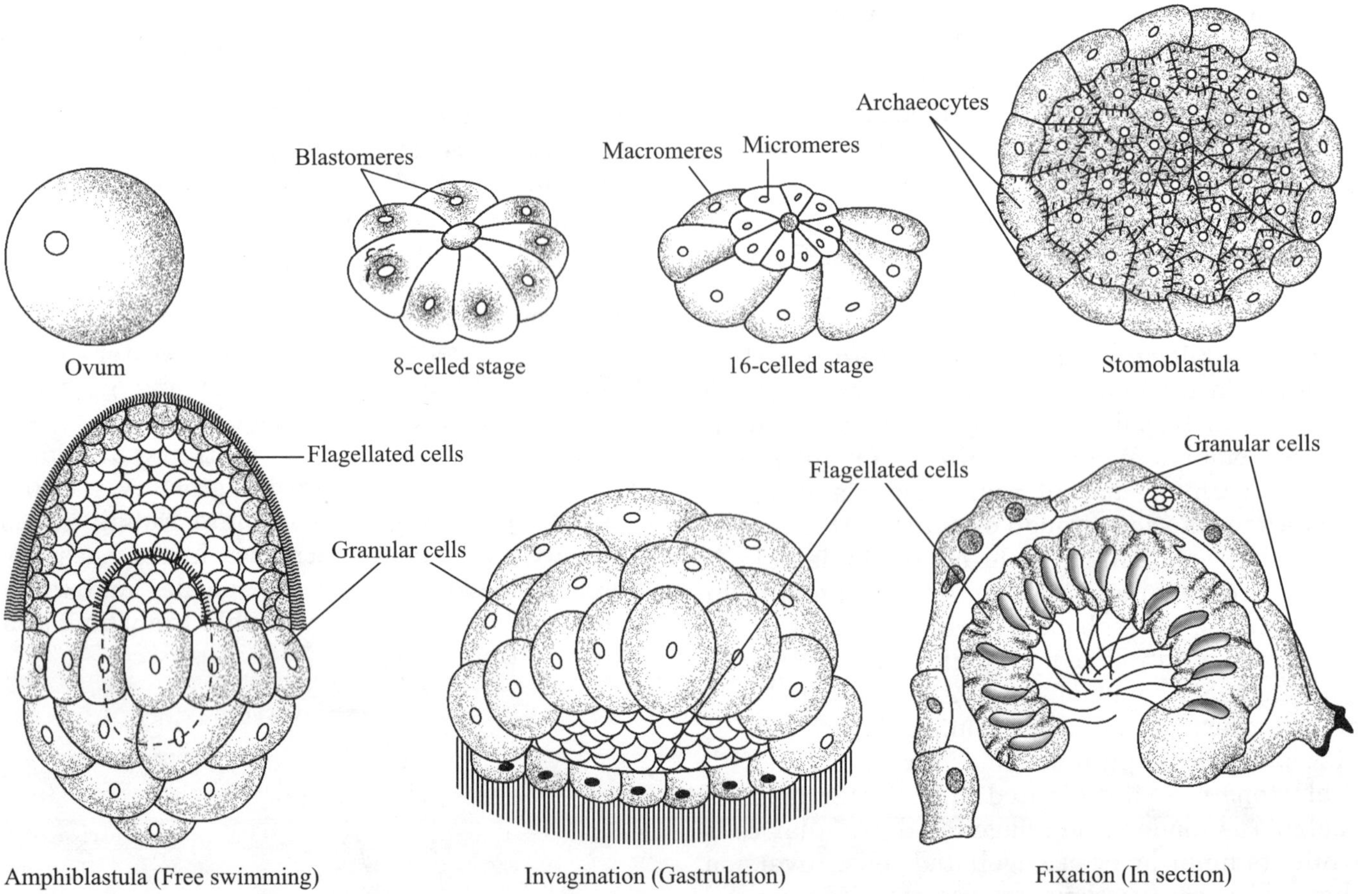


**Figure 4.19** Development of *Scypha* from ovum through subsequent stages.

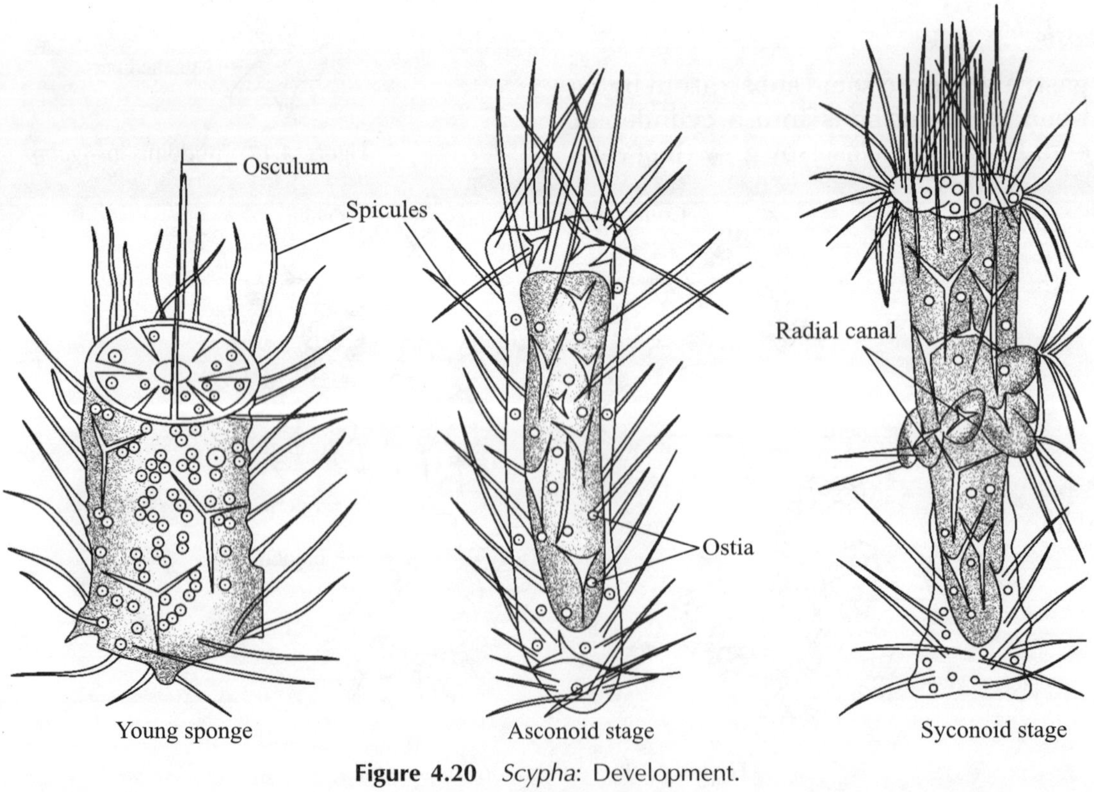


**Figure 4.20** *Scypha*: Development.

## 4.15 ECONOMIC IMPORTANCE

Sponges are used for domestic use and in the arts. They have relatively little direct value. Many kinds of crustaceans, worms, and other animals find shelter within sponge's canals. Some animals are found only on the bodies of other animals like crabs, where they hold a symbiotic relationship. The spicules afford protection to the sponge. Some boring sponges injure the shells of molluscs into which they burrow for protection

## REVIEW QUESTIONS

### Short Answer Questions

1. What is osculum?
2. What the name Porifera signifies?
3. Give the scientific name of bath sponge.
4. Name the middle layer of body wall in sponges.
5. What type of canal system is found in *Leucosolenia?*
6. What is prosopyle?
7. Which cells are found only in sponges?
8. In which animal pinacocytes are present.
9. Which cells form the spicules in sponges?
10. What is paragastric cavity?
11. What cells line the paragastric cavity of sponges?
12. What is the most common composition of skeleton of sponges?
13. Name the larva of sponge.
14. What are the reproductive cells of sponges?
15. Who coined the name Porifera?
16. Name an example of a fresh water sponge.
17. Name the cells present in the mesenchyme.
18. What is the function of spongioblasts?
19. What are archaeocytes?
20. What are myocytes?
21. What are megascleres spicules?
22. What are microscleres spicules?
23. Comment on the desma.
24. Write two characters of spongin.
25. Give two examples of Class Calcarea.
26. Give two examples of Class Hexactinellida.
27. Give two examples of Demospongiae.
28. State two points of difference between Calcarea and Hexactinellida.
29. State two points of difference between Hexactinellida and Demospongiae.
30. Name the larval stage of Porifera.
31. What is apopyle?
32. What is prosopyle?
33. Name three variations of leuconoid type of canal system.
34. What are poriferans?
35. What is the typical shape of poriferans?
36. What is the function of the pores in porifera?
37. Do sponges have nervous, circulatory and excretory systems?
38. What do you understand by the term diploblastic?
39. What is stomoblastula?
40. Give one character of the order Homocoela.

### Long Answer Questions

1. Give the characteristics of Porifera.
2. Give an account of canal system in sponges with suitable diagrams.
3. Describe the structure, function and mode of development of various skeletal elements of sponges.
4. Give an account of reproduction in Porifera with suitable diagrams.
5. Give an account of nutrition, gas exchange in Porifera.
6. Give an account of various types of cells found in sponge with suitable diagrams.
7. Clssify Porifera up to orders with suitable examples.

CHAPTER 5

# Phylum Cnidaria

Cnidarians are diploblastic and radially symmetrical animals. They usually bear tentacles, provided with nettling cells and only one body cavity. The body form is usually referable to one of two types, the polyp form and the medusoid form. The two forms are clearly differentiated one from the other. The term polyp, applied to the hydroid individual, is derived from the Latin name of the cuttlefish (*Polypus*) due to the apparent resemblance between the two animals. In each, the tentacles grasp food and bring it into the mouth. The nettling cells are tubular threads and remain coiled within a small bladder like structure. These threads are discharged from the nematocysts with a force to carry them even through the chitinous body covering of small crustaceans. The Portuguese man-of-war, inflict painful injury even to man.

The body consists of a central blind sac, which is called the *gastrovascular cavity*. The body wall comprises the outer epidermis and the inner gastrodermis. A connective tissue layer, the mesoglea, lies between the two epithelia. The mouth is usually surrounded by tentacles and opens at one end of the coelenteron. It lies at the tip of the manubrium that elevates it above the oral surface. The defining features are possession of stinging cells, which contain the cnida. The cnida on proper stimulation inverts and ejects a slender toxic thread in the direction of prey. Three types of cnidae are nematocysts, spirocysts and ptychocysts. All toxic cnidae are nematocysts. Spirocysts are sticky. The everted tubules of ptychocysts are used in constructing the felt like tubes.

The basic body plan is manifested as medusa and polyp. In some taxa, only one generation is present. In others, both polyps and medusa are found. A life cycle characterized by alternation of sexual swimming medusae with benthic asexual polyps is exhibited in many cnidarians. Cnidarians are ammonotelic. Diffusion through the body and tentacle surface eliminates the ammonia from the body. The gas exchange occurs across the general body surface. The nervous system, a plexus of basiepithelial neurons, serves sensory and motor systems. The life cycle involves a planula larva.

## 5.1 GENERAL CHARACTERISTICS

1. The cnidarians have a relatively simple body construction.
2. Diverse, mostly marine group with above 10,000 living species.
3. Possess a relatively simple, diploblastic body plan.
4. The basic body plan is a sac with a central digestive compartment, the gastrovascular cavity.
5. A single opening to this cavity functions as mouth and anus.
6. Two variations, the sessile polyp and the floating medusa, exist in the life cycle.
7. The cylindrical polyps adhere to the substratum by the aboral end and extend their tentacles.
8. Medusae are flattened and move by drifting passively and by contracting their bell-shaped bodies.
9. The tentacles of a medusa dangle from the oral surface.
10. Some cnidarians exist only as polyp.
11. Others exist only as medusa.

12. Some pass sequentially through both a medusa stage and a polyp stage in their life cycle.
13. Cnidarians are carnivores.
14. The batteries of cnidocytes on the tentacles defend the animal or capture prey.
15. The cnidae called *nematocysts* are stinging capsules.
16. The muscles and nerves exist in their simplest forms.
17. The cells of the epidermis and gastrodermis have bundles of microfilaments which are arranged into contractile fibres.
18. The gastrovascular cavity acts as a hydrostatic skeleton against which the contractile cells can work.
19. The coral animals live as solitary or colonial forms and secrete a hard external skeleton of calcium carbonate.
20. Each polyp generation builds on the skeletal remains of earlier generations to form skeletons, which are commonly called the *coral*.
21. In the tropical seas, coral reefs provide habitat for a great diversity of invertebrates and fishes.
22. The coral reefs in many parts of the world are being destroyed by human activity.
23. Diploblastic metazoan with ectoderm and entoderm separated by mesoglea.
24. Possess primary radial symmetry, often modified as biradial, quadriradial or other form.
25. The primary body axis is oral-aboral.
26. Each cnida resides inside and is produced by a cell, the cnidocyte.
27. Musculature is formed largely of myoepithelial cells derived from ectoderm and entoderm.
28. Exhibit alternation of polypoid and medusoid generations; but there are many variations of this basic pattern.
29. Head and centralized nervous system are absent.
30. Nervous system is a simple nerve net(s), composed of naked and largely nonpolar neurons.
31. Typically have planula larvae.

## 5.2 HABITAT

They are chiefly marine, though **Hydra** and **Protohydra** inhabit fresh water. The anaemones and the colonial hydroids are found at the seashore. The free-floating jellyfish with their graceful pulsations are conspicuous among the free-floating organisms.

## 5.3 TISSUES AND ORGANS

The medusoid represents higher grade of organization. The structure often either lacking or of low organization in the hydroids, but well developed medusae are found in some species. The muscle originates due to partial differentiation of epithelial cells in hydroids of the Hydrozoa. The sheets and bundles of muscle tissue are found in the medusae. The specialized sensory organs are associated with the distinctly centralized nervous system of the medusae. In these, organs of equilibrium occur in extremely diverse developmental stages, representing conditions ranging from the simple modified tentacles to complicated statocysts with the accessory protective vesicles. The reproductive organs exhibit the lowest degree of specialization. Ducts or other accessory sexual organs of any type do not accompany the gonads.

Most species feed on captured microscopic organisms which are killed by the nettling cells. Fishes and crustaceans are the food of some jellyfish. The minute fresh water hydras can feed on newly hatched fish. In its primitive condition, the gastrovascular cavity is a simple bag with a single opening. In the medusae, diverticula from the central chamber of the digestive system provide the greater space for the digestion and serve for the delivery of digested food to the distant body parts. These diverticula may have the form of definite vessels, called the *radial canals*, which may be united at their distal ends by a common vessel called the *circumferential canal*.

In the Anthozoa, a definite infolding of ectoderm projects for some distance from the mouth and opens into the digestive chamber as an oesophagus and the cavity is divided by a series of partitions called the *mesenteries* or *septa*, which provide the additional surface for digestion and food assimilation. Cnidarians are carnivores and feed on live prey which they usually capture using the tentacles, which are armed with cnidocytes. The digestion occurs in the coelenteron, which generally possess ciliated canals for the distribution of the partly digested food.

## 5.4 CLASSIFICATION WITH CHARACTERS

The following characters are mentionable:

1. Aquatic, mostly marine.
2. Soft bodied.
3. Radially symmetrical in adult.
4. Diploblastic.
5. The ectoderm and endoderm are separated by a noncellular mesoglea.
6. Single opening of the coelenteron, the mouth but no anus.
7. Generally carnivorous.
8. The tentacles are found around the mouth.
9. Polypoid and medusoid forms are found.
10. The nematocysts are present.
11. The nervous system, as a net with receptor cells, is present.

Classification followed here is that of Marshall and William (1972).

### *Class Hydrozoa*

1. Two main forms of zooids—polyps and medusa are present.
2. The epidermis consists of epithelio-muscular, interstitial and sensory and gland cells.
3. Stinging capsules are present.
4. The gastrodermis consists of amoeboid gland cells and epithelio-muscular cells.
5. The reproductive products are discharged externally.
6. The horny perisarc is present.

#### *Order Hydroida*

1. The zoophyte stage is fixed.
2. The sense organs are ectodermal.

**Example:** *Hydra, Obelia.*

#### *Order Trachylina*

1. The fixed zoophyte stage is absent.
2. The sense organs are derived from the gastrodermis.

**Example:** *Petasus, Glossocodon.*

#### *Order Hydrocorallina*

1. A massive skeleton is present.

**Example:** *Millepora, Stylaster.*

#### *Order Chondrophora*

1. Represented by a single large polyp.
2. The polyp possesses a chitinous, gas filled, lamellar float.
3. The adult is a free medusa.

**Example:** *Velella, Porpita.*

#### *Order Pteromedusae*

1. The pelagic modified actinulae become sexually mature.
2. No attached stages, or medusae are known.
3. The adult possesses no tentacles.

**Example:** *Tetraplatia.*

#### *Order Siphonophora*

1. Pelagic.
2. The colony exhibits extreme polymorphism of zooids.
3. No free sexual medusae.

**Example:** *Halistemma, Physalia.*

#### *Order Graptolithida*

1. Palaeozoic forms.
2. Found as fossilized perisarc of the branched colonies.

**Example:** *Monographus.*

#### *Order Stromatoporoidea*

1. Fossil forms.
2. They were reef-builders in mid-Palaeozoic times.

### *Class Scyphozoa*

1. Medusoid forms.
2. Presence of the endodermal and gastric tentacles.
3. True velum is absent.
4. The tentaculocysts are present.
5. The gonads discharge their products into the digestive cavity.

#### *Order Stauromedusae*

1. The umbrella is conical or vase shaped.
2. Mostly attached to external objects by an exumbrella peduncle.
3. No tentaculocysts.

**Example:** *Lucernaria.*

#### *Order Coronatae*

1. The umbrella is divided by a horizontal coronary groove.

2. Four to sixteen tentaculocysts are present.

**Example:** *Periphylla, Nausithoe.*

***Order Cubomedusae***

1. The cup-shaped umbrella with four sides is present.
2. Four perradial tentaculocysts are present.

**Example:** *Charybdaea.*

***Order Semaeostomeae***

1. Flattened, saucer-shaped, or disc-shaped umbrella is present.
2. The tentaculocysts are usually 8, 4 per radials and 4 interradials.
3. The square mouth is guarded by the four long, oral arms.

**Example:** *Aurelia.*

***Order Rhizostomeae***

1. The mouth is obliterated by the oral arms.
2. The stomach is continued into the canals.

**Example:** *Pilema.*

***Class Anthozoa***

1. Polypoid forms.
2. The stomodaeum is present.
3. The mesenteries extend inwards.
4. The coiled mesenteric filaments are present.
5. The nematocysts are complex.
6. The sexual products are discharged into the coelenteron.
7. The alteration of generation is absent except in one instance.

***Subclass Hexacorallia***

1. The tentacles and mesenteries are arranged in multiple of six.
2. The tentacles are simple and unbranched.
3. The siphonoglyphs are usually two in number.

***Order Actiniaria***

1. Simple.
2. The tentacles and mesenteries are numerous.
3. The skeleton is absent.

**Example:** *Teglia, Edwardsia, Adamsia.*

***Order Madreporaria***

1. Colonial forms.
2. Presence of an ectodermal skeleton.

**Example:** *Madrepora, Astrea.*

***Order Zoanthidea***

1. Solitary or colonial.
2. Presence of one ventral siphonoglyph.
3. The tentacles are unbranched.

**Example:** *Zoanthus.*

***Order Antipatharia***

1. Colonial.
2. The tentacles and mesenteries are few.
3. The skeleton is present in the form of branched chitinoid axis.

**Example:** *Antipathes.*

***Order Ceriantharia***

1. Solitary form.
2. Single, dorsally placed siphonoglyph.
3. All the mesenteries are complete.
4. The marginal- and oral- tentacles are present.

**Example:** *Pachycerianthus.*

***Order Corallimorpharia***

1. Solitary or aggregated anaemon-like polyps.
2. The skeletal structures are absent.

**Example:** *Corynactis.*

***Order Ptychodactiaria***

1. Anaemone-like polyps.
2. Ciliated tract on mesenterial filaments is absent.

**Example:** *Ptychodactis.*

***Subclass Octocorallia***

1. The tentacles and mesenteries are 8 in number.
2. The tentacles are pinnate.
3. One ventral siphonoglyph is present.

***Order Stolonifera***

1. The polyps are connected by the creeping stolons.
2. The stolons are embedded in a mat of tissue.
3. The skeleton is made up of separate calcareous spicules.

**Example:** *Tubipora.*

***Order Telestacea***

1. The colony consists of elongated polyps.
2. The colony arises from a creeping base.

**Example:** *Telesto.*

### *Order Alcyonacea*

1. The polyps are fused in their proximal parts.
2. The skeleton has calcareous spicules.

**Example:** *Alcyonium.*

### *Order Coenothecalia*

1. The skeleton has crystalline fibres of aragonite.
2. The skeleton is perforated by the cylindrical cavities.

**Example:** *Heliopora.*

### *Order Gorgonacea*

1. Tree-like compound appear.
2. The skeleton is ectodermal.
3. The spicules are present in mesoglea.
4. The siphonoglyph is absent.

**Example:** *Gorgonia.*

### *Order Pennatulacea*

1. The colony is elongated.
2. The polyps are dimorphic.
3. A horny skeleton supports the stem.

**Example:** *Pennatula* (Figures 5.1–5.5).

## 5.5 THE BODY WALL

Cnidarian epithelia—the outer epidermis and inner gastrodermis—include myoepithelial cells and are viewed by many workers as the most primitive muscle cells in the Metazoa. These columnar cells bear flattened, contractile, basal extensions called *myonemes*. In the epidermis, these cells are referred to as epithelio-muscular cells and in the gastrodermis are called *nutritive-muscular* cells. The myonemes rest against the middle mesoglea or mesenchyme and the opposite ends of the cells form the outer body and gut surfaces. The myonemes run parallel to free surfaces and contain contractile myofibrils. Myonemes of neighbouring cells interconnect, often forming longitudinal and circular sheets capable of contracting like true muscle layers. Other sorts of myoepithelial cells occur in some other animals. Similar cells are known even among mammals, where they are found in association with certain secretory tissues.

Some cnidarians also possess sub epidermal mesenchymal muscles, apparently derived from the contractile elements of the myoepithelial cells. In anaemones, for example, cord-like sphincters are sunk below the epithelium and reside as distinct

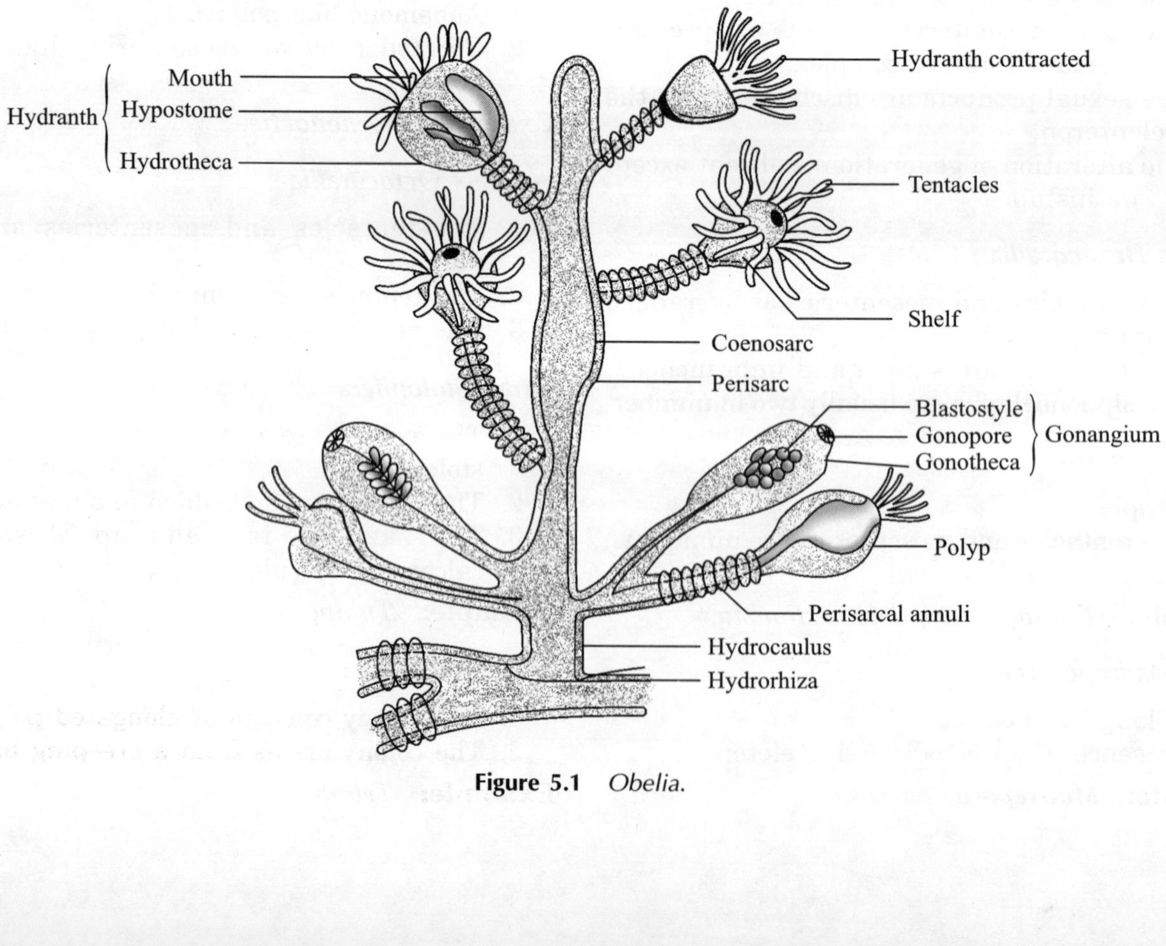


**Figure 5.1** *Obelia.*

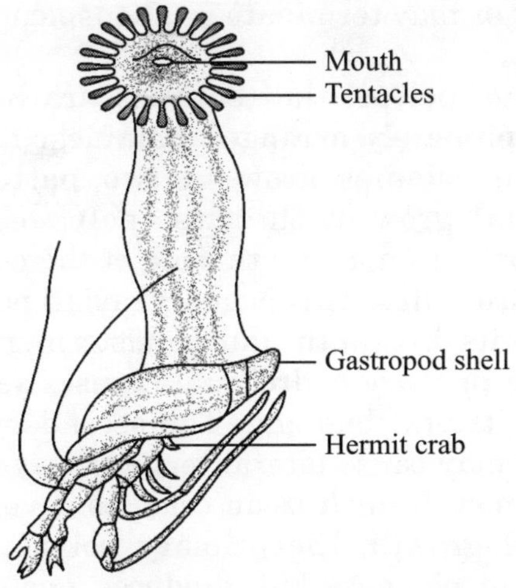


**Figure 5.2** *Adamsia.*

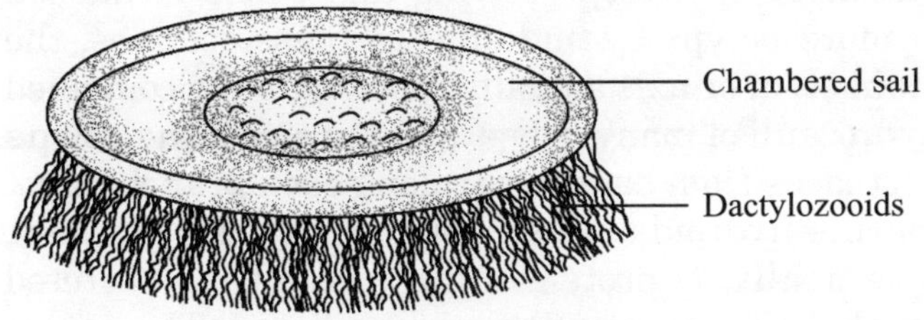


**Figure 5.3** *Porpita.*

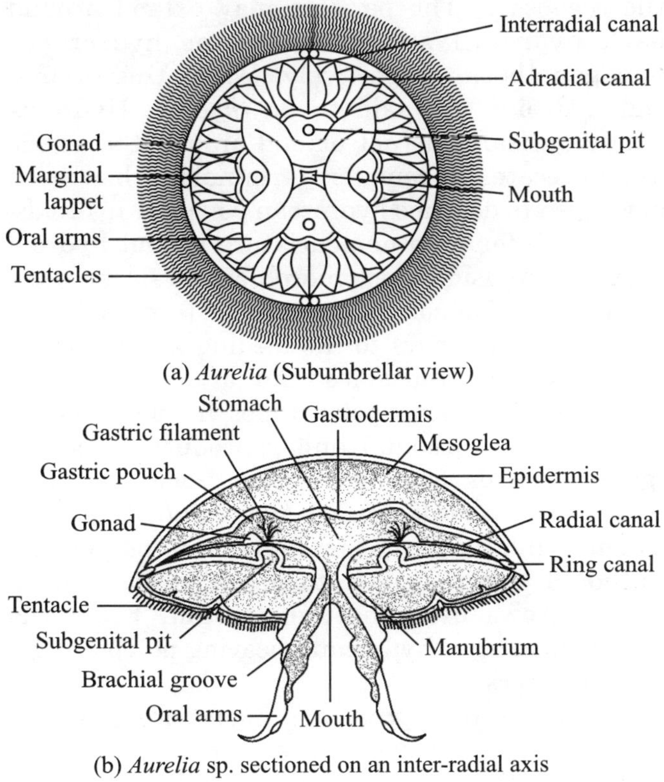


(a) *Aurelia* (Subumbrellar view)

(b) *Aurelia* sp. sectioned on an inter-radial axis

**Figure 5.4** *Aurelia.*

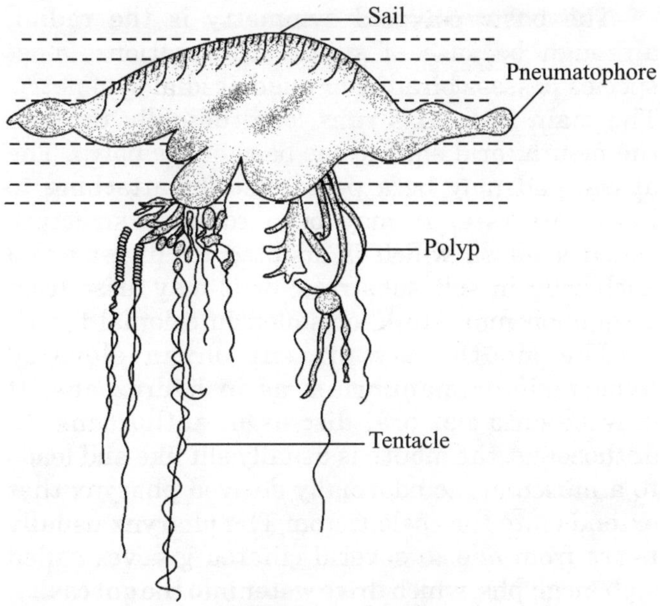


**Figure 5.5** *Physalia.*

muscles wholly within the mesenchyme. In addition to epithelio-muscular cells, the epidermis contains sensory cells, cnida-bearing cells called *cnidocytes*, *gland cells* and *interstitial cells*. The last are undifferentiated and capable of developing into other types of cells. The gastrodermis is histologically somewhat similar to the epidermis. Along with the nutritive muscular cells, it also contains cnidocytes (except in the Hydrozoa) and gland cells. In hydrozoans, the middle layer is a simple, gel-like, largely acellular mesoglea. Scyphomedusae and cubomedusae have very thick mesogleal layers with scattered cells. In anthozoans, the middle layer is often thick and richly cellular mesenchyme.

## 5.6 THE POLYPOID FORM

Polyps are much more diverse than medusae, largely as a result of their capacities for asexual reproduction and colony formation. These stages occur in four classes of cnidarians, although it is greatly reduced in the Scyphozoa and Cubozoa. Polyps are tubular structures with an outer epidermis, an inner gut sac (coelenteron) lined with gastrodermis and a layer of jelly-like mesoglea or mesenchyme in between. Most polyps are small, but some species of sea anaemones have large polyp; the largest is the tropical Indo-Pacific *Stichodactyla mertensii*, which can exceed a meter in diameter and the northeast Pacific *Metridium giganteum*, which can extend its column to a meter in height.

The basic polypoid symmetry is the radial, although because of subtle modifications, most species possess biradial or quadriradial symmetry. The main body axis runs longitudinally through the mouth (oral end) to the base of the polyp. The aboral end may be a pedal disc for attaching to hard substrata. It may be a rounded structure called a physa which is adapted for digging and anchoring in soft substrata; or it may arise from a common mat, stalk, or stolon in colonial forms.

The mouth may present on an elevated hypostome or manubrium as in hydrozoans. It may be on a flat oral disc as in anthozoans. In anthozoans, the mouth is usually slit like and leads to a muscular, ectodermally derived pharynx that extends into the coelenteron. The pharynx usually bears from one to several ciliated grooves called siphonoglyphs, which drive water into the gut cavity. In part, the presence of siphonoglyphs gives these polyps secondary biradial symmetry. The side of an anthozoan polyp that bears a single siphonglyph is called the *sulcal side* and the opposite side is called *asulcal side*. The coelenteron or gastrovascular cavity serves for circulation as well as digestion and distribution of food.

In hydrozoan polyps, the coelenteron is a single, uncompartmentalized tube. In scyphozoan polyps (scyphistomae), it is partially subdivided by four longitudinal, ridgelike mesenteries. In anthozoan polyps, it is extensively compartmentalized by mesenteries. Anthozoan mesenteries are projections filled with mesenchyme. They extend from the inner body wall toward the pharynx, some or all of them fuse with it as complete mesenteries. Those that do not connect to the pharynx are called incomplete mesenteries. In anthozoan polyps, the free inner edge of each mesentery below the pharynx has a thickened, cord-like margin armed with cnidae, cilia and gland cells and is called the *mesenterial filament*.

In some sea anaemones, these filaments give rise to the long threads, called acontia that hang free in the gastrovascular cavity. They function in defense and feeding. In most colonial anthozoans, the cellular mesenchyme unites individual zooids. In some, such as the soft corals, gastrovascular cavities are connected to one another by canals called *solenia*. The tentacles that surround the mouth contain hollow extensions of the coelenteron in anthozoans, whereas they house a solid core of packed gastro dermal cells in most hydrozoans. Tentacles may taper to a point (filiform tentacles) or may terminate in a conspicuous knob of cnidae.

In some polyps, the tentacles are branched, often as pinnately arranged pinnules. Branched hydrozoan colonies grow in two patterns. In monopodial growth, the first polyp elongates continuously from a growth zone at the distal end of the hydrocaulus. This primary (axial) polyp may even lose its hydranth and persists merely as a stalk. The primary hydrocaulus causes secondary polyps by lateral budding. The secondary polyps grow and may cause lateral tertiary polyps in the same fashion. In hydrozoan colonies developed by sympodial growth, the primary polyp does not continue to elongate but produces one or more lateral polyps by budding and then stops growing. The new polyps extend the colony upward to some distance, then stop growing and cause formation of more polyps by budding. In these colonies, the main stem or axis actually represents the combined hydrocauli of many polyps and the age of the polyp's decreases from base to tip along each branch. Most marine hydroids are surrounded, at least in part, by a nonliving protein-chitin exoskeleton secreted by the epidermis called the *perisarc*. This outer covering is absent in freshwater hydroids.

The living tissue inside the perisarc is termed the *coenosarc*. The perisarc may extend around each hydranth and gonozooid as hydrotheca and gonotheca, respectively. When this occurs, the hydroids are said to be *thecate*. Hydroids whose perisarcs do not extend around the zooids are athecate. A complex terminology has been developed to describe hydrozoan polyps or hydroids. One cause for such nomenclature is that hydroid colonies are usually polymorphic, containing more than one kind of polyp, or zooid. The term *hydranth* or *gastrozooid* refers to the feeding zooids, which typically bear tentacles and a mouth. Other commonly occurring polyp types include defense polyps (dactylozooids) and reproductive polyps (gonozooids or gonangia). Each zooid typically arises from a stalk, called *hydrocaulus*. In most colonial hydrozoans, the individual polyps are anchored in a root like stolon called *hydrorhiza*, which grows over the substratum. From the hydrorhiza, arise hydrocauli bearing polyps singly or in clusters.

Gastrozooids capture and ingest prey, provide energy and nutrients to the rest of the colony, including all the nonfeeding polyps. Dactylozooids occur in a variety of sizes and shapes and are heavily armed with cnidae. Often several dactylozooids surround each gastrozooid and serve for both defenses and food capturing. Gonozooids which produce medusa buds are called *gonophores* that are either released or retained on the colony. Whether released as free medusae or retained as gonophores, they produce gametes for the sexual phase of the hydrozoan life cycle. The living tissue (coenosarc) of the gonozooid is called the blastostyle, the gonophores arise from this tissue. When a gonotheca surrounds the blastostyle, the zooid is called a *gonangium*. The most dramatic examples of polymorphism among polyps are seen in the hydrozoan order Siphonophora and the anthozoan order Pennatulacea. Siphonophorans are hydrozoan colonies composed of both polypoid and medusoid individuals with up to thousand zooids in a single colony. This large order includes a variety of unusual and poorly understood forms, including the man-of-war, *Physalia*. The gastrozooids of siphonophorans are highly modified polyps with a large mouth and one long, hollow tentacle that bear many cnidae. This long feeding tentacle reaches lengths up to 13 metres in the Atlantic species *Physalia physalis*. The nonfeeding dactylozooids also bear one long (unbranched) tentacle. The gonozooids are usually branched. They produce sessile gonophores that are never released as free medusae.

Siphonophorans use a swimming bell (nectophore) or a gas-filled float (pneumatophore), or both, to help maintain their position in the water. The nectophore is a true medusoid individual with many of the structures common to free-swimming medusae. It was once thought to be a modified medusa, presently known to be derived directly from the larval stage and probably represents a highly modified polyp. Pneumatophores are double-walled chambers lined with chitin. Each float houses a gas gland, which consists of a mitochondrion-laden glandular epithelium-lining pit or chamber. The gland secretes a gas usually similar to air in composition, although in *Physalia*, it apparently includes a high proportion of carbon monoxide. Many siphonophorans have mechanisms by which they regulate gas in their floats.

Siphonophorans are grouped into three suborders on colony structure—the suborder Calycophorae includes colonies with swimming bells but no float, members of the suborder Physonectae have a small float and a long train of swimming bells; and those of the suborder Cystonectae have a large float and no bell. Calycophorans have a long tubular stem extending from the swimming bell, from which various types of zooids bud in groups called *cormidia*. Each cormidium acts as a colony within a colony and is usually composed of a shield like bract, a gastrozooid and one or more gonophores that may function as swimming bells. The cormidia commonly break loose from the parent colony to live independently, when they are termed *eudoxids*.

The physonectans have an apical float with a long stem bearing a series of nectophores followed by a long cormidia. The cystonectans including *Physalia* usually have a large pneumatophore with a prominent budding zone at its base, which produces the various polyps and medusoids. The hydrozoan order Chondrophora is composed of colourful oceanic organisms that drift about on the sea surface, occasionally washing ashore to coat the beach with their bluish-purple bodies. These animals are known as large, solitary, athecate hydranth polyps floating upside down at the surface instead of sitting on stalk attached to the bottom. The aberrant medusae of chondrophorans are short-lived and do not possess functional mouth or gut, probably relying instead on their symbiotic zooxanthellae for nutrition.

The aboral sail in *Velella* (the "by-the-wind-sailor") has no counterpart in sessile hydroids. In its ability to sail at an angle to the wind, *Velella* resembles the *Physalia*, a similarity attributed to convergent evolution. The pennatulaceans are the most complex and polymorphic members of the class Anthozoa. The colony is built around a main supportive stem, which are actually the primary polyp and buds lateral polyps in a regular fashion. The base of the primary polyp is anchored in sediment, but the upper, exposed portion (the rachis) produces polyps in whorls or rows, or sometimes united in crescent-shaped leaves.

Often these polyps are of two distinct types. Autozooids bear tentacles and function in feeding. Siphonozooids are small with reduced tentacles and serve to create water currents through the colony. In sea pens, the rachis is elongated and cylindrical. In sea pansies, it is flattened and shaped like a large leaf. In the deep-sea genus *Umbellula*, the secondary polyps radiate outward to give the colony the appearance of a pinwheel. The first deep benthic photos of *Umbellula* arose question about its systematic position. Gorgonians are also colonial anthozoans. Some grow in bushy shapes, others are planar and size and shape of the colony are often mediated by the hydrodynamics of the local currents. Where prevailing currents are about in one plane, the branches of the colony tend to grow largely in one plane perpendicular to the flow. In regions of mixed currents, the same species tends to grow in two planes.

## 5.7 THE MEDUSOID FORM

Free medusae occur in all cnidarian classes except the Anthozoa. Although variation in form exists, medusae are far less diverse than polyps and it is much easier to generalize about their anatomy. The relative uniformity of medusae is a result in part of their similar lifestyles in open water and their inability to form colonies by asexual reproduction. They are participants in colonial life. Some remain attached to hydrozoa colonies as sessile gonophores. Sessile benthic forms are rare. Even though the body walls of medusae and polyps are similar and both adhere to the general cnidarian bauplan, their gross morphologies are adapted to their different lifestyles.

Medusae are bell-, dish- or umbrella-shaped and usually provided with a thick, jelly-like mesogleal layer (hence the name "jellyfish"). The convex upper (aboral) surface is called the *exumbrella*. The concave lower (oral) surface is the *subumbrella*. The mouth is located in the centre of the subumbrella, often suspended on a pendant, tubular extension called the *manubrium*, which is almost always present on hydromedusae, but usually reduced or absent in scyphomedusae.

The coelenteron or gastrovascular cavity occupies the central region of the umbrella and extends radially via radial canals. In most hydromedusae, a marginal ring canal within the rim of the bell connects the ends of the radial canals. The presence of four radial canals and tentacles or in multiples of four (in hydromedusae) and the division of the stomach by mesenteries into four gastric pouches (in scyphomedusae) give most jellyfish a quadriradial symmetry. Most hydromedusae have a thin circular flap of tissue, the velum within the margin of the bell. Such medusae are termed *craspedote*. Those lacking a velum such as scyphomedusae are said to be acraspedote. Like that of polyps, the external surfaces of medusae are covered with epidermis and the internal surfaces (coelenteron and canals) are lined with gastrodermis. The bulky, gelatinous middle layer is either a largely cellular mesoglea or a partly cellular mesenchyme.

## 5.8 SUPPORT

Cnidarians employ a wide range of mechanisms for support. Polypoid forms rely substantially on the hydrostatic qualities of the water-filled coelenteron constrained by circular and longitudinal muscles of the body wall. The mesenchyme may be stiffened with fibres, particularly in the anthozoans. Colonial anthozoans may incorporate bits of sediment and shell fragments onto the column wall for further support. Many colonial hydrozoans produce a flexible, horny perisarc, composed largely of chitin secreted by the epidermis. In medusae, the principal support mechanisms, the middle layer, which ranges from a fairly thin and flexible mesoglea to an extremely thick and stiffened fibrous mesenchyme, which may be almost cartilaginous in consistency.

In addition to soft or flexible support structures, there are hard skeletal structures of three fundamental types—horny or wood-like axial skeletal structures, calcareous sclerites and massive calcareous frameworks. Horny axial skeletons occur in some groups of colonial anthozoans such as gorgonians, sea pens and antipatharian corals. Amebocytes in the coenenchyme secrete a flexible or stiff internal axial rod as a supportive base embedded in the coenenchymal mass. Axial rods are protein-mucopolysaccharide complexes (called *gorgonin* in Gorgonacea), but little is known of their chemistry. In the antipatharians (black coral), the axial skeleton is so hard and dense that it is ground and polished to make jewellery. In most octocorals, mesenchymal cells called *scleroblasts* secrete calcareous sclerites of various shapes and colours. Usually these sclerites give gorgonians their characteristic colour and texture.

In many species, the sclerites become quite dense and may even fuse to form a more or less solid calcareous framework. The precious red coral. *Corallium* is a gorgonian with fused red coenenchymal sclerites. In the stoloniferan organ-pipe corals (*Tubipora),* the sclerites of the body walls of the individual polyps are fused into rigid tubes. Invertebrate calcium carbonate skeletons do not usually have collagen in their framework, as occurs in vertebrates. However, in some gorgonians (e.g. *Leptogorgia*) the calcareous spicules include collagen component. Massive calcareous skeletons are found in certain groups of Anthozoa and Hydrozoa. The best known are the true or stony, anthozoan corals (order Scleractinia), in which epidermal cells on the lower half of the column secrete a calcium carbonate skeleton.

The thin layer of living epidermis that secretes it covers the skeleton and thus it might technically be considered an internal skeleton. Because the coral colony generally sits atop large nonliving calcareous framework, most authors consider the skeleton as external. The entire skeleton of a scleractinian coral is termed the *corallum*, regardless of whether the animal is solitary or colonial. The skeleton of a single polyp is called a *corallite*. The outer wall of the corallite is the theca; the floor is called the *basal plate*. Rising from the centre of the basal plate, there is often a supportive skeletal process called the *columella*. The basal plate and inner thecal walls give rise to numerous radially arranged calcareous partitions, the septa, which project inward and support the mesenteries of the polyp. Polyps occupy only the uppermost surface of the corallum. The skeletal thickness increases as polyps grow and the bottoms of the corallites are sealed off by transverse calcareous partitions called *tabulae*, each of which becomes the new basal support of a polyp.

The corallum can assume a great variety of shapes and sizes, from simple cup-shaped structures in solitary corals to large branching forms in colonial species. The corallum of colonial forms may be upright, low and massive, or even encrust on the other hard substrata. Members of the hydrozoan families Milleporidae and Stylasteridae also produce calcareous exoskeletons and they are often referred to as the *hydrocorals*. Like true corals, milleporid colonies may assume a variety of shapes, from erect branching forms to encrustations. Pores of two sizes that accommodate two kinds of polyps perforate the milleporid exoskeleton, termed a *coenosteum*. The gastrozooids live in large holes or gastropores and are surrounded by a circle of smaller dactylopores, which house the dactylozooids. Canals, lead downward from the pores into the coenosteum and are closed off below by transverse calcareous tabulae.

As growth proceeds and the colony thickens, new tabulae are formed, keeping the polyp pores at a more or less fixed depth. Hydrocoral colonies thus differ from scleractinian colonies in having the skeleton penetrated by living tissue. The stylasterine skeleton is similar to the milleporid skeleton, but the margins of the gastropore soften, bear notches that serve as dactylopores and the gastrozooids and dactylozooids are supported by calcareous, spine-like gastrostyles and low ridges called *dactylostyles*, respectively. Stylasterine gonophores arise in chambers called *ampullae*, which connect to the feeding zooids through the coenosteum.

## 5.9 MOVEMENT

The contractile elements of cnidarians are derived from their myoepithelial cells. In spite of the epithelial origin of these elements, the terms *muscles* and *musculature* for the sets of longitudinal and circular fibrils are used for convenience. In polyps, these two muscle systems work with the gastrovascular cavity as an efficient hydrostatic skeleton as well as a means of movement. However, unlike the fixed-volume hydrostatic skeletons of many animals (many worms), water can enter and leave the coelenteron of cnidarians, adding to its versatility as a support device. Polyp body musculature is most highly specialized and well developed in the anthozoans, particularly in the sea anaemones and many muscles lie in the mesenchyme. In anaemones, the muscles of the column wall are largely gastrodermal, although epitheliomuscular cells occur in the tentacles and oral disc.

Bundles of longitudinal fibres lie along the sides of the mesenteries and act as retractor muscles for shortening the column. The circular muscles derived from the gastrodermis of the column wall are also well developed. In most anaemones, the circular muscles form a distinct sphincter at the junction of the column and the oral disc. The circular fibres also occur in the tentacles and the oral disc and circular muscles surrounding the mouth can close it completely. When an anaemone contract, the upper rim of the column is pulled over, to cover the oral disc. In many sea anaemones, a circular fold—the collar or parapet—occurs near the sphincter to further cover and protect the delicate oral surface upon contraction. Most polyps are sedentary or sessile. Their movements consist mainly of food-capturing actions and the withdrawal of the upper portion of the polyp during body contractions. These activities are accomplished primarily by the epidermal muscles of the tentacles and oral disc and by the strong gastrodermal muscles of the column. Circular muscles work with the hydrostatic skeleton to distend the tentacles and body.

The variety of locomotor methods has evolved among polyps. Most can creep about slowly by using their pedal disc musculature. In some solitary hydrozoa polyps (*Hydra*), the column can bend far enough to allow the tentacles to contact and temporarily adhere to the substratum, where upon the pedal disc releases its hold and the animal somersaults or moves like an inchworm. Few sea anaemones can detach from the substratum and actually swim away by rapid flexing or bending of the column (*Actinostola, Stomphia*). Others swim by thrashing the tentacles (Boloceroides). These swimming activities are temporary behaviours, generally elicited by contact of a predator. In a few species of sea anaemones, the basal disc may detach and secrete a gas bubble, permitting the polyp to float away to a new location. Many small anthozoan species float hanging upside-down on the sea surface by using water surface tension forces (*Epiactis, Diadumene*). Sea anaemones of one family (Minyadidae) are pelagic and float upside- down in the sea by a gas bubble enclosed within the folded pedal disc. *Hydra* float upside-down by a mucus-coated gas bubble on the bottom of its pedal disc. One of the oldest forms of polyp locomotion is that of the sea anaemone *Liponema brevicornis* inhabiting the Bering Sea, which is capable of drawing itself into a tight cylindrical form that can be rolled around the sea floor by the bottom currents. Even colonial sea pansies (Pennatulacea) are motile and use their muscular peduncle to move to different depths.

Ceriantharians are burrowing, tube-building organisms. They differ with the true sea anaemones (*Actiniaria*) in several important ways. They have no sphincter muscle and their weak longitudinal gastrodermal muscles do not form distinct retractors in the mesenteries. As a result, cerianthids cannot retract the oral disc and tentacles as they withdraw into their tubes. In contrast to other anaemones, however, they possess complete layer of longitudinal epidermal muscles in the column, which allows a very rapid withdrawal response. The mere shadows of a passing hand will cause a cerianthid to pull rapidly itself deep into its long, buried tube.

In medusae, epidermal and subepidermal musculatures predominate and the gastrodermal muscles that are so important in polyps are reduced or lacking. The epidermal musculature is best developed around the bell margin and over the subumbrellar surface. Here the muscle fibres usually form circular sheets called *coronal muscles* that are partly embedded in the mesenchyme or mesoglea. Contractions of the coronal muscles produce rhythmic pulsations of the bell, driving water out from beneath the subumbrella and moving the animal by jet propulsion. The stiffened cellular collenchyme of Scyphomedusae and Cubomedusae includes elastic fibres that provide the antagonistic force to restore the bell shape between contractions.

Many medusae also possess radial muscles that aid in opening the bell between pulses. In craspedote forms, the velum reduces the size of the subumbrellar aperture, thus increasing the force of the water jet. The velarium of the fast-swimming Cubomedusae has the same effect and the evolutionary forces that produced these two convergent features were probably similar. Most medusae spend their time swimming upward in the water column, then sinking slowly down to capture prey by chance encounter, thereafter to pulsate upward once again. Some medusae change direction as they swim and many are strongly attracted to light (especially, those harbouring symbiotic zooxanthellae).

At least some medusae house their zooxanthellae in small pockets that expand during the day, exposing the algae to light. The pockets contract at night. Medusae can be abundant in certain localities. Some, such as the *Aurelia*, are known to aggregate at temperature or salinity discontinuity layers in the sea, where they feed on small zooplanktons, which also concentrate at these boundaries. A few unusual groups of medusae are benthic. Some hydromedusae (*Eleutheria, Gonionemus*) crawl about on algae or sea grasses by adhesive discs on their tentacles. Members of the scyphozoan order Stauromedusae (*Haliclystis*) develop directly from the polypoid stage and affix to algae and other substrata by an aboral adhesive disc.

## 5.10 CNIDAE

Before considering feeding and other aspects of cnidarian biology, it is necessary to consider the structure and function of cnidae. Cnidae, often referred to collectively as nematocysts in older works, are unique to cnidarians. They have a variety of functions, including prey capture, defense, locomotion and attachment. They are produced inside cells called *cnidoblasts*, which develop from interstitial cells in the epidermis and in many groups in the gastrodermis. Once the cnida is fully formed, the cell is called *cnidocyte*. During formation of a cnida, the cnidoblast produces a large internal vacuole in which a complex but poorly understood intracellular reorganization takes place. Cnidae may be complex secretory products of the Golgi apparatus of the cnidoblast. There is also some evidence that cnidae might have originated symbiogenetically from some ancient protist(s) and cnida like structures have been reported from diverse groups as dinoflagellates, sporozoans and microsporans.

Cnidae are the most complex intracellular structures. When fully formed, they are cigar- or flask-shaped capsules, 5–100 μm or more long, with thin walls composed of a collagen-like protein. One end of the capsule is turned inward as a long, hollow, coiled, eversible tubule. The outer capsule wall consists of globular proteins of unknown function. The inner wall is composed of bundles of collagen-like fibrils having a spacing of 50–100 μm, with cross-striations every 32 μm (in the nematocysts of *Hydra*). The distinct pattern of mini-collagen fibres provides the tensile strength necessary to withstand the high pressure in the capsule. The entire structure is anchored to adjacent epithelial cells (supporting cells) or to the underlying mesenchyme.

When sufficiently stimulated, the tube everts from the cell. In members of the classes Hydrozoa, Scyphozoa and perhaps Cubozoa, the capsule is covered by a hinged lid or operculum, which is thrown open when the cnida discharges. In members of these three classes, each cnida bears a long cilium-like bristle called a *cnidocil*, a mechanoreceptor that elicits discharge when stimulated. The cnidocil responds to specific water-borne vibration frequencies. Chemoreceptor on the adjacent supporting cells may actually tune the cnidocil to the proper reception frequency for available prey. Anthozoan cnidae lack a cnidocil and have a tripartite apical flap instead of an operculum. Cnidocytes are most abundant in the epidermis of the oral region and the tentacles, where they often occur in clusters of wart like structures called *nematocyst batteries*.

About 30 kinds of cnidae are known which can be assigned to three basic types. True nematocysts have double-walled capsules containing a toxic mixture of phenols and proteins. The tubule of most types is armed with spines or barbs that aid in penetration and anchorage in the victim's flesh. The toxin is injected into the victim through a terminal pore in the thread or is carried into the wound on the tubule surface. Spirocysts have single-walled capsules containing mucoprotein or glycoprotein. Their adhesive tubules wrap around and stick to the victim rather than penetrating it. The capsule tubules of spirocysts never have an apical; spirocysts occur only in the zoantharian Anthozoa. The third kind of cnidae, the ptychocyst, differs morphologically and functionally from both nematocysts and spirocysts. The capsule tubule of a ptychocyst lacks spines and an apical pore and is strictly adhesive in nature. In addition, the tubule is folded into pleats rather than coiled within the capsule. Ptychocysts occur only in the ceriantharians and function in forming the unique tube in which these animals reside.

Cnidae have usually been viewed as independent effectors and indeed, they often discharge on direct stimulation. Evidence suggests that the animal have at least some control of the action of its cnidae. For example, starved anaemones seem to have a lower firing threshold than satiated animals. It has also been demonstrated that stimulating discharge of cnidae in one area of the body results in discharge in stimuli, initially perceived by the cnidocil or a similar structure, cause most cnidae to fire. Cnidarians are known to discharge their cnidae in the presence of various sugars, low-molecular-weight amino compounds and occasionally glutathione—the latter being a chemical that is liberated when animals are injured or tissue breaks down. Reduced glutathione also causes feeding tentacles and gastrozooids to become active, writhe about and prepare for feeding. The ejection of the tubule from a cnida is called *exocytosis* and an individual cnida can be fired only once. The following three hypotheses have been proposed to explain the mechanism of firing:

- The discharge is the result of increased hydrostatic pressure caused by a rapid influx of water (the osmotic hypothesis).
- Intrinsic tension forces generated during cnidogenesis are released at discharge (the tension hypothesis).
- Contractile units enveloping the cnida cause the discharge by squeezing the capsule (the contractile hypothesis).

## 5.11 TYPE STUDY

### *Obelia* sp.

The *Obelia* is a sedentary, marine, colonial form, commonly known as *sea-fur*. This is cosmopolitan in distribution, which forms a whitish or high brown plant like fur in sea. They remain attached on the surface of seaweeds, molluscan shells and rocks in shallow water. Individual colony is called the *zooid*. The *obelia* is a trimorphic colony having three kinds of zooids, viz. polyp, blastostyles and medusae.

### *Polyp*

The polyp is constituted of two portions. The horizontal portion is called the *hydrorhiza*. The vertical portion is called the *hydrocaulus*. The hydrorhiza and hydrocaulus are hollow tubes called the *coenosarc*, which are covered externally by the perisarc. The coenosarc is made up of ectoderm and endoderm. Between these two layers, the noncellular mesoglea exists. The coenosarc contains a tubular cavity, coelenterons that is continuous and is filled with a fluid. A kind of nematocyst called the *basitrichous isorhizas* is present.

The free distal end of polyp produces a conical elevation, hypostome or manubrium, which is surrounded by a circle of about 24 tentacles. The tentacles are longer, tapering and filiform. The apex of hypostome bears a terminal aperture, called the *mouth*. Below the hypostome is the stomach. The polyp is nutritive zooid, carnivorous and feeds upon the aquatic crustaceans, nematodes and other worms. The tentacles help in catching the prey to mouth. The digestive juice is secreted in gland cells of gastric juices. The digestion is extra cellular as well as intracellular. The hydrorhiza gives mechanical encourage to the whole colony. The hydrocaulus arises from the hydrorhiza having short alternate branches bearing zooid. Two types of zooids are observed. The gastrozooid has the nutritive function and the blastostyles have the reproductive function.

### *Medusa*

Either the medusae are sexual or reproductive zooids and are with four testes or four ovaries in the subumbrella. A gonad have an ectoderm, an endoderm, mesoglea and a small diverticulum of radial canal. The medusa is a modified zooid and form as a hollow bud from the coenosarc of blastostyle in the spring and summer. When fully formed, it breaks free and emerges from the mouth of the gonotheca. It is circular, tiny umbrella-like in shape, having convex outer surface, the exumbrella and concave inner surface, the subumbrella. From the centre of the sub umbrella arises a short projecting manubrium (*L. manus* = handle). The apex of manubrium bears a square mouth which is surrounded by the four oral lobes.

The mouth leads into an enteric cavity, which gives rise to four radial canals. These canals are delicate ciliated tubes, run to the margin of bell to join a ciliated circular canal running near margin. The enteric cavity and canals represent enteron, which distributes food. Four gonads project from the middle of the radial canals. The gonads mature after the release of medusa from the gonotheca. The edge of the bell are produced inwards as a thin fold, called the *velum*. The velum is insignificant in the *Obelia*. The medusa with a velum is called the *craspedote* (Figures 5.6–5.8).

From the edge of the bell, numerous small solid tentacles hang downwards. The tentacles have swollen bases due to the accumulation of interstitial cells, which are absent in the other regions. Such basal swellings are called the *bulbs*. The nematocysts are formed continuously in the bulbs. Near the bulbs, there are ocelli with pigment granules and nerve cells. The ocelli are sensory to light. Eight marginal sense organs, called the *statocysts* lie at regular intervals. The statocyst

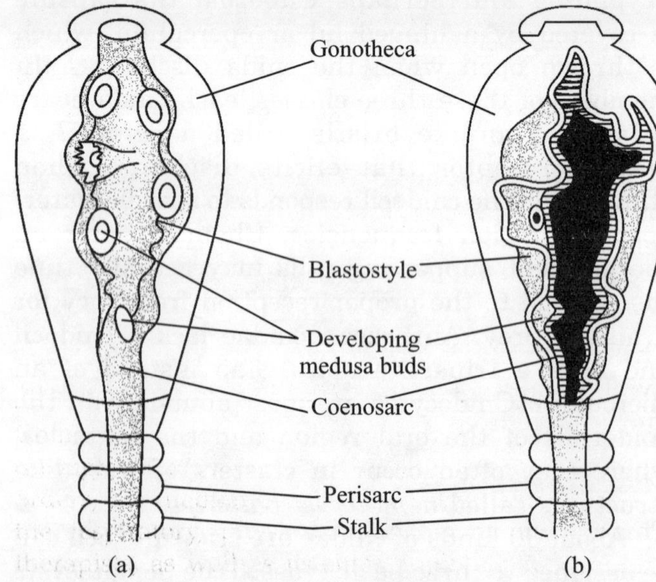


**Figure 5.6** Blastostyle of *obelia* (a) Entire (b) L.S.

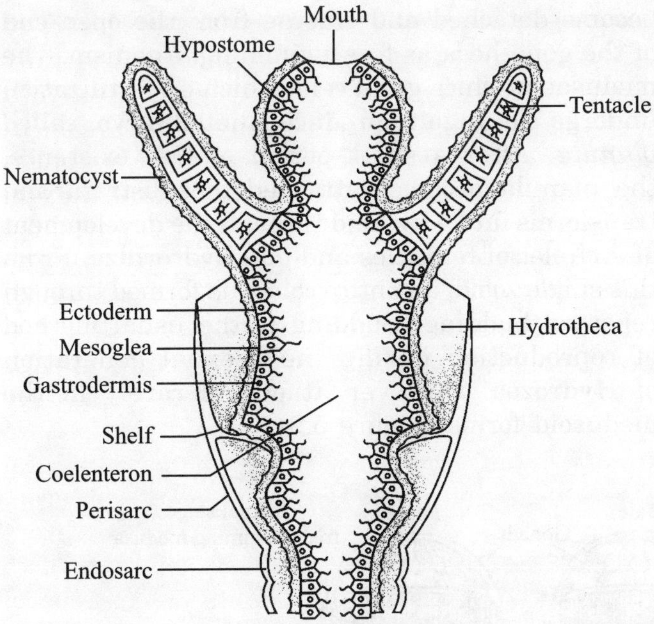


**Figure 5.7** Vertical section of hydranth of *obelia*.

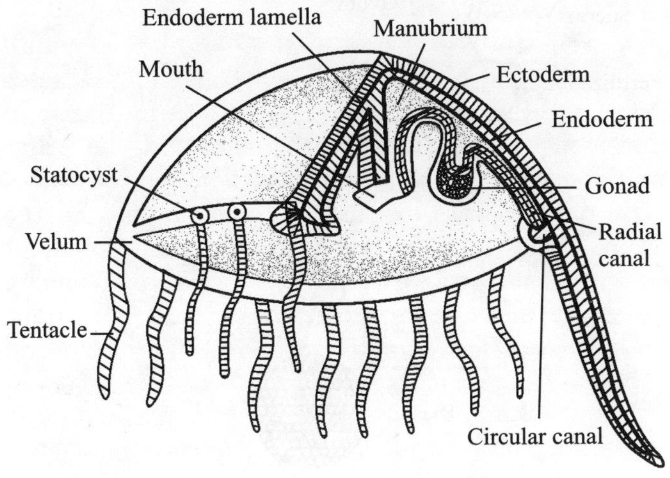


**Figure 5.8** Medusa in section.

is filled with a fluid containing the calcareous granules, called the *otolith*, which lie in a special cell, called the *lithocyte*.

### Nutrition

The medusae are carnivorous. The food consists of living animals. The digestion is both extracellular and intracellular.

### Muscular system

This system is restricted to the epidermal layer and is best developed around the bell margin and subumbrellar surface. Some epitheliomuscular cells of medusa have their contractile extensions, which are oriented to form a powerful circular band of striated fibres.

### Nervous system

In bell margin, epidermal nerve cells are organized and become concentrated into the two nerve rings. The nerve rings connect with the fibres innervating tentacles, musculature and sense organ.

### Sense organs

The bell margin is richly supplied with the sensory cells, contains ocelli and statocysts. The ocelli consist of patches of pigment and photoreceptor cells, which are organized within either a flat disc or pit. The ocelli are typically located on side of tentacular bulbs. The statocysts are located between the tentacles and act as an organ of the equilibrium. When the bell tilts, statocysts respond to the pull of gravity and stimulate the sensory bristle to which they are attached.

### Reproductive organ

The medusae are dioecious. The germ cells do not arise in the gonads but arise from the interstitial cells of ectoderm of blastostyle and then migrate into the medusa. Then through radial canals, they take up their position in ectoderm of gonads. When germ cells mature, the gonads rupture releasing spermatozoa and ova into the water.

### Reproduction and metagenesis

In asexual reproduction at water temperature above 20°C, the buds, which normally form the gonangia, break free from the colony and settle down. A stolon arises from the lower end of bud and produces a new colony of *Obelia* asexually. This is a special mode of asexual reproduction.

### Life history and development

The life cycle of *Obelia* serves well to illustrate metagenesis. The hydroid colony starts as a single polyp which by continued budding produces a series of individuals alternating along a common stalk. Such a colony is attached to the object by a root like structure, the hydrorhiza. In a young colony, all individuals are alike and function as the vegetative zooids. Starting near the base of the colony, reproductive individuals appear. These are sacs- or vase-shaped zooids, termed the *gonangia*, which budded off in the axis between the vegetative individuals and the common stalk. These are club-shaped, cylindrical form, which remain covered by a gonotheca that opens at its distal end by the gonophores. They contain blastostyles on which lateral buds form, develop into the medusae or gonophores.

The blastostyles has no mouth and tentacles. They end distally into a flattened disc. Gonotheca, blastostyles and gonophores together constitute a gonangium. Each gonangium is composed of a central core upon which buds are formed. As development proceeds, these buds display medusae characters and on reaching full development become detached and emerge from the open end of the gonotheca as free-swimming organism. The medusae produce germ cells which on fertilization undergo cleavage to produce ciliated larva, called *planula*. After a short period of free existence, the planula becomes attached to substrata and transforms into a hydroid through the development of a circlet of tentacles and of a hydrorhiza. From this single zooid, an entire colony is formed through repeated budding. Budding is the usual method of reproduction during the asexual generation of Hydrozoa. However, it occurs rarely in the medusoid forms (Figure 5.9).

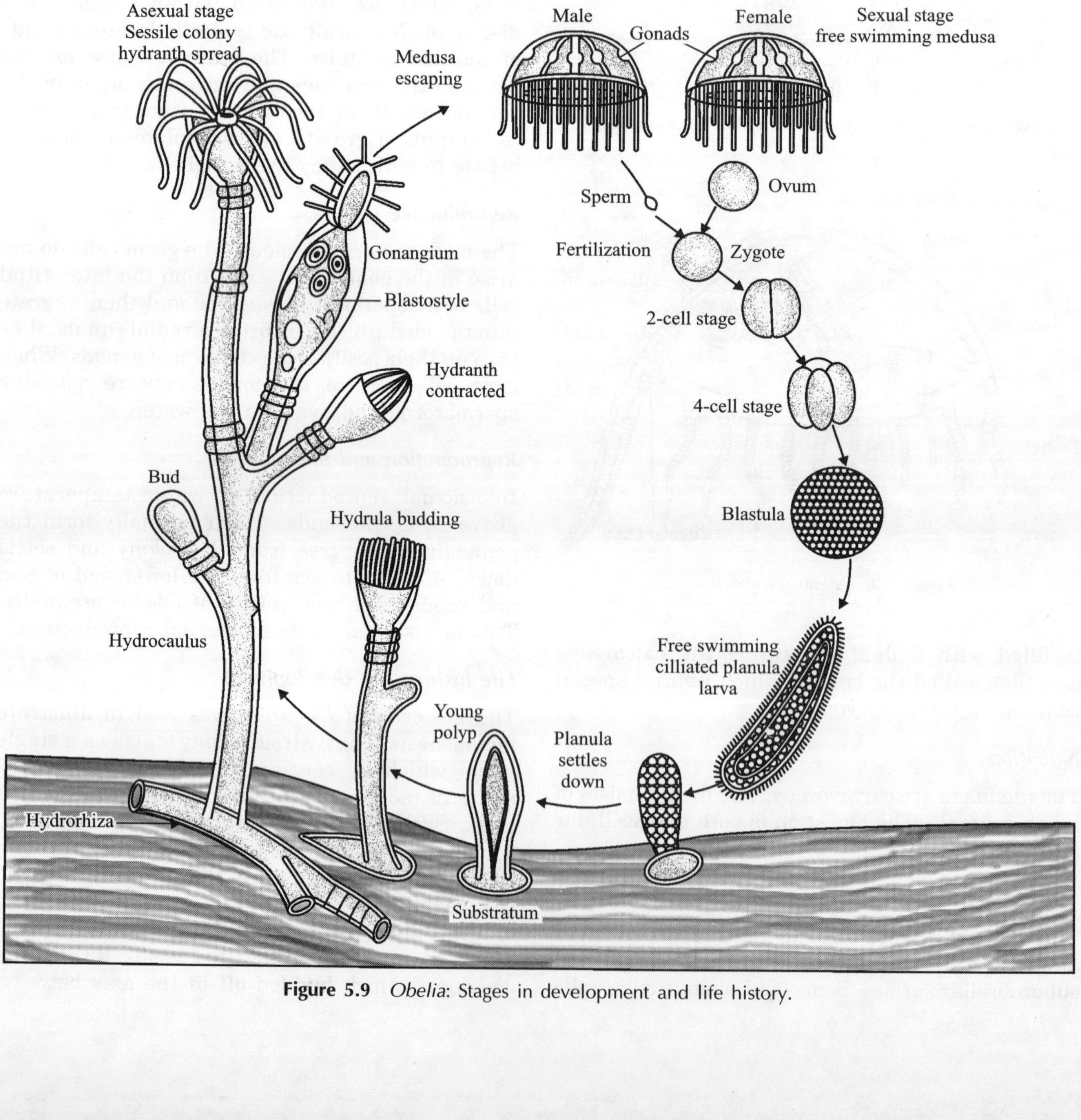


**Figure 5.9** *Obelia*: Stages in development and life history.

***Reduction of gonophore***

The cnidarians bears the gonads termed the *gonophore*, which is ordinarily a free-living medusa with gonads either on manubrium or on radial canals of the gastrovascular system. In some hydrozoans, the medusoid generation is lacking as a result of gonophore reduction. Usually, the gonophores are products of asexual reproduction and they undergo development to a stage while still attached to the hydroid and are then released. Sometimes the gonophores remain permanently attached to the parent hydroid, and organs essential to the independent existence of the gonophore do not develop to a functional stage.

A reduced gonophore comprises little more than an ectodermal bag within which the germ cells are arranged about a core of entoderm called the *spadix*. The spadix represents the remains of a degenerated manubrium. The term sporosac is coined to designate such a reduced gonophore. Fertilization usually takes place in the open sea water. The gametes are set free there. The sperms may be carried into the female medusa with the water currents and there they fertilize eggs *in situ*. The parent medusae die soon after liberating their respective gametes. The zygote undergoes complete or holoblastic and equal cleavage to form a single-layered blastula with the blastocoel. Some cells migrate into the blastocoel, eventually filling it completely to form a solid gastrula, known as the *stereogastrula*. Its outer cell layer becomes the ectoderm and inner cell mass the endoderm.

The gastrula elongates, ectoderm cells become ciliated and now it is called the *planula*. A cavity, called *enteron* develops in the solid endodermal mass by a process of delamination. The planula thus becomes a two-layered larva. The planula, after a short free-swimming existence, settles on some solid object by its broader end. The free end forms a manubrium with a mouth and a circlet of tentacles. Thus, a simple polyp or hydrula is formed, which grows a hydrorhiza from its base, from which an *Obelia* colony is formed by budding.

*Metagenesis:* There is an alternation of polypoid and medusoid generations in the life history of *Obelia*. The asexual polypoid generation produces polyps and blastostyles by the asexual budding. The medusae form the gametes, which after fertilization develop into a polypoid colony, from which medusae are produced again by the budding. An asexual polypoid generation thus alternates with a sexual medusoid generation.

The polyp and medusa both bears $2n$ number of chromosome. This phenomenon is known as the alternation of generations until recently. Alternation of generations means that individual exists in two distinct forms. Such forms alternate each other regularly in the life history. One individual possesses the power to reproduce the other by asexual reproductions, which again by the sexual reproduction causes the next generation. So, a true alternation of generation occurs always between a diploid asexual and haploid sexual generation. In *Obelia*, the medusae are modified zooid capable of free-swimming existence. They are not produced directly from a zygote, but are budded off from the blastostyle. The gonads found in the medusa are not formed in it. They are formed in the ectoderm of blastostyle, which later on migrate into the medusa and situated on its radial canals. Thus, it is difficult to distinguish between the sexual and asexual generations. Hence, the term metagenesis is used to replace "alternation of generation" in *Obelia*. In *Obelia,* there is a regular alternation between the polypoid and medusoid phases, both of the phases being diploid. Such an alternation of generations between two diploid phases is known as the metagenesis.

## 5.12 ANTHOZOA

This heterogeneous taxon includes the stony corals, sea fans, anaemones and sea pens. Anthozoa is the largest cnidarian taxon, with more than 6000 species of exclusively marine polyps. Anthozoans are always polyps. Medusa does not exist in their life cycle.

## 5.13 *Aurelia*

*Aurelia aurita* commonly known as the *moon jelly* inhabits coastal waters and is not dangerous. This is cosmopolitan and consists of a large, relatively flat disc called the bell with the mouthparts hanging below. The oral surface of the bell bears the mouth and is called the *subumbrellar* surface. The opposite aboral surface is the exumbrellar surface. The oral-aboral axis extends from the centre of the aboral surface to the centre of oral surface and constitutes the axis of symmetry.

The bell margin is shallowly indented at eight positions. Sensory complices, the rhopalia are located in the indentations between the scallops. The outer surface is covered with the epidermis. The coelenteron is lined by the second epithelium, the gastrodermis. Thick mesoglea, which makes about all of the mass lies between these two thin epithelial layers. The mesoglea is made up of protein fibres, a watery ground substance and a few amoeboid cells.

Four long oral arms of the manubrium lie in the centre of the oral surface. The oral arms are oriented at the right angles to each other and form a 'X' across the oral surface. The two axes that pass through the oral arms, perpendicular to the oral-aboral axis and to each other, are the perradial axes. The axes bisecting the angles between the adjacent oral arms are the interradial axes. The surface of each oral arm possesses two ribbons like ridges that run along its length and enclose a deep ciliated brachial groove. The grooves of the four oral arms join at the mouth. The margins of grooves bear the small brachial tentacles. The four oral arms radiate outward from the short, vertical column called the *manubrium* which remains attached to the centre of oral surface.

The bell margin bears numerous, short marginal tentacles bearing the cnidocytes. Tentacle is a circular veil of tissue, known as the *velarium*. Eight sensory rhopalia are distributed around the bell margin. *Aurelia* feeds on the zooplanktons that become entangled in the mucus of the aboral surface. The mucus is moved by the monociliated epidermal cells to the bell margin where it accumulates until removal by the tips of oral arms. Cilia in brachial grooves transport both the mucus and food to mouth. The coelenteron is elaborate arrangement of chambers and channels. This is lined by the ciliated gastrodermis and extending through the mesoglea and opens to the outside via the mouth. It is both digestive and fluid transport system, physically divided into the two regions to serve two functions.

Extracellular digestion occurs in a central set of chambers by the digestive enzymes. From these chambers, ciliated canals are extended which distribute the partly digested food to the body, where phagocytosis and intracellular digestion occurs. The digestive region possesses the stomach and gastric pouches. It opens into a central stomach. Four semicircular gastric pouches (Figure 5.10) open from the sides of the stomach are the sites of extracellular digestion. Each pouch contain a horseshoe-shaped gonad on its floor. The opaque gonads are easily visible than the transparent one. The edge of the gastric pouch is visible around the outside margins of gonad.

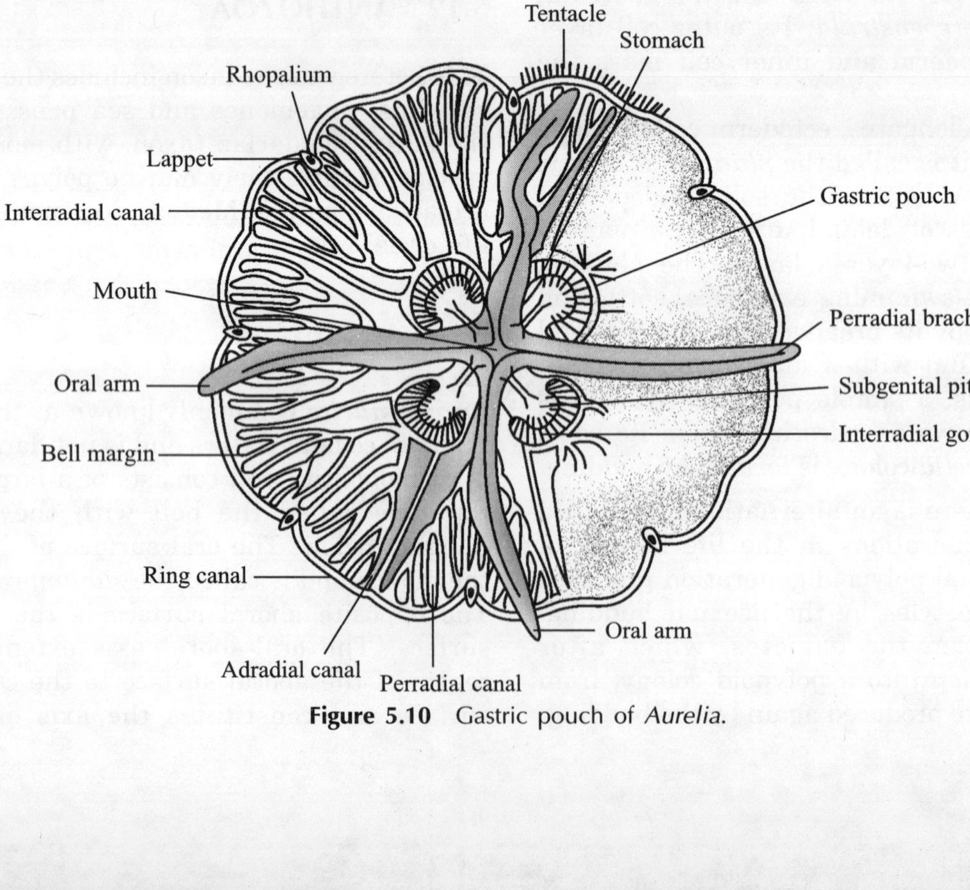


**Figure 5.10** Gastric pouch of *Aurelia*.

The gastric pouches are interradial in position. The passage way between the stomach and each gastric pouch is separated to form three short, side-by-side, ciliated channels. The channels move food between the stomach and gastric pouch. A fringe of short, threadlike gastric filaments is located on the floor of each gastric pouch along the inner margin of each gonad. These filaments secrete enzymes into the gastric pouches. The distributive portion of the coelenteron consists of a set of narrow, ciliated canals for transport of partially digested food to and from the periphery. The system comprises a ring canal in the margin of the bell and a set of 16 radial canals that extend from the stomach to the ring canal. Numerous short blind canals emerge from the ring canal and extend into the marginal tentacles and sense organs.

Some radial canals are branched and some are not. Four perradial and four interradial canals are located on the perradial and interradial axes respectively. Canals are branched and move materials from the ring canal back toward the stomach and gastric pouches. The perradial canals emerge from the stomach. The interradials and adradials canals emerge from the gastric pouches. Between each perradial axis and its neighbouring interradial axis, there is an adradial canal. These eight canals are unbranched and flow in them is away from the gastric pouches. The septal funnels are four invaginations in the oral surface of bell, one below each gastric pouch. The pits extend up into the mesoglea almost to gonads. The pits probably provide the oxygen to the developing gametes.

### *Sensory complex*

The sense complices are the eight rhopalia, which are evenly spaced around the bell margin. Each rhopalium possess a statocyst for gravity detection in addition to serve as mechanoreceptors, chemoreceptors and an ocellus for photoreception. Two small, triangular, side-by-side rhopalial lappets are located beside the rhopalium (Figure 5.11). The complex is provided by a branch of ring canal. The lithostyle bears sensory receptors including a pair of tiny dark ocelli, two obscure chemosensory epithelium and a spherical statolith containing tiny, opaque granules of calcium sulphate. The statolith is a device for detection of gravity. A short branch of ring canal called the *rhopalial canal* extends into the rhopalium to supply food.

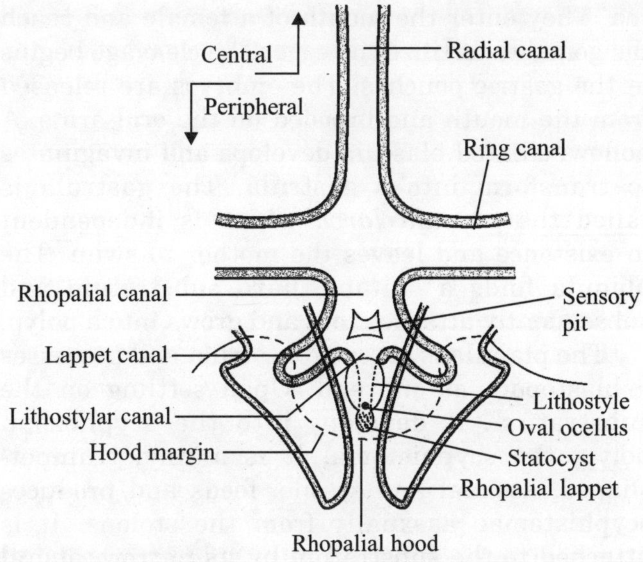


**Figure 5.11** Oral view of rhopalium of *Aurelia*.

The canal branches into two lappet and a central lithostylar canal.

### *Nervous system and musculature*

*Aurelia* has a sub epithelial nerve plexus but no nerve ring. A concentration of neurons called the *rhopalial ganglion* is associated with each rhopalium. The musculature comprises a ring of circular epidermal muscles near the bell margin. Pulsations of the bell margin result from the contractions of ejecting water from the oral surface and propel the animal in the aboral direction. The recovery is affected by elastic recoil of mesoglea.

### *Symmetry*

*Aurelia* has a distinct tetramerous radial symmetry, which is found to be arranged around the oral-aboral axis. Four oral arms, four gonads, four sub genital pits, four gastric pouches and four or eight of each type of radial canal are found. There are eight sense organs.

### *Life history*

The life cycle includes the alternating generations of the medusae and polyps. Large, sexual, free-living, pelagic, motile medusa generation produces the asexual, sessile and benthic polyp generation. *Aurelia* is gonochoric. Gametes are released from the gastro dermal gonads into the gastric pouches in medusae. The ova remain in the coelenteron. The spermatozoa come out via the mouth into the sea. They enter the mouth of a female and reach the gonad to fertilize the eggs. The cleavage begins in the gastric pouches. The embryos are released from the mouth and brooded on the oral arms. A hollow, ciliated blastula develops and invaginates to transform into a gastrula. The gastrula is called the *planula larva* which is independent in existence and leaves the mother to swim. The planula finds a suitable hard substratum and subsequently attaches to it and grows into a polyp.

The planula is an ovoid gastrula and possesses a blastopore at one end. Upon settling on the substratum, it develops into the scyphozoan polyp, the scyphistoma. It is a tiny, trumpet-shaped, asexual and sessile, feeds and produces scyphistomae asexually from the stolons. It is attached to the substratum by its narrow, aboral end. The wider free oral end bears a central mouth which remains surrounded by a ring of 16 long tentacles. The mouth sits atop a low mound called the *manubrium*. The scyphistoma undergoes an asexual process of transverse fission called *strobilation* to form a stack of tiny discs which mature to form medusae. A scyphistoma undergoes strobilation is called *strobila*. The young medusae, the ephyrae accumulate in a stack at the oral end of scyphistoma. The ephyra farthest from the scyphistoma is the oldest and is released first. The tiny ephyrae swim away from the scyphistoma and develop into sexually mature medusae in less than a year. Adult features are mostly present in ephyrae but in different proportions. Their obvious feature is disproportionately large eight pair of lappets.

The lappets are inconspicuous in adult medusae. They dominate the ephyra (Figures 5.12–5.14).

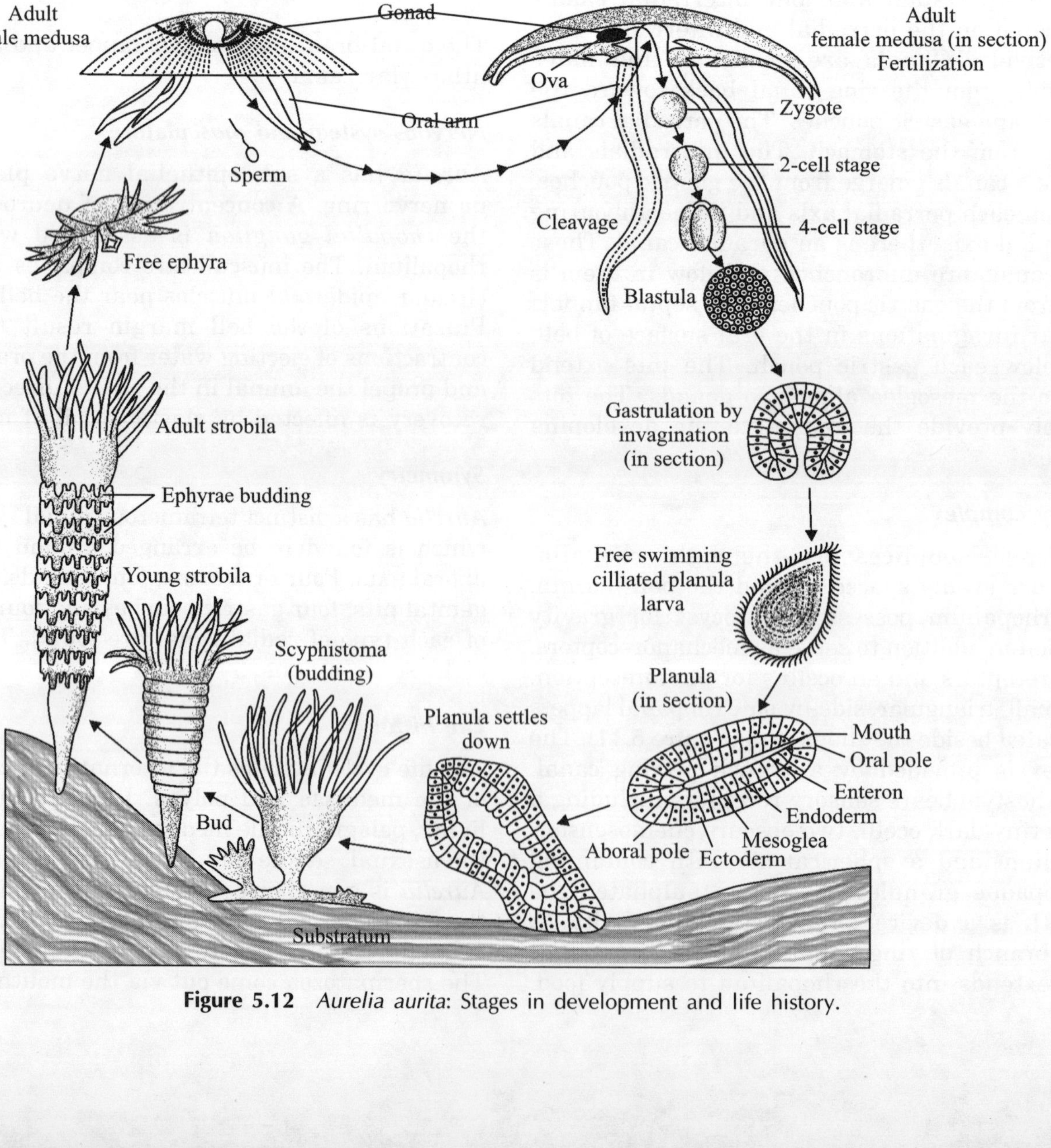


**Figure 5.12** *Aurelia aurita*: Stages in development and life history.

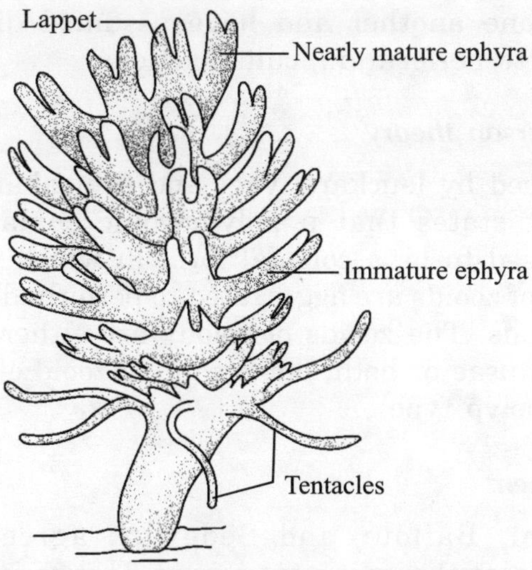


**Figure 5.13** Strobila of *Aurelia*.

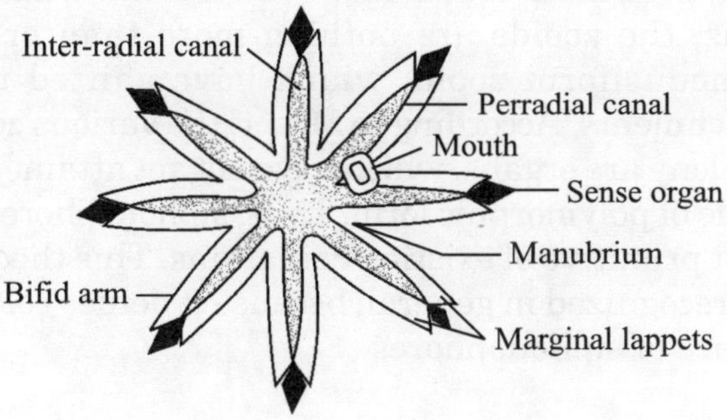


**Figure 5.14** Early ephyra (oral view).

## 5.14 POLYMORPHISM IN SIPHONOPHORA

The original plan of two body forms changing in the life cycle of the Hydrozoa has modified in members of Siphonophora. These are free-floating or swimming colonial forms in which the parts have become so interdependent and so highly specialized for limited functions that it is difficult to distinguish whether the individuals of the colony are hydroid or medusoid. Polymorphism reaches its peak in siphonophores which are common in warm seas. Ocean currents and storms sometimes bring large numbers of siphonophores into cool water. Many are attractively coloured and dangerous because of their powerful nematocyst. A continuous tube of the alimentary system connects all the individuals of a colony and each performs a limited service for the entire colony. One individual consists of an air-filled bladder, otherwise called *pneumatophore* which regulates the position of the colony at the surface of the ocean. The pneumatophore serves as a sail. The gastrozooids consist of feeding zooids, the dactylozooids serve as the sensory organs and the bracts are organs of defense. When single organism is represented by more than one kind of individuals, which differ in form and function, it is said to be polymorphic and the phenomenon is called as the *polymorphism*.

### *Zooids*

A polymorphic colony consists of more than one type of individuals, called the *zooids* or *persons*. These individuals are reducible in two main types, polyp and medusa.

1. *Polypoid zooids*

#### *A. Gastrozooids*

The gastrozooids (nutritive polyps, siphons) ingesting food is tubular with a large mouth and a contractile, hollow tentacle. The tentacle bears numerous fine lateral contractile branches, called the *tentilla*. The tentilla ends in a complicated knob or coil of nematocysts. The tentacles do not occur in *Velella*. The endoderm of gastrozooids is pigmented and projects as villi into the coelenteron.

#### *B. Dactylozooids*

The protective polyps of the colony are termed the dactylozooids, palpons, feelers or tasters. They are long, tubular, tentacle like zooids without a mouth or shorter zooids with a long unbranched tentacle. They are provided with the nematocysts.

#### *C. Gonozooids*

The reproductive zooids, termed the gonozooids or blastostyles are devoid of the mouth and produces the medusoids by the asexual budding. Generally, the gonozooids takes the form of a branched stalk, called the *gonodendron*, bearing grape like clusters of the gonophores and often provided with a gonopalpon. It develops no tentacle.

2. *Medusoid forms*

#### *A. Pneumatophore*

This is an apical air sac of hydrostatic functions and appears under the two forms. In Physonectae and Cystonectae, it is probably a specialized medusoid which is formed typically as a swelling at the upper extremity of the coenosarc. The ectodermal cavity of the coenosarc is distinguishable into a central part, the air gland which secrets a gas that passes through a pylorus into a distal part. The air sac is lined by a chitinous secretion. Many classes of radial pouches and septa of varying numbers, which perhaps correspond with radial canals of a medusoid are found between the air gland and outer wall of pneumatophore. In Disconectae, the pneumatophore comprises the chitinous plate, which bear a crest or sail. The air tubes forms pneumatophore and penetrate the centradenia.

#### *B. Aurophore*

This is characteristic of the Auronectae and is placed at side of pneumatophore. This is ovoid in shape and is traversed by a minute canal, which leads from the cavity of pneumatophore to exterior. Round this canal lies the pistillum, a mass of muscle enveloped in a strong chitinous tube.

#### *C. Nectophore, nectocalyx, nectozooid or swimming bell*

This is a medusoid which devoid tentacles, manubrium and mouth but retains velum, circular and radial canals. The musculature is well developed. The nectophore has a locomotor function.

#### *D. Hydrophyllium (bract, phylozooid)*

This is shield shaped and protective in function. It consists of a curved plate of thick mesoglea and is covered externally by the ectoderm and containing a solid endodermal core (phyllocyst).

#### *E. Gonophore*

This occurs singly or in clusters. They are degenerate medusae with no mouth, tentacles or sense organs. They have a velum, canals and a manubrium, which bears gonads. The gonophores produce the germ cells. (Figures 5.15–5.20)

## 5.14.1 Origin of Polymorphism

#### *Polyorgan theory*

Originally proposed by Huxley, Eschscholtz and Metschnikoff, this theory states that a polymorphic colony is supposed to be a single medusa. Its various components are modified organs of this medusa. The parts of zooid, i.e. manubrium, tentacles, umbrella, etc. multiply independently from one another and have assumed different forms to perform difficult functions.

#### *Polyperson theory*

Proposed by leuckart, Vogt and Genenbaur, this theory states that a polymorphic cnidarian is supposed to be a colonial form in which various types of zooids are aggregated to perform different functions. The zooids of colony are either polyps or medusae or both, but primitive zooid of colony is of polyp type.

#### *Comment*

Hackel, Balfour and Sedgwick agreed that polyperson theory is more correct. They viewed that primitive zooid of colony was probably medusoid that produced other medusae by the budding. Thus, the zooids are nothing more than organs of medusiform zooid, which have shifted their attachments. According to Moser, the various zooids of colony are organs, which have not yet attained the grade of polymorphic form. Thus, siphonophores are most primitive of existing cnidarians. This theory is not recognized in general, because it denies colonial nature of siphonophores.

#### *Significance of polymorphism*

In siphonophores, the asexual colonies lead to such specialization that the individual polyp comes to act as organ. Cnidarians are excellent examples of a common phenomenon among invertebrates—extensive use of asexual reproduction to form the colonies. Asexual colony formation allows mimicking the organ system, with individual polyps specializing for the specific functions. Asexual reproduction allows for the exponential growth that allows quick utilization of ephemeral resources. The survival of colony is size-independent. Any number of individuals can ensure this, depending on the level of resources available. Asexual colony formation also reduces competition between the individuals of a species by allowing them to share rather than fight over resources.

Finally, sexual reproduction brings about the rapid cloning of phenotypes which are proven to be affective by virtue of their survival. It also imparts a high degree of the morphologic flexibility and adaptability to allow expression of seasonal and environmental polymorphism. The Hydrozoa usually have an alternation in generations in which the hydroid is often the more conspicuous. The hydrozoan polyp lacks the longitudinal folds of the entoderm. The mouth opens into the coelenteric cavity. The hydromedusa, when present, possesses a smooth bell margin. The concave surface is partially covered by a membrane called the velum, which extends inward from the edge of the bell. In the members of this class, the gonads may be borne either on the radial canals or on the manubrium. The germ cells are released from the gonads through the surface of the body.

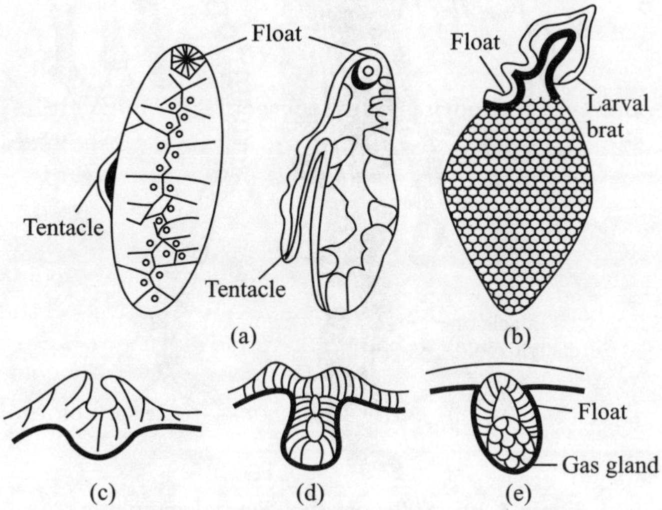


**Figure 5.15** Development of float in Agalmids.

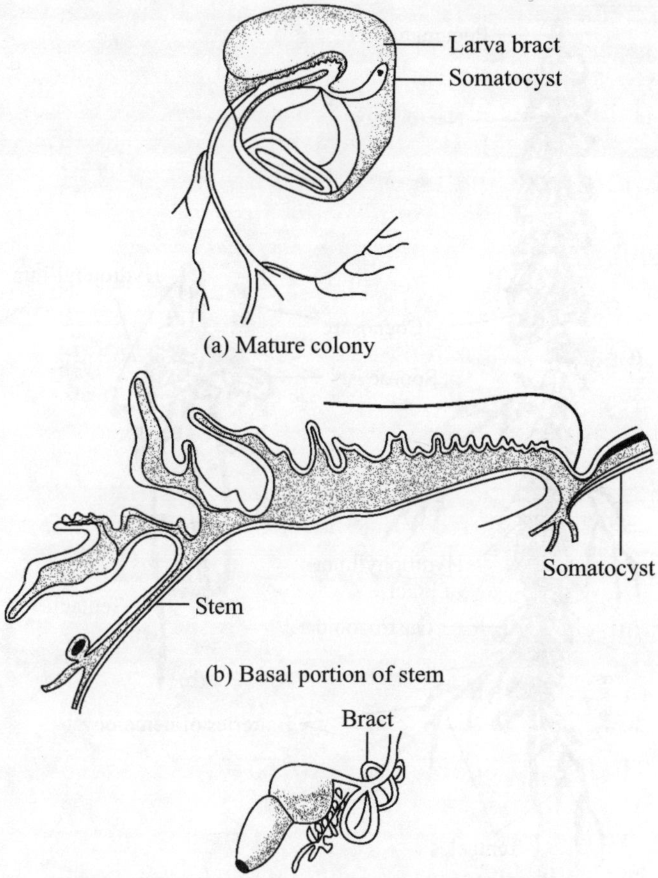


(c) Terminal cormidium with bract, gonophore and siphon

**Figure 5.16** Development of cormidia in Sphaeronectes.

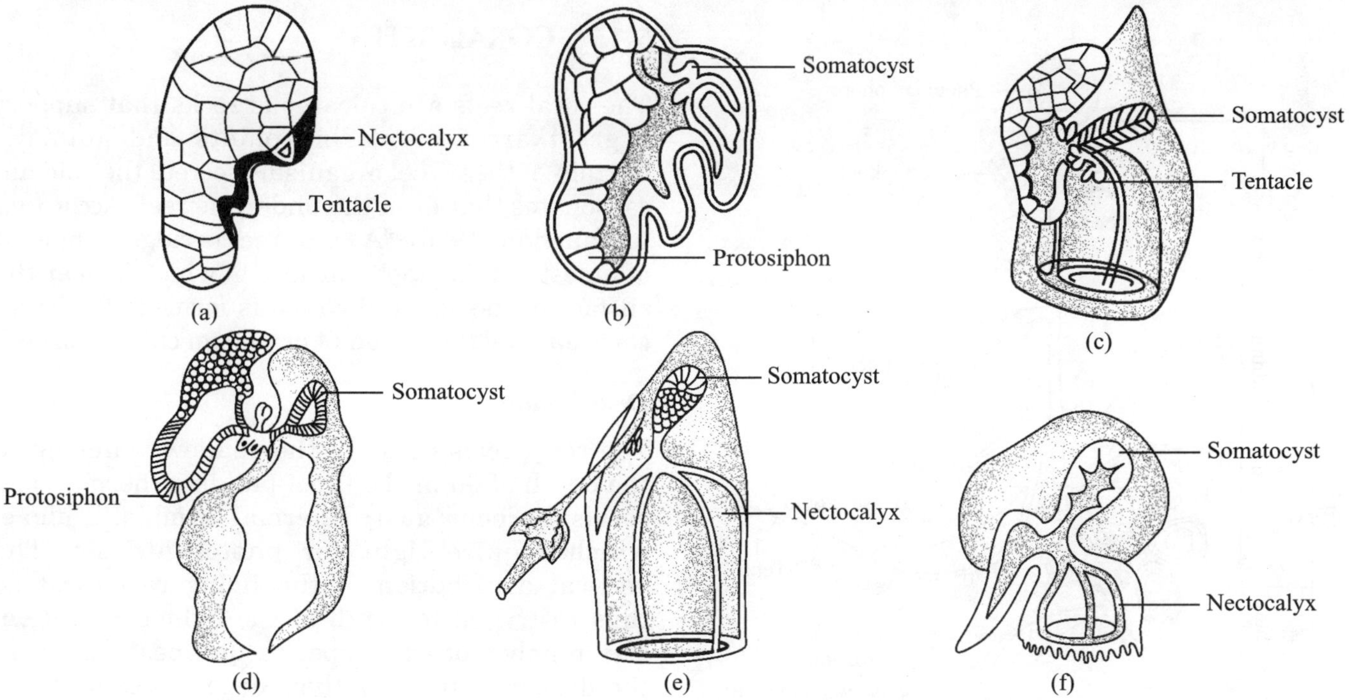


(a), (b), (c): *Galeolaria quadrivalvis*; (d), (e): Same in sagittal section; (f) Metschnikoff's larva of *Sphaeronectes truncata*

**Figure 5.17** Development of calycophorida.

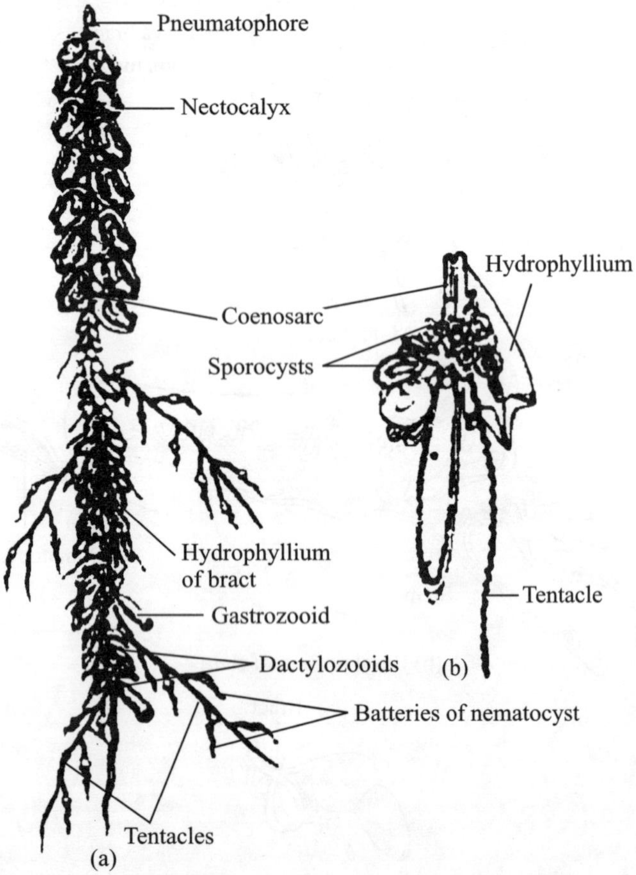


Figure 5.18 *Halistemma* (a) Entire colony; (b) A cormidium.

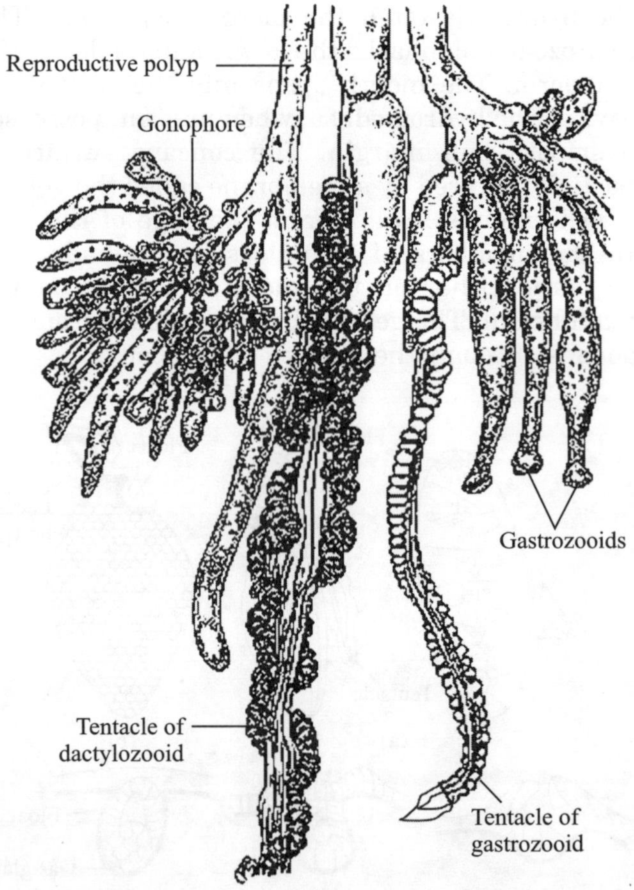


Figure 5.20 *Millepora* showing polymorphism.

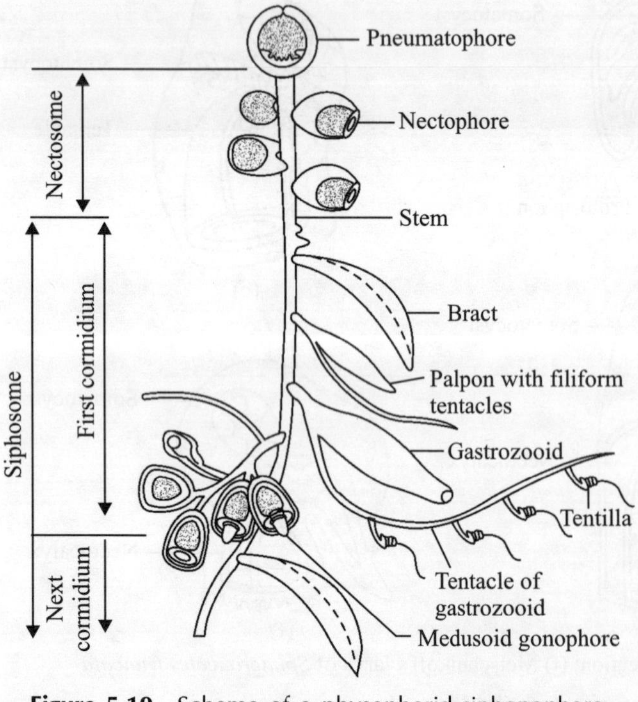


Figure 5.19 Scheme of a physophorid siphonophore.

## 5.15 CORAL REEFS

The coral reefs are calcareous rocks that support a great array of marine plants and animals. Certain of these reef organisms secrete the calcium carbonates that form the underline reef. According to Vaughan (1970), "A coral reef is ridge or mound of limestone the upper surface of which is near the surface of the sea and which is formed of calcium carbonate by the action of organism chiefly corals".

### *Distribution*

The coral reefs occur in the shallow water up to the depth of 60 m. The reef-building (hermatypic) corals harbour gastro dermal symbiotic algae, which require light for photosynthesis. The vertical distribution of the living reef coral is, thus restricted to depth of the light penetration. The number of coral species declines rapidly in the deeper water. For their dependence on light, the reef corals require clear water and found only where the surrounding water contains the small amounts of the suspended material, i.e. water of low turbidity and low productivity. The coral reefs are further restricted in their distribution by the water temperature and occur only in tropical seas with average minimum water temperature (not less than 20°C). They are confined to the tropical parts of Atlantic, Indian and Pacific oceans and inhabit the regions as follows:

1. The Caribbean and related water, including, Florida, Bermuda, Bahamas and West Indies.
2. The indo-pacific region from the east coast of Africa, including Madagascar, through the Indian ocean, including Maldives, Laccadive, Cocos, Chagos, etc. and throughout western pacific, including Malayan archipelago, Philippines and innumerable pacific islands scattered to east of Australia as far as Hawaii.
3. The pacific ocean northeast of Australia is called the *Coral sea*, where most islands are formed entirely of the coral rock.

The reef corals diversity decreases in the higher latitudes to about 30° north and south, beyond which they are usually not found. Bermuda at 32° north is an exception to this rule. Stag horn and elkhorn corals, which are conspicuous on West Indian reefs, are absent in the Bermuda because of the lower water temperature in winter. The greatest concentration occurs near the equator, where the water temperatures are highest.

### 5.15.1 Kinds of Coral Reefs

The coral reefs are usually divided into three primary types—fringing reefs, barrier reefs and atolls.

#### *Fringing (shore) reefs*

Fringing reefs grow in the shallow waters, closely bordering the shores of some volcanic islands or parts of some continent. They are small in breadth and extend from the coast up to 1/4 mile out, as a bench or platform with its surface more or less level. It consists of the seaward slope, reef edge, seaward flat, boulder zone and inner flat. The active coral growth occurs at the outer or seaward edge of the reef plateau, called the *reef edge* or *front*. It is a narrow belt, 20–40 yards in breadth and subject to continuous surf. Beyond the outer edge, there is a steep seaward slope down to the sea bottom. The corals grow on this slope, up to a depth of 50 fathoms. Between the reef edge and shore is a slightly lower and about flat surface, called the *reef flat* or *seaward flat*, composed largely of the coral, sand, mud, dead coral and other debris. The boulder zone is characterized by the dead coral masses and reef-fragments, torn off by breakers and carried across the flat by the surfs.

#### *Barrier reef*

Barrier reefs are parallel to the coast and separated from the shore by a lagoon. It consists of the same part as fringing reef, but the place of inner flat is taken by the lagoon flat. The latter has a thin covering of sand and shows a luxurious coral growth. The lagoon may be so broad that besides its distant barrier reef, the shore may possess a narrow fringing reef as well. There may be a fringing reef on one coast and a barrier reef on the other and two often merging into each other, as reefs of Nagau island of Fiji. A barrier reef entirely encircles the island of Tahiti. The great barrier reef of Australia is the longest in the world, stretching over 1000 miles along the Australia's northeast coast. This is not a single reef, but a long string of separate reefs, which are not always in a line. Inside this, all sorts of coral growth are found. Another long barrier reef is located in the Caribbean coast of Belize between Mexico and Guatemala.

#### *Atolls*

Atolls lie above the old submerged volcanoes. They are ring-like reefs containing an interior lagoon. The lagoon varies from a few 100 yards to 50 miles in diameter. The rim of the ring is narrow, but a few 100 yards wide. It may be complete or broken by several gaps or channels. The reef may be so low that the sea waves dash over it into the central lagoon. The rim of the reef bears a linear series of little islets. The outer side of the atoll reef slopes off steeply into depths of the ocean. Thousands of atolls dot indo-pacific region, forming nearly 90% of all the coral formations of that region. The atolls are hundreds or thousands of miles from the nearest land. The largest atoll of the world is the Suvadiva of Maldives, covering an area of 42 by 32 miles, with circumference of about 121 miles and not less than 102 separate little islets on its rim. The atoll of Bikini with atomic and hydrogen bomb test fame has a lagoon area of 280 square miles but a land area of only 287 square miles.

### *Reef formation*

The reef building involves the constructive and the destructive phases. The scleractinian corals and larger skeletons of some other organisms from framework material, or the bricks of reef platform, finer skeletal material forms 'mortar'. When a living colony die, underlying skeleton becomes riddled with the boring organisms. Eventually, a large coral is fragmented to the chunks. The deeper surfaces exposed by the borers may be colonized by the bryozoans and coralline algae, both depositing the calcium carbonate. Destructive phase duration is determined by the deposition of fine, excavated material produced by the borers. This material accumulates; eventually borers are smothered in their own debris and boring stops. The more rapidly the boring is halted, the larger will be the chunks of coral left intact. The debris produced by the boring, as well as fine skeleton and shells of gorgonian corals, calcareous algae, sponges, molluscs and other organisms settles into the crevices and holes, slowly filling the spaces between the larger framework pieces, with a calcareous mortar. The fusion or lithification of this material actually occurs within the 10–15 cms of surface.

In the agitated water, magnesium calcite particles (4–30 μ) precipitate and act as a cement. The reefs may grow seaward, but the light and water depth limits the vertical growth. The core sampling on the most modern reefs reveals a platform thickness of 6 to more than 1000 m. The lower parts of many platforms are located below the photic zone. This can be explained by the fluctuations in the sea level and by the subsidence of bedrock, upon which the reef is resting. Both having occurred at Semisweet. The geologic history of Eniwetok began during the early Cenozoic period, when a fringing reef developed around an emergent volcanic cone. This oceanic peak gradually subsided to below sea level. The coral deposition occurred at a rate equalizing the subsidence and the original fringing reef was transformed into an atoll, resting on vertical walls almost one mile in height. The fluctuating sea level periodically caused the emergence and submergence.

Rise in the sea level from the last Pleistocene ice age low of 120 m below the present level, gave Eniwetok its present emergent form. Although, the existing coral reefs have been formed by the scleractinian coral, the reefs in their past have been produced by the other organisms. Two extinct groups of the anthozoans, tabulate and rugose corals, formed great reefs during the Palaeozoic. The reefs have also been formed in the past by the algae that secrete the calcium carbonate. The coralline algae are still prevalent today and are common members of the reef communities, but there are no existing reefs composed primarily of the living coralline algae. Many theories have been put forward to explain the formation of the coral reefs. Some such theories are:

**Darwin–Dana subsidence theory:** This theory states that the fringing reefs gave rise to the other types of the coral reefs. The fringing reefs are formed on the shores, which are inclined downward. Sometimes, the shore between the developing fringing reef and the land mass sinks. The fringing reefs thus, turn into a barrier reef. The barrier reef and main land mass always have the sea water between them. In case, the main land mass is an island and itself subsides completely under the water, the reef turns into an atoll.

**Semper–Murray solution theory:** Murray advanced the view that the corals grow on the highest peaks of the ocean bottom. Deposition of the sediments brings them to an optimum level. The barrier reefs and atolls are produced by the better growth of the corals in edges of the coral deposition and by the solution of its inner part.

**Submerged bank theory:** The coral reefs are formed on the flat surfaces, which at first remain in the lesser depths. Such surfaces with the growing reefs, later submerged slowly and go down to the greater depths. This theory denies the transition of one common type of reef into the different types. It advocates an independent origin of the different types of reefs. Further, regarding the shape of an atoll, it lays importance on winds and water currents.

**Daly Glacial–Control theory:** This theory advocates that during the last glacial period, the formation of ice caps lowered the level of water of the ocean. At that time, the extreme cold temperature prevailed. Subsequently, the ice melted and the temperature rose. The corals began to grow upon the flat platform of the ocean bottom and kept place with the rising level of the ocean. The support of Daly Glacial theory was given by calculating that almost all the existing coral reefs had been formed since the ice age.

### *Problem and prospect of conservation*

Hurricanes and unusually low tides cause deaths of the many corals and other organisms, but in time, the reef recovers. Disasters produced by the man are much more serious and recovery may be prolonged. The reefs in many parts of the world are in jeopardy from a variety of conditions. Construction causes the silting and changes the patterns of water flow. The patch reefs in castle harbour, Bermuda, have never been recovered from the construction of air base there, during the Second World War. The silt and fertilizer in run off from farmland have destroyed or damaged fringing reefs in many places. Oil pollution from the tank washing and spills, is taking a toll in certain areas. Removal of the large corals for construction or for making cement is a common practice in undeveloped countries. The damage is compounded, when quarries drag the large coral heads over the bottom, destroying the smaller corals in process. Fishermen, sometimes dynamite reefs, in order to collect the stunned fish. Although dynamiting is outlawed, the enforcement is difficult in many countries. Corals bear great importance and are very beneficial to man. They have been noted for their island building operations. Many coral islands have been of great strategic value in Pacific Ocean, for use as military bases and landing fields. Some corals are very beautiful and certain species such as 'precious red coral', are made into the attractive jewellery. Others are used in the indoor aquariums and rock gardens, because of their unusual form and design (Figures 5.21–5.28).

*Conclusion:* The entirely aquatic phylum Cnidaria is represented by the polyps like sea anaemones and corals and by medusae like the jellyfish. A polypoid or a medusoid is a radially or biradially symmetrical and uncephalized with a single body opening, the mouth which is surrounded by the tentacles studded with stinging capsules known as the *nematocysts*. Nematocysts serve for offense and defense. Intrinsic nematocyst is the characteristic feature of the phylum. The body and tentacles in cnidarians consist of endoderm and ectoderm and between the two cell layers lies the mesoglea. Thats why cnidarians are diploblastic. The coelenteron communicates with the surrounding medium through the mouth and serves for gas exchange and digestion. All cnidarians are carnivorous. The cnidae and tentacles are active in prey capture. Polyps are typically sessile. Some medusae possess sensory structures. Cnidarians

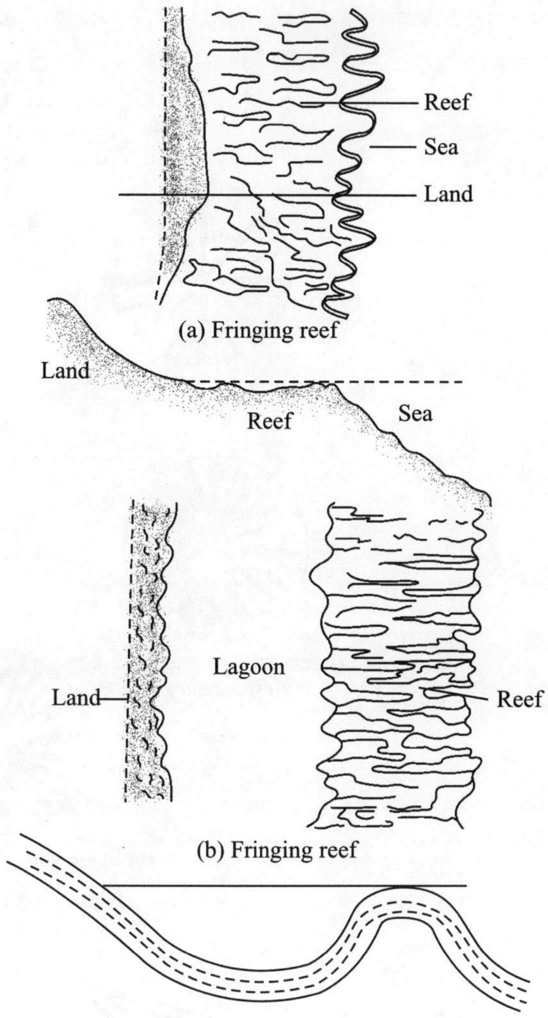


**Figure 5.21** Fringing and barrier reefs.

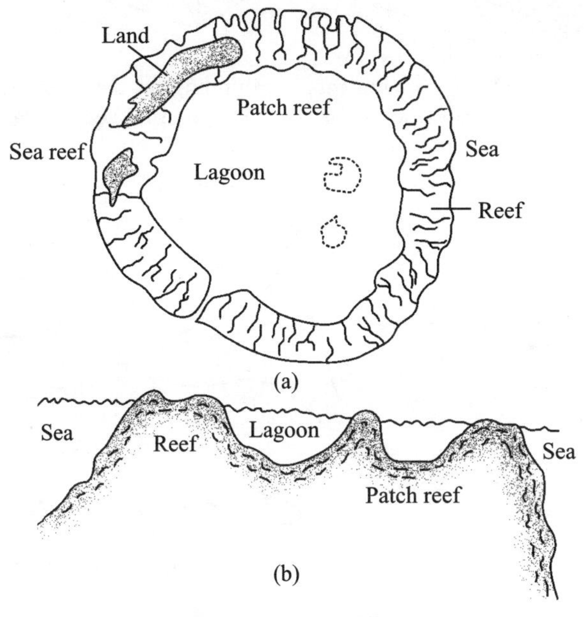


**Figure 5.22** Atoll.

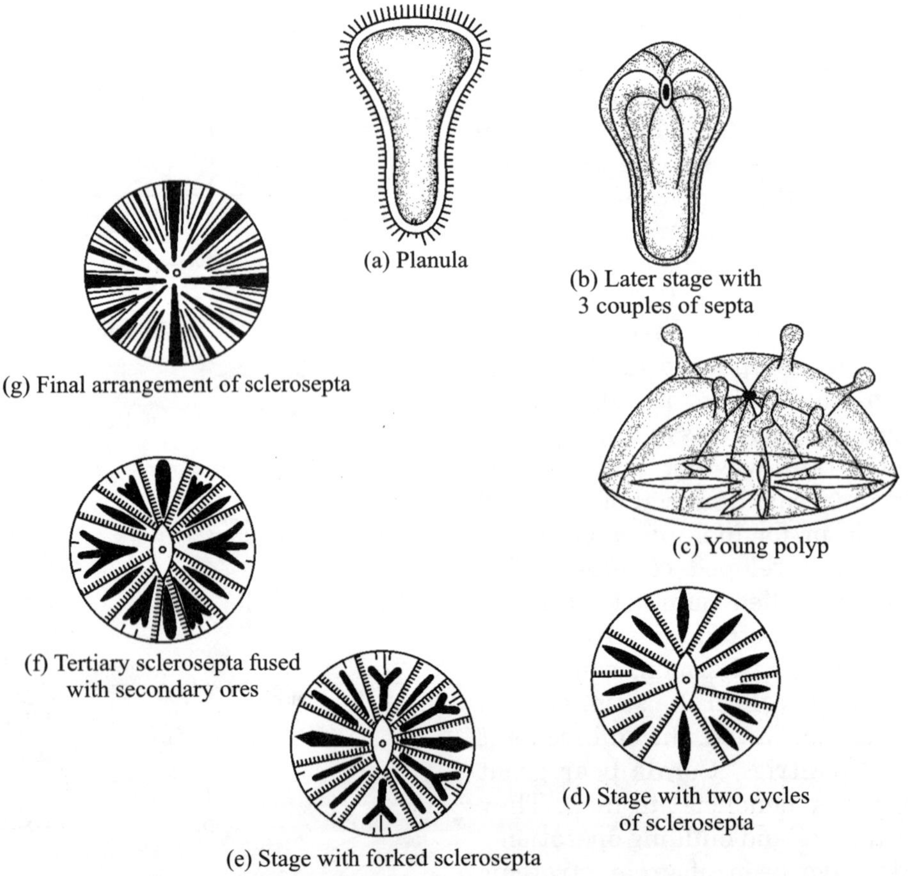


**Figure 5.23** Formation of coral reef.

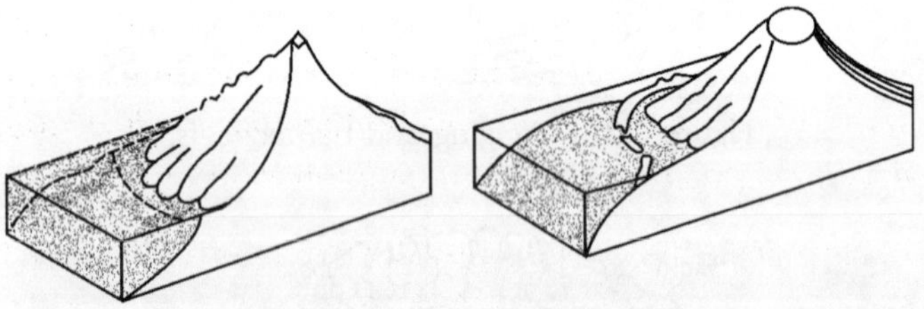

**Figure 5.24** A bringing reef surrounding an active volcano-subsidence of sea floor or rising sea levels.

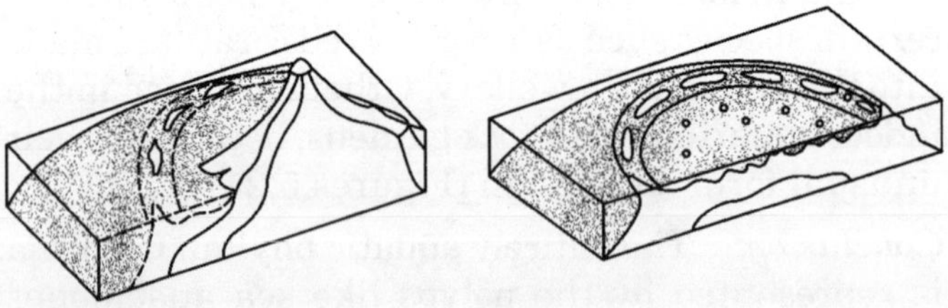

**Figure 5.25** Shrinkage of the island; formation of atoll with the disappearance of the island.

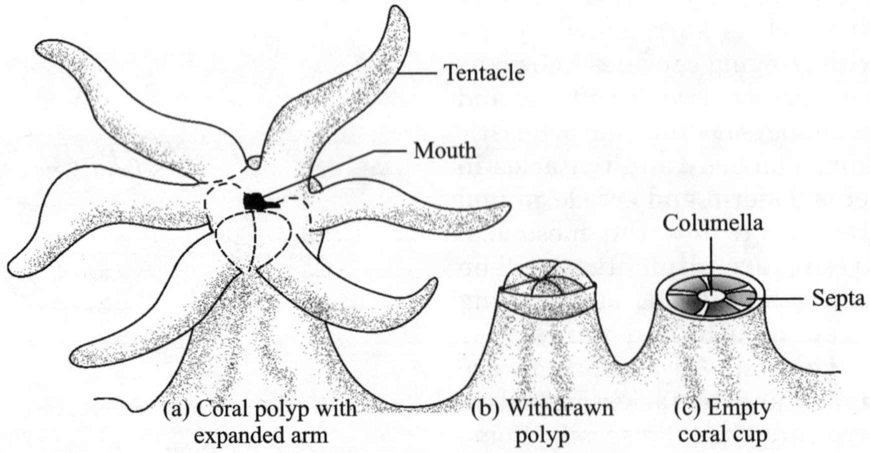


**Figure 5.26** Coral.

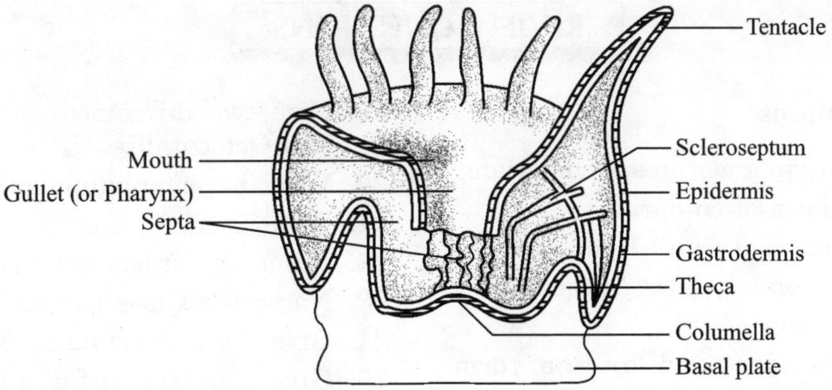


**Figure 5.27** One-half of a simple coral.

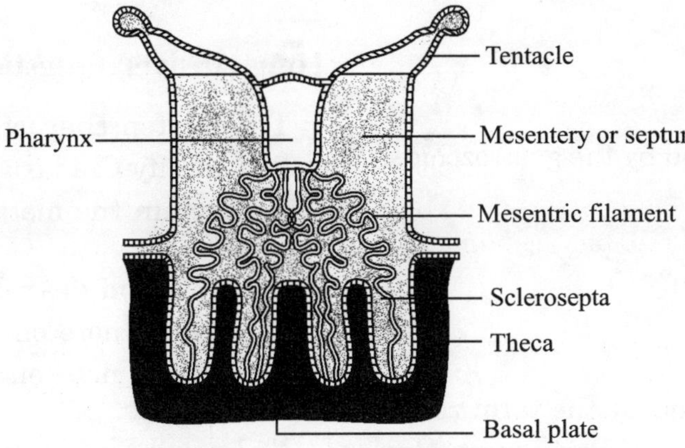


**Figure 5.28** V.S. through coral in its theca.

are generally passive predators, feeding on prey that blunders into their tentacles. Some members absorb dissolved organic matter from sea water. The typical cnidarian life cycle comprises an alternation between a medusa and a polyp which is termed *metagenesis*, the former being the sexually reproductive stage and the latter the asexual stage. In the entire class Anthozoa, medusa is absent. Some anthozoan species form gametes. Some species may reproduce vegetatively as well.

Life cycles of the Hydrozoa are the diverse. The polyp is the more conspicuous and persistent stage in most taxa. However, some lack the medusa phase, whereas others lack the polyp phase. The cnidarian larva is called the *planula* which is a pear-shaped ciliated animal. In the "typical" cnidarian life cycle, male and female medusae spawn freely into the sea water, where fertilization occurs and a planula develops ultimately. The planula settles on and attaches to the substratum and metamorphoses into a polyp. The cnida or nematocyst is secreted by the Golgi apparatus of the cnidoblast. The appropriate physical and/or chemical stimulus causes a cnida to fire, everting a tubule which is many times the length of the capsule. The tubule delivers a toxin which may stick to a prey item, or may entangle an object. A cnidae can fire once. Nematocysts occur in all classes of Cnidaria, but more than 30 varieties of nematocysts are limited to members of certain classes. Spirocysts which are adhesive in nature are found only in Anthozoa. Ptychocysts are the taxonomically limited in distribution, occur only in the order Ceriantharia of Anthozoa. They serve to entangle bits of mud in their robust tubules to form the felt work for constituting the tube of these burrowing animals. Polyps in a number species propagate vegetatively and form colonies or clones. Polymorphism is encountered in colonies of some hydrozoans and anthozoans where the polyps specialize for feeding, defense and sexual reproduction. Some polyps form skeleton within external to their tissues. Some skeletons are mineralic, others are the organic and some are both. Spheroidal or discoidal medusae are solitary and most is pelagic. Medusae assumes all orientations in the water and few species are able to propagate vegetatively.

## REVIEW QUESTIONS

### Short Answer Questions

1. Mention two important features of cnidarians.
2. Name a larval form of cnidarians.
3. Give two characters of Class Hydrozoa.
4. Give two Characters of Class Anthozoa.
5. What are nematocysts?
6. What do you understand by the term metagenesis?
7. What are medusae?
8. Define polymorphism.
9. What is velum?
10. What is coelenteron?
11. What do you understand by the gastrozooids?
12. What are gonozooids?
13. What is aurophore?
14. What is hydrophyllium?
15. Define coral reef?
16. Comment on the atoll.
17. What do you understand by the term zooid?
18. Give the scientific name of Portuguese Man of War.
19. Give two differences between Hexacorallia and Octocorallia.
20. What is pneumatophore?
21. What is mesoglea?
22. Name the order to which *Velella* belongs.
23. Name the Class to which *Porpita* belongs.
24. Give the generic name of sea fan.
25. Give the scientific name of mushroom coral.

### Long Answer Questions

1. Give ten characters of Phylum Cnidaria.
2. Classify Cnidaria up to class with examples.
3. Explain the metagenesis taking life cycle of *Obelia*.
4. Draw and describe the life cycle of *Aurelia*.
5. Write a note on coral reef formation.
6. Write a note on the types and distribution of corals.
7. Describe polymorphism in Siphonophora with suitable diagrams.

# CHAPTER 6

# Phylum Ctenophora

The name ctenophore comes from the Greek word *ktenos* meaning 'comb bearer'. They are simple, live in marine waters from the polar to tropical, inshore to offshore and from near the surface to the deep ocean. About 150 species are known throughout the world's ocean. The best-known members occur near-shore. They are typically planktonic, transparent and not pigmented. They swim by synchronous beating of the eight rows of comb plates. They are common members of the plankton in most coastal areas. They may be abundant in the spring and early summer. Their small size and transparency make them relatively inconspicuous. Ctenophores possess true tissues. Between the epidermis and the gastrodermis, there is a cellular mesenchyme. The nervous system is simple, noncentralized nerve net. The locomotive structures called *ctenes* arranged in comb rows are arranged radially. Retractile tentacles, anal pore, adhesive prey-capturing structure called *colloblast* and an apical sense organ containing a statolith are the important features. Most ctenophores are spherical or ovoid in shape, although some are flattened.

The mouth is situated at the oral pole. The aboral pole bears the apical sense organ. Each comb plate or ctene is composed of a transverse band of long, fused cilia. The tentacle sheath produces the inner wall. The tentacles are typically very long and contractile and bear lateral branches called *filaments* or *tentillae*. The epidermis of both tentacle and lateral tentillae is richly armed with colloblasts. The stomodeum lies on the oral-aboral axis of the body and is flattened in one plane of body symmetry, the stomodeal plane. Bisecting the animal along the stomodeal plane separates the two tentacular halves of the body. The second plane called the tentacular plane is defined by the position of the tentacle sheaths.

## 6.1 SUPPORT AND LOCOMOTION

Elastic mesenchyme provides the structural support, makes up most of the body mass and contains both elastic supportive cells and muscle cells. Tension in the looped muscles tends to maintain the spherical shape. Radial muscles diminish the radius and also open the pharynx. Most ctenophores are pelagic. The gelatinous body and low specific gravity maintain relatively neutral buoyancy, allowing them to float about with the ocean currents. Neutral buoyancy appears to be maintained by passive osmotic accommodation. Ctenophores may temporarily accumulate at discontinuity layers in the sea, where a water mass of one density overlies water mass of a slightly different density.

The beating of the ctenes provides locomotor power for moving up and down in the water column and to locate richer feeding grounds. Each ctene consists of a transverse band of hundreds of very long, partly fused cilia that beat together as a unit. Each cilium has a typical 9 + 2 microtubule structure, but each also possesses unique set of lamellae at the 3 and 8 doublets. The mesenchymal musculature is used to maintain body shape and helps in feeding. It is involved in prey swallowing, pharyngeal contractions and tentacle movements. Usually both longitudinal and circular muscles are present just beneath the epidermis. In the platyctenids, stomodeal musculature facilitates the creeping locomotion. In the cestids, body muscles may generate swimming undulations. The lobate ctenophores swim by muscular flapping of their two oral lobes. The lobate species *Leucothea* swim either by typical slow or by rapid ctene propulsion. Giant smooth muscle fibres are found in *Beröe*.

## 6.2 FEEDING AND DIGESTION

The long tentacles of cydippids have a muscular core with a colloblast-laden epidermal covering. The tentacles trail passively or are 'fished' by various movements of the body. On contact with (zooplankton) prey, the colloblasts burst and discharge an adhesive material. Each colloblast develops from a single cell and consists of a hemispherical mass of secretory granules attached to the muscular core of the tentacle by a spiral filament, which coils around a straight filament. The straight filament is the modified nucleus of the colloblast cell. The spiral filament adheres to the prey by the sticky material produced in the secretory granules. As the tentacles accumulate prey, they are periodically wiped across the mouth by muscular contractions. Members of the orders Lobata and Cestida bear short tentacles and zooplankton are trapped in mucus on the body surface and then carried to the mouth by ciliary currents. Some ctenophores prey on larger animals, especially gelatinous forms.

The cydippid, *Lampea* lives embedded in the body of pelagic tunicates, *Salpa* on which it feeds. The pharyngeal epithelium is richly endowed with gland cells that produce the digestive juice. Large food items are tumbled within the pharynx by ciliary action. Digestion takes place extracellularly, mostly in the pharynx. The digested food passes via a small chamber from the pharynx into a complex system of radiating gastrovascular canals. The anal pores serve as a primitive anus, helping the mouth in the voiding indigestible wastes. They may also serve as an exit for metabolic wastes. Minute pores lead from the various canals into the mesenchyme. Surrounding these pores are circlets of ciliated gastrodermal cells called *cell rosettes*, which appear to regulate the flow of the digestive soup and perhaps play a role in excretion.

## 6.3 CIRCULATION, EXCRETION, GAS EXCHANGE AND OSMOREGULATION

Independent circulatory system is absent. The gastrovascular canal system distributes nutrients to most body parts and picks up metabolic wastes from the mesenchyme for expulsion out of the mouth or anal pores. The cell rosettes may also transport wastes to the gut. Gas exchange occurs across the general body surface and walls of the gastrovascular system. All these activities occur by diffusion through the mesenchyme. The movement of water over the body surface is enhanced by the beating of the comb plates. The extensive canal system and the ciliary bands help to overcome the problem of long diffusion distances.

## 6.4 SENSE ORGAN

The apical sense organ, the calcareous statolith functions in balance and orientation and is supported by four long tufts of cilia called balancers. The whole structure is enclosed in a transparent dome that is derived from cilia. From each balancer arises a pair of ciliated furrows, each of which connects with one comb row. Each balancer innervates the two comb rows of its particular quadrant. Tilting the animal causes the statolith to press heavily on the downside balancers and the resulting stimulus elicits a vigorous beating of the corresponding comb rows to right the body. The two comb rows in each quadrant innervated by single ciliated furrow beat synchronously. If a ciliated furrow is cut, the beating of the two corresponding comb rows becomes asynchronous. The normal direction of ciliary power strokes is to the aboral pole, so that the animal is driven forward oral end first. The beat in each row begins at the aboral end of the comb row and proceeds in metachronal waves toward the oral end. Stimulation of the oral end reverses the direction of both wave and power stroke. The removal of apical sense organ or statolith results in an overall lack of coordination of the comb rows and the injured ctenophore loses its ability to maintain a vertical position.

## 6.5 REPRODUCTION

### *Asexual reproduction and regeneration*

Ctenophores can regenerate any lost part, including the apical sense organ. Entire quadrants and even whole halve regenerate. Platyctenids reproduce asexually by a process that resembles pedal laceration in sea anaemones. Small fragments break free as the animal crawls and each piece can regenerate into a complete adult.

***Sexual reproduction and development***

Most ctenophores are hermaphroditic but a few are gonochoristic. The gonads arise on the walls of the meridional canals. Pelagic forms shed their gametes via the mouth into the surrounding sea water, where fertilization takes place. Special sperm ducts occur in some platyctenids. The eggs are centrolecithal and formed in association with nurse cell complexes. Polyspermy is common. Those that spawn freely produce embryos that grow quickly to cydippid larvae. Species belong to the order Beroida lack this larval phase. Development is thus indirect, although growth to the adult is gradual rather than metamorphic. In the *Coeloplana* and *Tjalfiella*, fertilization is internal and embryos are brooded until a cydippid larva is formed and released.

## 6.6 CLASSIFICATION WITH CHARACTERS

1. Eight rows of cilia are used for locomotion.
2. The cilia in each row are arranged to form the comb plates.
3. Members of the order Cydippida are round or oblong, usually less than 3 cm in diameter. They have solid bodies distinguished by the eight radial comb rows for locomotion and two usually branched tentacles. *Pleurobranchia* is the most common.
4. Members of the order Lobata are usually larger than the coastal cydippids, with a pair of highly expandable lobes that are used as sticky surfaces to capture prey, accompanied with the short tentacles that remain inside the lobes. They feed primarily on small planktonic prey.
6. The third group composes the members of the order Beroida. They are elongate or cylindrical and are often more translucent. They may be colourless or light pink or yellowish.
7. Planktonic, carnivores.
8. The accidental introduction via a ship's ballast water of an American ctenophore, *Mnemiopsis leidyi*, into the Black Sea in the early 1980s, caused a full ecosystem fishery collapse in less than 10 years.
9. Life cycle is simple.
10. Except for the Platyctenida, ctenophores complete their entire life cycle in the plankton and they are considered holoplanktonic.
11. Most species are hermaphroditic. They spawn eggs and sperm freely into the sea water, where the sperm find the eggs and the eggs are then fertilized.
12. Fertilized eggs develop through a larval stage. Larva hatches out of the free-floating fertilized egg grows into an adult.
13. Hermaphroditic ctenophore species are probably self-fertile in few species.
14. They reproduce at a young age, whereas most medusae grow to the adult size before beginning to produce eggs or sperm.
15. They produce small numbers of gametes before attaining adult size and may have very rapid generation times.
16. After attaining the adult state, they spawn eggs and sperm daily, since there is sufficient food available in the plankton.
17. The starved ctenophores do not produce the gametes and shrink in the body size.
18. When food becomes available, they grow again to some predetermined size and then attempt for the gamete production.
19. They have a pair of anal pores.
20. They possess a third tissue layer.
21. Although most ctenophores swim, one group creeps along the bottom of the seas.
22. Most species live on the other animals, for instance with the echinoderms, sponges or benthic cnidarians.
23. Many ctenophores like other planktonic organisms are bioluminescent.
24. Their bodies are made up mostly of water.
25. Two fossil ctenophore had been found in the Late Devonian, in the famous Hunsrück-kscheifer slates of the southern Germany. Both of them owe their preservation due to the rapid precipitation of pyrite in the tissues. Both are similar to the living ctenophores of the order Cydippida. Ctenophore-like forms have been found in the Cambrian-age Burgess Shale of the Canadian Rocky Mountains and Chengjiang Formation of southern China.
26. Ctenophores possess eight comb rows of fused cilia arranged along the sides of the animal. These cilia beat synchronously and propel the animal through the water.

27. Some species move with a flapping motion of their lobes or undulations of the body.
28. Many have two long tentacles, but some lack tentacles completely.

***Class Tentaculata***

1. Ctenophores with tentacles.

***Order Cydippidea***

1. Presence of an unbranched meridional and stomodaeal vessel.
2. Body may be circular in section or is slightly compressed in the lateral plane.

**Example:** *Hormiphora*

***Order Cestidea***

1. Band like in appearance.
2. Presence of numerous lateral tentacles.
3. Union or branching of the meridional and stomodaeal vessel occurs.

**Example:** *Cestus*

***Order Lobata***

1. Presence of numerous nonretractile lateral tentacles.
2. Presence of four pointed process or auricles.

**Example:** *Deiopea*

***Order Platyctenea***

1. A pair of retractile lateral tentacle is present.
2. Meridional canal is absent.

**Example:** *Ctenoplana*

***Class Nuda***

1. Presence of the tentacles.

***Order Beroidea***

1. Mouth is very wide.
2. The gullet occupies the major part of the interior of the body

**Example:** *Beroe*

## 6.7 OCEANIC CTENOPHORES

Besides relatively well-known coastal ctenophores, many species are found only far offshore near the surface. Each species is found in only a small portion of the water column, depending on water temperature, light availability, depth and other factors. Many oceanic species found under similar conditions worldwide and tend to be more fragile as they do not tolerate wave action and sediment load of coastal waters.

Open ocean near-surface ctenophores include the *Cestum veneris*. Lesser known species include the *Ocyropsis*. A few oceanic species are large, about a metre in maximum dimension. Most ctenophores live near the surface of the ocean and are colourless, like the transparent *Deiopea kaloktenota*. Oceanic species which inhibit deep sea include *Bathyctena*.

## 6.8 BENTHIC CTENOPHORES

Platyctene ctenophores found in warm water are usually only a few cm in diameter. They are planktonic with comb plates only in larvae. After planktonic larvae settle down, the often brightly-coloured benthic adults appear like nudibranchs than ctenophores. They are often coloured similarly to their substrate organism. A pink starfish may support pink platyctene. These ctenophores possess branched tentacles. The presence of expanded tentacles during feeding is the interesting feature.

## 6.9 BIOLUMINESCENCE

Light-scattering is produced by beating of the eight rows of locomotory cilia. Most ctenophores are bioluminescent, but the light can be visible in darkness. Bioluminescence is also found in many marine animals.

## 6.10 CTENOPHORES: ROLE IN THE FOOD WEB

All ctenophores are carnivores. They feed using a pair of highly extensible, branched, sticky tentacles that adhere to the small zooplankton prey. The mouth is located at one end. The prey of cydippid and lobate ctenophores are the copepods, amphipods, euphausiids and larvae of clams and snails. The principal preys of beroid ctenophores are other ctenophores. Ctenophores are eaten by many medusae and by sea turtles. Tentacles of most cydippid and lobate ctenophores are covered by the colloblasts that help adhere to the prey.

After prey capture, the tentacle with its adhering captured-prey is contracted and brought across the ctenophore's mouth, where it is released as the prey is ingested. *Euplokamis* possess tentacles with the unusual musculature that allows them to coil around the prey items to control captured prey on its way to the mouth. *Haeckelia* carry nematocysts on their tentacles instead of the colloblasts. These nematocysts are sequestered from the medusae that *Haeckelia* feeds on and are then recycled for use in feeding by such ctenophores.

## 6.11 CTENOPHORES AND MEDUSAE

Both ctenophores and medusa were once placed together in the phylum Coelenterata. However, this concept has been replaced. Recent molecular evidence suggests separation in the evolution of the Cnidaria and the Ctenophora. Jellyfish is a term that refers to medusae. The bell-shaped jellies found in coastal waters which swim like an umbrella that opens and closes. They lack stinging cells. To capture prey, they possess sticky cells called *colloblasts*. In a few species, special cilia in the mouth are used for biting gelatinous prey.

## 6.12 *Pleurobrachia pileus*

Ctenophores or comb jellies are delicate, transparent, mostly pelagic, marine and carnivores. They have biradial symmetry, oral-aboral axis of symmetry and triploblastic organization, with a thick cellular mesoglea. Some true organs are present. About 80 species ranges in size from millimetres to over a metre. There is no alternation of generations, no sexual dimorphism and no colonial organization. Cnidocytes are absent and cilia rather than muscles are used for locomotion. The gut is complete, with openings at both ends. The gut is divided into the digestive and distributive regions. The cilia are well-developed and used for various purposes. Locomotion is done by the eight longitudinal rows of paddles. Each paddle is composed of thousands of cilia. Cilia are involved in sensory reception. Some of them function in a mechanical, rather than neuronal or communication system. The connective tissue compartment is a mesoglea which contains several types of cells, especially muscle. *P. pileus* is a small, about 1 cm in diameter. The tentacular plane coincides with the plane of the paper. The pharyngeal plane is perpendicular to it.

### External anatomy

*Pleurobrachia* is a slightly ovoid sphere. This is weakly flattened on the two opposite sides. At the two ends of the ovoid, there are the oral and aboral poles. The mouth is a transverse slit located at the oral pole. Two anal pores are located at the aboral pole. The oral-aboral axis is the axis of symmetry around which is arrayed the biradial symmetry. Two planes, including the axis of symmetry and at right angles to each other, are the planes of symmetry. One tentacular plane passes via the two tentacle sheaths on opposite sides of the body. The pharyngeal plane includes the plane of the flattened pharynx. These two planes divide the jelly into equal, mirror image halves.

### Comb rows

The most unique feature is its eight comb rows, which are arranged like meridians. Each row is composed of several successive plates of large, fused cilia. Each plate is known as a *ctene* which functions as a paddle.

### Tentacles

The two tentacle sheaths are deep ectodermal invaginations which open on the surface of the aboral hemisphere. They run into the interior to end on the opposite sides of the pharyngeal region. The sheaths and tentacles lie in the tentacular plane. A single tentacle arises from a large, opaque tentacle bulb at the bottom of each sheath. The tentacles run from the sheath and trail through the water. The tentacles are branched and the branches are known as *tentilla* which bear colloblasts. Colloblasts are cells reminiscent of cnidocytes and arise from the epidermal interstitial cells. They release mucus. The mucus is discharged on the prey. The tentacles are muscular and can be retracted into sheaths. The completely extended tentacles of *Pleurobrachia* may be up to 25 cm in length.

### Coelenteron

The mouth opens into a large, wide, flat, epidermal pharynx where extra cellular digestion begins. The pharynx is strongly flattened in the pharyngeal plane. The pharynx runs aborally from the mouth and empties into the small stomach near the centre of the animal. The coelenteron including the stomach is endodermal. This is lined with ciliated, secretory, phagocytic gastrodermis. Ciliated endodermal canals arise from the stomach and distribute food to the tissues.

It includes 8 meridional canals, 2 tentacular canals, 2 pharyngeal canals to opposite sides of the pharynx. A pair of transverse canals and an aboral canal runs vertically to open to the exterior via 2 anal pores at the aboral pole. The gut is divided into a central digestive region for extracellular digestion and a peripheral distributive portion. The pharynx is the major extra cellular digestive chamber. The canals are the fluid transport system. The gastrodermis lining the distributive canals is composed of ciliated cells and phagocytic cells. Extra cellular digestion begins in the pharynx with the hydrolytic enzymes. Partially digested food moves into the stomach and from there into the canal system. In the canals, the partially digested material is phagocytized or pinocytosed. Finally, the digestion is completed intracellularly. Undigested material passes through the aboral canal and is voided via one of the two anal pores at the aboral pole. The epithelial walls of the canals include the cells specialized for light, osmoregulation, endocytosis and intracellular digestion, gametocytogenesis and fluid transport.

### *Aboral organ*

The aboral organ containing a calcareous statolith is a centre for gravity detection and controls the comb rows. The statolith rests on four tufts of balancer cilia. Two ciliary grooves run from each balancer, one each to the aboral ends of each pair of comb rows. Tilting the body away from its vertical orientation increases the pressure of the statolith on the balancers. Pressure of the statolith on a balancer changes the beating rate of its cilia. The ctenes of those comb rows beat faster and return the animal to vertical position.

### *Reproduction and development*

Ctenophores are hermaphroditic. Gonads of both sexes are located in the lining of the meridional canals. Gametes are shed to the sea water through numerous gonopores in the comb rows. Fertilization is mostly external. Development is direct without metamorphosis but involves a cydippid larva (Figures 6.1–6.3).

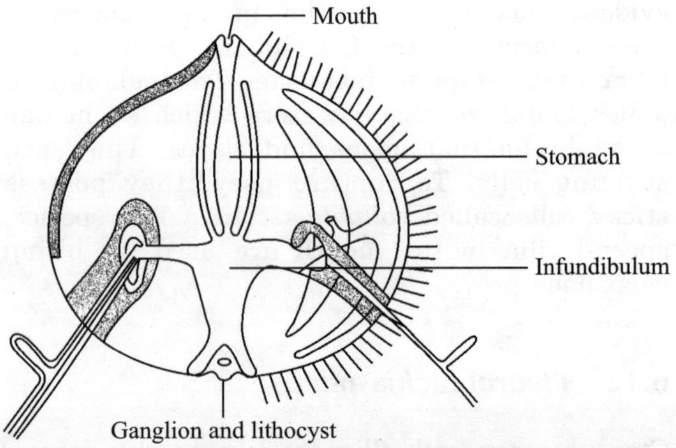


**Figure 6.1** *Pleurobrachia.*

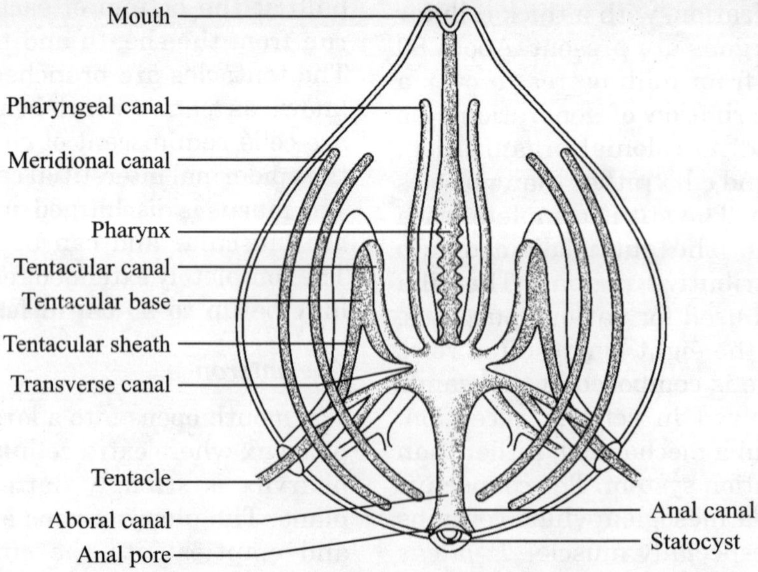


**Figure 6.2** Digestive system of a Ctenophora.

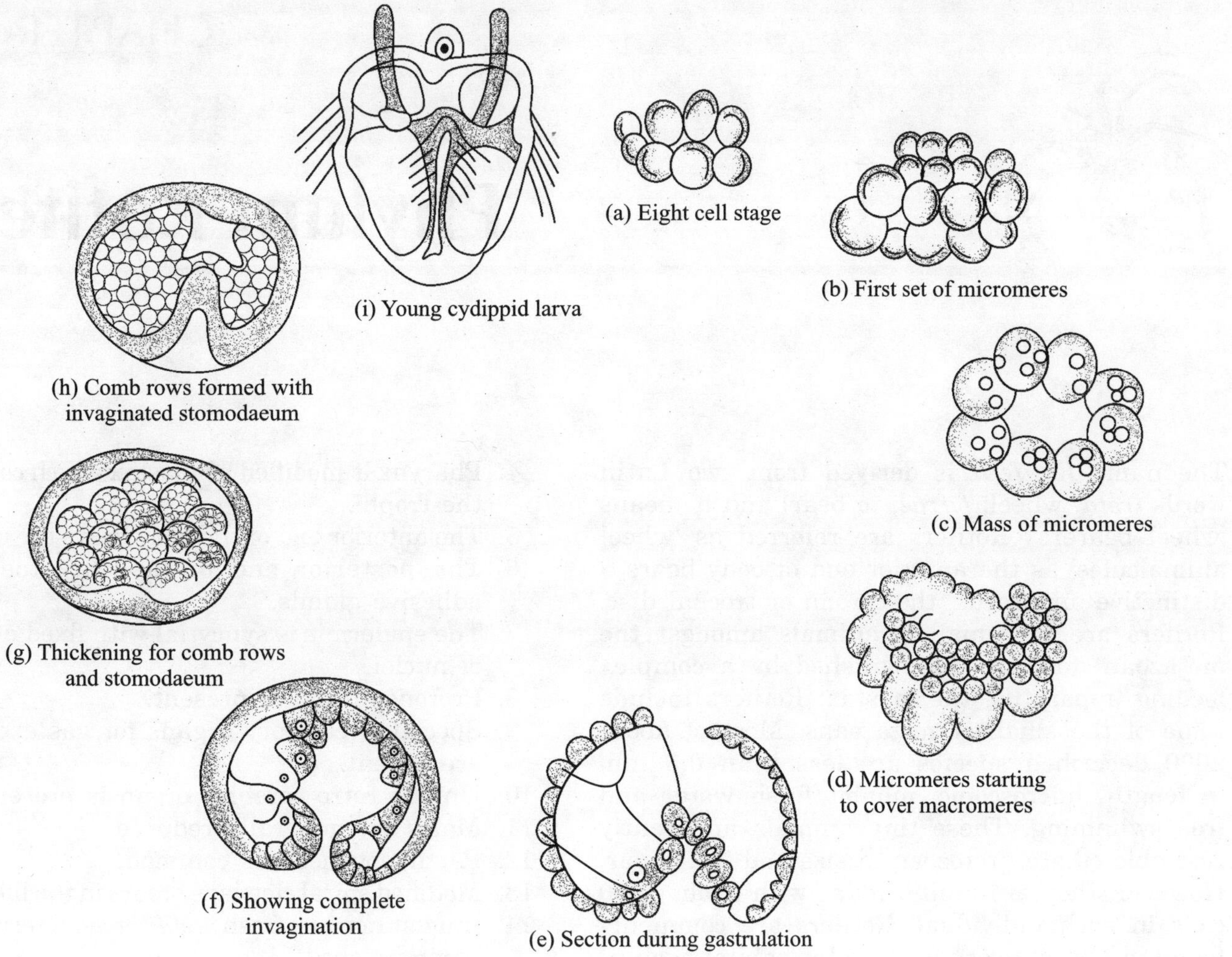


**Figure 6.3** Development in *Ctenophora*.

## REVIEW QUESTIONS

### Short Answer Questions

1. Name the sense organ of *Pleurobrachia*.
2. What is stomodaeum?
3. Comment on per-radial canal.
4. Give two characters of Ctenophora.
5. What is colloblast?
6. Mention the order to which *Beroe* belongs.
7. What do you understand by the term "bioluminescence"?
8. Give one distinguishing character of class Tentaculata.
9. Give one character of class nuda.
10. What is tentilla?
11. What is statolith?
12. What is comb plate?

### Long Answer Question

1. Describe the salient features of *Plerobrachia* with suitable diagrams.

CHAPTER 7

# Phylum Rotifera

The name *Rotifera* is derived from two Latin words (*rota*, wheel; *ferrae*, to bear) and it means 'wheel bearers'. Rotifers are referred as 'wheel animalcules' as the anterior end of body bears a distinctive apparatus, the corona or trochal disc. Rotifers are the smallest animals amongst the metazoan and are distinguished by a complex feeding apparatus, the mastax. Rotifers include some of the smallest metazoans. Most of about 2000 described species are less than 0.5 mm in length, microscopic, inhabit fresh water and free swimming. These tiny animals apparently resemble ciliate protozoan in size and behaviour. However, they are multicellular with about 1000 cells in each individual. Rotifers are commonly found in almost every fresh water habitat such as lakes and rivers besides places that remain moist only periodically, such as moss, soil and leaf litter. They are the most abundant in fresh water benthic habitats. Rotifers inhabit all habitats except polar latitudes. About 50 species are the marine. Two genera are solely marine. Rotifers have a fluid filled body cavity which contains a loose syncytium of amoeboid cells. The body cavity derives from the blastocoel of the embryo. Mesothelium is absent. They are eutelic. Individuals of each species have a characteristic and constant number of cells. Tissues are often syncytial. Growth occurs through increase in the cell size, rather than through addition of the new cells.

## 7.1 CHARACTERS OF ROTIFERS

1. Triploblastic and bilateral.
2. Unsegmented blastocoelomates.
3. Gut is complete and regionally specialized.
4. Pharynx is modified as mastax which contains the trophi.
5. The anterior end of the body bears the corona.
6. The posterior end often bears toes and adhesive glands.
7. The epidermis is syncytial with fixed number of nuclei.
8. Protonephridia is present.
9. Special circulatory organs for gas exchange are absent.
10. Unique retro cerebral organ is present.
11. Males are generally reduced.
12. Parthenogenesis is common.
13. Modified spiral cleavage occurs in the life cycle.
14. Inhabit marine, fresh water or semi terrestrial environments.
15. The contractile, wormlike body is elongated and composed of 16 rings.
16. Bdelloid rotifers can undergo cryptobiosis, withstand extended periods of desiccation and return to the metabolic activity when wetted.
17. This ability makes them to inhabit bryophytes (mosses) in terrestrial habitats that are only periodically wet.
18. During wet conditions, the rotifers inhabit the thin film of water on the surface of the moss. Here, they feed and complete the parthenogenetic life cycle.
19. The well-developed foot and toes maintain contact with the moss substratum. In dry weather, the film of water dries out and the rotifers start the process of forming the desiccation- resistant tun. This is not a cyst as there is no secreted covering. The tun is the dried rotifer enclosed in its wrinkled and withered integument.

20. The formation of the tun is not rapid. Rotifers require the presence of interstices and crevices in which trapped water evaporates more slowly than from the surface of the leaves.
21. Rotifers artificially deprived of these interstices will not survive the drying process. In wet conditions, the rotifer absorbs water and resumes its normal morphology, metabolism and activity.
22. Rehydration occurs in as little as 10 minutes or may take up to a day.
23. No special respiratory or haemal systems are present.
24. Gas exchange is done over the unelaborated body surface.
25. Fluid transport occurs by diffusion and movement of the blood in the haemocoel.
26. Rotifers are covered by an intraepidermal lorica. This lorica may be thick or thin.
27. In bdelloids, the lorica remains in the rings.
28. The trunk is the thick central region of the body.
29. The head and the foot are narrow.
30. The ciliated corona is located at the anterior end of the head.
31. The corona is used for swimming through the water and for creating a feeding current into the mouth.

## 7.2 HABITAT

Rotifers inhabit all kinds of water. Most species are limited to fresh or slightly brackish water. Some species have distinct ranges of tolerance for certain environmental variables and therefore occur in special habitats. Rotifers mostly live in the small water bodies or in the littoral zone of the larger water bodies. Parasitic rotifers live inside or on the outer surface of the other animal. Few rotifers are the marine. Some inhabit mosses or lichens. Rotifers are ecologically divided into three groups as follows:

1. Alkaline fauna, those found in hard water.
2. Acidic fauna, those found in soft water.
3. Transcursion fauna, the dormant eggs of such fauna can tolerate both the alkaline and acidic conditions. Bdelloids inhabit fresh water, mostly in intermittently moist terrestrial mosses and soil. Some are benthic in the aquatic habitats. Species inhabiting the terrestrial habitats are tolerant of desiccation.

## 7.3 SHAPE, SIZE AND COLOURATION

1. They are small and aquatic.
2. Many are not larger than protozoan.
3. They are elongated, typically may be slender and worm-likes (*Rotaria*), broad (*Polychaetus*), flattened sacciform (*Asplanchna*) or even spherical (*Trochosphera*).
4. Usually, they are transparent and colourless.
5. Brown red and orange coloured rotifers are seen in alpine and polar regions.

## 7.4 EXTERNAL ANATOMY

1. The body surface is annulated allowing flexibility.
2. They often bear spines, tubercles, other sculpturing and may have thickened casing or lorica.
3. Rotifers bear single dorsal- and paired lateral-sensory antennae.
4. The foot is elongated with cuticular annuli.
5. It often bears one to four projections called *toes* at the distal end through which the ducts from pedal glands pass.
6. Foot is narrower than trunk region.
7. End of foot usually bears *toes*.
8. In both crawling and sessile forms, foot contains pedal glands that open by ducts at tips of toes, or other parts of foot.
9. The posterior end of bdelloid rotifers bear small toes connected with pedal glands in the foot.
10. The glands secrete an adhesive which the rotifer uses to attach to the substratum.
11. Spurs look like toes are dorsal and relatively large.
12. In planktonic rotifer, foot is usually reduced or disappeared altogether (Figures 7.1–7.5).
13. The secretion of the pedal glands attaches them temporarily to the substratum.
14. Foot is absent in some swimming forms. This is modified for permanent attachment in sessile types.
15. The corona is the remarkable external feature with variable morphology.

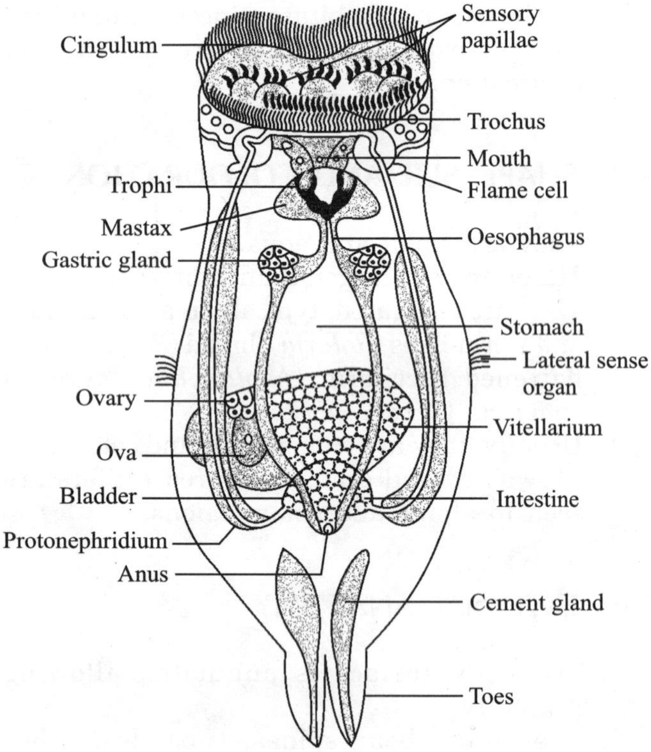


**Figure 7.1** *Epiphanes.*

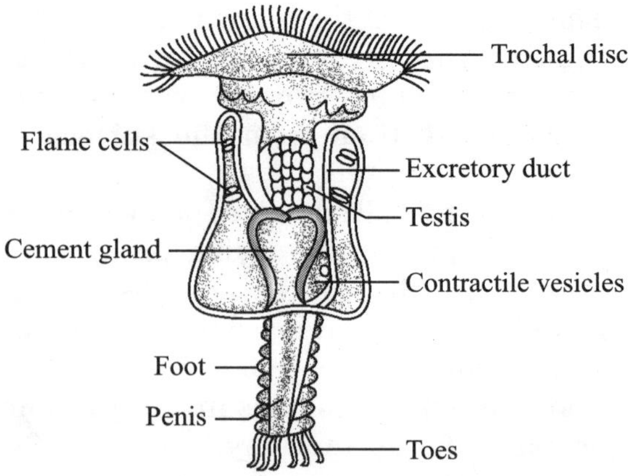


**Figure 7.3** *Brachionus* male in dorsal view.

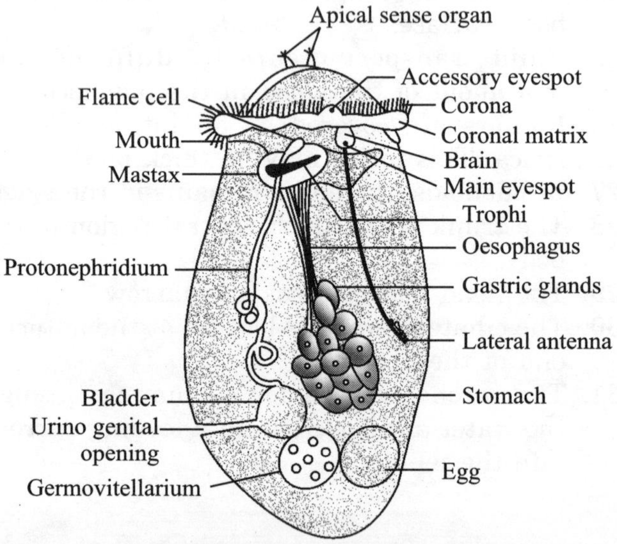


**Figure 7.4** *Asplanchna.*

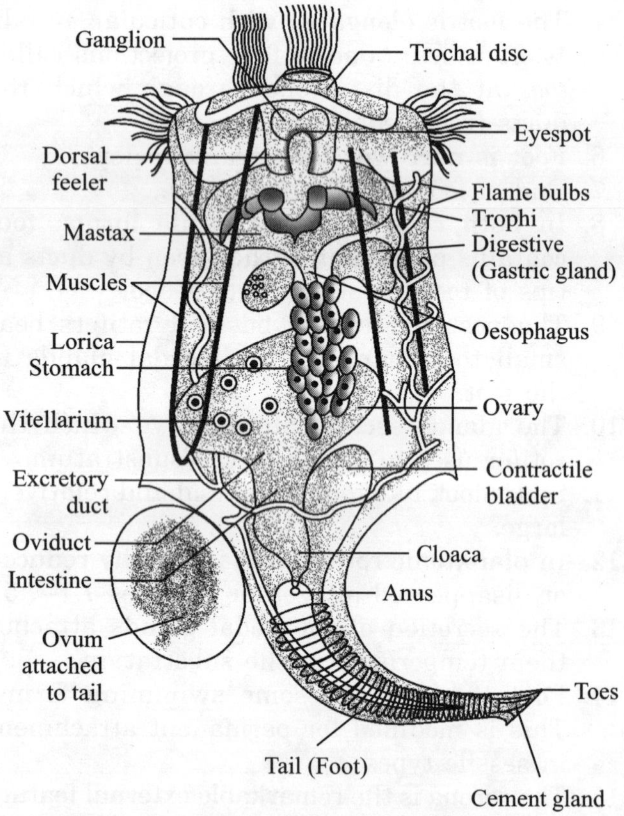


**Figure 7.2** *Brachionus* female in dorsal view.

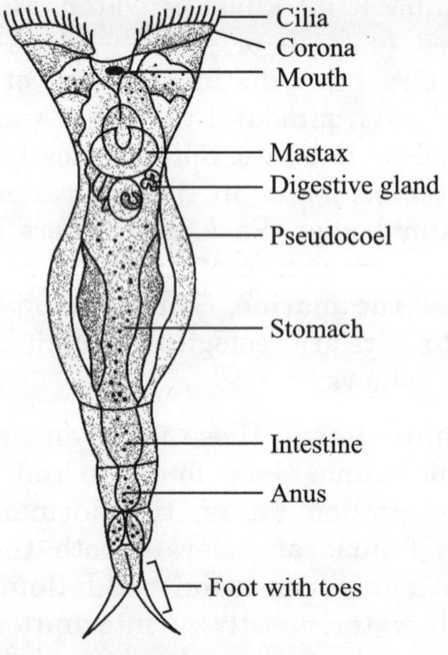


**Figure 7.5** *Philodina.*

16. Well-developed patch of cilia constituting the buccal field or circumoral field that surrounds the mouth.
17. The buccal field extends dorsally around the head as a ciliary ring called the circumapical field.
18. The extreme anterior part of the head bordered by this ciliary ring is the apical field.
19. In the familiar corona, the buccal field is reduced and the circumapical field is separated into two ciliary rings, one slightly anterior to the other.
20. The anterior most ring is trochus, the other is called *cingulum*.
21. The sessile or free-swimming trochus is a pair of well-defined anterolateral ring of cilia which are called *trochal disc*.
22. Trochal discs retract or extend for locomotion and feeding.
23. Metachronal ciliary waves along these trochal discs impart the impression of rotating wheels.
24. Many organs and tissues display eutely.
25. Constancy is also observed in cell or nuclear number.

## 7.5 BODY WALL, BODY CAVITY, SUPPORT AND LOCOMOTION

1. Mostly possess a gelatinous cuticle outside the syncytial epidermis.
2. The skeletal lamina of the epidermis produces the lorica and other surface structures.
3. Beneath the epidermis, various circular and longitudinal muscle bands are found.
4. Sheets or layers of body wall muscles are absent.
5. The internal organs lie within a blastocoelom.
6. Body support and shape are also maintained by the intro-epidermal skeletal lamina and the hydrostatic skeleton.
7. Often, the cuticle is flexible to allow slight changes in shape. Thus, increases in hydrostatic pressure in the body cavity are used to protrude body parts.
8. Body parts protract and retract by various muscles.
9. Swimming is performed by beating the coronal cilia, forcing water posteriorly along the body and moving the animal forward.
10. During creeping movement, a rotifer attaches its foot with secretions of the pedal glands. It then elongates its body and extends forward.
11. It attaches the extended anterior end to the substratum, releases its foot and draws its body forward by muscular contraction.

## 7.6 CORONA

1. Corona or wheel organ, the most striking feature, is an elongated ciliated area around mouth in its simplest development.
2. It varies greatly within the group in relation to swimming and feeding.
3. A large ventral ciliated area called *buccal field* surrounds the primitive corona.
4. From buccal field, cilia extend around anterior margin of head to form a crown like ring called the *circumapical band*.
5. Area inside the ring without cilia is called *apical field*.
6. Corona generates the feeding and locomotory currents.
7. Corona shows variation in structure as follows:
   - (i) In a common corona, cilia are arranged as two discs. Two discs of cilia beat in a circular manner, one clockwise and one counter clockwise and look like two wheels spinning, from which name rotifer is derived.
   - (ii) In *Euchlanis*, some cilia are modified to cirri or bristles.
   - (iii) The buccal field in *Collotheca* and related forms is modified into a funnel with mouth at bottom with reduced ciliation
   - (iv) In *Polyarthra* and related forms, corona is divided entirely to form circumapical band, which is transformed into two circlets of modified cilia—an anterior band called *trochus* and a posterior band called *cingulum*, which passes below the mouth. The head of bdelloid and notommatid rotifers carries a mid-dorsal projection called the *rostrum*. This little projection bears the cilia and sensory bristles at its tip.
   - (v) Corona is absent in adult females of *Atrochus, Cupelopagis* and *Acyclus* but present in males and young females.

Corona of male usually differs from that of females of same species, being usually less modified. Anterior structures include the eyes, one or two short antennae and mucus secreting retrocerebral organ. Lorica may be divided into distinct plates and is usually ornamented with ridges and spines. Spines may be long and moveable.

## 7.7 BODY WALL AND PSEUDOCOEL

1. The body wall consists of a syncytial epidermis, containing an intrasyncytial lamina, known as the *lorica* and underlying by individual circular and longitudinal muscles.
2. Distinct cuticle may be ringed, sculptured or ornamented in various ways.
3. Cuticle often forms an encasement called *lorica* in trunk
4. Cuticle is intracellular. It is part of epidermis which is thin, syncytial and possesses a constant number of nuclei.
5. Body muscles lies beneath the epidermis.
6. Muscle fibres are either circular or longitudinal but musculature is not organized into distinct circular and longitudinal sheaths.
7. Pseudocoel lies beneath body wall and surrounds the gut and internal organs.
8. Pseudocoel is filled with fluid and a syncytial network of branching amoeboid cells (Figure 7.6).

## 7.8 DIGESTIVE SYSTEM

1. The pharynx contains a unique mastax composed of several cuticular jaws which are used for manipulating food.
2. The mastax can be found near the junction of the head with the body.
3. The gut is usually complete and the muscular pharynx includes an elaborate, cuticular mastax with up to seven hard jaws or trophi.
4. Some part surrounds the ventral mouth of corona. In suspension feeders, a ciliated buccal tube may be present between the mouth and pharynx.
5. At the lower end of pharynx, there is an enlargement, the mastax or gizzard. Following account of mastax is based on Harmen and Shipley (1952).
6. Mastax is a strong muscular sac containing trophi or hard chitinous chewing organs with an antero-ventral inlet from pharynx and a postero-dorsal outlet through which food passes into stomach either directly or through a slender gullet.
7. In ventral wall of gizzard of most *Ploima*, there is a median piece, called the *fulcrum*, from which two pieces of rami run forwards and upwards and are hinged on fulcrum. Y-shaped structure formed by these there pieces is called *incus*.
8. At either side of gizzard and at a higher level, there is a paired piece, the malleus, so called for its resemblance to a hammer, in

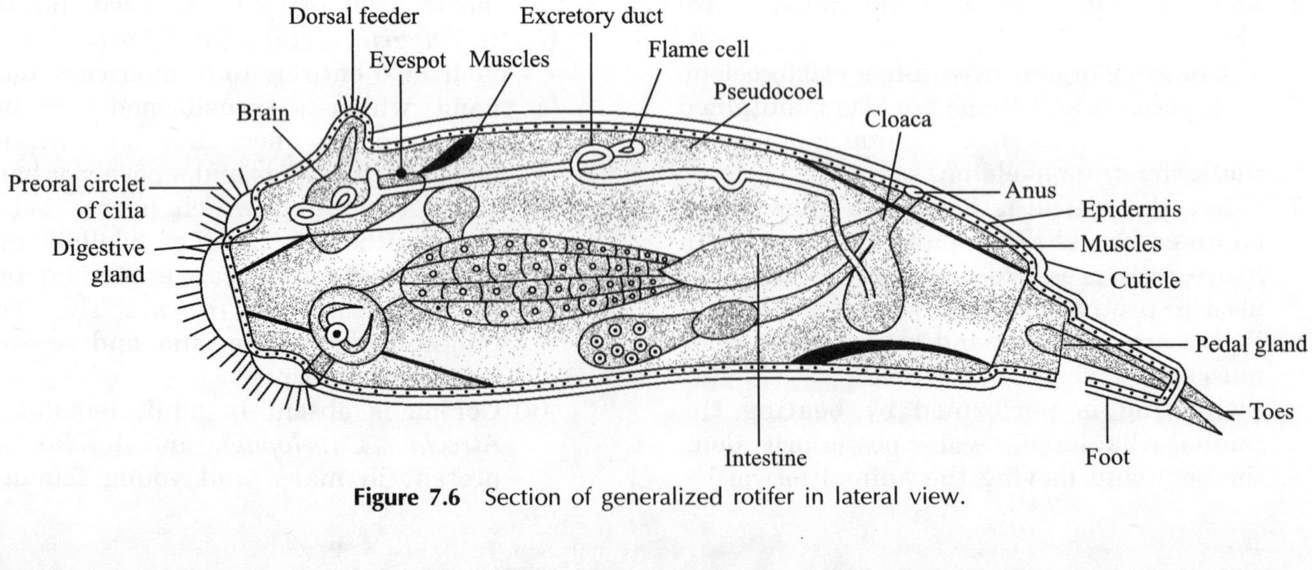


**Figure 7.6** Section of generalized rotifer in lateral view.

which manubrium (handle) runs backwards and is embedded in side walls of mastax, while toothed claw or uncus runs forwards and inwards and is hinged at its inner side with tip of ramus.

9. As unci and rami are strongly toothed, the gizzard forms a very efficient chewing apparatus.

   (i) In malleate type mastax, pharynx is short, dilatable. Junction point of unci and rami may be protruded for biting, or for clinging to host (in *Albertia, Drilophagus*) or for pretension of food (Rattulidae). All bdelloids have a malleate mastax with two large plate like unci that grind food particles.
   (ii) In virgate type mastax, trophies are slender, scarcely toothed and frequently asymmetrical. In submalleate type mastax, only mallei are slender.
   (iii) In forcipate type mastax, both unci are slender and sharply pointed.
   (iv) In malleoramate type mastax, manubrium is a looped structure, while uncus is formed of several parallel, slender elongated teeth. This characterizes the family Melicertidae and genera *Triarthra, Ptcrodina* and *Pedalion.*
   (v) In unicinate mastax, mallei are in curved hooks with a few teeth at free end. Rami are simple or absent and there is no fulcrum as occurs in Flosculariaceae only.
   (vi) In Asplanchnidae, rami are large and hooked and constitute incudate mastax, but reduced rami may be present.
   (vii) In *Asplanchnopus*, they are almost as well-developed as in Melicertidae, affording a transition to malleormate type. Here, mastax is very peculiar being divided into dorsal and ventral chambers. Dorsal chamber forms a great purse-like sac or crop with a frame work of four longitudinal bags into the gullet and open pharynx. Ventral pouch is smaller and in its base large rami are inserted. Therefore, they can be protruded into the crop. This ventral sac with rami may be everted through crop and mouth to swallow small rotifers and Entomostraca which from food of these groups, or to eject undigested remains of food. Two lateral sacs open at the junction of the ventral pouch and crop, but their role in deglutition of food is uncertain. Trophi in bdelloids are represented by rami in form of a sphere, excavated on curved sides and transversely ridged on two flat sides for attachment of muscles. This gizzard is called *ramate*. Glandular masses called *salivary glands* are located in the mastax walls of most rotifers. There is a pair of gastric glands at the junction of oesophagus and stomach. Stomach is a large sac or tube that passes into a short intestine. Excretory organs and oviduct open in the terminal end of intestine, sometimes called *cloaca*. Anus opens to the dorsal surface near the posterior trunk (Figures 7.7–7.10).
   (viii) The mastax in trapping forms is often reduced.
   (ix) The ectodermal mastax is lined with cuticle.

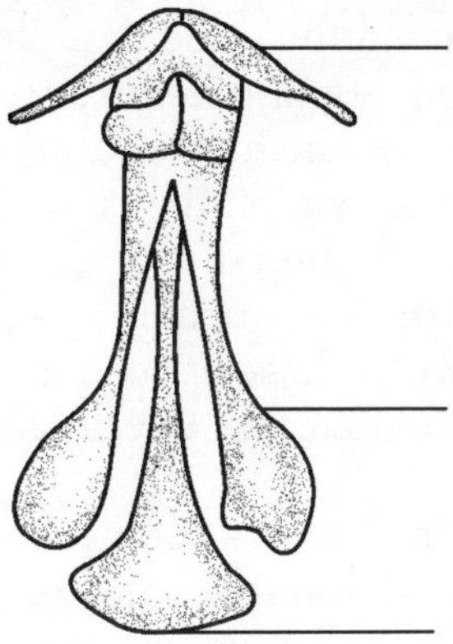


**Figure 7.7** Fulcrate type of mastax.

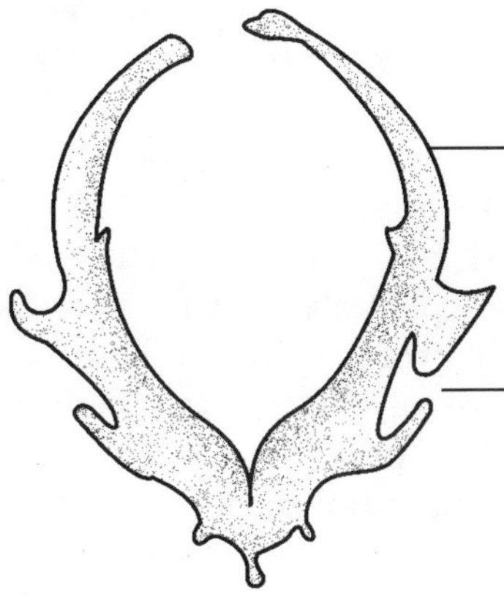


**Figure 7.8** Incaudate type of mastax.

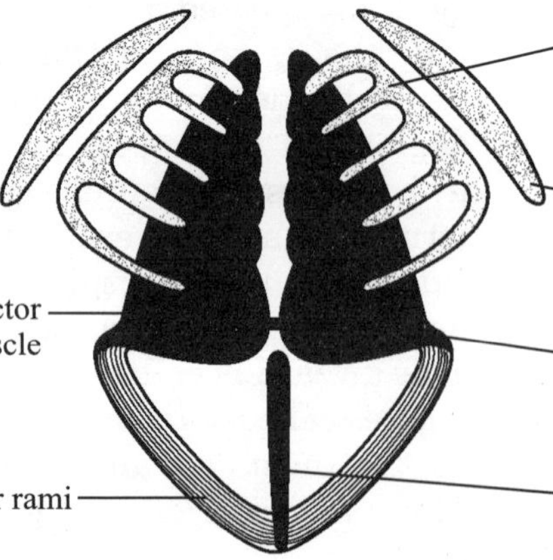


Figure 7.9 Malleate type of mastax.

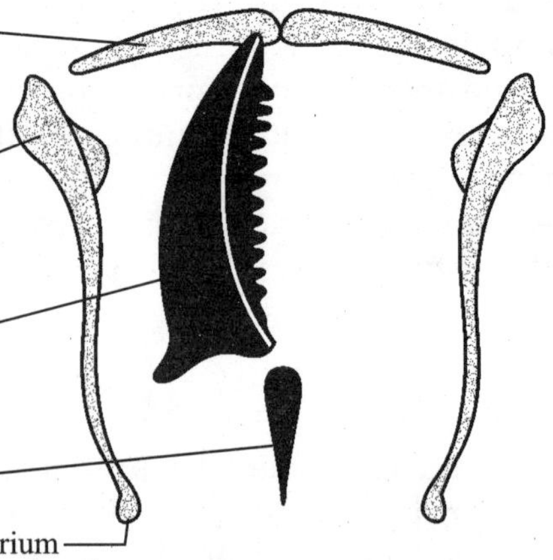


Figure 7.10 Forcipate type of mastax.

## 7.9 FEEDING AND DIGESTION

1. Ciliary suspension feeders possess well-developed coronal ciliature and a grinding mastax. These forms include the bdelloids with trochal discs and a ramate mastax. Some monogonontan rotifers have separate trochus and cingulum, and a malleate mastax which feed on organic detritus.
2. The feeding current is produced by the cilia of the trochus, which beat in a direction opposite that of the cilia of the cingulum.
3. Particles are drawn into a ciliated food groove that lies between these opposing ciliary bands and are carried to the buccal field and the mouth.
4. Raptorial feeding is common in many Monogononta. Coronal ciliation in such rotifers is reduced or used solely for locomotion. Raptorial feeders obtain food by grasping it with protrusible, mastax jaws; most possess either a forcipate mastax (nonrotating) or incudate mastax (rotating 90–180 degrees during protrusion)
5. Raptorial rotifers feed mainly on small animals and ingest plant materials as well. They ingest their prey in whole and grind it to smaller particles within the mastax, or they may pierce the plant body or animal with the tips of the mastax jaws and suck fluid from the prey.
6. Some monogonontan rotifers adopt a trapping method of predation in which the spines or setae are arranged as a funnel-shaped trap in the corona. The mouth in such trappers is located in the middle of the ring of spines, thus, captured prey are drawn to it by contraction of the trap elements.
7. Few rotifers lead symbiotic lifestyles.
8. Seisonids live on marine *Nebalia*. Seison crawl around the base of the legs and gills of their host, feeding on detritus and on the host's brooded eggs.
9. Some bdelloids live on the crustacean gills.
10. The digestive tract of most rotifers is complete and more or less straight.
11. The mouth leads inward to the pharynx (mastax) either directly or via a short, ciliated buccal tube.
12. Depending on the feeding method and food sources, swallowing is accomplished by various means, including ciliary action of the buccal field and buccal tube, or a piston-like pumping action of certain elements of the mastax apparatus.
13. Just posterior to the mastax, there are ducts of the salivary glands.
14. There are usually two to seven such glands. They are believed to produce digestive enzymes and perhaps lubricants.
15. A short oesophagus connects the mastax with the stomach.
16. The paired gastric glands open into the posterior end of the oesophagus.
17. The walls of the oesophagus and gastric glands are generally syncytial.
18. The thick-walled stomach may be cellular or syncytial, usually comprising a specific number of cells or nuclei in each species.
19. The short intestine leads to the anus, which is

located dorsally near the posterior extremity of the trunk.

20. Except for *Asplanchna* which lacks hindgut, the expanded cloaca joins the intestine and anus.
21. Oviduct and the nephridioducts empty into the cloaca.
22. Digestion probably begins in the lumen of mastax and is completed in the stomach extracellularly.
23. Absorption also occurs in the stomach.

## 7.10 CIRCULATION, GAS EXCHANGE AND EXCRETION

1. Special organ for internal transport between tissues and the environment is absent.
2. The blastocoelomic fluid serves for circulation, which is aided by general movement and muscular activities.
3. Small body size reduces the diffusion distances, facilitates the transport and exchange of gases, nutrients and wastes.
4. Gas exchange probably occurs over the general body surface where the cuticle is sufficiently thin. Most rotifers possess one pair of flame bulb.
5. The protonephridia is located far forward in the body.
6. A nephridioduct from each flame bulb leads to a collecting bladder, which empties into the cloaca via a ventral pore.
7. In the bdelloids, the ducts open directly into the cloaca, which is transformed to act as a bladder. The protonephridial system is primarily osmoregulatory.
8. The excess water from the body cavity and probably from digestion is pumped out via the anus with the help of muscular contractions of the bladder. Thus urine is significantly hypotonic relative to the body fluids.
9. Protonephridia eliminates nitrogenous excretory products from the body.
10. Some rotifers withstand extreme environmental stresses by entering a state of metabolic dormancy. They remain in a dormant condition for while four years-reviving on the addition of water. Some have survived freezing experimentally in liquid helium at –272°C and other severe stresses.
11. Typically two protonephridia are present in pseudocoel.
12. Protonephridia is of flame cell type, but nucleus lies at some distance from terminal ends of branched tubules. Two collecting tubules empty into a bladder which opens into ventral side of cloaca. Contents of bladder are emptied through anus by contraction.
13. The U-shaped metanephridia lined by ciliated epithelium act as gonoducts.
14. The metanephridia opens to the exterior through nephridiopore on side of the anus.

## 7.11 NERVOUS SYSTEM AND SENSE ORGANS

1. A nerve layer consists of nerve fibres and nerve cells are differentiated into nerve ring at anterior end.
2. Lateral nerve innervates the trunk musculature. Neurosensory cells are present dorsally.
3. The cerebral ganglion is located dorsal to the mastax in the neck region.
4. Several nerve tracts arise from the cerebral ganglion, some of which bear additional small ganglionic swellings.
5. Two major longitudinal nerves positioned either ventrolaterally or one dorsally and one ventrally.
6. The corona bears touch sensitive bristles and often a pair of ciliated pits which are thought to be chemoreceptor.
7. The dorsal and lateral antennae are probably tactile.
8. Some rotifers bear sensory flosculi, which are arranged as cluster of micro papillae Flosculi may be tactile or chemosensory.
9. Errant rotifers possess at least one simple ocellus in the cerebral ganglion. This cerebral ocellus may be accompanied with one or two pairs of lateral ocelli on the coronal surface and sometimes by a pair of apical ocelli in the apical field.
10. The lateral and apical ocelli are multicellular epidermal patches of photosensitive cells. Associated with the cerebral ganglion, there is retrocerebral organ. This glandular structure causes ducts that lead to the body surface in the apical field. It may secrete mucus to aid in crawling.

## 7.12 REPRODUCTIVE SYSTEM

1. Rotifers are gonochoric; depending on taxon, males are rare, unknown, or produced occasionally. Reproduction is usually parthenogenetic but sometimes sexual in monogonont species with both sexes.
2. Males when present is smaller than females, short-lived, aberrant or degenerate and usually have no functional gut and do not feed.
3. Females have a gonad consisting of a combined germarium (ovary) and vitellarium (yolk gland).
4. Most rotifers are oviparous. Few planktonic species are viviparous. In mature individual, two large germovitellaria, or combined ovaries and yolk producing organs, on the right and left sides of the trunk are found. Each may be dominated by a large yolky egg.
5. A single sac like testis and ciliated sperm duct are present.
6. Sperm duct runs directly to a gonopore.
7. Parthenogenesis is probably the common method of reproduction.
8. The male reproductive system includes single testis, a sperm duct and a posterior gonopore whose wall is usually folded to form a copulatory organ.
9. Prostatic glands may be found in the wall of sperm duct.
10. The short-lived males possess a reduced gut unconnected to the reproductive tract.
11. The female system includes paired (Bdelloidea) or single (Monogononta) syncytial germovitellaria. Eggs as a rule are formed in the ovary.
12. Eggs receive yolk directly from the vitellarium before passing along the oviduct to the cloaca. In species without intestinal portion of the gut, the oviduct passes directly to the outside via a gonopore.
13. Copulation may take place by insertion of the male copulatory organ into the cloacal area of the female
14. Sometimes the males attach to females at various points of the body and are believed to inject sperm directly into the blastocoelom.
15. The sperm find their way to the female reproductive tract for fertilization.
16. The number of eggs produced by an individual female is usually 20 or less.
17. Once fertilized, the ova produce a series of encapsulating membranes. They are then getting attached to the substratum or carried either externally or internally by the brooding female.
18. Parthenogenesis is clearly the rule among the bdelloids. However, it is also common and usually occurs seasonally in the monogonontans, where it tends to alternate with sexual reproduction.
19. During favourable conditions, females reproduce parthenogenetically through production of mitotically derived diploid ova also called *amictic ova*. These eggs develop into more females without fertilization.
20. When ova from amictic females are exposed to particular environmental conditions (so-called *mixis stimuli*), they develop into the mictic females, which then produce mictic (haploid) ova by the process of meiosis.
21. The exact stimulus may change in day length, temperature, food resources or increases in population density. These cycles are commonly termed *summer cycles* and *autumn cycles*.
22. However, this is misleading because mixes can also occur during warm weather and many populations have several periods of mixis each year.
23. The resting zygote is extremely resistant to low temperatures, desiccation and other environmental extremities. In favourable conditions, the zygotes develop and hatch as amictic females. The development of rotifers involves modified spiral cleavage.
24. The isolecithal ova undergo unequal holoblastic early cleavage for producing stereoblastula. Gastrulation occurs by epiboly of the presumptive ectoderm and the involution of the entoderm and mesoderm.
25. The gastrula gradually hollows to produce the blastocoel. The mouth appears in the area of the blastopore.
26. The *larva* eventually settles and attaches to the substratum. Always, cell division during postembryonic life does not occur.
27. Many rotifers exhibit developmental polymorphism. It is the expression of alternative morph types under various ecological conditions. The alternative adult morph types are perhaps the products of flexible developmental pathways,

triggered by environmental cues and often mediated by internal factors like hormonal activities.

28. The specific molecular form of vitamin E-alpha tocopherol constitutes such a stimulus. *Asplanchna* obtains tocopherol from its food source such as algae or other plant material. The chemical acts directly on the developing tissues of rotifers and stimulates the differential growth of the syncytial hypodermis after ceasation of cell division. Predator-induced morphologies are also found among the rotifers.

## Conclusion

Rotifers have drawn attention for more than 3 hundred years. Well studied rotifer is *Brachionus*, commonly called the white mouse. In many rotifers, the corona is developed as two concentric rings of cilia namely trochus and cingulum. Another feature is a mastax with trophi. Trophi with 7 articulating pieces process food in a number of ways. Sexual reproduction in monogononts (mixis) occurs due to specific environment signals. Mixis produce mictic females which produce haploid eggs or, if fertilized, diploid embryos (resting eggs or cysts). Resting eggs undergo obligatory diapause, ultimately hatch as amictic females. The ecological and reproductive roles of amictic and mictic females vary. Females in species like *Asplanchna*, *Conochilus*, and *Sinantherina* produce diploid and haploid eggs. These amphoteric females produce males and resting eggs, or females and resting eggs. Pseudosexual eggs are known in *Keratella*. In culture, *Synchaeta pectinata* produces diapausing amictic eggs and non-diapausing eggs. Rotifers are eaten by invertebrate predators and fish fry, transferring their energy available to higher trophic levels. Species abundance can differ remarkably from year to year.

Rotifers are used in pollution monitoring as bioindicator or as part of saprobic assessment system. The latter aims to integrate a number of abiotic and biotic features into a single variable to assess eutrophication. Rotifers are becoming important in ecotoxicological testing. With about 25% of the total number of cells in body, their nervous system is simple, consisting of a concentration of ganglia at the brain, some additional ganglia serving the mastax, body, and foot, paired ventral neurons, and mechano-, chemo-, photo-receptors. Many species has a retro cerebral organ, comprising paired sub cerebral glands, an unpaired retro cerebral sac, and ducts leading to coronal surface. Neuro-pharmacological studies show that cholinergic systems function in at least 12 species and that adrenergic and catecholaminergic systems are reported in other species including both bdelloids and monogononts, Such studies indicate that distribution of neurons in brains is bilateral with distinct species-specific patterns. Bipolar neurons are typical, uni- and multipolar neurons are found in small numbers, and catecholaminergic neurons comprise ≈5% of the total number of brain cells. Spine formation is probably a defense mechanism to reduce predation by tactile predators.

# REVIEW QUESTIONS

## Short Answer Questions

1. Mention the significance of the term *rotifera*.
2. Mention two examples of rotifers.
3. What is corona?
4. Distinguish between trochus and cingulum.
5. What is buccal field?
6. What is caudovascular ganglion?
7. Comment on the pedal glands of rotifers.
8. What is fulcrum?

## Long Answer Questions

1. Describe various types of mastax of rotifers with suitable diagrams.
2. Describe the anatomical peculiarities of rotifers in brief.

CHAPTER 8

# Phylum Phoronida

Phoronida is a small taxon of 14 worm-shaped lophophorates. Phoronids, the members of the Phylum Phoronida, are tubicolous, hermaphrodite, vermiform, coelomate bilaleria with a horseshoe-shaped (usually) or circular lophophore, a dorsal anus and one pair of metanephridia. They are benthic, infaunal or epifaunal, suspension feeders.

In favourable conditions, their populations densities may be high. They secrete and inhabit a stiff chitinous tube. Sand, shell fragments, debris, sponge spicules and other particles are known to adhere to the chitinous tube. The body consists mainly of two regions—the mesosome and the metasome—each containing a coelomic space. The mesosome with its mesocoel is large and is the region of the lophophore. It surrounds the mouth as a coelomic ring with extensions into the lophophoral tentacles. Most of the body is the metasome and its cavity is the metacoel or the trunk coelom. The metacoel is partitioned by four longitudinal mesenteries. The small anterior epistome overhangs the mouth. The long gut is complete and runs from the anterior, terminal mouth to the sub terminal anterior anus. The anus is found on the dorsal midline near the anterior end and just above the mouth. The closed haemal system contains haemoglobin in corpuscles. The lophophore serve as the major respiratory structure. Two metanephridia open through the nephridiopores beside the anus. The nervous system includes a nerve ring and a longitudinal nerve cord on the left side. Phoronids may be gonochoric. However, most are hermaphroditic. Fertilization is internal. The larva is an actinotroch which is a type of trochophore. Phoronids undergo asexual reproduction which causes aggregations of individuals. The individuals in the aggregation get entangled with each other (Figure 8.1).

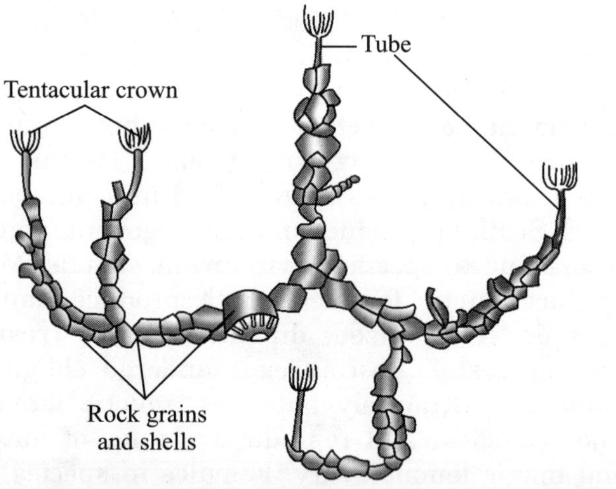


**Figure 8.1** Aggregation of tubes of *Phoronis prammaphila.*

## 8.1 HABIT AND HABITAT

They are exclusively marine, sedentary, benthic and worldwide in distribution. They inhabit shallow waters of tropical and temperate zones, but absent in polar and sub polar water. They live (only loosely fastened) inside chitinous, leathery tube secreted by them during the adult stage. Solitary, cylindrical and vertical tubes are generally embedded in the sandy or muddy bottom.

## 8.2 DISTINCTIVE CHARACTERS

1. Small, slender, solitary, tubicolous and excusively marine.
2. Unsegmented and coelomate.
3. Body vermiform and cylindrical.

4. The anterior end is provided with usually spirally coiled lophophore.
5. Presence of two rows of the ciliated tentacles.
6. The mouth is located inside the lophophore.
7. The alimentary canal is U-shaped.
8. The RBCs are rounded with nuclei.
9. The closed type of circulatory system is present.
10. The nervous system is epidermal.
11. Presence of a circum-oesophageal ring.
11. Hermaphroditic.
12. The life cycle involves a free-swimming actinotroch larva.

## 8.3 EXTERNAL ANATOMY

Body is differentiated into an anterior lophophore and a posterior trunk.

### *Lophophore*

The horseshoe shaped tentacular crown with numerous, slender, hollow, ciliated, nonretractile tentacles is present at the anterior oral end. The mesodermal exoskeleton supports the tentacles. The tentacles are arranged in two rows. The inner row is incomplete in the middle line. The tentacles of each row are basally fused to form a membrane. The lophophore consists of the outer and inner ridges between which is a groove leading into the median mouth. Over hanging the mouth, there is a broad lobe called the upper lip or epistome. A short space separates the mouth from the anus and is situated at the tentacular extremity. Two ciliated nephridial tubes open near the anus [Figures 8.2(a) and 8.2(b)].

### *Trunk*

This is narrow, slender, cylindrical and without appendages. Trunk is demarcated from the lophophore by a slight grove. The trunk is uniform in diameter throughout except the posterior end, which enlarges to form end bulb. The trunk shows slight annulations.

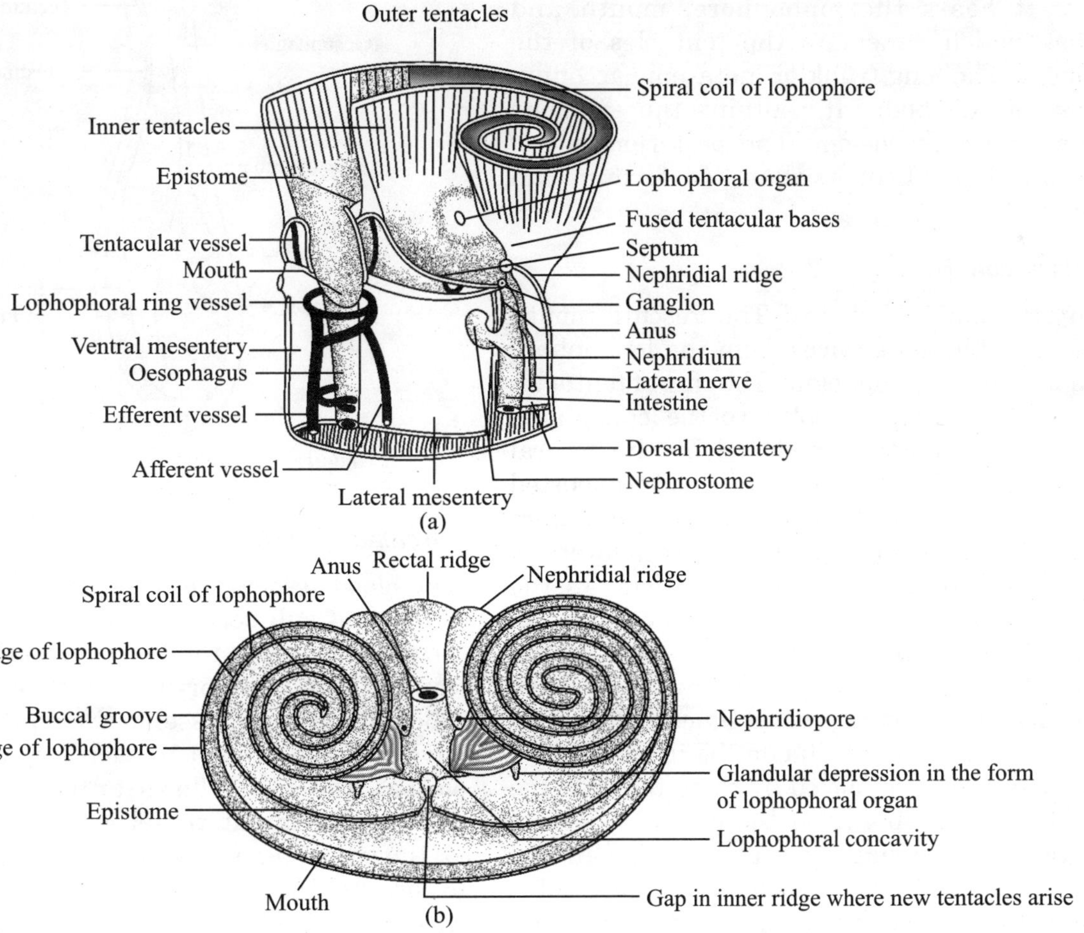


**Figure 8.2** (a) Anterior end of body showing internal organization (b) Anterior end or lophophore magnified in *Phoronis australi.*

## 8.4 INTERNAL STRUCTURES

### *Body wall*

The cuticle is well-developed in anterior region. Columnar cells and neurons are present between the epidermis and the muscular layer. The muscular layer consists of thin outer circular muscles and thick inner longitudinal muscles. The coclomic epithelium is syncytial in nature.

### *Coelom*

The coelom is spacious, lined by coelomic epithelium and with two unequal regions. External openings are absent in the anterior region. Posterior region occupies the whole length of the trunk and is divided into four longitudinal compartments by the mesenteries. The posterior coelom communicates with the exterior through the nephridia. Colourless, albuminous coelomic fluid contains the coelmocytes, RBCs, spindle bodies and pigment cells. The mesosome is the extreme anterior most of the two phoronid body regions. It bears the lophophore, mouth and mesocoel, which runs into the tentacles of the lophophore. The long trunk or metasome accounts for most of the body. It contains the spacious metacoel or trunk coelom. The posterior end of the trunk is slightly swollen for forming an ampulla.

### *Alimentary canal*

Alimentary canal is U-shaped. The cresentic mouth receives the ciliated grooves from the lophophore and leads into the oesophagus. The ring of tentacles encircles the mouth at the bottom of the lophophore. The feeding current is produced by the lateral tentacular cilia. The food particles are transported to the mouth by the frontal cilia. Water enters the open end of the lophophore and exits laterally between the tentacles. The food particles in the exiting current strike the tentacles for initiating a local reversal of beat in the lateral cilia on the sides of the tentacles, which throws them back into the bell. This occurs several times, before they finally strike the frontal cilia on the inside edge of the tentacles and are moved by them towards the mouth. Cilium is absent on the outside edges of the tentacles. Oesophagus with the thick wall and internal folds leads into a long descending tube, prestomach or proventriculus. The gut is a long U-shaped with one limb descending from the mouth down into the trunk to the ampulla. The other limb ascend back to the anus. The glandular oesophagus runs from the mouth down the trunk to the stomach, which is located in the posterior end just before the gut reverses its direction. The intestine exits from the stomach at the bottom of the "U" and is the ascending limb of the gut tube. It ends at the anus.

Muscles are absent in prestomach, but is provided with an internal mid-dorsal band and in end bulb. The end-bulb is dilated to produce a stomach. Stomach is provided with an inner ciliated epithelium in the middle region but at two sides epithelium becomes syncytial. A constriction marks the stomach from the intestine. Intestine is differentiated into the proximal and distal region, passes into rectum and opens through the anus (Figure 8.3).

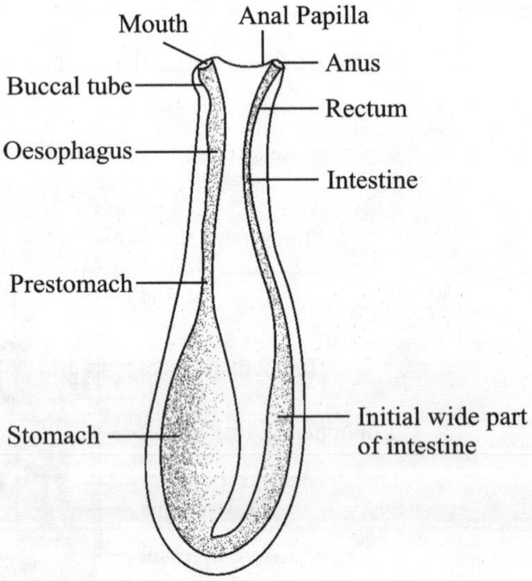


**Figure 8.3** Digestive system of phoronid.

### *Circulatory system*

The blood vessels are closed tubes with contractile walls and sub serves the functions of the heart. Situated between the two limbs of the alimentary canal, there is a dorso-median or afferent vessel. The afferent ring vessel in the region of lophophore gives off one branch to each tentacle. In afferent vessel, the blood flows anteriorly. The ventro-lateral or efferent vessel runs along the left side of the oesophagus in which blood flows posteriorly. Aborally, the dorso-median and ventro-lateral vessels communicate with the haemal plexus of the stomach wall. The blood contains only RBC, which is supported by colourless plasma.

The red colour of their corpuscular haemoglobin makes them easily observable in living specimens. They are joined to the circular blood vessels at the base of the lophophore. The tentacular vessels extend into each tentacle of the lophophore from the circular vessels. At the posterior end of the trunk, the dorsal vessel and ventral vessels are connected to form a plexus or capillary bed in the wall of the gut and gonads.

#### *Excretory system*

The U-shaped metanephridia lined by ciliated epithelium act as gonoducts. The metanephridia opens to the exterior through nephridiopore on the side of the anus.

#### *Nervous system*

A nerve layer consists of nerve fibers and nerve cells are present which is differentiated into nerve ring at anterior end. Lateral nerve innervates the trunk musculature. Neurosensory cells are present dorsally.

#### *Reproductive system*

*Phoronis* is a hermaphrodite, although some species are dioecious. Gonads are loose masses situated posteriorly. Usually testis occurs on the ventral side and ovary on the dorsal side of the lateral vessels. The specialized cells of inner coelomic wall produce the sperms and ova. The fertilization takes place in the body cavity called the *lophophoral cavity*. The sperm, exit the coelom via the nephridiopores, is organized into spermatophores by the lophophoral organs. The spermatophores are transported by the water currents to other *Phoronis*. In the lophophore of the partner, the sperm becomes amoeboid. It penetrates the body wall of the metacoel to fertilize the eggs. The resulting zygotes exit the metacoel via the nephridiopores. Some *Phoronis* brood their embryos. *P. architecta* is not one of them. *P. architecta* is gonochoric. It spawns freely in the sea. The hermaphrodite, *P. hippocrepia*, is a brooding species.

The gonads are found in the metacoel at the posterior end of the oesophagus. The gametes are released into the metacoel from the gonads where they mature and remain stored. Finally, they are fertilized and release through the metanephridia and nephridiopores. The trunk coelom contains eggs or zygotes. The eggs are large, immobile and yolky. The sperms are tiny, flagellated and mobile.

## 8.5 DEVELOPMENT

The lophophoral cavity serves as a brood pouch containing ova and embryos in all the development stages. The development is indirect. The cleavage is complete, slightly unequal and results in the formation of coeloblastula. The vegetal pole of coeloblastula becomes flattened and invagination starts in a typical embolic fashion. A free-swimming ciliated larva when formed is called *actinotrocha larva*. The planktonic actinotrocha has a gelatinous body, which ranges between 1 and 5 mm in length. Overhanging the larval mouth, there is a hook-like hollow pre-oral lobe and a circlet of ciliated tentacles. An apical plate with eyespots is situated in preoral lobe of larva. A pair of excretory organs, solenocytes, post oral ciliated ridge below the mouth and a girdle of slender tentacles arising from preoral lobe are characteristics features of actinotrocha. Actinotrocha ultimately settles to the bottom and transforms into the adult.

## 8.6 ACTINOTROCHA LARVA

The planktotrophic actinotrocha is the remarkable phoronid larva. The preoral hood at the anterior end overhangs the mouth. The anus is situated at the posterior end and remains surrounded by a ciliary ring, the telotroch. Just posterior to the mouth and oral hood, there is a girdle of ciliated larval tentacles which make the actinotrocha distinctive. The larval tentacles transform into the adult tentacles or they may be replaced by a new set of adult tentacles at the time of metamorphosis. There is an invagination of the body wall, called the *metasomal sac* in the midway along the ventral surface of the larva. The actinotrocha has a more or less straight gut. Metamorphosis from the actinotrocha to the adult is rapid requirring only about 30 minutes. In metamorphosis, the major axis of the larva, which is originally the antero-posterior axis, transforms into the dorso-ventral axis. The animal shortens along the antero-posterior axis and elongates along the dorso-ventral axis. This occurs due to the evagination of metasomal sac from the ventral surface and its growth into a long wormlike body. A loop of the gut is pulled down into the everting metasomal sac and causes the lengthening of body. Finally, this becomes the largest part of the animal containing the ascending limb and the descending limbs of the gut. The lophophore and the anus remain in the old anterior end. The anterior end of the animal is the anterior and posterior ends of the larva and the so-called posterior end of the adult is the ventral end of the larva (Figures 8.4–8.6).

## 8.7 AFFINITIES

Phoronida, Brachiopoda and Ectoprocta are collectively called *lophophorate coelomates*. Hyman suggested the term *lophophorate* for these animal groups because of the presence of lophophore.

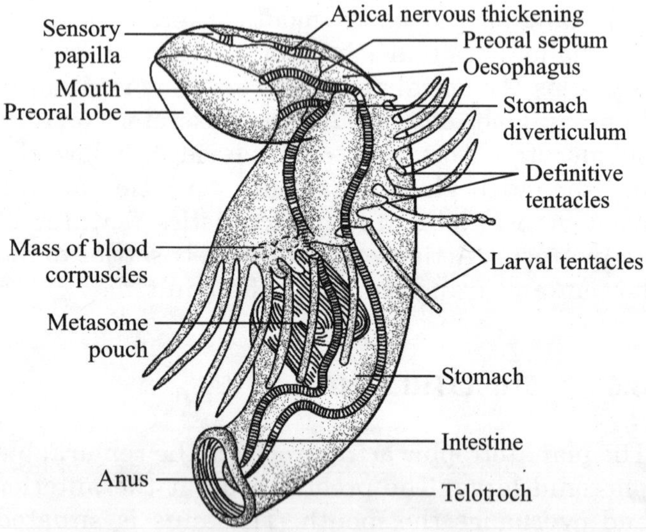


**Figure 8.4** Mature actinotrocha a with buds of tentacles.

### *With Brachiopoda*

*Similarities*

1. In the presence of a lophophore.
2. An epistome is present.
3. A U-shaped alimentary canal is present.
4. A coelom is present in the septum.
5. A sub epidermal nerve plexus form a nerve center in the mesocoel.
6. A pair of metanephridia is present.

*Dissimilarities*

1. The nerve centre is supra enteric in phoronida but subenteric in Brachiopoda.
2. The larval tentacle is present in phoronids but absent in brachiopods.
3. The correlation between brachiopod shell and phoronid exoskeleton is difficult to establish.
4. In phoronids, cleavage pattern is spiral but this is not spiral in brachiopods.

### *With Ectoprocta*

*Similarities*

1. There is a septum between mesocoel and metacoel.
2. In the presence of (a) horseshoe-shaped lophophore (b) an epistome and (c) an U-shaped alimentary canal.

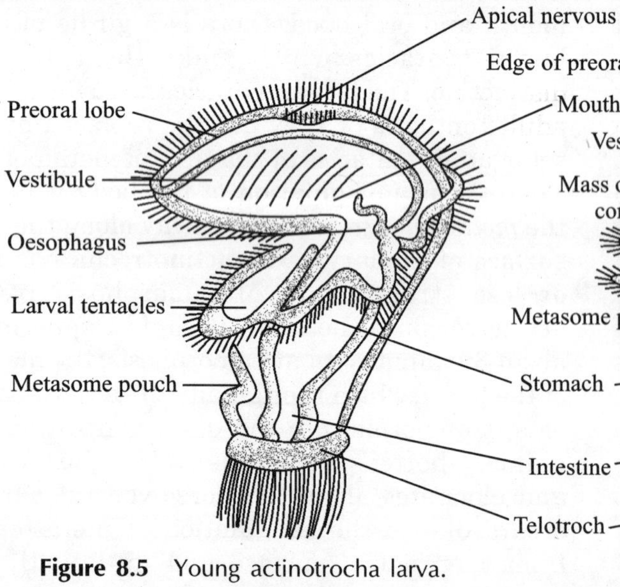


**Figure 8.5** Young actinotrocha larva.

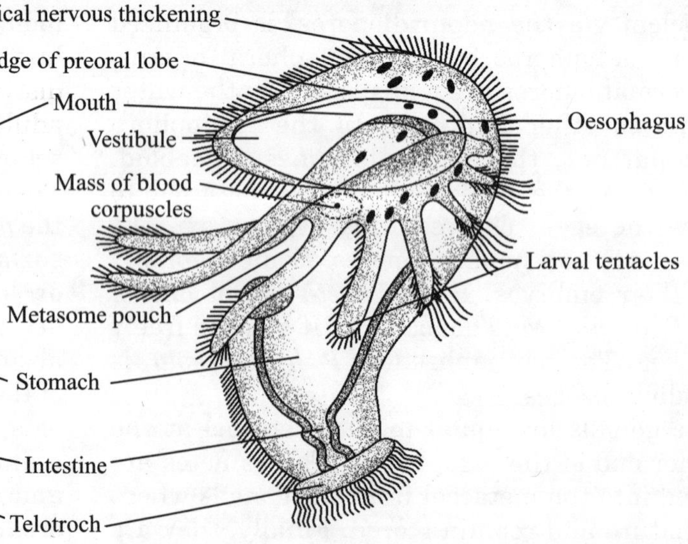


**Figure 8.6** Later stage with trunk and tentacles elongating.

*Dissimilarities*

1. The coelom is endomesodermal in Phoronida while ectomesodermal in Ectoprocta.
2. The region between the mouth and the anus is dorsal in Phoronida but ventral in Ectoprocta.
3. The circulatory system and nephridia are absent in Ectoprocta but well formed in Phoronida.

*Comment:* The relationship between ectoprocts and phoronids cannot be justified because of major structural differences between them.

**With Annelida**

*Similarities*

1. In passing out of mature germ cells through nephridia.
2. The actinotrocha and tornaria larva have many common features such as:
   (a) Both are free swimming, ciliated, pelagic forms with distinct pre-oral lobe.
   (b) Cilla bordering pre-oral lobe of actinotrocha represent metatroch, prototroch and trochophore, respectively.
   (c) Similar disposition of telotroch. Presence of a pair of solenocytic nephridia.

*Dissimilarities*

1. In Phoronida, body is unsegmented but in annelid the body is segmented.
2. In actinotrocha, mesoderm occurs as mesenchyme alone, but it is found in teloblastic bands in trochophore.

*Comment*

1. The relationship between phoronids and annelids cannot be justified critically.
2. The similarities between their larvae may be due to adaptive convergence.

**With Hemichordata**

*Similarities*

1. Proboscis, collar and trunk of hemichordates are similar to epistome, mesosome and trunk of phoronids.
2. A pair of glandular pockets opening into the stomach of phoronids is supposed to be paired notochord of the hemichordates.
3. Presence of septum between the middle and posterior parts of body forms.

*Dissimilarities*

1. Body divisions in Phoronida could not be justified by embryological studies.
2. The mentanephridia present in Phoronida but is absent in hemichordata.
3. The notochord nature of glandular pockets in Phoronida is difficult to establish.

*Conclusion:* It appears that phoronids form a sort of connecting link between protostomous and deuterostomous phyla.

## REVIEW QUESTIONS

### Short Answer Questions

1. What is lophophore?
2. What is epistome?
3. What do you understand by proventriculus?
4. Name the larval forms found in Phoronida.
5. What are lophophoral organs?
6. Name the excretory organ of *Phoronis*.
7. What is mesocoel?
8. What is lophophoral cavity?
9. What is metasomal sac?
10. Name the phyla of lophophorate coelomates.
11. What is mesosome?
12. What is metasome?

### Long Answer Questions

1. Give an account of *Phoronis* with suitable diagrams.
2. Describe the affinities and systematic position of Phoronida.

CHAPTER 9

# Phylum Ectoprocta

The name Ectoprocta derived from two Greek words *ektos* meaning without and *proktos* meaning anus. It refers to the position of the anus outside the tentacular crown of the animal surrounding the mouth. Ectoprocts are microscopic, sessile, colonial, unsegmented and coelomate animals. They are fastened permanently in the exoskeletal cases or gelatinous secretion of their own. The name *moss animal* and old *zoophyte* aptly suggest the superficial resemblance of an ectoproct colony to a moss. Another popular name *Corallines* refers to their superficial resemblance to the hydroid coelenterates and corals.

Impreato first casually asserted the animal nature of some zoophytes. He gave the name pores to the ectoprocts. Several workers like Termbley and Baker, Pallas, Cuvier and Grant have studied the various aspects of ectoprocts. Brain and Papyn have ended the separation of Bryozoans into two phyla, namely, Ectoprocta and Entoprocta. The Stenolaemata and Gymnolaemata include the stoloniferous or nonstoloniferous colony. Stoloniferous colonies are found in some members of the order Ctenostomata in which the zooids arise separately from the horizontal 'runners'. Nonstoloniferous colonies are either encrusting or arborescent or discoidal, but the zooids are always compacted. Zooids do not arise separately and remain at some distance from one another. Zooids of many gymnolaemates and some stenolaemates are polymorphic.

Lophophore-bearing autozooids serves for feeding and digestion. Other individuals of a colony are collectively called the *heterozooids*, which are incapable of feeding. Kenozooids are modified for attachment. Many gymnolaemates possess avicularia, each of which bears a mandible which articulates against a rigid rostrum. Zooids possessing avicularia defend the colony against small organisms.

## 9.1 CLASSIFICATION WITH CHARACTERS

1. Trimeric and enterocoelic.
2. Colonial lophophorates.
3. Unsegmented and sedentary.
4. Epistome and protocoel are mostly absent.
5. Lophophore may be circular or U-shaped.
6. The presence of cicumoral ring, otherwise called the *lophophore*.
7. The gut is U-shaped.
8. The anus is located close to the mouth and outside the lophophore.
9. Triploblastic.
10. Most of the members have the two distinct body parts—zoecium and the polypide.
11. Life cycle involves a ciliated free-swimming larva.
12. Typical circulatory and excretory structures are absent.
13. Colonies are produced by asexual budding.
14. Zooids within a colony are often polymorphic and usually hermaphroditic.

### *Class Phylactolaemata*

1. Inhabit fresh water.
2. Lopophore is horse-shoe shaped.
3. Presence of an epistome and statoblasts.

**Example:** *Plumatella, Cristatella*

### *Class Gymnolaemata*

1. Mostly marine

2. Lophophore is circular
3. Epistome is absent
4. Various types of trochosphere larva are found in the life-history.

*Order Cyclostomata*

1. The zooecia is tubular
2. Operculum does not cover the aperture

**Example:** *Crisia*

*Order Cheilostomata*

1. The aperture of zooecioum is covered by an operculum.

**Example:** *Bugula, Flustra*

*Order Ctenostomata*

1. Aperture of zooecioum is closed by an operculum.

**Example:** *Alcyonidium* (Figures 9.1–9.3).

## 9.2 HABITAT

All stenolaemates and almost all gymnolaemates are the marine. A few gymnolaemates have also penetrated brackish water. One or two ctenostomes (belonging to the order Gymnolaemata) are found in fresh water. Gymnolaemates live permanently attached to the rock, shells, seaweeds, other animals, or any other kind of stable substrate. They are strictly benthonic and pelagic only in larval stages. They occur in all seas at all latitudes and at all depths to about 6000 m. Phylactolaemata include excursively fresh water forms. They occur everywhere in the shallow water, rivers and streams as well as in the ponds and lakes. They are particularly abundant in the clear, well-oxygenated water and are found attached to submerge logs, twigs, rocks, shells and vegetation.

## 9.3 SHAPE, SIZE AND COLOURATION (COLONY FORMS)

Ectoprocts form colonies, which may be fenestrated or lobulated. Most colonies are small, but some reach considerable dimensions. Individuals comprising colony are usually less then 0.5 min in length. Most colonies are white or pale. Some colonies are brown, blue or violet.

## 9.4 EXTERNAL STRUCTURE

A complete ectoproct colony is termed *zoarium*, which is composed of several individuals or units called the *zooids*. Zooids may be chitinous, gelatinous or calcareous in nature and tubular, ovoid or polyhedral in shape. When zooid is a typical feeding individual, it is called an *autozooid*, which makes up bulk of colony. The reduced or modified zooids are known as *heterozooids*. The

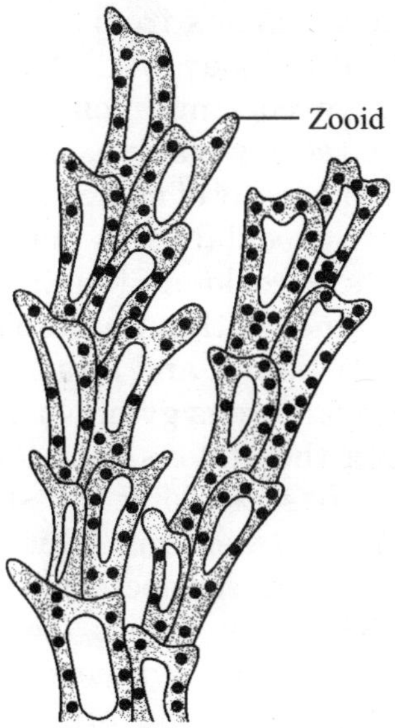


**Figure 9.1** *Bugula.*

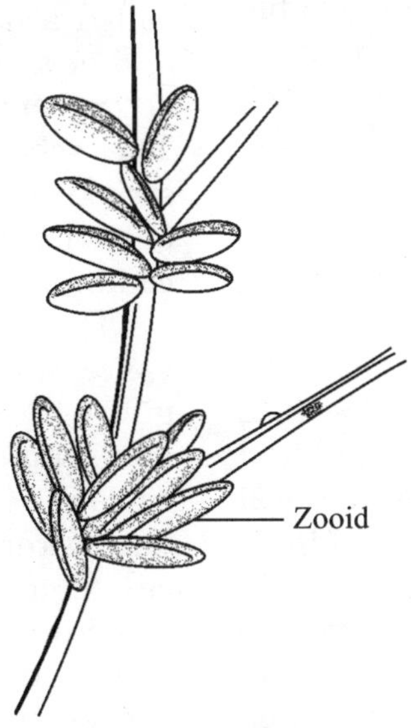


**Figure 9.2** *Bowerbankia.*

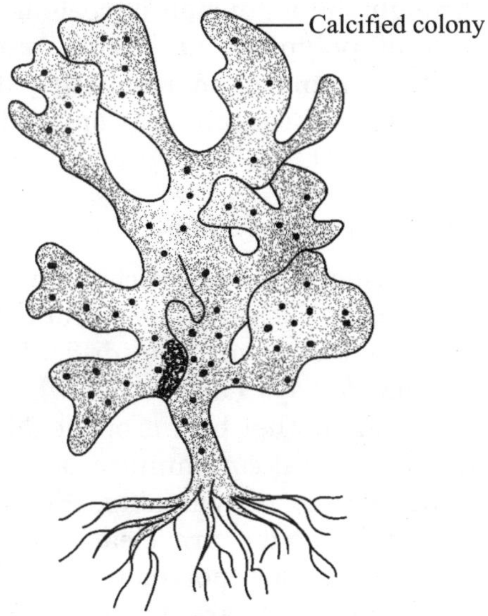


**Figure 9.3** *Flustra.*

common type of heterozooid is modified to form the stolon, attachment discs, roof like structures and other such vegetative parts of the colony.

### *Zoecium*

In gymnolectes, zoecium consists of a protein-cuticle or cuticle overlying the calcium carbonate. The shape and arrangement of zoecia varies in different genera. In *Bugula,* calcification is light. An opening, the orifice enables lophophore to protrude. In many marine bryozoans (ectoprocts), orifice is provided by a movable, hinged operculum rim, or may be provided by spine-like projections *(Aeverrillia)*, lobes or lips *(Farrella)* and is closed by a movable, hinged operculum *(Beannia)*. The body wall is generally interned at orifice for some distance to form a chamber, called the *vestibule*. A restricted opening lie at the bottom of the vestibule, through which the lophophore is protracted or retracted. This opening is provided with a sphincter or diaphragm

### *Introvert*

Anterior part or introvert bears the mouth opening surrounded by a circlet or ciliated tentacles. The disc bearing the mouths and tentacles is called the *lophophore*. In gymnolaemates, the circular lophophore consists of a simple ridge which carries 8–30 or more tentacles. The lateral surface and inner median surface of each tentacle bear a longitudinal tract of the multiciliated cells. Within the tentacle, longitudinal muscle fibres and nerves pass through the centre of a collagen tube and through the extension of coelom. It is immovable part of polypied, i.e. living parts within body wall. It remains permanently fastened in the zoecium and contains coelom and internal organs (Figure 9.4).

### *Zooid interconnections*

Some aspects of ectoproct colony organization are interesting. Clear definitions of the term colony are somewhat elusive. This difficulty arises because it is not always easy to tell where one individual ends and another begins or the degree of structural and functional communication among individuals is uncertain or variable. Ectoproct zooids, at least autozooids, are clearly demarcated by the elements of the polypide. However, the way in which the zooids are interconnected differs among various groups. In phylactolaemates, the metacoel

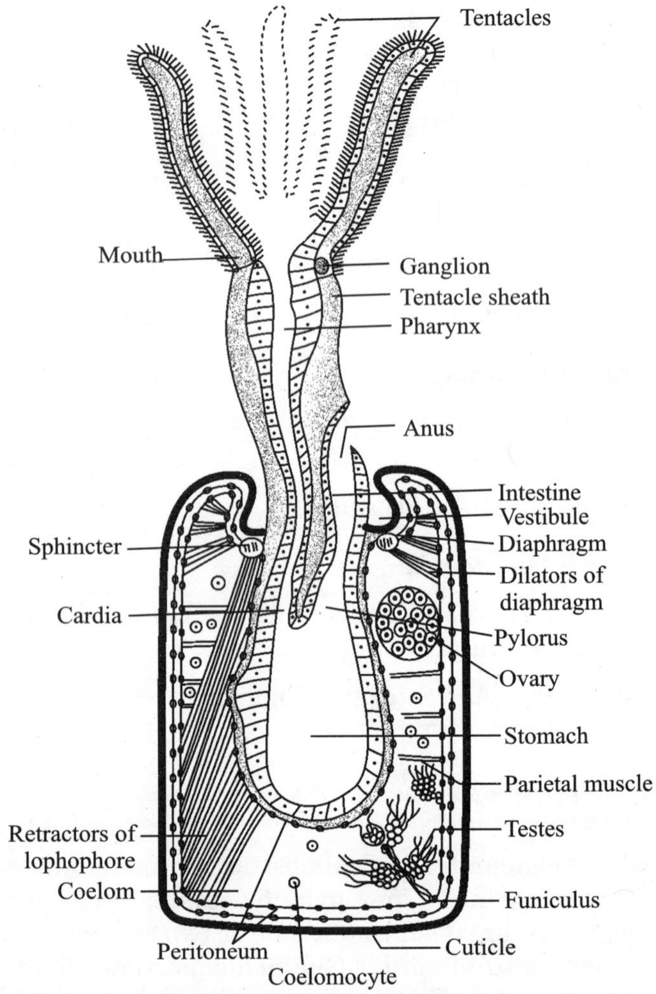


**Figure 9.4** Extended zooid of a gymnolaemate ectoproct.

is continuous among zooids and uninterrupted by septa. Each zooid bears a tubular tissue cord, called a *funiculus*, which extends from the inner end of the curved gut to the body wall. All other ectoprocts lack extensive coelomic connections. The zooids are separated by various sorts of structural components. The walls of adjacent zooids of stenolaemates bear interzooidal pores that allow communication of exosaccalcoelomic fluid.

The funiculus is contained within the entosaccal space with the rest of the viscera and attaches the gut to the body wall. Stoloniferous gymnolaemates have septa spaced along the stolons between the zooids. A cord of tissue passes along the stolons and through pores in each septum. This cord, called a *stolonal funiculus*, connects with the funiculus of each zooid which arises from the stolon. In most nonstoloniferous gymnolaemates, the cystid walls of adjacent zooids are pressed tightly together producing what are called *duplex walls*. These double walls bear pores with tissue plugs, which again, usually connect with the funiculus of associated zooids. Ectoproct zooids are clearly interconnected structurally, either by direct sharing of coelomic spaces or by funicular tissue. Functionally, these connections provide a means of distributing materials through the colony and perhaps other communal activities as well. Some workers even suggest that the funiculus is homologous to the blood vessel. Perhaps, these tissues are remnants of the circulatory system in other lophophorates.

## 9.5 BODY WALL, COELOM, MUSCLES AND MOVEMENT

The body wall comprises the outer zoecium, the underlying epidermis and the peritoneum. Circular and longitudinal muscle sheets are generally present, absent or greatly reduced in other groups and are replaced by muscle bands. Ectoprocts can retract their lophophores into their zoecial casings. Many ectoprocts possess species-specific surface sculpturing including spines and pits. Some ectoprocts produce chemicals for defense. The mechanisms of lophophore retraction and protraction depends largely on the arrangement of muscles, the degree of rigidity of the zoecium and the hydraulic qualities of the coelomic compartments.

The main coelom provides fluid filled spaces on which muscles act directly or indirectly to increase hydraulic pressure for the protraction of the lophophore. The epistome and protocoel play no part in this process. The septum between the mesocoel and metacoel is perforated. When the metacoel is compressed, the polypide is partially forced out of the cystid and protract the lophophore. At the same time, the coelomic fluid moves into the mesocoel and erects the tentacles. Phylactolaemates protract their lophophores by contraction of the circular muscles of the flexible body wall around the metacoel. These ectoprocts possess a ring-shaped, muscular diaphragm just internal to the orifice. The diaphragm dilates as the lophophore is protracted and serves to close partially off the orifice after the lophophore is withdrawn by retractor muscles.

Stenolaemate ectoprocts have erect, tubular zooids surrounded by heavily calcified zoecia. The inflexibility of the body wall and absence of well-developed sheets of muscles preclude use of the direct compression action of the phylactolaemates. Stenolaemates have evolved a mechanism of lophophore protraction which is unique among the ectoprocts. The key structure is a membranous sac attached by ligaments to the body wall and formation of a thin epithelial layer separating the metacoel or entosaccal coelom, from an outer exosaccal cavity. Distally, the exosaccal space lies between the outer body wall and a thick layer of atrial dilator muscles. When a zooid is retracted, the atrial dilators are relaxed and a special atrial sphincter effectively closes the inner end of the atrium. The retracted polypide presses against the membranous sac and force the fluid into the distal region of the exosaccal cavity.

Protraction of the lophophore involves relaxation of retractor and atrial sphincter muscles and contraction of atrial dilators. The contracted muscles press outward against the atrial wall and force fluid into the basal region of the exosaccal cavity, thereby protracting the polypide. The wall of the membranous sac houses numerous, separate bands of circular muscles, which may aid in lophophore extension. Members of the order Ctenostomata possess an uncalcified, flexible zoecium composed of gelatinous, chitinous, or leathery material. Retraction of the lophophore is done by the usual retractor muscle, which is aided by longitudinal parietal muscles that pull in the atrial chamber. When the lophophore is fully retracted, a sphincter contracts, the orifice is closed, and in some species folding a pleated collar over the end of the zooid occurs.

Contraction of the transverse parietal muscles pulls the cystid walls inward causing an increase in coelomic pressure that then protracts the lophophore. Cheilostomate ectoprocts are housed in zoecia with calcium carbonate which is deposited between the epidermis and an outer chitinous zoecial layer. Creating changes in coelomic pressure has been solved by the retention of special uncalcified parts of the cystid wall on which muscles can act. Each zoecium is more or less box-like, the outer surface of which bears the orifice, called the *frontal surface*. In many cheilostomates, the frontal membrane is uncalcified and flexible. Contraction of parietal protractor muscles pulls the frontal membrane inward, thus increasing coelomic pressure and pushing out the lophophore.

Nearly all cheilostomates possess a calcified operculum that closes over the orifice when the polypide is retracted, but the exposed frontal membrane presents weakness in their defense against predation. Many species have evolved additional protective devices. Cribrimorph ectoprocts bear hard spines that project over the membrane and in some cases actually meet and fuse to form a cage above the vulnerable area. In others, a calcified partition, called the *cryptocyst*, lies beneath the frontal membrane, separating it from the soft parts within. The cryptocyst bears pores through which the protractor muscles extend. In ascophoran cheilostomates, the entire frontal surface is calcified except for a small opening. This opening, called the *ascopore*, leads inwards to a blind sac called the *compensation sac* or *sacs*. This structure is an inwardly pouched, flexible portion of the body wall on which the protractor muscles insert. Contraction of these muscles pulls the wall of the compensation sac inward as the ascopore allows water to enter the sac. Pressure is exerted on the coelom and the lophophore is protracted.

## 9.6 LOPHOPHORE, FEEDING AND DIGESTION

Ectoprocts are typically suspension feeders, apparently feed largely on protists and invertebrate larvae of appropriate size. The lophophore of phylactolaemates bears at its base a food groove that leads to the mouth. On protraction, the tentacles of the circular lophophores of stenolaemates and gymnolaemates are erected in a bell-shaped arrangement around the mouth. Each tentacle bears three ciliary tracts along its length, one frontal tract and two lateral tracts. During suspension feeding, the lateral cilia create a current that enters the open end of the funnel, flows towards the mouth and then out between the tentacles. Some food particles are carried directly to the area of the mouth by the central flow of water. Other potential food moves peripherally with the current toward the intertentacular spaces.

When a particle contacts lateral cilia, a localized reversal of power stroke direction is started in those cilia and the particle is tossed onto the frontal edge of the tentacle. The particle is repeatedly bounced in this fashion from lateral to frontal and is moved towards the mouth. Many ectoprocts capture relatively large food particles including live zooplankton. Many species engage in flicking movements of individual tentacles with which a food particle has come in contact. *Bugula neritina* is capable of trapping zooplankton by folding its tentacles over the prey and pulling it to the mouth.

In some ectoprocts, the zooids of the colony function together in the feeding and rejection of waste, or nonfood materials. In many, groups of zooids cooperate to produce general currents that bring water to several clustered zooids and then flow away via excurrent chimneys between the clusters. Such currents move larger amounts of water over the lophophores which moved by individual zooids. The generation of strong excurrent water flow away from the colony surface helps to push nonfood material and faeces far enough to reduce the possibility of recycling. In *Cauloramphus spiniferum*, large particles are actually passed from zooid to zooid and then dumped into an excurrent chimney.

The digestive tract of ectoprocts is U-shaped. The mouth lies within the lophophoral ring and in the Phylactolaemata it is overlaid by an epistome. Ciliary tracts lead into the mouth from the surrounding peristomial field. Internal to the mouth, there is a muscular pharynx. A valve separates the lower end of the pharynx from the descending portion of the stomach, which is called the *cardia*. In some species, this is modified as a gizzard. The cardia leads to a central stomach from which arises a large caecum and the funiculus attaches to the caecum. The ascending portion of the stomach or pylorus also arises from the central stomach and leads to a proctodeal rectum and the anus. The flow of material from the pylorus to the rectum is controlled by a sphincter.

In phylactolaemates, an oesophagus precedes the stomach and the hindgut is elongated as an intestine. Ingestion is done by the sweeping action of the peristomial and oral cilia and by muscular contractions of the pharynx. Digestion begins extracellularly in the cardia and central stomach and is completed intracellularly in the caecum. Food is moved through the gut by peristalsis and cilia. Undigested material is rotated and formed into a spindle-shaped mass by the cilia of the pylorus and then passed to the rectum for expulsion.

## 9.7 CIRCULATION, GAS EXCHANGE AND EXCRETION

Circulation of metabolites in single zooids occurs by diffusion. Intrazooid diffusion distances are small. The coelomic fluid provides a medium for passive transport. Interzooid circulation is facilitated by the confluent coelom in phylactolaemates, the cystid pores in stenolaemates and the funicular cords of most gymnolaemates. Gas exchange occurs across the walls of the protracted parts of the polypide, particularly the lophophore. The tentacles provide very high surface area. Ectoprocts contain no respiratory pigments and gases are carried in solution. Metabolic waste products are transported by phagocytic coelomocytes which apparently occurs in part by the formation of brown bodies. The appearance of brown bodies is usually associated with the degeneration of polypides. In most gymnolaemates, a brown body is left within the cystid following polypide degeneration. In some cheilostomates, the new polypide regenerates in such a way that the brown body is housed within the gut of the new zooid and is then expelled via the anus.

## 9.8 NERVOUS SYSTEM AND SENSE ORGANS

The nervous system and sense organs are predictably reduced. A neuronal mass or cerebral ganglion lies dorsally in the mesosome near the pharynx. A circumenteric nerve ring arises from this structure. Nerves run from the ring and ganglion to the viscera, and motor and sensory nerves runs into each tentacle. Interzooidal nerve fibres occur. The only known receptors are tactile cells on the lophophore and on avicularia. Some ectoprocts exhibit a marked negative geotaxis prior to settling, which is a direct response to gravity. Larvae also usually have well developed ocelli and are positively phototactic, while free swimming. Settlement is often accompanied with a shift to a negative phototaxis.

## 9.9 REPRODUCTION AND DEVELOPMENT

### *Asexual reproduction*

Asexual reproduction causes colony growth and regeneration of zooids. Except for the unique cases of polyembryony in stenolaemates, each colony begins from a single, sexually produced, primary zooid called the *ancestrula*. The ancestrula undergoes asexual budding to form daughter zooids in a chainlike series, a plate or a disc. The budding pattern determines the growth form of the colony and is highly variable among species. Budding involves only elements of the body wall. In most gymnolaemates, a partition forms that isolates small chamber, the developing bud, from the parent zooid. The bud initially includes only components of the cystid and an internal coelomic compartment. A new polypide is then generated from the living tissues of the bud. The epidermis and peritoneum invaginate. The former produces the lophophore and the gut.

The peritoneum produces all of the new coelomic linings and the funiculus. Budding in phylactolaemates and stenolaemates is similar except that the polypide develops first and is then encased by a new cystid wall. In addition to budding, fresh water ectoprocts reproduce asexually by the formation of statoblasts. Statoblasts generally form on the funiculus of an autozooid and include peritoneal and epidermal cells plus a store of nutrient material. Some statoblasts sink to the bottom. Others float by enclosed gas spaces. Some bear surface hooks or spines and are dispersed by passive attachment to aquatic animals or vegetation (Figures 9.5–9.8).

### *Sexual reproduction*

Most ectoprocts are hermaphroditic. Each zooid is capable of producing sperm and eggs. The colonies of dioecious species may consist of zooids of one sex, or more commonly may include both male and female individuals. Gametes usually arise from transient patches of germinal tissue. Only in the stenolaemates, there is any real organ present; that is the testis remains surrounded by a discrete cellular lining. Gametes are proliferated into the metacoel and migrate to the mesocoel before release. Sperm migrates into the coelomic lumina of the tentacles. In some species, the sperm escape through special coelom pores on the particular tentacles. A few cheilostomates exhibit free spawning of eggs as well as sperm. Fertilization and development are fully external. In other ectoprocts, the ova are retained by the parental zooids and brooded at least during early ontogeny. In gymnolaemates, that release their ova

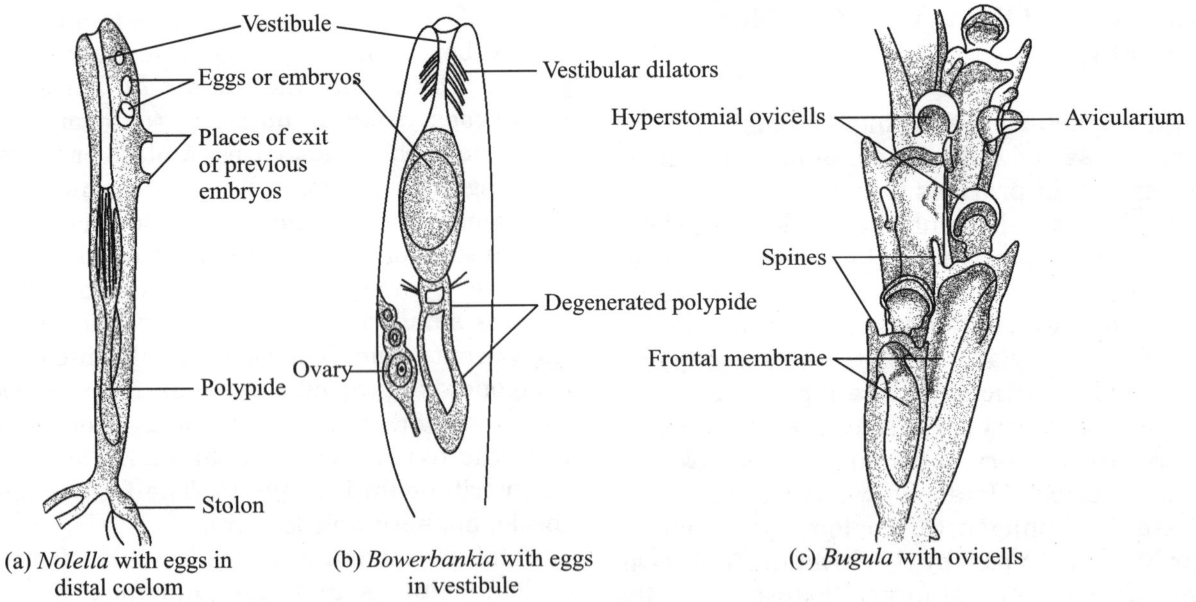


**Figure 9.5** Brooding chambers in ectoprocts.

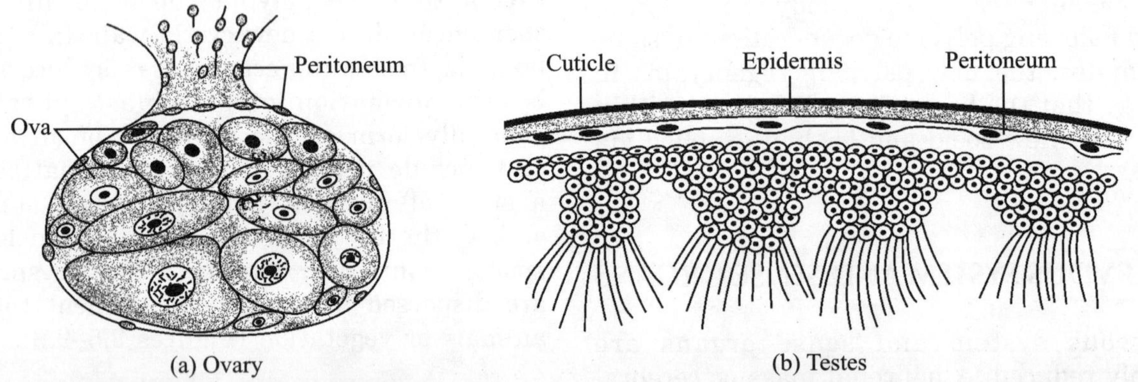


**Figure 9.6** Reproductive organs of ectoprocts.

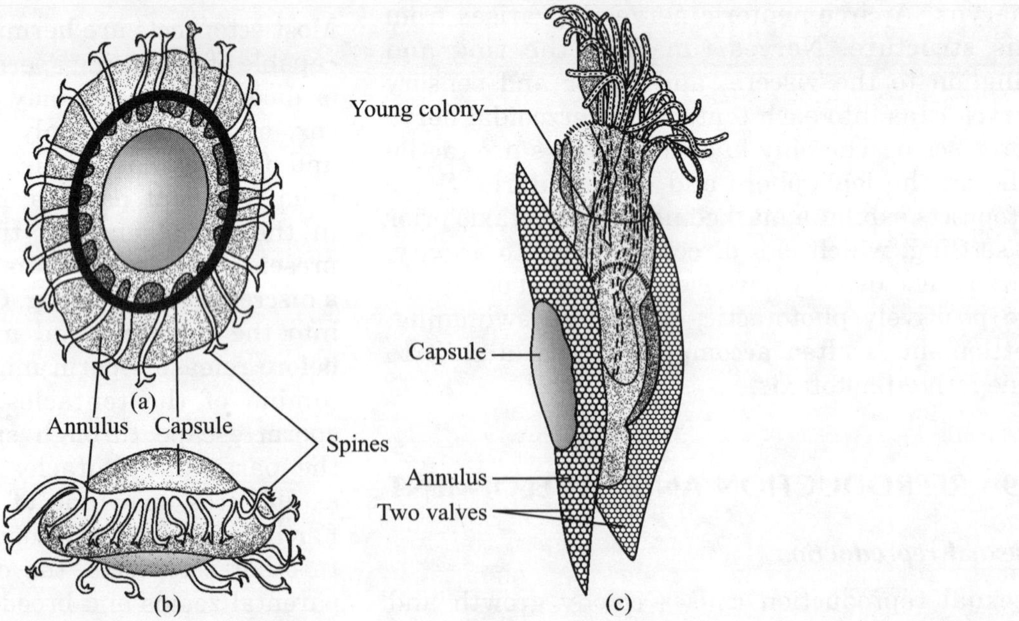


**Figure 9.7** (a) Spinoblast of *Cristatella* in ventral view, (b) Same in lateral view, (c) Germinating statoblast of *Lophopus*.

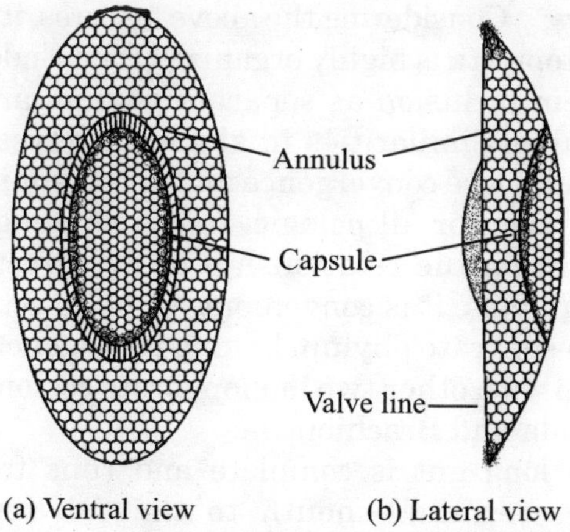


**Figure 9.8** Floating statoblast of *Hyalinella punctata*.

to the sea water or to some external brooding area, the eggs are shed from the mesocoel through an opening called the *super neural pore*. This pore is located between the bases of two tentacles.

Stenolaemates and phylactolaemates brood their embryos. The former in special individuals are called the *gonozooids* that are modified by loss of the polypide and the latter in embryo sacs produced by invaginations of the body wall. A variety of brooding methods occurs among gymnolaemates, usually involving the formation of an external brooding area called an *ovicell* or *ooecium*. This phenomenon of polyembryony is unique in these animals and may represent heterochronic displacement of the usual asexual budding process of other ectoprocts. Each embryo develops cilia and escapes as a simple larva, which settles and undergoes a metamorphosis similar to that of gymnolaemates.The coeloblastulae gym mnolaemates undergo gastrulation by delamination.

Free-swimming larvae are eventually produced. Many species that freely spawn have a flattened, triangular larva called a *cyphonautes*. These larvae have a functional gut and may remain in the plankton for months. The larvae of brooding species lack a digestive tract and lead very short, pelagic lives. Despite these differences, gymnolaemate larvae have some fundamental similarities. They possess a sensory pyriform organ complex and a pouch like adhesive sac, both of which are important in settling and metamorphosis. Some ctenostomate ectoprocts produce nonfeeding vesiculariform larvae.

Ectoproct larvae are at first positively phototactic and most possess pigment spots that are thought to be light sensitive. Following a planktonic phase, the larvae usually become negatively phototactic and swim toward the bottom. Once in contact with the substratum, the pyriform organ complex is apparently used to test for chemical and tactile cues reflecting the suitability of the substratum for settling. Once a proper surface is selected, the adhesive sac everts and secretes sticky material for attachment. After attachment, there is a remarkable reorganization of tissue positions accompanied by histolysis of various larval structures. The metamorphosed larva then generates the primary zooid, or ancestrula..

## 9.10 AFFINITIES

Systematic position of ectoprocts is controversial due to their structural alliances with other groups of animals.

### *Relationship with Phoronida*

Caldwell (1888) emphasized the relationship between Phoronida and Ectoprocta based on the presence of the following similar features:

1. Both are provided with horseshoe shaped lophophore.
2. Presence of episome.
3. U-shaped alimentary canal.
4. Similar disposition of coelom and the presence of septum which separates mesocoel and metacoel.

However, there are the following important differences between them:

1. Origin of coelom is different.
2. Region between mouth and anus is dorsal in Phoronida and ventral in Ectoprocta.
3. Circulatory system and nephridia are absent in Ectoprocta but in Phoronida both systems are present.

*Remarks:* Relationship between Ectoprocta and Phoronida cannot be established, because of differences between them. Phoronida is nearer to lophophorate ancestor and Ectoprocta occupies a subsequent stage.

### *Relationship with Brachiopoda*

*Similarities*

1. Both have similar body construction.
2. Bivalve shell of cyphonautes larva of Ectoprocta is comparable to the shell of brachiopod.
3. U-shaped alimentary canal.

*Differences*

1. The brachiopod shell could not be compared to the exoskeleton of ectoprocta.
2. Shell is laterally placed in Ectoprocta but in Brachiopoda the shell is dorso-ventrally placed.
3. Chitinous setae are present in Brachiopoda but absent in Ectoprocta.

*Remarks:* Because of lack of specific relationship between them, Ectoprocta and Brachiopoda are placed in two separate phyla having remote phylogenetic connection.

### Relationship with Endoprocta

*Similarities*

1. Presence of looped alimentary canal.
2. Presence of ciliated tentacular circlet.
3. Presence of similarities in larval stages.

*Dissimilarities*

1. Ectoprocts possess true coelom but this is absent in endoprocts.
2. Tentacular crown surrounds only mouth in ectoprocts, but in endoprocts both mouth and anus are enclosed by crown of tentacles.
3. Nephridia and gonoducts are absent in Ectoprocta, but in Endoprocta both of them are present.

*Remarks:* Considering the above features, it is clear that Ectoprocta is highly organized then Endoprocta and their exclusion as separate phylum are more reasonable. Similarities in alimentary system are due to adaptive convergence and larval similarities are common for all pelagic, free swimming larval forms. From the relationship of Ectoprocta with other animals, it is convenient for place ectoprocts under a separate phylum having phylogenetic relationship with other two lophophorate coelomates—Phoronida and Brachiopoda.

The long gut is complete and runs from the anterior terminal mouth to the sub terminal, anus. The anus is situated on the dorsal midline near the anterior end and just above the mouth. The closed haemal system contains haemoglobin in blood corpuscles. The lophophore is the major respiratory structure. Two metanephridia open via nephridiopores beside the anus. The nervous system includes a nerve ring at the anterior end and a longitudinal nerve cord on the left side. Phoronids may be gonochoric. However, most are hermaphroditic. Fertilization is the internal. The larva is an actinotrocha which is like a trochophore larva of annelids. Phoronids often reproduce by asexual methods. Such asexual reproduction may result in aggregations of individuals. The individuals in the aggregation become entangled with each other.

## REVIEW QUESTIONS

### Short Answer Questions

1. Mention the habit of Bugula.
2. What is Zoecium?
3. What is avicularium?
4. What is trunk coelom?
5. What is funiculus?
6. Give two important characters of ectoprocts.
7. Distinguish between the classes Phylactolaemata and Gymnolaemata.
8. Give an example of order Ctenostomata.
9. What are zooids?
10. What are heterozooids?
11. What is "duplex wall"?
12. What is phototaxis?

### Long Answer Questions

1. Give a generalized account of the Phylum Ectoprocta.
2. Discuss the affinities of the Ectoprocta with other Phyla.

# CHAPTER 10

# Phylum Platyhelminthes

Members of the phylum Platyhelminthes have a wormlike shape with a third layer of cells, the mesoderm that produces muscular layers, excretory system and an elaborate reproductive system. They have bilateral symmetry. Cephalization, a specialization of the anterior end, results in the clustering of the sensory structures and clumps of nerves. Many elaborations are related to increase in the mobility, higher activity and general increase in responsiveness of the animals. The body cavity is more complex with the specialized internal organs. The coelomic fluid loosely filled with mesoderm cells called the *parenchyma* act as a hydrostatic skeleton, a site for the collection of wastes or gametes and a primitive circulatory system. The flatworms exhibit the simplest body organization, an acoelomate body plan. Platyhelminthes is also known as the *flatworms* due to their almost paper-thin structure. Their digestive system is incomplete and they possess only a mouth. The flame cells are often present as an osmoregulatory system. Turbellarians are free-living, while the other two classes (Trematoda and Cestoda) are obligate parasites.

The flatworms have no body cavity, no lungs for respiration. They receive oxygen via diffusion. For keeping body parts ventilated with the oxygen, they need maximum surface area. This is why flatworms are flat. The need for diffusion restricts them to the moist habitats, such as water, leaf litter and most notoriously as parasites like tape worm. The flatworms are the most ancestral animal with centralized ganglion. They do not possess a true gut but provided with a gastrovascular cavity. They have the most prolific reproduction system. The ending segments of the tapeworm are generally filled entirely with the eggs. The same pharyngeal opening both takes in food and expels waste. In larger flatworms, the gut is often very highly branched to transport the food to all parts of the body. The flukes, like other parasitic flatworms, have complex life cycles often involving two or more host organisms. The planarians are free-living flatworms, less than a centimetre long with a simple life history. They inhabit fresh water and are carnivores or scavengers. Platyhelminthes have practically no fossil record. A few fossil is reported. Eggs of fossil trematode have been found in the Egyptian mummies and in the dried dung of the Pleistocene ground sloth. Trematode larvae that parasitize molluscs may leave the pits or thin spots on the inside of the shell. These pits may be recognized on fossil shells.

## 10.1 GENERAL CHARACTERISTICS

1. Live in the marine, fresh water and damp terrestrial habitats.
2. They may be free-living or parasitic.
3. Thin bodies range in size from nearly microscopic to more than 20 m long.
4. Triploblastic, with a middle embryonic tissue layer, a mesoderm.
5. The presence of gastrovascular cavity with only one opening.
6. They lack a coelom.
7. They have no specialized organs for the gas exchange and circulation.
8. Simple excretory apparatus functions mainly to maintain the osmotic balance.
9. Excretory apparatus consists of the ciliated cells, called the *flame bulbs*.
10. They are divided into four classes: Turbellaria, Monogenea, Trematoda and Cestoidea.

11. Nearly all turbellarians are free-living and mostly marine.
12. The monogeneans and the trematodes live as parasites in or on other animals.
13. The trematodes parasitize a wide range of hosts and most species have the complex life cycles with alternation of sexual and asexual stages.
14. Most monogeneans are external fish parasites.
15. Their life cycles are simple, with a ciliated, free-living larva that starts an infection on a host.
16. Members of the class Cestoidea are also parasitic. Adults are parasites of vertebrates, including human.

## 10.2 CLASSIFICATION WITH CHARACTERS

The following are remarkable characters of platyhelminthes

1. Acoelomate, bilaterally symmetrical.
2. Metameric segmentation is absent.
3. Hard general body skeleton is absent.
4. Organ systems are being separated by a parenchymatous connective tissue.
5. Blood vascular system is absent.
6. Well-developed alimentary canal or absence of gut.
7. Rarely anus is present.
8. The flame cells are generally present.
9. Almost hermaphrodite.
10. A shell surrounds the complete egg.

The scheme of classification followed here is according to Marshall and William (1972).

### *Class Turbellaria*

1. Mostly nonparasitic, with the ciliated epidermis and rhabdites.
2. Gut cavity is present in all except Acoela.
3. Excretory pores are variable in number.
4. Life cycle is usually direct.

### *Subclass Archoophora*

1. The ovary is undivided.
2. Exclusively marine.
3. The eggs are entolecithal.

### *Order Acoela*

1. The pharynx is absent.
2. The gut cavity is absent.
3. The gut consists of a loose digestive syncytium.

**Example:** *Convoluta.*

### *Order Polycladida*

1. Gut with the pharynx and intestinal caeca.
2. The caeca are branched, radially arranged.

**Example:** *Thysanozoon.*

### *Subclass Neophora*

1. Marine, fresh water or terrestrial.
2. The ovary is divided into germarium and vitellarium.
3. The eggs are ectolecithal.

### *Order Rhabdocoela*

1. Mainly free-living.
2. Gut is provided with pharynx and intestine.

**Example:** *Actinodactylella.*

### *Order Temnocephala*

1. The posterior sucker and anterior tentacle are present.
2. The epidermis is not provided with cilia in some species.

**Example:** *Temnocephala.*

### *Order Tricladida*

1. Free living.
2. The pharynx and three intestinal limbs are present.

**Example:** *Bipalium*, *Polycelis.*

### *Class Digenea*

1. Endoparasite.
2. The epidermis of adult lacks cilia and tentacles.
3. The gut possesses oral sucker, pharynx and bifurcated intestine.
4. Single excretory posterior pore is present in the adult.

**Example:** *Fasciola.*

### *Class Aspidogastrea*

1. Endoparasite.
2. The epidermis of adult lacks cilia and rhabdites.
3. The gut does not possess oral sucker.
4. Pharynx and intestine are present.

5. A large ventral sucker or several distinct suckers are present.
6. Single excretory pore is present in the adult.

**Example:** *Aspidogaster.*

***Class Monogenea***

1. Ectoparasites or endoparasites.
2. The epidermis of the adult lacks cilia and rhabdites.
3. The gut possesses pharynx and intestine.
4. A pair of eversible suckers is present.
5. A pair of lateral excretory pores is present.

***Order Monopisthocotylea***

1. Feeding occurs mainly on the epidermis, rarely on the blood.
2. The intestinal epithelium is continuous and cuboid, or columnar.
3. The genito-intestinal canal is absent.

**Example:** *Entobdella, Gyrodactylus.*

***Order Polyopisthocotylea***

1. Blood feeders.
2. The intestinal epithelium is discontinuous.
3. The genito-intestinal canal is present.
4. The bucco-intestinal canal is present.

**Example:** *Polystoma.*

***Class Cestoda***

1. Endoparasites.
2. Epidermis of the adult lacks cilia and rhabdites, but provided with the microvilli.
3. The gut is absent.
4. The suckers are armed with the tentacles or hooks.
5. The life cycle involves free-swimming larva.
6. The six-hooked, or ten-hooked larva occur in the life cycle.

***Order Caryophyllidea***

1. Scolex is provided with or without suckers.
2. The body is not strobilated.
3. The vitellaria are well developed.
4. The uterus is present with external opening.
5. The single set of genitalia is present.

**Example:** *Caryophyllaeus, Archigetes.*

***Order Spathebothridea***

1. Scolex is absent.
2. Genitalia are replicated.
3. Vitellaria are well developed.

**Example:** *Cyathocephalus.*

***Order Diphyllobothridea***

1. Strobilization occurs but no true apolysis.
2. The scolex is present with two sucking grooves.
3. Three-host life cycle occurs.

**Example:** *Diphyllobothrium, Amphilina.*

***Order Bothriocephalidea***

1. Strobilization and true apolysis present.
2. The scolex is present with two sucking grooves.

**Example:** *Eubothrium.*

***Order Tetraphyllidea***

1. The scolex is present with the sucking cups.
2. The uterus lacks primary external opening.

**Example:** *Acanthobothrium.*

***Order Trypanorhyncha***

1. Strobilization occurs.
2. The scolex is present with four small suckers and four eversible spiny tentacles.

**Example:** *Tetrarhynchus.*

***Order Proteocephalidea***

1. Strobilization occurs.
2. The scolex is present with four acetabulate suckers.

**Example:** *Ophiotaenia.*

***Order Cyclophyllidea***

1. Strobilization occurs.
2. The scolex consists of four acetabulate suckers, usually with terminal armed rostellum.
3. Three-host life cycle occurs.

**Example:** *Taenia, Hymenolepis* (Figures 10.1–10.5).

## 10.3 *Fasciola*

*Fasciola* is a well-known parasite of herbivores with a worldwide distribution in reservoir hosts. Sheep, goats, cattle, buffaloes, horses and rabbits

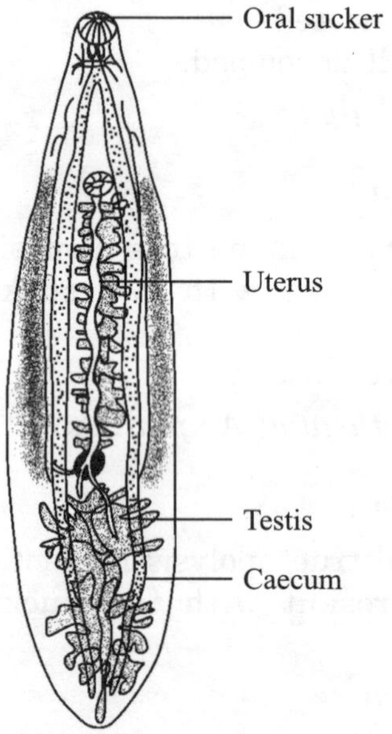


**Figure 10.1** *Opisthorchis.*

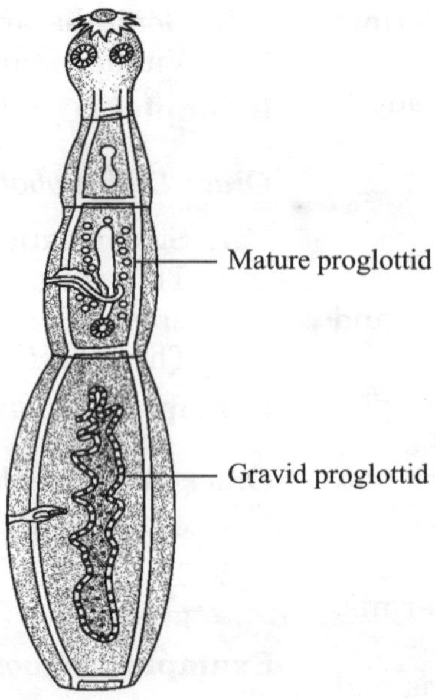


**Figure 10.2** *Echinococcus.*

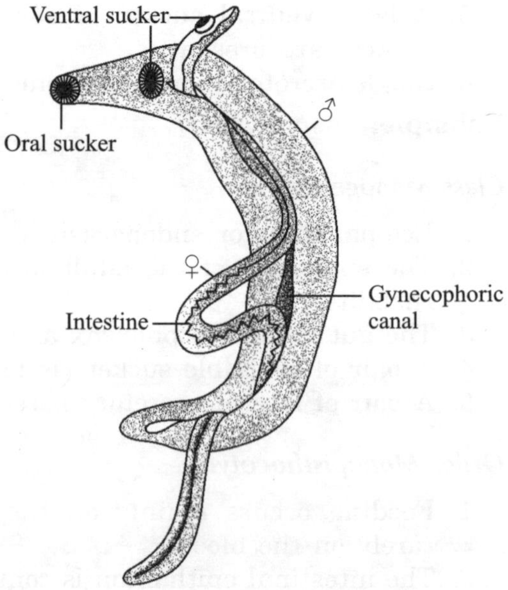


**Figure 10.3** *Schistosoma.*

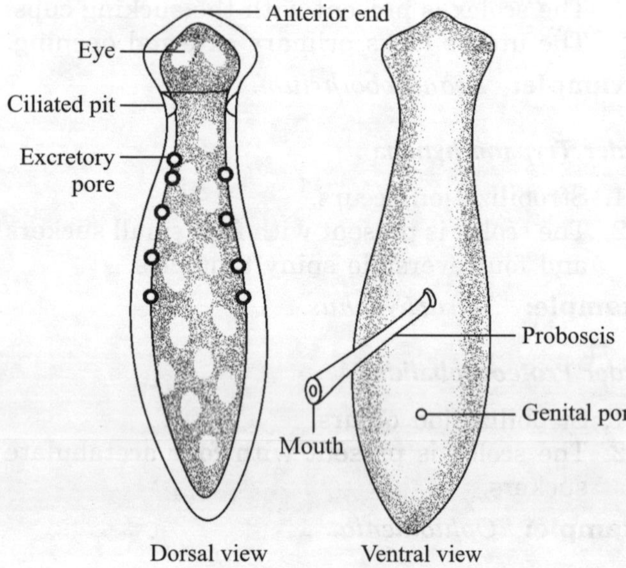


**Figure 10.4** *Planaria.*

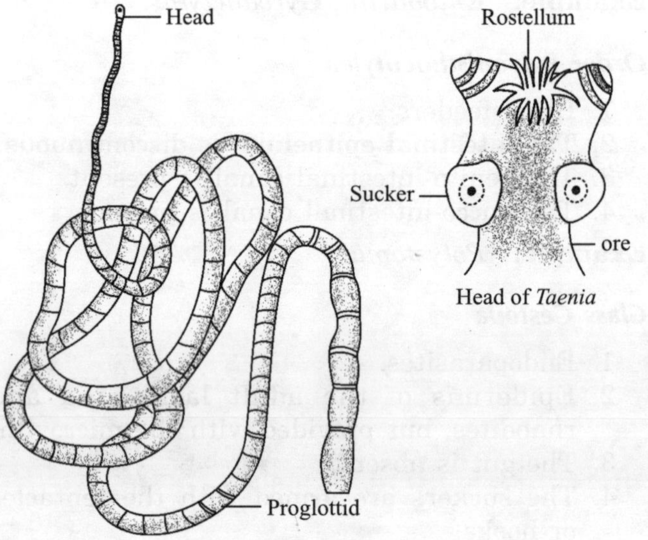


**Figure 10.5** *Taenia* (Adult).

show infection rates up to 90% in some areas. Human infection was very sporadic until 1978. The estimated number of infected people is 2.4 million in 61 countries. The number of people at risk is more than 180 million throughout the world. Infected people are reported mostly from the Bolivia, China, Ecuador, Egypt, France, Islamic Republic of Iran, Peru and Portugal. In the Eastern Mediterranean Region, only in Libyan Arab Jamahiriya, no confirmed case of human infection is reported.

In the Iraq, Lebanon, Morocco, Tunisia and Yemen fewer than 100 cases are documented, as the problem has not received enough attention in these countries. Population at risk in the Egypt is considered 27 million as reported by WHO. *F. hepatica* is hermaphroditic. It was not known to infect man. Human infection is reported in many countries including Europe and the USA. *F. hepatica*, commonly called the *sheep liver fluke* was the first trematode discovered as early as 1819 by de Bill. It is the largest and the most common liver fluke found in the man. Its primary host is the sheep and to a less extent the cattle. It is worldwide in distribution being found in the sheep leaning areas. It causes the economically important disease, the liver rot in sheep.

### Morphology

This is a large leaf-shaped fluke, measuring 3 cm in length by 1.5 cm in breadth and brown to pale gray in colour. The anterior end has a conical projection and at its apex is situated at the oral end. The anterior sucker perforates the mouth. On the ventral surface, a little behind the head lobe, the posterior sucker is situated. It is bigger in size than the anterior sucker. Between the two suckers and close to the posterior sucker, there is genital opening through which the penis sometimes protrudes. The suckers are cup-like muscular organ, which serves for attachment to the host. The excretory aperture is single and lies at extreme posterior tip of the body. The Canal of Laurer opens on the middle of the dorsal surface. The body surface is marked by the presence of several conical projections, spinules or papillae, which are extensions of the cuticle surrounding the body. The outer tegument is covered by tiny spines, which face backwards enabling them to attach themselves along with their suckers to tissues (Figure 10.6).

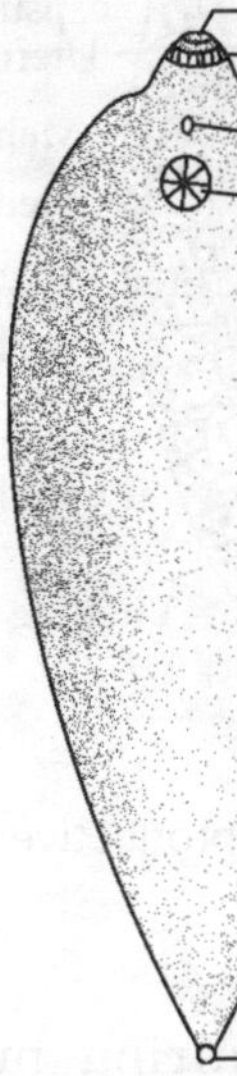


**Figure 10.6** *Fasciola hepatica* (Adult in ventral view).

### Body wall

The body wall consists of the following succession of layers:

(i) An outermost region, the epidermis bearing backwardly directed spines.
(ii) A basement membrane.
(iii) Circular, longitudinal and diagonal muscles.
(iv) The cells with protoplasmic processes connected to and continuous with the outermost spine bearing layer. Internally, the spaces between organs are filled by the parenchyma.

### Digestive system

The mouth is situated in the middle of the anterior sucker and opens into the rounded bulb-like pharynx, which is muscular and suctorial. Next to the pharynx, the oesophagus is located, which is very short and is followed by the intestine. Intestine is bifurcated into two limbs, the right and the left, and runs to the posterior side of the body. Each limb ends blindly and gives off numerous blind branches or caeca. The caeca of the inner side are short and simple, while those of the outer side are large and branched. Practically, every region of the body is traversed by the caeca. The intestine and its branch are very prominent, as they remain filled with the biliary matters, desquamated epithelium and blood on which the animal feeds. The anus is absent.

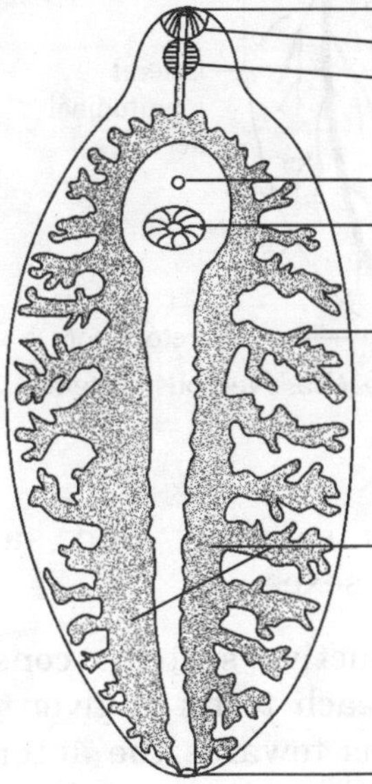


**Figure 10.7** *Fasciola hepatica:* Digestive system.

### Excretory system

The excretory system or system of water vessels consists of a median longitudinal excretory canal, or protonephridia, which opens to the posterior by the excretory pore situated at posterior tip of the body. From the anterior region of the main canal, four large canals are given off. Each canal gives repeated branches. These branches finally end in fine microscopic vessels or capillaries, which enter the inconspicuous flame cell of flame bulb. The excretory canals normally contain the fat droplets.

### Nervous system

The nervous system consists of a pair of prominent nerve ganglia situated one on either side of the oesophagus. A nerve ring around the oesophagus joins the ganglia together. From the nerve ganglia, the nerves are given off to head lobe and to the posterior part of the body. Amongst these posterior nerves, a pair becomes larger and stouter than rest and run posteriorly as lateral nerve cords and bears the branches. In association with parasitic mode of life, conspicuous sense organs are absents (Figure 10.8).

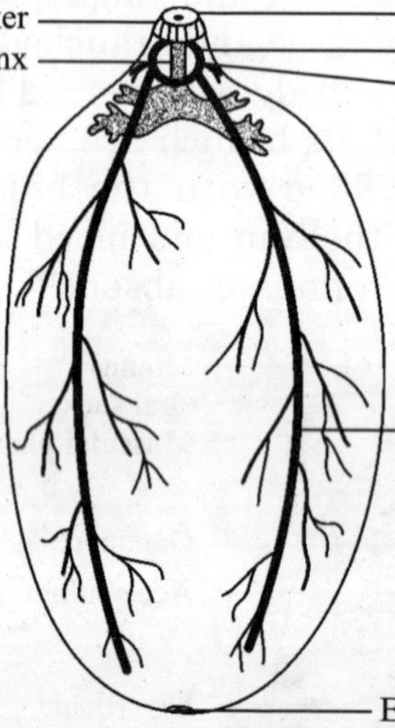


**Figure 10.8** *Fasciola hepatica:* Nervous system.

### Reproductive system

The animal reproduces sexually and same individual bears both the sexes.

**Male:** The male reproductive systems consist of a pair of testis. From each testis is given off, a vas deferens, which runs towards the anterior region. Two vasa differentia unites and forms a median coiled and dilated vesicular seminalis. From it, a narrow tube, the ejaculatory duct leads to the male aperture situated at the tip of cirrus or penis. The penis opens into the genital atrium.

**Female:** The female reproductive organ is a single ovary, or germarium. The tubules of ovary open into a single median oviduct. The vitelline glands are numerous, minute, round and open by the small ducts into a large duct. There are two such ducts. Two ducts meet and form a lateral vitelline duct, which runs transversely and opens into a small chamber called the *yolk reservoir*. From it comes out a median vitelline duct, which opens into the oviduct. Around this junction, there are groups of unicellular shell glands, or accessory female glands, which open by the small ducts into the lumen of oviduct. The lumen of oviduct at this region is dilated and called the *ootype*. The uterus is a wide convoluted tube, formed by the oviduct and median vitelline duct and it opens in the genital atrium near the base of citrus. A canal, termed *Laurer's canal* leads from junction of oviduct and median vitelline duct and opens externally on the dorsal surface. The common genital aperture is situated on the mid-ventral line between the suckers and close to the ventral suckers (Figure 10.9).

**Figure 10.9** *Fasciola hepatica:* Reproductive system.

### Life cycle

The life cycle is complex requiring more than one intermediate host. In *F. hepatica*, cross fertilization is common. Fertilization is followed by the copulation. The copulation takes place in the bile duct of the sheep and during which the cirrus of one *Fasciola* is inserted into the Laurer's Canal of other *Fasciola*. In the process, sperms are deposited into the oviduct. In self fertilization, sperms from the same *Fasciola* enter the female genital aperture and pass down the uterus to fertilize the egg in the oviduct. The eggs enter the duodenum with the bile and leave the host in faeces. The eggs require 9–10 days to complete development before releasing miracidium.

The completion of life cycle depends on the transference of parasite from one host to the other through an intermediate host. The eggs are fertilized in the lower part of the oviduct and then pass onto the uterus. Each egg receives an amount of yolk from the yolk cells and vitelline secretions, and finally becomes enclosed in a proteinaceous shell or capsule secreted by the shell glands. The shell becomes hard when it enters the uterus. The hardening is caused by the action of quinine. One pole of the eggshell bears a small lid or operculum for exit of the future larva. The typical eggs are large, 130–150 μm by 63–90 μm and are laid before complete development of the enclosed miracidium and pass into the alimentary tract via the bile ducts, eventually pass out of host in the faeces. The egg can survive if it falls on the damp soil. The hatching of eggs depends on the temperature but at 26°C eggs hatch in about 10–12 days in a small miracidium larva. Light stimulates the miracidium.

### *Miracidium*

The completely developed miracidium ultimately escapes from the egg shell. The ciliated larva bears the eye spots. A pair of eye spots of miracidium produces an enzyme that changes permeability of the vitelline membrane and the membrane surrounding the viscous cushion. This allows water to enter. The viscous cushion, composed of the mucopolysaccharide, absorbs water and swells, increase pressure inside the egg (miracidium is temporarily compressed), until operculum flies open. Two vesicles swell to help push the miracidium out of egg. The epidermal cells, sub-musculature and sub-epithelium together from the body wall of the miracidium. Anteriorly, the miracidium is provided with a conical, apical papilla and attached to it is a glandular sac, called the *apical gland*. The penetration gland is found on each side of the apical gland. A pair of protonephridia is also present.

The miracidia only live for about 24 hours and must find a snail host of the genus *Lymnea*. Throughout the greater part of the world, *F. hepatica* is transmitted by *Lymnaea truncatula*. *Lymnaea bulimoides techella* and *L. tomentosa* are also implicated in the transmission of *F. hepatica*. Inside the snail, the miracidia produce sporocyst. The miracidium is conical and bears a distinct head lobe. Within the body just below the epidermis lie delicate layers of the circular and longitudinal muscle of fibres, mesenchyme, one pair of flame bulbs with ducts and a sac-like intestine. The germ cells in blocks are present on the posterior part of the body. The miracidium swims freely in the water, or crawls over damp surface for sometime and dies in case it does not find its second host (*Lymnaea*, *Succinea* and *Fossaria*). The miracidium is broad anteriorly. The cuticle of miracidium is ciliated (Figure 10.10).

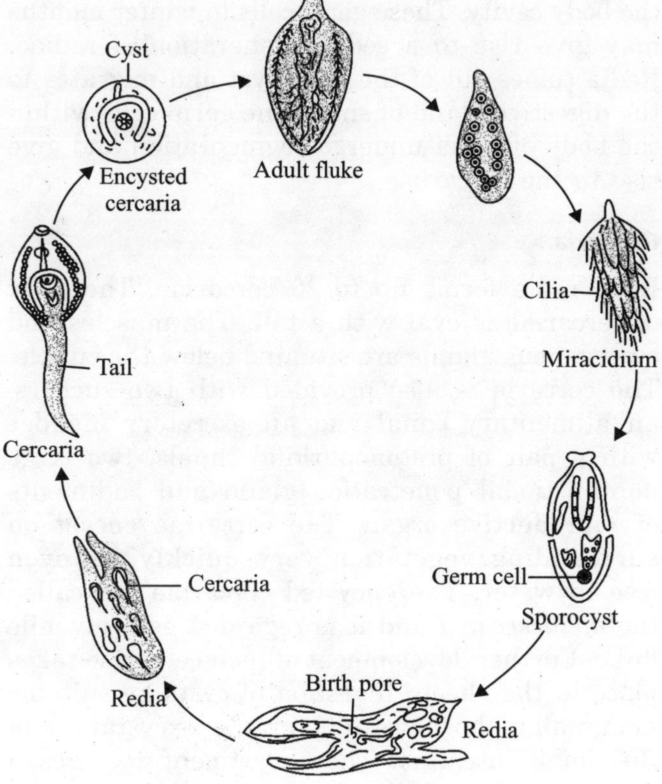


**Figure 10.10** Life cycle of *Fasciola*.

### *Sporocyst*

On meeting the snail, miracidium bores into it by the operation of penetration glands and reaches the internal organs, especially the pulmonary sac. It metamorphoses into the sporocyst by casting off the ciliated covering. The sporocyst is elongated in appearance and its internal cavity is lined with the cells. It also contains the germ cells. The germ cells and lining cells are budded off which undergo segmentation and produce the redia (first generation). The penetration gland, apical papilla and eye spots disappear in the sporocysts. The sporocyst is elongated, about 0.7 mm long and is covered with a thin cuticle.

### Redia

The sporocysts transform into the first generation redia, which in turn produce the second generation rediae. The second generation rediae transform into the cercariae, which emerge from the snail host 5–7 weeks after the initial infection. A redia is provided with the mouth, pharynx and a simple intestine. It bears a circular ridge or collar a little behind the anterior end. The bulging of the body wall forms it. A redia is elongated in appearance and bears a pair of short muscular projections at the posterior side. Many germ cells are found in the body cavity. These germ cells in winter months may give rise to a second generation of rediae. Redia comes out of the sporocyst and migrates to the digestive gland of snail. The germ cells within the body of redia undergo segmentation and give rise to the cercaria.

### Cercaria

Each redia forms up to 20 cercariae. The body of cercariae is oval with a tail. The muscles and cystogenous glands are situated below the cuticle. The cercaria is also provided with two suckers, an alimentary canal and an excretory bladder with a pair of protonephridial canals, two large nonfunctional penetration gland and rudiments of reproductive organ. The cercariae encyst on surrounding vegetation very quickly, or even free in water. The encysted cercariae are called the *metacercaria* and it is regarded as a juvenile fluke. Further development of metacercariae takes place in the sheep. Ingestion of cysts by suitable mammalian hosts results in the excystation in the small intestine. The excystment requires a complex series of stimuli.

The outer cyst is dissolved by trypsin and pepsin secreted by the host. Escaping from inner cyst wall requires other factors such as the temperature of about 37°C, 5% $CO_2$, a low redox potential and presence of the specific bile salts. Newly excysted fluke burrow through the intestinal wall, creep over viscera to reach liver. In liver, they burrow into the parenchyma and wander about for about 8 weeks to enter the bile duct.

During migration through liver, metabolism of liver fluke switches from aerobic to anaerobic. The adult fluke could live for 10 or more years. The cercaria is provided with a long tail, oral and ventral suckers. The alimentary canal is well-developed and consists of the mouth, pharynx and a bifid intestine. The paired excretory tubules with flame cells, germ cells and peripheral cyst-forming cells are also found. The cercaria comes out of the body of the snail. It swims for sometime and then takes the refuge on green leaves. Ultimately, the tail is shed and it becomes encysted. Cercaria in encysted condition is called the *metacercaria* (Figures 10.11–10.13).

### Transmission

In the past, *F. gigantica* was the species infecting herbivorous animals throughout the Egypt. At present, both the species co-exist in the domestic animals with infection rates varying between 30% and 90%. *F. hepatica* infection is common than

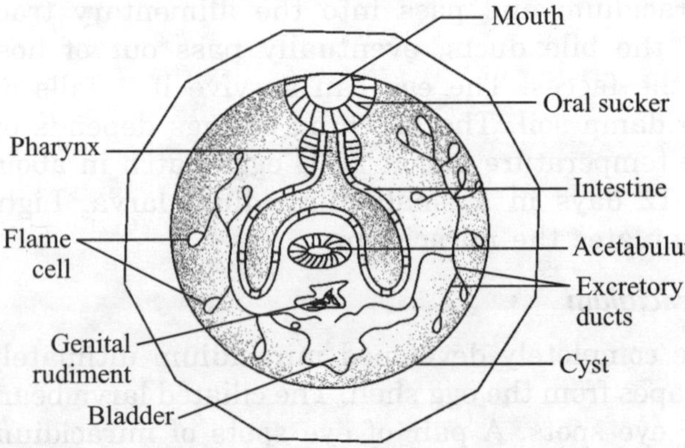


**Figure 10.11** Metacercaria larva.

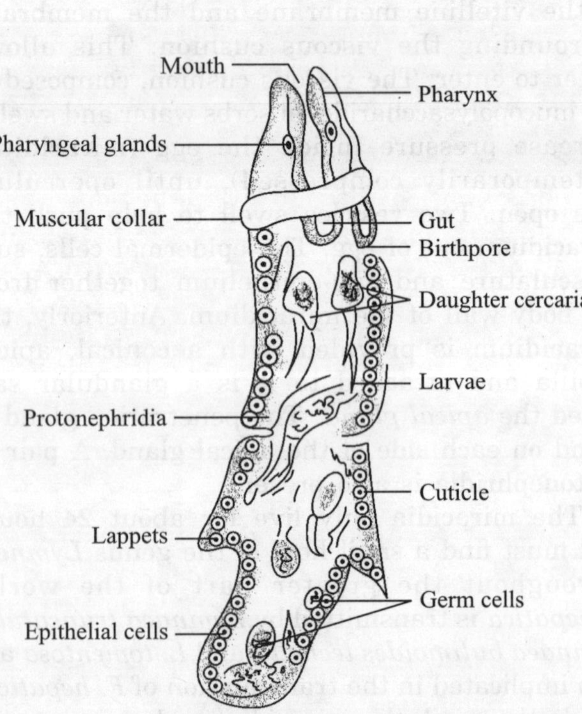


**Figure 10.12** Redia larva.

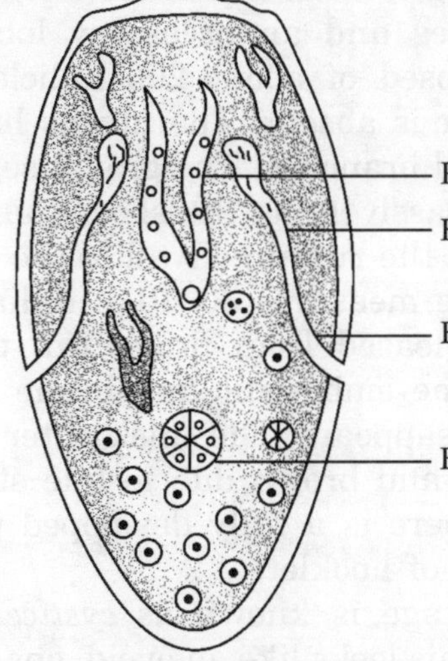


**Figure 10.13** Sporocyst larva.

*F. gigantica*. Only *Lymnaea cailliaudi* is detected in water bodies of the Nile Delta. The overall rates of snail infection vary between 10% and 40%. The highest infection rate was found in summer and this may be a factor responsible for lowering snail density in this season. *L. cailliaudi* is responsible for transmission of both the *F. hepatica* and *F. gigantica*. *Biomphalaria alexandrina* is found to harbour *F. gigantica* in areas reclaimed from the desert (see Mandal, 2011). Metacercariae swallowed by the definitive host during their feeding on the infected green leaves excyst, once they reach the duodenum. They then penetrate the intestinal wall and are found in the coelomic cavity. From the body cavity, they penetrate the liver, migrate through the parenchyma and become established in the bile ducts, where they mature. During migration through liver, the young adult actively feed on the host's liver cells. Rate and extend of the development of *F. hepatica* within the snail host depend on the degree of infection and on the availability of nutrient within snail, primarily those stored in the digestive gland.

## 10.4 TAPEWORMS

Tapeworms have been known for generations. Individual proglottids or segments of the worms are obvious. Prehistoric man must have noticed them and wondered about their origin. Tapeworms were prevalent in the ancient Egypt and were probably mentioned in the Papyrus Ebers, dating around 1550 BC. Parasites may have been *T. saginata*. According to Herodotus, the ancient Egyptians did not eat pork. Intestinal parasite infection by tapeworms is becoming more common in the US. Tapeworms are a type of parasites that need a human to continue their life cycles. They belong to a group known as the *Cestoda*. Several varieties of human tapeworms are *Taenia solium*, *T. saginata*, *Diphyllobothrium latum*, *Echinococcus granulosus* and the dwarf tapeworm, *Hymenolepsis nana*. *T. saginata*, also known as the beef tapeworm is the most common in North America.

*T. solium*, or the pork tapeworm, infects millions in Latin Americas. *D. latum* or the fish tapeworm along with the pork tapeworm is prevalent in Asia. These parasites occur in the various sizes ranging from 6 inches to up to 26 feet long as ribbon-like flat creatures. Their body is divided into three parts—head, neck and segments of the body known as the *proglottids*. The head contains hooks by which it attaches itself to the intestinal wall. This parasite can survive and form a new tapeworm as long as the head remains attached to intestinal mucosa. The proglottids contain eggs, which detach themselves and travel through the body. They absorb partly digested food through their skin from their host. The adults infest the digestive tracts of a wide range of hosts such as fish, cats, dogs, cows, pigs and humans. Most have both, the female and male, reproductive organs and can produce and fertilize their eggs.

### *Life cycle*

The adults of both species live in the small intestine of man, the definitive host. The gravid segments are very active and escape through the anus, releasing large numbers of eggs in the perianal region or on ground, where they can survive for long periods. When ingested by the pigs or cattle, the eggs hatch and release an oncosphere that migrates through intestinal wall and blood vessels to reach the striated muscle within which it encysts to form the cysticerci. When man eats inadequately cooked meat containing cysts, the oncospheres excyst, settle in small intestine and develop there into the adult cestodes over the next 3 months or so. The segments of *T. solium* are somewhat less active than those of beef tapeworm but its eggs, if released in upper intestine, can invade host (auto-infection), setting up potentially dangerous larval infection known as the *cysticercosis* in muscle of any other site (see Mandal, 2011).

Both humans and cattle or pigs are necessary to complete the life cycle of *Taenia*. The eggs ingested by the intermediate hosts usually contain the oncospheres which hatch out in the duodenum, pass into the intestine where they penetrate the intestinal wall and are then carried by circulation to be deposited in the tissues (usually muscle). There they develop into cysticerci larva, which is the white, ovoid and measuring approximately 8 × 5 mm. Humans become infected by ingesting inadequately, cooked beef or pork with the cysticerci containing an invaginated protoscolex. The protoscolexes evaginate and pass into the small intestine where they attach themselves to the mucosa and develop into the adult worms. The eggs and proglottids are passed out in the faeces, then eaten by the intermediate host and perpetuate the life cycle. According to estimates, about 50 million people in the world are affected by cycticercosis. The cysts are usually only revealed by the autopsy. In a small number of cases, the central nervous system is infected causing neurocysticercosis, a potentially deadly condition. Once the undeveloped tapeworm infects a host, it does not develop into a fully grown tapeworm. It drills through the host's abdominal lining and enters the blood stream. Through arteries and veins they travel to all major muscles and organs. Once the undeveloped tapeworm finds a home, the liver or any large muscle, it forms a fluid sac, called a *cyst*, into which it encloses itself. The embryo does not leave this sac until some other animal or human eats the host (cow's flesh). This is the main mode of infection. Most people notice tapeworm in their stool. Sometimes infections may last a long time depending on the organ that is affected.

### 10.4.1 *Taenia solium*

#### *Morphology*

The adult measures up to 3 metres in length. The 'Pin-head' size scolex is globular, provided with the four circular suckers, a rostellum which is armed with a double row of alternating large and small hooklets. The neck is short and up to 10 mm in length. Total average number of proglottids is estimated as 800. The common genital pore is present laterally at middle of each segment alternating between the right and left side. The ovary is located in the posterior third of each proglottid and consists of two large symmetrical lobes and an accessory lobe. The testes are composed of about 200 follicles. The vaginal sphincter is absent. The uterus has only 5–10 thick lateral branches. The gravid segments are passed out passively as the short chain. The eggs escape from the ruptured wall of the uterus. The spherical egg measures 30–40 μ in diametre. When freshly released from proglottid, the egg has a thin hyaline embryonic membrane around it, which soon disappears. The thick outer wall is radially striated and brown due to bile staining. In the centre, there is a fully developed embryo with three pairs of hooklets.

The larval stage is known as *cysticercus* or *bladder worm*. This looks like an ovoid, opalescent, milky white bladder or vesicle surrounded by the fibrous capsule. It contains a thick fluid rich in protein and salt. The scolex of larva, with its sucker, lies invaginated within bladder and can be seen as a thick white spot. It measures usually about 5 mm by 10 mm and remain viable for several months.

#### *Life cycle*

The eggs and sperms of the same proglottids or of different proglottids unite to produce the fertilized eggs or zygote. The embryo is formed from the fertilized egg and remains surrounded by an envelope, called the *embryophore*. The embryophore and eggshell constitute the structure called the *oncosphere*. The terminal proglottids containing oncosphere detach from the worm body and pass out of the human body along with the faecal matter. When the pig consumes human faeces containing the oncosphere, the development of worm takes place and results in the hexacanth larva. The hexacanth penetrates the human gut wall with the help of its six hooks, reach into the blood stream and finally muscle.

On reaching the muscles, the larva forms a cyst wall around its body. The hooks of the worm are lost in muscle, takes the shape of a balloon and a balloon shaped larval stage is called the *cysticercus* or *bladder worm*. Infection in human occurs through the consumption of undercooked pig flesh. When undercooked pork is eaten, the cyst wall of bladder worm disintegrates within the human intestine and adult worm is formed. Man may also become an intermediate host and develop larval stage of disease in one of the ways:

(i) By heteroinfection, the most common route, in which the eggs present in food contaminated by faeces of the *Taenia* carriers are ingested
(ii) By exogenous autoinfection, due to ano-oral contamination in patients harbouring the adult worm
(iii) By endogenous autoinfection in which the eggs of the adult tapeworm living in the small intestine return to stomach due to the reverse peristalsis

The last two modes of autoinfection are far less frequent as it is uncommon to find patients having simultaneous infestation with the cysticercosis and taeniasis in more than 10–15% cases (Rabiela-Cervantes et al. 1982) (Figures 10.14–10.16).

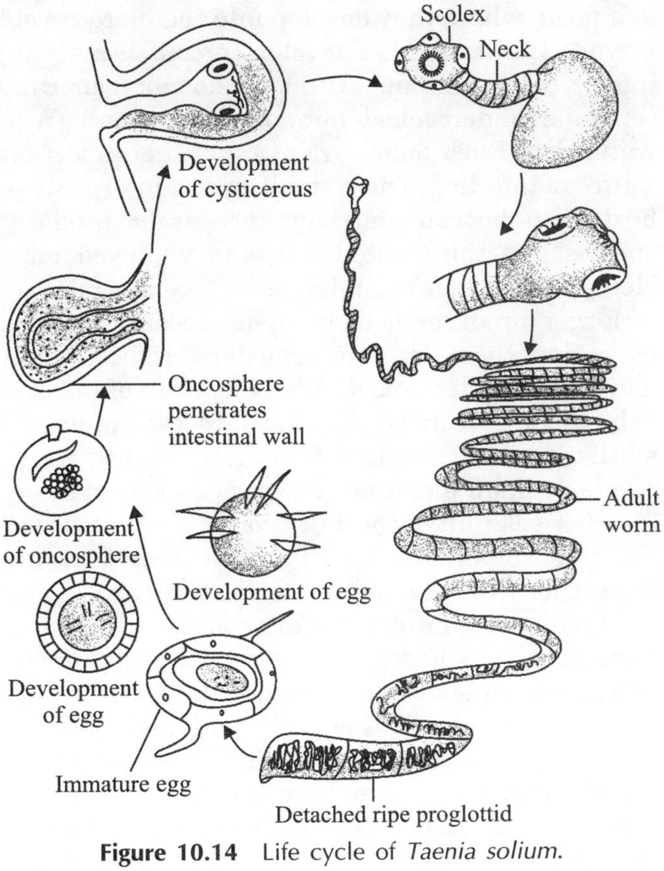

**Figure 10.14** Life cycle of *Taenia solium*.

## 10.4.2 *Taenia saginata*

*Taenia* species are the most common cestode parasites of humans. More than 60 million people are infected with *T. saginata* (beef tapeworm) worldwide. The parasite could eventually stretch 25 m (82 feet) and live a quarter of a century and during this time it releases billions of infective eggs

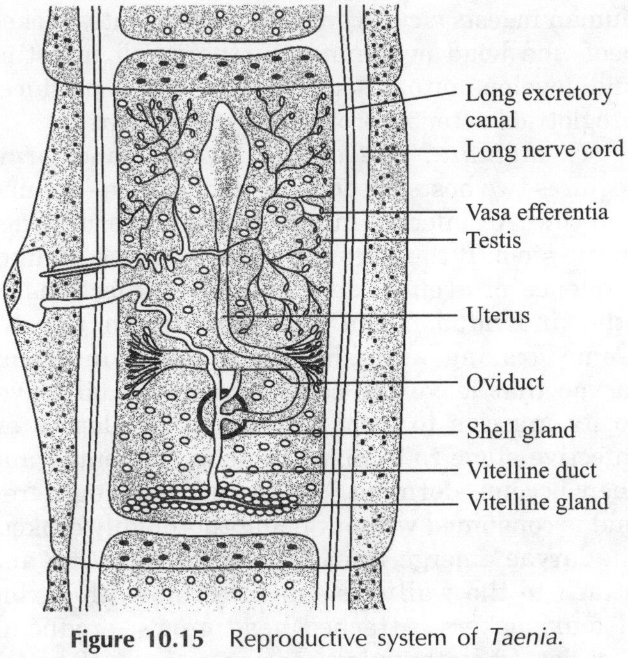

**Figure 10.15** Reproductive system of *Taenia*.

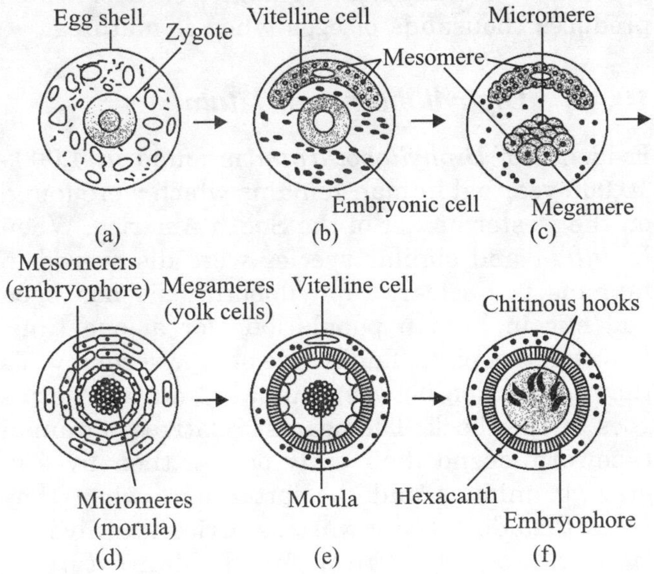

**Figure 10.16** Segmentation of zygote in *Taenia*.

into the environment through its host's stool. The adults are usually about 5 m in length, live in the small intestine with head attached to the mucosa. The eggs and gravid proglottids are passed in the faeces. When ingested by the cattle, the eggs hatch in the small intestine. Each released larva (oncosphere) penetrates the mucosa, passes via bloodstream to the tissues, especially the muscles, subcutaneous tissues and central nervous system, where it vesiculates to form a bladder worm (*Cysticercus bovis*). These usually reach 10 mm in size and contain an invaginated head. When a human ingests a cysticercus in insufficiently cooked beef, the head evaginates in the small intestine and develops into an adult worm, which produces proglottids after approximately 2 months.

*T. saginata*, one of the largest tapeworms requires two hosts to complete its life cycle—human and cow. An infected human excretes worm's eggs in the stool. If the eggs end outside on the ground, sequence of events cause spread of infections in other individuals. The eggs are infective to cows (some grazing animals). Eggs hatch into tiny larvae that leave the cow's intestines and travel to its muscles to form tiny cysts, develop to an infective stage to humans in about 2 months and then become dormant. If the cow is slaughtered and is consumed without being thoroughly cooked, the larvae emerge from cyst in the intestine and attach to the wall of small intestine with a ring of four suckers. Attached head grows, producing a series of rectangular segments, each of which produces thousands of eggs when it matures.

### 10.4.3 *Diphyllobothrium latum*

Evidence of *Diphyllobothrium* is found in 4,000–10,000 year old human remains which were found on the western coast of the South America. When *D. latum* and similar species were discovered in humans is unclear. Diphyllobothriasis has been endemic in human populations for a long time. Investigation of the family tree of tapeworm reveals that association between humans and tapeworms goes further back. The nearest relatives of human tapeworm spend their early part of their lives in grazing animals and the latter part when they reach the adulthood in various African carnivores. Hence, tapeworms first infected ancestors of human species when they began consuming the worm-infested meat as hyenas and big cats. These studies connect human tapeworm infestations with the increased scavenging of meat and hunting by human beings.

*D. latum* is an important intestinal worm, known as the human broad tapeworm. The term broad refers to the fact that proglottids are generally wider than their length. This has a wide distribution and occurs in countries bordering the Baltic Sea, also in Russia, Switzerland and North America where humans eat uncooked or partly cooked fish. The fish tapeworm is a broad, long worm; often grow to lengths of 3–7 feet at maturity. They may attain 30 feet consisting of up to 4,000 proglottids. The main body is filled with male and female reproductive organs. The female produces an incredible number of eggs, more than 1,000,000 a day. The ivory coloured adult attaches to the intestinal wall by the two sucking grooves which are located in the scolex. They live in humans, the final host for 20 years. The immature eggs are passed in the faeces with human stool. Under suitable conditions, the eggs mature (about 18 to 20 days) and produces oncospheres which develop into a coracidia. After ingestion by a suitable fresh water crustacean, the coracidia develop into the procercoid larvae.

Following ingestion of the copepod by a second intermediate host, typically minnows and other small fresh water fish, the procercoid larvae are liberated from the crustacean and migrate into fish flesh, where they develop into the plerocercoid larvae (sparganum). The plerocercoid larvae are infective for humans. Humans do not generally consume undercooked minnows and small fresh water fish. Such fishes are also not the important source of infection. These small second intermediate hosts can be consumed by the larger predator species like the trout, perch and walleyed pike. Here, the sparganum migrate to the musculature of larger predator fish. Humans become infected by eating these later intermediate infected host fish raw or undercooked. After ingestion of infected fish, the plerocercoid develop into the immature adults. Later it develops into mature adult which reside in small intestine. The eggs appear in faeces 5 to 6 weeks after the infection.

It is sometimes found in the fish that are migrating from the salt water to the fresh water for breeding. One can be infected by consuming raw, lightly cooked migratory species of salmon, perch, pike, pickerel and turbot. Eating raw fish dishes such as Japanese sushi and sashimi loads to spread this disease. The fish tapeworms are found in places where humans, bears and fish-eating mammals defecate and streams from which this fish are obtained. Most infected people do not show any symptoms.

### 10.4.4 *Hymenolepis nana*

Hymenolepiasis is caused by two cestodes. *H. nana*, the dwarf tapeworm, the smallest cestode and most common cause of all cestode infections is encountered worldwide. This is the only cestode that parasitizes humans without requiring an intermediate host. In temperate areas, its incidence is higher in children and institutionalized groups. The adult *H. nana* resides in the ileal portion of the small intestine. They produce proglottids that disintegrate in the small intestine and release eggs that are immediately infective. A morphologically identical variant, *H. nana var. fraterna*, infects the rodents and uses insects as intermediate hosts.

#### *Morphology*

The egg containing the oncosphere bears three pairs of hooklets and remains surrounded by a membrane. This membrane has two polar thickenings from which arise threadlike filaments extending into the space between the membrane and colourless hyaline shell, unlike those of *H. diminuta* which do not possess any filaments. The adult worm is normally 2.5–4 cm long. The scolex is knob like in shape, has a rostellum with hooklets and 4 suckers. The segments are wider than they are long. The ova are spherical or ovoid measuring 30–47 μm in diametre.

#### *Life cycle*

The life cycle does not require an intermediate host, complete development occurring within the villi of a single host, resulting in a direct life cycle. It can also use an insect as an intermediate host. The eggs that are released from the mature proglottids in the upper ileum are usually passed out in faeces. If swallowed by another human, they develop into the hexacanth oncospheres and burrow into villi of the small intestine. This is where they develop into the tailless cysticercoids and then migrate towards the ileum and attach, to begin the formation of the proglottids. The eggs, which are ingested by the insects, such as fleas, beetles or cockroaches, hatch to form the tailed cysticercoids, which remain unmodified as long as they are inside the insect. If they are accidentally swallowed by a human they pass down to the ileum and establish themselves (see Mandal, 2011).

The eggs passed with the stool cannot survive more than 10 days in the external environment. When an egg is ingested in contaminated food or water or from hands contaminated with the faeces, oncosphere contained in egg is released. It invades the intestinal villus and develops into a cysticercoid larva. On rupture of villus, the cysticercoid returns to the intestinal lumen, evaginates its scolex, attaches to the intestinal mucosa and develops into an adult. An alternate mode of infection consists of internal autoinfection, where eggs release their oncospheres directly into the intestine without passage through the external environment. The life span of an adult worm is 4–6 weeks, but internal autoinfection allows infection to persist for years. When insects ingest eggs, they develop into the cysticercoids, which can infect humans or rodents on ingestion.

### 10.4.5 *H. diminuta*

*H. diminuta*, the rat tapeworm, is a cestode of rodents which is sometimes diagnosed in humans and has an adult length of 20 to 60 cm. *H diminuta* is reported from the various areas of the world and requires an arthropod as an intermediate host. The eggs ingested by arthropod develop into the cysticercoid larvae. Ingesting the arthropods infects rodents and humans can be accidentally infected through the same mechanism, by ingesting insects in precooked cereals. *H. diminuta* infections are most often asymptomatic. The diagnosis depends on the demonstration of eggs in stool specimens. The concentration techniques and repeated examinations increase the likelihood of detecting light infections.

### 10.4.6 *Diphylidium caninum*

*D. caninum*, the double-pored dog tapeworm mainly infects the dogs and cats and occasionally found in humans. Dog is the principal definitive host. Other potential hosts include the cats, foxes and people, mostly children. This is found worldwide and human infections are reported in Europe, Philippines, China, Japan, Argentina and the United States. The adult worms, up to 60 cm in length and 3 mm in width, reside in the small intestine of host, where they attach by their scolex. They produce proglottids with two genital pores. The proglottids mature, become gravid, detach from the tapeworm and migrate to the anus or are passed in stool. Subsequently they release typical egg packets. Following ingestion of an egg by intermediate host, the larval stage of dog or cat flea, *Ctenocephalides*, an oncosphere is released into the flea's intestine.

The oncosphere penetrates the intestinal wall, invades the insect's haemocoel and develops into a cysticercoid larva. Ingesting adult flea containing cysticercoid infects vertebrate host. In the small intestine of vertebrate host, the cysticercoid develops into the adult, which reaches maturity about one month after the infection. It occurs in the intestines of carnivorous mammals and can be acquired by humans. It has two sets of reproductive organs per proglottid. The uterus breaks down into the egg capsules, each containing several eggs with the oncospheres. The gravid proglottids break free and pass out in faeces, releasing egg capsules. Biting lice or larvae of fleas can ingest the egg capsules and serve as intermediate hosts. In lice, oncospheres hatch, penetrate into the body cavity and develop to cysticercoids. In fleas, the oncosphere remains in the body cavity until larval flea metamorphoses to adult. Then parasite completes development to cysticercoid stage. When fleas or lice are ingested during grooming by the definitive host, cysticercoid is freed, evaginates and attaches and matures to the adult. Characteristic structures include suckers and hooked rostellum on scolex, cirrus within cirrus pouch, vas deferens, testes, uterus, ovary, ootype and vitellaria. In gravid proglottids, the egg capsules contain several eggs with oncospheres.

Most infections are asymptomatic. The pets may exhibit behaviour to relieve anal pruritis such as the scraping the anal region across grass or carpeting. Mild gastrointestinal disturbances may occur. The most striking feature in animals and children consists of passage of the proglottids, which can be found on the perianal region, in faeces, on diapers. The proglottids are motile, when freshly passed and may be mistaken for maggots or fly larvae. Demonstrating the typical proglottids or egg packets in stool or environment confirms the diagnosis.

### 10.4.7 *Echinococcus granulosus*

Human echinococcosis or hydatidosis is caused by the larval stages of the genus *Echinococcus*. *E. granulosus* causes the cystic echinococcosis, the form most frequently encountered. *E. multilocularis* causes the alveolar echinococcosis. *E. vogeli* causes the polycystic echinococcosis. *E. oligarthrus* is an extremely rare cause of human echinococcosis. Cystic hydatid disease is primarily a dog tapeworm. Sheep, cattle, deer and other grazing animals are the intermediate hosts. Dogs and other carnivores become infected from the wild rodents or their faeces. Shepherds who use dogs are the most common victims of hydatid disease. The alveolar hydatid disease is primarily the fox tapeworm. *E. granulosus* is endemic in the sheep raising areas of the Mediterranean, USSR, Australia, South America and Africa. It is found in immigrants to the North America in Canada and in the United States, along the Mississippi and in Alaska. *E. multilocularis* is endemic in regions where foxes are found, central Europe, Siberia and North America in the north central United States and Canada and in Alaska.

The adult *E. granulosus*, 3 to 6 mm long, resides in the small bowel of the definitive hosts, dogs or other canids. The gravid proglottids release eggs that are passed in the faeces. After ingestion by a suitable intermediate host, the eggs hatch in the small bowel and release oncospheres that penetrate the intestinal wall and migrate through the circulatory system into the various organs, particularly liver and lung. In such organs, the oncospheres develop into the cysts. The cysts enlarge gradually, producing protoscolices and daughter cysts that fill the cyst interior. Definitive host becomes infected by ingesting cyst-containing organs of infected intermediate host. After ingestion, the protoscolices evaginate, attach to the intestinal mucosa and develop into the adult stages in 32 to 80 days. Same basic life cycle occurs in *E. multilocularis* which is 1.2 to 3.7 mm. There are a few small differences. Definitive hosts are the foxes and to a lesser extent the dog, cats, coyotes and wolves. Intermediate hosts are small rodents. The larval growth in liver continues indefinitely in the proliferative stage resulting in the invasion of surrounding tissues.

*E. vogeli* is up to 5.6 mm long. Its definitive hosts are bush dogs and domestic dogs. Intermediate hosts include rodents and larval stage in the liver, lungs and other organs, develop both externally and internally, resulting in multiple vesicles. *E. oligarthrus* is up to 2.9 mm long with a life cycle that involves wild cats as definitive and rodents as intermediate hosts. Hydatid disease is caught by humans from the infected dogs. Dogs and dingoes carry the worm in their gut without becoming ill. Eating with the infected hands or other hand-to-mouth contact after patting a dog is enough for the eggs of hydatid worm to be swallowed and cause infection. When swallowed, the hydatid eggs are transported by blood to other parts of the body with resulting release of the oncospheres in the intestine and the development of cysts in various organs. In early stages of hydatid disease no symptoms may be felt. Symptoms depend onsite of parasitic cyst which is cause of disease. The symptoms due to a large liver cyst may be a sense of weight, vomiting, feeling overly full after meals, or pain, indigestion and jaundice. The cysts may also occur in lungs.

Early symptoms may be coughing, chest pain or coughing blood. The first symptom may be coughing up salty fluid after rupture of a cyst. This may lead to shock from allergy, itching of the skin or chest infection. The cysts in other body organs may cause seizures, blindness, deafness and kidney pain or heart problems. All these forms are potentially deadly. The rupture of a cyst at any site can cause death from allergic shock. The only effective treatment is surgery to remove the cysts, sometimes with antiworm drugs. Some cysts in vital organs cannot be readily removed by the surgery.

### 10.4.8 *Clonorchis sinensis*

Three major liver trematodes or flukes that infect humans are *Clonorchis sinensis*, *Opisthorchis* species and *Fasciola hepatica*. The adult fluke measure 11 μm–20 μm (by 3 μm–4.5 μm) and are lanceolate in shape, translucent and brownish in colour, hermaphroditic, possess two suckers. The ova contain the fully developed miracidia and possess prominent opercular shoulders (flask shaped egg). They are bile stained and measure 29 μm by 16 μm. In wet mounts, they are transparent and it is quite easy to see their anatomy. There can be up to 6,000 worms present and a daily egg output of 1,000 eggs per microlitre of bile or 600 per gram of faeces. The cercariae possess eye spots, well developed penetration and cystogenous glands.

### 10.4.9 *Opisthorchis spp.*

*Opisthorchis felines*, the cat liver fluke occurs in the Southeast Asia and in Central and Eastern Europe, particularly in the Siberia and other parts of the former Soviet Union. Prevalence rates of 40 to 95% in some areas and infection of over 16 million people is estimated by WHO. The adults are 7 mm to 12 mm in length and 2 mm to 3 mm in breadth. They inhabit biliary and pancreatic ducts of mammalian host, attaching with mucosa. The fully developed eggs deposited by the female are passed in the faeces. After ingestion by the first intermediate host, the eggs release miracidia, which pass through the developmental stages, *viz.*, sporocysts, rediae and cercariae in snail. The cercariae released from the snail penetrate the fresh water fish, the second intermediate host and encysts as metacercariae in the muscles or under scales of fish. Ingesting undercooked fish containing the metacercariae infect cats, dogs and various fish-eating mammals including the humans. After ingestion, the metacercariae excyst in the duodenum and ascend via the ampulla of Vater into biliary ducts, where they attach and develop into adults.

Adults lay eggs after 3 to 4 weeks. Most infections are asymptomatic. In mild cases, manifestations include dyspepsia, abdominal pain, diarrhoea or constipation. In infections of long duration, symptoms are severe hepatomegaly and malnutrition may be present. Sometimes cholangitis, cholecystitis and cholangiocarcinoma may develop. Acute infection resembling Katayama fever, schistosomiasis, with symptoms such as fever, facial edema, lymphadenopathy, arthralgias, rash and eosinophilia. Chronic infection presents the same manifestations as *O. viverrini*, with additional involvement of pancreatic ducts. Diagnosis is based on microscopic identification of eggs in stool specimens. Patient outcome depends on the early detection and treatment. Human cases of opisthorchiasis may affect liver, pancreas and gall bladder. If not treated in early stages, opisthorchiasis may cause cirrhosis of liver and increased risk of liver cancer, but may be asymptomatic in children. Two weeks after flukes enter the body, parasites infect biliary tract. Severe anaemia and liver damage may also incapacitate infected person for 1–2 months. Treatment is generally with a single dose of praziquantel. Detection of fecal *Opisthorchis* antigen is reported.

*O. viverrini* are dorso-ventrally flattened, lancet-shaped, thin and transparent. Adults are monoecious and of average size 7.0 × 1.5 mm. The oral sucker is sub terminal. The ventral sucker lies about anterior one-fifth of the body length. Two testes are deeply lobed, diagonal. They are situated near the posterior extremity. The testes are morphologically different from the *C. sinensis*. The slightly coiled seminal vesicle ends in ejaculatory duct, which open via the genital pore just in front of the ventral sucker. The cirrus sac and cirrus are absent. Multilobated ovary is located in front of anterior testes and nearby is seminal receptacle and Laurer's canal. Vittellaria consist of numerous follicles which is disposed as columnous groups in lateral fields between the ventral sucker and testes. Excretory bladder is long, sac-like tube, s-shaped and runs between two testes.

### 10.4.10 *Spirometra mansonoides*

The North American human infections are caused by *S. mansonoides*. Other species can also infect humans. *S. mansonoides* is found in the domestic and feral dogs and cats in the southern United States. *S. erinacei*, *S. mansoni* and *S. ranarum* are found in feral and domestic dogs and cats in Asia. Cats and dogs are the primary hosts of adult parasite. A variety of fish, reptiles, amphibia, birds and mammals are second intermediate and paratenic hosts. These hosts maintain the third stage larvae that infect people. The stage living in the copepods can also infect people. People get infected by drinking water containing infected copepods, by consuming raw or poorly cooked meat or by using poultices of raw snake or frog tissue. Person-to-person transmission does not occur. In still fresh or brackish water, the infective larval stages only occur within copepods in ponds. The filtration of relatively large copepods breaks the cycle. Most infected copepods survive a few weeks in wild. Third stage larvae likely survive as long as their host.

The life cycle is complex and includes first and second intermediate hosts, paratenic hosts and adult worm hosts. The eggs passed in faeces develop into the embryos in water within several weeks. The ciliated first stage larvae called the *coracidia hatch* and swim. They are eaten by the *Cyclops* and other copepods. In several weeks, the larvae develop into third procercoid stage in the copepod. When a tadpole, fish, snake eat the copepod, frog, or mammal, larvae migrate into the tissues to develop into final larval stage, a sparganium that may stay a long time. In the

**Medically Important House fly**

**—Fatik Baran Mandal**

House fly is very common insect. Some insects are friendly to human society, while others are dangerous in transmitting human diseases. House fly is medically important. They belong with the Order Diptera, which means two winged flies.

House flies are 6–9 mm in length. Four stages in their life cycle are the egg, larva, pupa, and adult. In summer, the eggs generally hatch within 3 hours into larva. The average hatching time of egg is 48 hours. Larval period lasts up to 2 days, at the end of which larva transforms into pupa, Pupa stages lasts for 4 to 5 days in summer, but may be prolonged in winter. The adult emerges from pupa through a circular slit. Male and female mates within 24 hours after emergence from pupal case. The female lays 75–150 eggs per batch at an interval of 3 to 4 days.

On an average, the fly live for 15 days. The maximum life span is known to be 70 days. Usually, the adult tend to remain within a radius of 100–500 meter from their breeding places. They can fly up to 20 miles with the help of wind, and are active at daytime. At night, they take shelter in the dark corners of the room, under the roof, and behind the clothes. They prefer milk, fruit, cough, pus, blood, sugar, and stool, and always in search of food. Horse manure, cow manure, hog manure, chicken manure, human excreta, garbage, organic manure, fishmeal, blood and bone meal, sewage, and waste food behind trough attract house flies to oviposit there. The fly feeds intermittently, and defecates at the interval of 5 minutes, and vomits frequently. The fly sits on sputum, pus, and stool for purpose of feeding, or not and indiscriminately feeds on feces, sweets fruits, sugar, and milk.

Owing to this habit, the fly can contaminate human food with the micro-organisms form the filth. The whole body including the legs of a housefly is studded with numerous hairs. Microorganisms get attached to these hairs, and carried over to different food substances. The fly transmits germs by vomiting the ingested material, and by passing the ingested pathogens along with stool.

House flies can mechanically transmit bacterial infections such as *Salmonella typhosa*, *S. paratyphi*, *S. enteritidis*, *Shigella dysenteriae complexa*, *Vibrio chlolerae*, *Mycobacterium tuberculosis*, and *M. leprae*; viral diseases such as poliomyelitis, and infectious hepatitis. They also spread protozoan like *Entamoeba histolytica*, *E. coli*, and *Giardia lamblia*. The egg of *Ancylostoma duodenale*, *Ascaris lumbricoides*, and *Echinococcus granulosus* can be carried both internally and externally by this fly. The ova of helminths like *Taenia solium*, *Enterobius vermicularis* can be carried externally.

*Article source*: http://health.ezinemark.com/medically-important-house-fly-7d2cd037bc94.html

US, paratenic hosts are usually the frogs and water snakes. Dogs and cats become infected by eating paratenic or second intermediate hosts and in them larvae migrate to the gut where they mature into the adults who shed eggs. An Asian folk remedy for wounds is to apply a poultice of raw snake or frog tissue from which *Spirometra* larvae may migrate into human tissues. Immature worms wander through cutaneous tissues for days to months causing swellings. They may invade eye and central nervous system, brain where they can cause serious damage. Surgical removal is only treatment.

### 10.4.11 *Paragonimus westermanni*

*P. westermanni* is a lung fluke found in both humans and animals. The adults are 12 mm long and are found in capsules in lung. They are hermaphroditic. It is necessary for worms to be present in the cyst for fertilization to occur. Disease is seen in the Far East, China, South East Asia and America.

Adult is an ovoid, reddish brown fluke about 12 mm long. Eggs are the ovoid, brownish yellow, thick shelled and operculated. They measure 80–100 μm by 45–65 μm and may be confused with ova of *D. latum*. Humans become infected by ingestion of insufficiently cooked crayfish or crabs containing metacecariae, which excyst in intestine, penetrate through wall into peritoneal cavity and make their way through diaphragm and pleura into lungs. Lung cysts in which worms most commonly occur usually contain 1–3 flukes.

Eggs become freed into bronchial tubes and pass out with sputum. But they may also appear in faeces in large numbers as a result of being swallowed. Sporocyst and 2 redia generations occur, giving rise to creeping cercariae. These penetrate several fresh water crustaceans in which they encyst in various sites such as gills, muscles, heart and liver. Encysted metacercariae are not immediately infective and undergo further maturation. Mammalian hosts ingest cysts, which are then ingested in duodenum and the freed metacercariae penetrate through intestinal wall into body cavity to reach pleural cavity in about 4 days and the lungs in about 14–20 days. In lungs, a fibrous capsule is formed by host after about 6 weeks the worms mature and produce eggs, which characteristically appear in sputum.

### 10.4.12 Schistosomes

Schistosomes are the only blood trematodes. They differ from other trematodes in that they have separate sexes. Male worms resemble a rolled leaf where they bear longer and slender female in a ventral canal. They complete their life cycle using definitive and intermediate hosts. *S. mansoni*, *S. haematobium* and *S. japonicum* with *S. mekongi* and *S. intercalatum* are responsible for human disease. Blood is rich in glucose, amino acids and along with plasma and blood cells. It is suitable environment for egg producing trematodes. More than 200 million people are reported to be infected over at least 75 countries with 500 million or more people exposed to infection. Disease caused by *Schistosoma* is called *schistosomiasis* or *Bilharzia*. This is the most important of helminth diseases.

#### *S. mansoni*

*S. mansoni* occurs in West and Central Africa, Egypt, Malagasy, the Arabian Peninsula, Brazil, Surinam, Venezuela and the West Indies. Intermediate host is an aquatic snail of genus *Biomphalaria*. Man is the most common definitive host, occasionally baboons and rats are infected. Adults live in smaller branches of inferior mesenteric vein in lower colon. Adults live in plexus of veins draining rectum and colon and in branches of portal vein in the liver. Adults migrate to mesenteric and portal veins. Life cycle is similar to that of *S. haematobium*. Adult male measure up to 15 mm in length and females measure up to 10 mm.

Schistosomes remain in copula throughout their life span. Male surrounds the female with his gynaecophoric canal. Male is actually flat but sides roll up forming groove. Cuticle of male is covered with minute papillae. Female only possess these at anterior and posterior end as middle section being covered by male body. Oral and ventral suckers are present, with ventral one being lager holding worms in place, preventing them from being carried away by circulatory current. Ova are 114–175 μm long by 45–68 μm wide, light yellow-brown, elongate and possess a lateral spine. A nonviable egg is dark coloured. It does not show internal structural detail or flame cell movement. Eggs become calcified after treatment. They are usually smaller, black. Lateral spine of ova is a distinguishing factor. Eggs are not operculate and infection is acquired by penetration of cercaria through skin.

### S. Japonicum

*S. japonicum* is prevalent in China, Japan, the Philippines and Indonesia. It occurs in the portal and mesenteric vessels of man as well as in other mammals. Cats, dogs and cattle play major roles in transmission. Intermediate host is *Oncomelania hupensis* in south and southeastern china, *O. quadrasi* in the Philippines, *O. formosana* in Japan. Adults are longer and narrower than the *S. mansoni*. Male measures 12 to 26 mm long and about 0.3 mm thick. Sucker is situated close together near the anterior end. Cuticle is spiny on the suckers. Oesophagus is surrounded by several glands. Intestine bifurcates before reaching the level of ventral sucker, again joining in the last quarter of body. Six to eight lobes is found in the testes. Testes lie behind the genital pore and opens directly posterior to ventral sucker. Ovary lies behind the middle portion of the body and the vitelline gland fills the posterior quarter. Ova are about 55–85 μm by 40–60 μm, oval with a minute lateral spine or knob. Eggs are passed in the faeces of the host. Eggs may be seen in the faeces as quickly as 5 weeks. Sexual maturity is reached in about 4 weeks.

Female penetrates into the small vessels of the mucosa or submucosa of intestine, laying eggs in capillaries. Some eggs are transported to the liver and other organs via blood stream. Eggs are passed in the intestinal lumen and pass out through faeces. In snails, there occur two generations of sporocysts. Ultimately the cercariae is formed which swarm out of the snail and swim about in the water. Infection occurs when the host comes in touch with infected water. Cercariae penetrate through skin and ultimately reach the systemic circulation. Those get the chance to reach the abdominal vessels and pass to portal veins develop further and mature in about 4 weeks.

## 10.5 GENERAL NOTES

### Excretory system in Taenia solium

Flame cells and excretory ducts together constitute the excretory system. There are four excretory ducts, two are dorsal vessels and two are ventral vessels. Dorsal vessels are short, thin and small and present only in the anterior proglottids. They are also known as the *dorsal excretory ducts* or *dorsal collecting vessels*. Ventral vessels are situated one on each side of the body and run along the entire length. These are known as the ventral excretory ducts. All four vessels join in the scolex to form a ring vessel. Dorsal vessels carry the excretory materials to the scolex and ventral vessels are not uniform in diametre and are joined by a transverse excretory canal in the posterior margin of each proglottid. Excretory canals are lined by the cuticle and nonciliated. They are connected with the small ciliated branches which ramify into fine capillaries. Each capillary terminates into a flame cell. Flame cells are distributed all over the parenchyma.

### Flame cells

Under light microscope, each flame cell is found to consist of a large nucleus. Plasma membrane produces many pseudopods like processes to increase the absorbing surface area. A row of basal granules are present in front of the nucleus. From each basal granule, a cilium originates and ultimately a bunch of cilia is produced. Inner margin of plasma membrane forms the inverted funnel like tube within which the bunch of flagella flickers like flame of a candle. Hence the name flame cells. Each flame cell is connected with a capillary branch. Flame cell cytoplasm contains many excretory granules as found under electron microscope. Each flame cell (protonephridium) appears to be a complicated structure.

According to Wilsow and Webster (1974), inner closed part of a protonephridium forms a very complicated organ which is known as *terminal organ*. Terminal organs formed by the union of two different cells and have finger like projection. One of them is flame cell containing a bunch of cilia representing flame. Other cell is tubular which is formed by coiling extra cellular cytoplasm. Marginal parts to cytoplasm are so arranged that it appears as a zip fashioned chain. Two sets of projection emerged from the cytoplasm of the flame cell. Their actual function is unknown. Of these two, one is the outer set and the other is the inner set. Each of the two ventral vessels after meeting with the large transverse vessel becomes swollen to form a caudal vesicle, which opens outside through the excretory pore. When last proglottids are shed, caudal vesicles are also shed with them. Thereafter no caudal vesicles are formed. Ventral vessels open outside directly.

### Physiology of excretion

Many flame cells are found all over parenchyma. In

cytoplasm of flame cell, nitrogenous wastes diffuse. These nitrogenous wastes from globules and are transported to capillary by flickering of cilia. Each capillary vessel opens into the small collecting tubules. Collecting tubules open into four excretory vessels. All excretory materials are collected in the caudal vesicle. Pulsating action of caudal vesicle expels excretory materials out through the excretory pore. Flame cells not only help in excretion, but also maintain the osmoregulation (Figure 10.17).

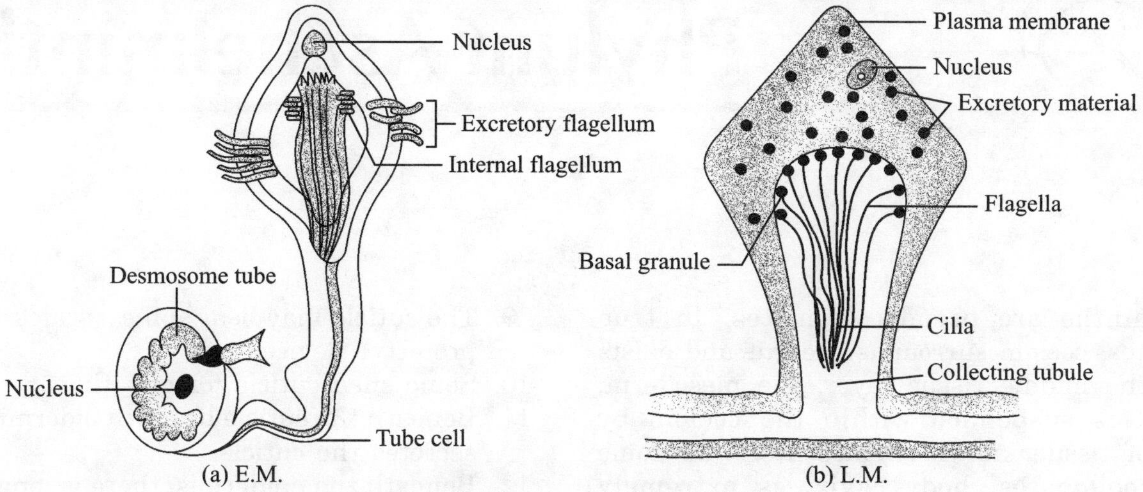


**Figure 10.17** Flame cell of *Taenia* under E.M. and L.M.

## REVIEW QUESTIONS

### Short Answer Questions

1. What is the common name of *Fasciola hepatica*?
2. Name the disease caused by *Fasciola*.
3. Mention the scientific name of intermediate host of liver fluke.
4. From which germ layer flame cells are derived.
5. Which are the organs of attachment in platy helminthes?
6. Comment on the Laurer's canal.
7. Comment on cirrus and cirrus sac.
8. What is ootype?
9. What is shell gland or mehlis gland?
10. Why fasciola is digenetic?
11. Mention two morphological adaptations of liver fluke.
12. Mention four anatomical adaptations of *Fasciola*.
13. Mention the physiological adaptation of liver fluke.
14. Mention adaptative features in the life cycle of *Fasciola*.
15. Which stages in the life history of liver fluke infects the primary host.
16. Mention the correct sequence of various larvae in liver fluke.
17. What are the excretory organs in Platyhelminthes?
18. Name the larval form of *Fasciola* which penetrates snail.
19. How is gas exchange done in flatworms?
20. Name an example of fresh water flatworm.
21. Are flatworms monoecious or dioecious?
22. What is the direct development?
23. Name the classes of Platyhelminthes.
24. Name a human disease caused by platyhelminthes?

### Long Answer Questions

1. Classify Platyhelminthes up to order with examples.
2. Describe the life cycle of *Fasciola hepatica* with suitable diagrams.
3. Describe the life cycle of *Taenia solium* with suitable diagrams.
4. Write a note on *Echinococcus granulosus*.
5. Describe the morphology and life cycle of *Schistosoma haematobium* with suitable diagrams.

CHAPTER 11

# Phylum Aschelminthes

Aschelminths are pseudocoelomates. In true coelomates, coelom surrounds the gut and exists within the middle tissue layer, the mesoderm. The gut is suspended within the coelom by sheets of tissue known as mesentery. In some pseudocoelomates, body cavity is extremely small. In others, body cavity is extensive, does not exist within the mesoderm, without true mesenteries, and their development in the embryo is different. The Aschelminthes is an obsolete Phylum of pseudocoelomate. The term *Aschelminth* is generally used as an informal name for several invertebrate groups. According to Marshall and William (1974), Phylum Aschel-minthes contains the Class Nematoda, Class Rotifera, Class Gastrotricha and Class Kinorhyncha. Rotifera has been described previously. In this chapter, Class Nematoda has been dealt with in details. A preliminary account of the Classes Gastrotricha and Kinorhyncha has also been given.

## 11.1 GENERAL CHARACTERS

1. Pseudocoelomate.
2. Most have complete digestive system.
3. The nematomophs lack a complete digestive system.
4. Most members have a muscular pharynx.
5. Protonephridia are found in the fresh water forms.
6. Primitive brains and some sensory structures are present.
7. Most are dioecious.
8. Life cycles are usually simple except the parasitic forms.
9. The cuticle may bear spines, scales, or other protectve devices.
10. Some shed cuticle for growth.
11. Beneath the cuticle lies the epidermis, which secretes the cuticle.
12. Beneath the epidermis, there is longitudinal muscle.
13. Most are fresh water animals, a few are marine.

## 11.2 CLASS KINORHYNCHA

1. Less than 1 mm long, elongate, bilateral worms, and found exclusively in the marine environment.
2. No external cilia, or locomotor appendages.
3. There are about 150 known species.
4. Body is composed of units called the *zonites*.
5. Head, the zonite 1 bears mouth and oral cone besides spines.
6. Neck, the zonite 2 has spines called the *scalids* and plates called the *placids*.
7. Head could be retracted into the neck.
8. Each trunk zonite possess pair of lateral spines and one dorsal spine.
9. Body wall is composed of the cuticle, epidermis and two pairs of muscles.
10. Digestive system is complete.
11. Feed on the diatoms, algae and organic matters.
12. Minute, cylindrical animals.
13. The head is covered with circlets of spines.

### *Order Cyclorhagida*

1. The closing apparatus works with the 2nd zonite.

**Example:** *Echinoderes.*

*Order Homalorhagida*

1. The closing apparatus works with the 3rd zonite.

**Example:** *Pycnophyes.*

*Order Conchorhagida*

1. The closing apparatus is not placed.

**Example:** *Semnoderes.*

## 11.3 CLASS GASTROTRICHA

1. Small group of free living, microscopic animals.
2. Found in marine and fresh water habitats.
3. The body is worm-like and unsegmented.
4. Adhesive glands are present.
5. Protonephridia are provided with flame bulb.

*Order Macrodasyoida*

1. Exclusively marine.
2. Male reproductive system is always present.
3. Protonephridia are absent.

**Example:** *Platydasys.*

*Order Chaetonotoida*

1. Almost exclusively fresh water
2. Male reproductive system is absent.
3. Protonephridia are present

**Example:** *Neodasys, Xenotrichula.*

## 11.4 CLASS NEMATOMORPHA

1. Superficially resembles the Nematoda
2. Cloaca is present in both the sexes.
3. Lateral cords and excretory system are absent.
4. Pseudocoelomate, filiform.
5. Digestive system degenerates at one end, or both ends.
6. Fresh water forms except *Nectonema*, which is marine.
7. Juveniles are parasitic on arthropods.

**Example:** *Gordius, Paragordius.*

## 11.5 CLASS NEMATODA

The nematodes constitute a very adaptable phylum. The term *nematoda* indicates the thread like appearance of the worms. Nematodes are generally tapered at two ends, unsegmented and commonly known as *roundworms,* or *threadworms*. Nematode infection in human is widespread. Parasitic nematodes are very successful and important group inhabiting vertebrates, invertebrates and plants, with the biomedical and the agricultural importance. About 100,000 species of nematode are described so far from vertebrates alone. Approximately, one out of four individuals in average of world's population suffers from a nematode infection of some sort. Infections have a wide distribution and found in both temperate and tropical climates.

The body is covered with a heavy cuticle. The body cavity is not a true coelom, but a pseudocoel. Highly developed sensory and locomotor organs are absent. During direct development, distinctive free larval stages are lacking. The nematodes include parasites of plants, animals and many nonparasitic free-living species. Free-living nematodes are exteremely abundant in the soils and sediments, where they feed on bacteria and detritus. Plant parasitic nematodes cause disease of economically important crops. Well-known parasitic nematodes include hookworms, pinworms, Guinea worm (*Dracunculus*), and intestinal roundworms (*Ascaris*).

Human-inhabiting nematodes found in the body fluids such as blood, lymph channels and intestine. They can withstand a wide range of external conditions of osmotic pressure, temperature and pH, utilize a wide range of food sources and survive under both aerobic and anaerobic conditions. In length, they range from a fraction of a millimetre to more than a metre. There are no appendages and no segmentation, though in some free-living forms striations, cuticular scales, or bristles may give a superficial appearance of segmentation.

## 11.6 CHARACTERS OF THE CLASS NEMATODA

1. External cuticle is modified as alae.
2. The cuticle may bear spines and bristles.
3. Presence of the stylet, or teeth, cutting plates, rasps, or denticles.
4. A valve between the oesophagus and intestine may be present.
5. Phasmids may be present on tail.

6. Presence of renette cells, or lateral canals.
7. Mostly bisexual, a few are the hermaphrodite.
8. Feed on organic material—rotting organic debris, living tissues of plants, animals, fungi, etc.
9. Microscopic to several metres long.
10. Many are parasitic, but most are free living.
11. Lack cilia and often move by the thrashing via longitudinal muscles.

**Example:** *Wuchereria, Trichinella, Ascaris.*

## 11.7 MORPHOLOGY

Majority of nematodes are small, typically spindle shaped, bilaterally symmetrical, with remarkable feature of stylized body plan and stylized life cycle. The body wall is composed of an external, noncellular layer, the cuticle which is elastic and smooth, directly beneath which lies the subcuticule. Cuticle plays an important role in its physiology and grows vectorially from a central growth zone. Spines, bristles, warts, papillae may be found on the cuticle. The cuticle consists of epi-, exo-, meso- and endo- cuticle. Thin, cellular, or syncytial hypodermis lies beneath the basal lamina. According to some workers, four main layers in cuticle are cortical-, fibrillar-, homogenous- and fibre- layer. The fibre layer is often reduced especially in the larvae. The nervous system modulates spontaneous activity of the muscle sheet. A layer of partially differentiated muscular cells lines the outer wall of the body cavity and comprises the chief bulk of the body wall.

In *Ascaris lumbricoides*, the cuticle contains 80% collagen, which is stabilized by the sulphydryl links. This muscular layer is not a continuous lining of the body cavity, as it is interrupted by the slight breaks in four regions, which are designated as the dorsal-, ventral-, and lateral-lines. The fairly conspicuous thickening of the subcuticula stands in the middle of each lateral surface is called the *lateral line*. Less pronounced intrusions of the subcuticula occur in the middorsal and midventral lines. Near the anterior extremity of the body, the ducts of the excretory system are united and communicate with the exterior through a single excretory pore on the ventral surface of the body.

## 11.8 MUSCLES

One, or more layers of longitudinally arranged muscle cells called the *somatic musculature* connect to the cuticle by the fibres. The circular, or diagonal muscle is absent. The longitudinal muscle cells are unique in connection with nerve rather than the muscle. The elastic components restore force in the cuticle, when the longitudinal muscles contract and the two are linked hydrostatically via the pseudocoel. During movement, internal hydrostatic pressures can be up to 200 mm Hg. Each muscle receives a dual innervation from an excitatory and an inhibitory nerve fibre. The excitatory fibres release acetylcholine for the muscle contraction, while inhibitory nerve fibres release 4-aminobutyric acid for inhibiting the muscle contraction. Other nerve fibres may release catecholamines. Several neuropeptides function as neurohormones, coordinate maturation, reproduction, and moulting.

## 11.9 DIGESTIVE SYSTEM

The digestive tract is complete and consists of mouth, buccal cavity, pharynx, intestine, rectum and sub-terminal anus. The gut comprises fore-, mid- and hind- gut. In the male, the cloaca is present. The rectum and vas deferens open into the same cavity. The pharynx, with a Y shaped lumen, pumps food into the intestine against the high internal pressure. It is believed that pharynx lining is a resilin type elastic protein, which works by a click mechanism. The pharynx is pulled open by the radial muscles, the kinetic energy for which is stored in the pharynx lining. When radial muscles relax, the pharynx snaps shut. The pharyngeal muscles are very peculiar electro physiologically, and are extremely fast relaxing. The intestine is one cell thick. The muscle coat around the intestine is absent.

The food moves primarily by the body movements. A sphincter lies between the intestine and rectum for preventing the contents of intestine being shot out, when the rectum is opened. The anus remains closed not by a sphincter muscle, but by internal pressure. A dilator muscle pulls the anus open. The high internal pressure expels the rectum contents. The digestive system is a straight tube with a mouth at the anterior extremity and the anus on the ventral surface slightly in front of the posterior tip. The mouth is usually surrounded by liplike organs. The parasitic forms frequently bears the chitinized structures for anchoring the host tissues. Behind the mouth, there occurs a muscular oesophagus with an especially conspicuous pharyngeal bulb. The stomach-intestine continues posteriorly as a flattened tube. In the male, the genital ducts open into the posterior region of the digestive tube which is transformed into the cloaca.

## 11.10 NERVOUS SYSTEM

Fairly simple nervous system possesses a nerve ring around the pharynx and associated with this are several small ganglia. Six, or 8 longitudinal nerves run posteriorly from these ganglia. The ventral nervė chord is the largest. The circumoesaphageal commissure encircles the oesophagus, and rectal commissure encircles the rectum. These commissures are two major nerve centres. The anterior longitudinal-, posterior longitudinal-, ventral longitudinal- and the peripheral nerves are present. Both mechanoreceptors and chemoreceptor are found. The papillae innervated by papillary nerves are mechanoreceptors. The prominent paired sense organ, the amphids lies on either side of head and innervated by amphidial nerves, are chemoreceptors. They are glandular, produce secretions to respond to sex pheromones. Amphids have a combined sensory/secretory function. Another pair of sense organs on tail is called the *phasmids*. The campaniform sensillae like sense organs called *deirids* as well as eye-spots are found in some species.

## 11.11 EXCRETORY SYSTEM

Two basic types of excretory system are glandular type and tubular type. The first type consists of a large cell called the *ventral gland* in pseudocoelomic cavity, which is connected to the excretory pore by a duct that ends in an ampulla. Several types of tubular excretory system are found. In *Ascaris*, it consists of two longitudinal canals in the hypodermal cords, connected anteriorly with each other by a cross canal, and connects to excretory pore through a medial duct. This forms an H system. The modification of H system results due to loss of one, or more arms of the system to produce Y, or M shaped systems. Walls of excretory canals are lined with microvilli. It is believed that like flame cell system, ultrafiltration occurs in the canal system followed by reabsorbtion. The driving force for ultrafiltration is provided by high internal pressure. Now it has shown that filtration does not occur in the canal system. Urine forms by secretion.

The chief end product of nitrogenous metabolism is ammonia. Some infective larvae excrete large quantities of amines. Amines are strongly alkaline, very toxic and may help to neutralize acidic end products. Parasitic nematode breaks down carbohydrate to acidic end products such as lactate, succinate and acetate. Amines disrupt the keratin layer of skin, or help to disperse mucus and aid in penetration. The excretory system also plays an important role in osmoregulation by excreting water and ions. The gut plays role in nitrogenous excretion and ionic balance. The parasitic nematodes often have relatively limited osmoregulatory capacity. They are osmoconformers with varying degrees of volume regulation. Gonad lies in the body cavity. In animal inhabiting nematodes, often 4 to 8 large stellate cells called the *pseudocoelomocytes* with unknown function remain in fixed positions in the body cavity. Probably, they are concerned in haem metabolism. One, or two renetts empty through an excretory pore. The respiratory, or circulatory organs are absent (Figure 11.1).

## 11.12 REPRODUCTIVE SYSTEM

The male is recognizable by its smaller size and more pointed and curved posterior extremity. From the cloaca of the male, the copulatory spicules are protruded. The reproductive organs are very simple. Normally, they consist of a gonoduct, which is directly continuous with the gonad. This tube in the male is usually single, while in the female it is often bifurcated with the opening of a single vagina occurring on the ventral surface. In some forms, the ducts and gonads are extremely long and coiled throughout the greater part of the body length, while in others, they are a single straight rod. Nematodes reproduce only sexually. The eggs are fertilized within the female oviduct and are

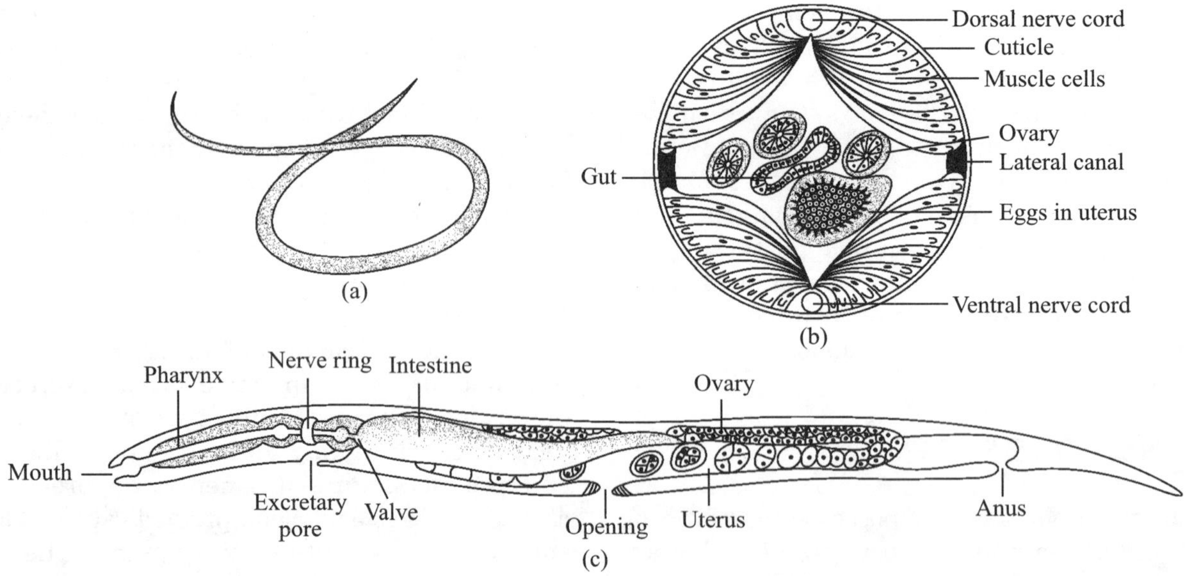


**Figure 11.1** The structure of nematodes: (a) External view, (b) Transverse section based on *Ascaris,* (c) Longitudinal section of female nematode.

discharged either before the development has proceeded far, or are retained until the young are fully developed. Sometimes, the eggs are not laid, but are retained in the body of the female until the young have left the egg membranes and are thus brought forth viviparously. Parthenogenesis occurs sometimes (Figures 11.2 and 11.3).

Both sexes are separate. Male is smaller than female. Male reproductive system consists of testes, vas deferens, seminal vesicle, and ejaculatory duct. The accessory sex organs like spicules allow internal fertilization. The copulatory bursa and genital alae may be present. The female possesses one, or two tubular ovaries, which lead to a uterus, or united uteri. The uterus opens through the vulva. The female are viviparous, oviparous, or ovoviviparous. Parthenogenic females and hermaphrodites occur in some species. The sperm is amoeboid. In some, sex ratio depends on the population density, males being produced when population density is high. The female reproductive system maintains a prodigious egg output. They are monarchic, when with one set of male reproductive system and diarchic, when with two sets of male reproductive system. They are called *monodelphic*, when with one set of female reproductive system and didelphic, when with two sets of female reproductive system.

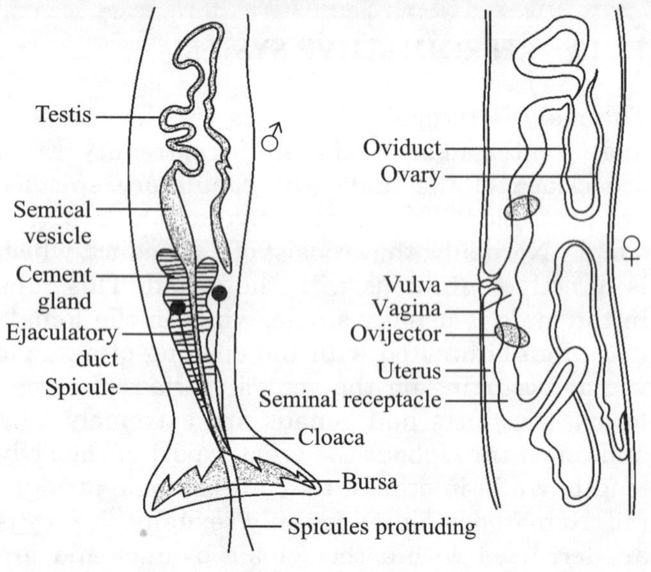


**Figure 11.2** Genital organs in hookworm (Adapted from Faust).

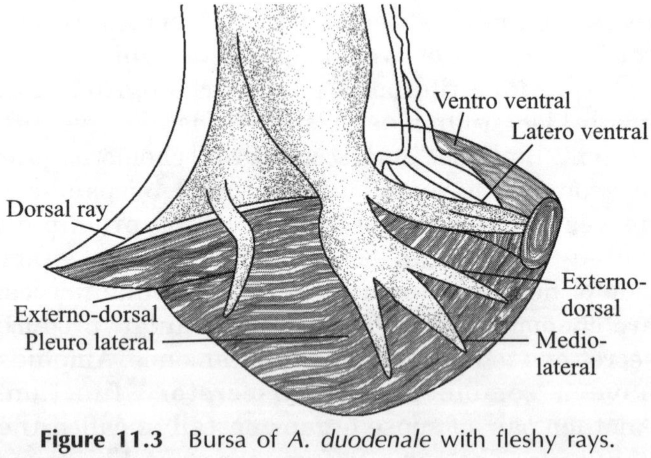


**Figure 11.3** Bursa of *A. duodenale* with fleshy rays.

## 11.13 LIFE CYCLE

In most intestinal nematodes, all developmental stages occur in a single host. These species are monoxenous. To be infective, the eggs of such species remain in the soil. These species are called *soil-transmitted nematodes*. Some nematodes need an intermediate host for the development. Most of these species live outside the digestive tract. These are heteroxenous. In another group, larvae of worm migrate to other organs of the same host and cause infection. Example is *Trichinella*, in which pig and human can act both as intermediate and final hosts. These are autoheteroxenous. Others that successfully invade intestine are generally larger, enter the tissues and grow to relatively enormous lengths.

One characteristic feature of some nematodes is migration within the tissue of host. *Ascaris* larvae penetrate the gut wall, migrate via the blood and lymphatic systems to the heart, and then lungs, where they break into alveoli, coughed up and swallowed again. Thus, they pass through the stomach twice. Such migration causes tissue damage. Invasion of lungs by the larvae result in parasitic pneumonia. It is suggested that originally *Ascaris* may have had a two-host life cycle, one of which has lost, but the migration remains. In an abnormal host, the infective larvae may undergo considerable migrations and in wrong host the larvae may finish up in aberrant sites. The larvae of *Toxocara canis* undergo aberrant migrations in humans and show a predilection for brain and optic nerve. In children, this causes blindness and is linked with epilepsy and asthma.

Skin penetration by the infective larvae is thought to be a primitive condition. Migration may be a relic of a previous host, which is now lost. Tissue migrations seem to be necessary for the normal development. L3 hookworm larvae placed directly in intestine do not develop. Larvae can migrate twice. If removed from intestine and injected back into tissues, they once again find their way to intestine. Migration may be necessary to provide developing larva with specific stimuli, or nutrients. Tissue may be the favourable site for parasites, but they are forced to leave the tissues and migrate under the influence of host immune response. Success of parasitic nematode is attributed to their stylized body plan and stylized life cycle. The life cycles of parasitic nematodes also involve three hosts as in Gnathostomatidae, the intestinal parasites of carnivores. The adult *Mermis* is free-living, while the larval stages are parasitic in insects. Gut of adult *Mermis* lacks anus and is just a blind ending sac.

The stages of development are usually called the *larvae*. After the fourth moult, the adult emerges and development of gonads occurs. The adult continues to grow in size without further moults. In many nematodes, third stage larva is often a resting, or resistant stage (dauer larva). Parasitic nematodes retain the basic life cycle of four moults and four larval stages, and in vast majority, the L3 is the infective stage. It is often a nonfeeding resistant stage. Often they retain a sheath formed by the shed second-stage cuticle. In some, the life cycle is shortened by retaining the larvae within the egg, but larva inside the egg undergoes the normal sequence of moults. They do not undergo asexual reproduction at any stage during their life cycle. Parasitism has arisen independently on at least seven separate occasions. It is possible to arrange nematode life cycles in order of increasing complexity.

## 11.14 INTESTINAL NEMATODES

### 11.14.1 *Trichuris trichiura*

*T. trichiura*, first described by Linnaeus, is commonly known as the *whip worm* due to whip-like form of the body. They have a cosmopolitan distribution though commonly found in the tropical climates and in areas with poor sanitation and occur in areas, particularly where the *Ascaris* and hookworm are found, as its eggs require the same conditions to allow for the embryonation. Both species can be found in the human together. *T. trichiura* cause trichuriasis. It is reported to infect up to 800 million people throughout the tropics and temperate regions.

#### *Morphology*

The anterior three-fifth of the worm is thin. The posterior two-fifth is thick and stout. The anterior end contains the oesophagus. The posterior portion contains intestine and reproductive organs. The posterior end is coiled ventrally in male, while in female it is comma shaped. The male is 30 to 40 mm in length and in case of female the length is 40 to 50 mm. The adults possess enormously elongated capillary-like oesophagus with anus situated in extreme tip. Elongated oesophagus remains embedded in the mucosa of intestine. There is no lip. The vulva lies at the junction of thread-like thickened region of the body. The posterior end is thicker and lies free in lumen of the large intestine. The ova are barrel shaped, bile stained with bipolar plugs. The ova measure 50–54 μm by 20–23 μm.

### *Life cycle*

The adult worms inhabit the caecum and upper part of the colon. In heavy infection, they are also found in the terminal ileum. They attach to the mucosa by embedding the anterior portion of body in the superficial tissues. They are oviparous. The fertilized female lays about 5,000 brown, barrel shaped eggs per day. The eggs require a warm, moist environment with the plenty of oxygen for the embryonation. In the embryonated condition, they are extremely resistant to the environmental conditions. The eggs are unsegmented at oviposition and are passed out along with the faeces (Figures 11.4–11.5).

The eggs undergo an embryonation period in soil, which may last up to the several months after which they become infective. When the human swallows embryonated eggs, the larvae are released in the upper duodenum. The larvae attach themselves to villi lower down the small intestine, or may invade the intestinal walls. After few days, the juveniles migrate down towards the caecum, attach them to the mucosa, and reach their final attachment site. The larvae reach maturity within a month after infection through four moulting. There is no lung migration. The time from the ingestion of the infective eggs to the development of adult is about three months. The infection is transmitted by swallowing contaminated soil or through the ingestion of contaminated food, or drinks. By the action of digestive enzymes, the eggshell dissolves and the larva comes out. The liberated larvae reach caecum and attain maturity in due course. The worm has a life span of 5 to 10 years. The children are commonly getting infected, as they are more likely to swallow soil.

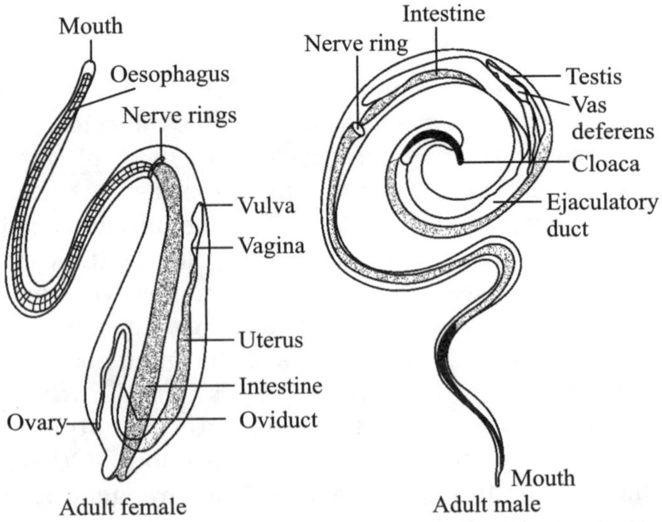


**Figure 11.4** *Trichuris trichura.*

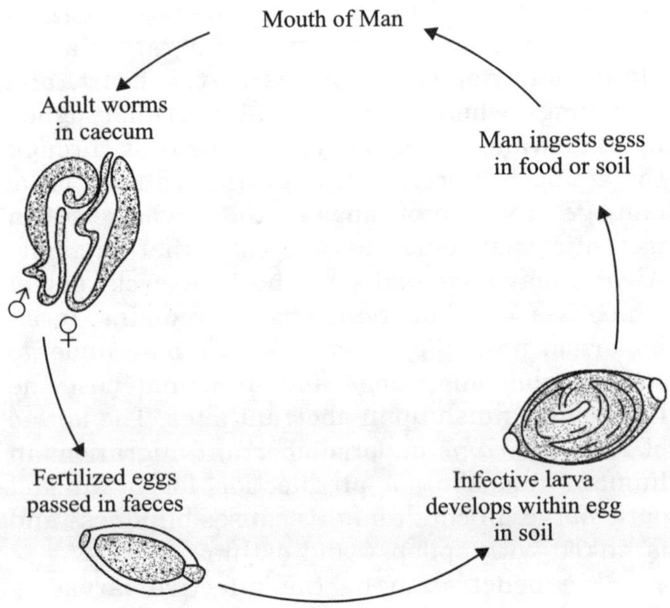


**Figure 11.5** Life cycle of *Trichuris trichura.*

## 11.14.2 *Trichinella spiralis*

*T. spiralis* was first observed by James Paget. This was named and described by Professor Richard Owen. The family Trichinellidae contains the only genus *Trichinella* and was originally thought to contain one species. *T. spiralis* causes the serious and often fatal disease called the *trichinosis* (trichinelosis). It is a parasite of carnivorous animals and is common in rats and in swine fed on uncooked garbage and slaughter house scraps. Human infection is caused mostly by eating raw pork with sausages. It is cosmopolitan and prevalent in many European countries with the highest incidence in China.

Adult female is about 2–3 mm long and 90 μm in diametre. Male is smaller, measuring 1.2 mm long by 60 μm in diametre. Adult females are ovoviviparous and up to 1,500 larvae may be released by a single worm. Four varieties of this species exist worldwide. *T. s. spiralis* found in the temperate zone is highly infective for pigs, rats and man. *T. s. nelsoni* found in tropics, shows low infectivity for pigs and rats, and shows high infectivity for lions and hyenas. *T. s. nativa* found in Arctic shows low infectivity for pigs and is found in polar bears and resistant to freezing. *T. s. pseudospiralis* found in New Zealand exhibits low infectivity for pigs, rats and mice. *T. spiralis* has long been recognized to cause a zoonosis which is transmitted to man by ingestion of infected pork (Figures 11.6–11.7).

### *Life cycle*

Infection in definitive hosts is acquired by eating raw or undercooked pork containing encapsulated larvae. Rats are known to be the most highly

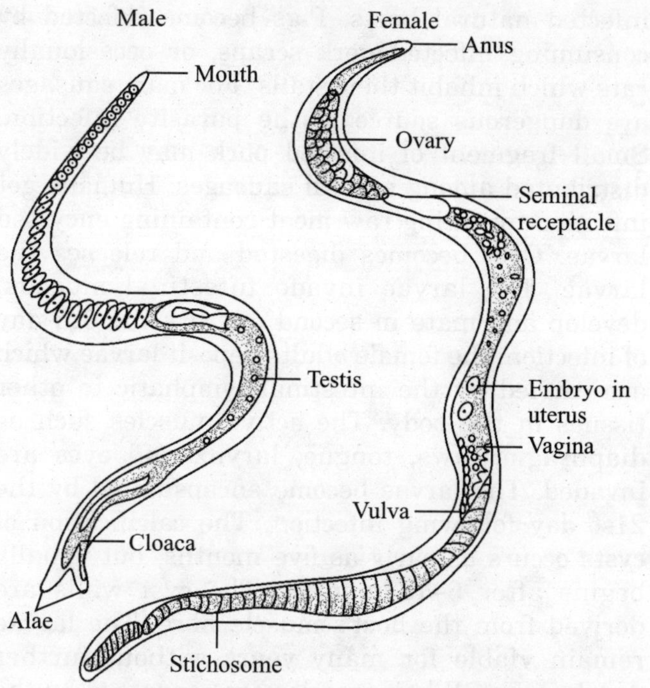


**Figure 11.6** *Trichinella spiralis.*

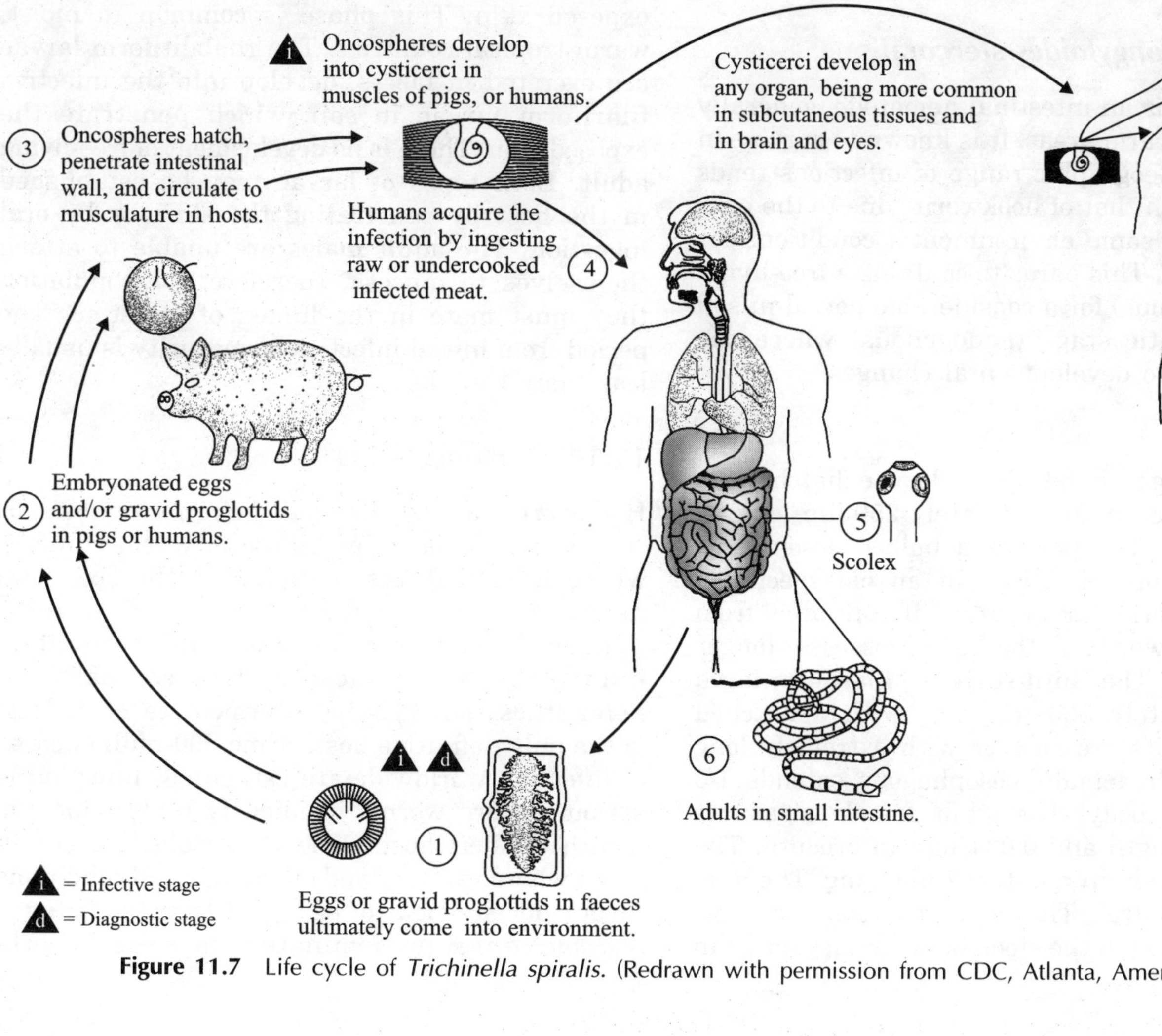


**Figure 11.7** Life cycle of *Trichinella spiralis.* (Redrawn with permission from CDC, Atlanta, America)

infected natural hosts. Pigs become infected by consuming infected pork scraps, or occasionally rats which inhabit their stalls. For man, sausages are dangerous source of the parasite infection. Small fragment of infected pork may be widely distributed among several sausages. Humans get infection by eating raw meat containing encysted larvae. Cyst becomes digested and releases the larvae. The larvae invade intestinal mucous, develop and mate in second day. By the 6th day of infection, the female adults deposit larvae which are carried by the intestinal lymphatic to other tissues in the body. The active muscles such as diaphragm, jaws, tongue, larynx and eyes are invaded. The larvae become encapsulated by the 21st day following infection. The calcification of cysts occurs as early as five months, but usually begins after 6–18 months. The cyst walls are derived from the host's muscle fibre. The larvae remain viable for many years without further development. When muscles are consumed by the definitive host, sexual maturation rapidly occurs.

## 11.14.3 *Strongyloides stercoralis*

*S. stercoralis* is an intestinal nematode, generally found in the warm areas. It is known to survive in sub-tropics. Geographic range of infections tends to overlap with that of hookworm, due to the eggs requiring the same environmental conditions for embryonation. This parasite contains a free-living stage (exogenous) for a considerable period in soil and a parasitic stage (endogenous) where the larvae undergo developmental changes.

### *Morphology*

The first stage rhabditiform larvae hatch from the eggs measure approximately 250 μm long by 20 μm wide. They possess a bulbed oesophagus and a short buccal cavity. In an old specimen, the rhabditiform larvae are differentiated from those of hookworm, as the latter possess a longer buccal cavity. The third stage, or filariform larva is approximately 500 μm long with a notched tail. The adults are slender with extremely long oesophagus. In female, oesophagous extends 1/3 to 1/2 of the body. The adult female measures 2.2 mm in length and 0.034 mm in breadth. The male is much shorter, only 0.7 mm long. The anal opening is ventral. The tail is pointed. The eggs are rarely found in the stool, because they hatch in the intestine. The eggs are oval and thin shelled, measuring 50–58 μm by 30–34 μm.

### *Life cycle*

The worm is ovoviviparous and bears the resemblance with the life cycle of hookworm. The life cycle is complex. After attaining maturity in intestine, the female begins egg production. The eggs hatch into rhabditiform larvae (noninfective) and are passed out in the faeces. The rhabditiform larvae transform into the filariform (infective) larvae, which penetrate the skin that is exposed to soil. The larvae migrate via blood to the heart, alveoli of lungs, trachea and are ultimately swallowed. The filariform larvae in intestine can reinfect host by penetrating the intestinal mucosa, perianal, or perineal skin. The larvae undergo direct development (homogonic), or indirect development (heterogonic). The noninfective (rhabditiform) larvae develop into the free-living adults in soil within 2–5 days and produce infective third stage, or filariform larvae, which could penetrate the exposed skin. This phase is common in moist, warm tropical countries. The rhabditiform larvae are excreted in faeces, develop into the infective filariform larvae in soil, which penetrate the exposed skin. There is no development of free-living adult. Both types of larvae may be established in the host by penetrating the skin, or by oral ingestion. The adult males are unable to attach themselves to mucosa, therefore, for copulation they must mate in the lumen of intestine. The period from initial infection to maturity is usually less than 4 weeks.

## 11.14.4 Hookworms

Hookworm disease, also called *ancylostomiasis*, or "miners anaemia" is a disease of warm climates where human faeces contaminate the soil. Two nematode (round worm) species, *Ancylostoma duodenale* and *Necator americanus* inhabits human. They are most destructive parasitic helminthes. There is no intermediate host. Man is the only definitive host. Some 900 million cases of infection worldwide are estimated. Infection is serious where worms insidiously undermine the health of their hosts. They particularly occur in areas where sanitary and environmental conditions favour development of eggs and larval infections. *N.americanus* predominately is a new world hookworm and was introduced from the Africa to the Western Hemisphere. It is also recorded in the Far East, Asia, Africa, South America and Oceania.

*A. duodenale* is an old world hookworm and is the only species of Europe and areas bordering the Mediterranean. It is also recorded in the Middle East, North China, Africa, Asia and South America. Adult worms inhabit the small intestine, mostly the jejunum, less often the duodenum and infrequently the ileum by their powerful buccal armature. Both species have the similar morphology; female measures approximately 10–13 mm and male 8–11 mm. They are about twice the size of the rat hookworm. The male has a posterior copulatory bursa, which is absent in female. A vulvar opening of female is found almost one third of body length from the posterior end. The females have two ovaries. Most of the female body is filled with the eggs. The mouth shows a pair of chitinous plates on the dorsal surface. The buccal cavity bears two hooks like teeth on top and two triangular cutting plates on bottom. The mouth of *N. americanus* has four cutting plates, two on ventral and two on dorsal surfaces. The head is curved in both the species.

In *Necator* adults, the head is more pronounced forming a definite hook at anterior end. The buccal cavity attaches worms to the mucosa of small intestine. The teeth and cutting plates pierce the mucosa. The bursa of male is well developed. Adult *Necator* is distinguished from *Ancylotstoma* by the split of dorsal rays and close arrangement of lateral rays. Hookworm disease is rare outside the tropics as larvae cannot survive in the colder climates. Cold, acidic, or alkaline conditions, flooding, exposure to direct sunlight, dry conditions, and salt kill hookworm larvae. Relatively stable warmth and humidity allow hookworm larvae to survive in mines, even outside the tropics, which signifies the name "miners anaemia."

*A. duodenale* causes more severe illness with lower number of worms than *N. americanus*. Exposure to hookworms moderately affects the human immune system, reducing the risk of allergies and some other diseases. Hookworm disease could be mild, when only a few worms are present. People with long-term infection with higher number of worms become thin, tired, and weak, and the associated anaemia causes a pale yellowish skin color. Symptoms of infection include abdominal pain and malnutrition through bloody diarrhoea, vomiting. Symptoms vary widely and depend on the nutritional status and overall health of the human host, as well as the species of worm and number of individual worms involved.

### 11.14.5 *Ancylostoma duodenale*

The name *Ancylostoma* is derived from two Greek words, *ankylos* meaning hook and *stoma* meaning mouth, which signifies the curved anterior end of worm that give it a hook-like appearance. It is a blood feeding intestinal nematode. It was discovered in 1838 by the Italian physician Angelo Dubini. Looss studied the pathogenesis and modes of transmission. It is found in all tropical and subtropical countries, where humidity and temperature favours the larval development. It is prevalent in the Europe, North Africa especially in Egypt, South America, Caribbean islands, India, Ceylon, Central and North China and Pacific islands.

**Habitat:** The adult worms inhabit the small intestine of the infected person, mostly in the jejunum, less often in the duodenum, and infrequently in the ileum by their powerful buccal armature.

**Adult worm:** They are relatively stout, cylindrical, pale pink, or grayish white in colour, but may be reddish brown due to the ingested blood. The body is curved with the concave dorsal aspect and the convex ventral aspect. The somewhat constricted anterior ends bents dorsally. The mouth is not at the tip, but directed dorsally. The large and conspicuous buccal capsule is lined with a hard substance and is provided with 6 teeth, 4 hook-like on the ventral surface and 2 knob-like on the dorsal surface. Out of five glands, the oesophageal gland remains connected with the digestive system and secretes a ferment, which prevents blood clotting.

**Male:** The adult males are 8 to 11 mm in length and 0.4 mm thick. The posterior end is expanded into a compulatory bursa, which consists of 3 lobes, 1 dorsal and 2 lateral. The lobes are provided with the chitinous rays. The dorsal lobe contains 3 rays, while the 2 lateral lobes contain 10 rays. The total number of rays is thus 13. The cloaca into which rectum and genital canal open is situated within bursa. The two long, retractile bristle-like copulatory spicules are present, tips of which project from the bursa.

**Female:** They are relatively larger, 10 to 13 mm long and 0.6 mm thick. Conoid hind end with a sub terminal anus is situated ventrally. Vulva opens ventrally at the junction of middle and posterior thirds of the body. The vagina leads to two intricately coiled ovarian tubes, which occupy the hind and middle parts of the worm. During copulation, the male attaches its copulatory bursa with the vulva. The copulating pair thus presents a Y-shaped appearance. The life span of an adult worm in the human intestine is estimated to be about 4 years.

**Eggs:** Eggs are oval, or elliptical, measuring 60 μ by 40 μ, colourless, not bile stained and having a thin transparent hyaline shell membrane. When released in the intestine, eggs contain an unsegmented ovum. During their passage through the intestine, ovum develops. When passed in the faeces, egg contains a segmented ovum usually with the 4 blastomeres. A clear space between the segmented ovum and the eggshell is present. Eggs float in the saturated salt solution. A single female lays about 20,000 eggs per day.

**Life cycle:** The man is the only definitive host. No intermediate host is required. Multiplication of the worm does not occur inside the human body. Adult worms live in small intestine, attached firmly to mucous membrane of gut lining, and feed on blood and tissue. Adult females deposit their eggs whilst in gut. They can produce up to 20,000 eggs per day. Eggs are then passed out in faeces. Rhabditiform larvae hatch from eggs in warm, damp soil and feed on bacteria. After about one week and passing through 2 moults, they become infective filariform larvae, climb into a suitable position and wait for a suitable host to pass by. Larvae enter the host by penetrating the unbroken skin, reach subcutaneous tissue, consequently enter lymphatic, or small venules, then into general blood stream and are carried to heart, lungs, trachea and larynx. They crawl over the epiglottis to back of pharynx and are swallowed. Larvae then enter into the stomach, oesophagus and lastly reach their final abode, the duodenum, or jejunum. They are then swallowed and develop into adults in the small intestine. Larvae that are initially swallowed may not show migration. Larvae live up to 6 weeks in the tropics. Various life cycle stages may be summarized as follows:

- **Passage of eggs from infected host:** Eggs containing segmented ova with 4 blastomeres are passed out with human faeces.
- **Development in soil:** The egg hatches into a rhabditiform larva (250 μ in length) in soil, in about 48 hours. Rhabditiform larva, which feed on bacteria, moults twice on third and fifth day. It then develops into a filariform larva (infective stage). The time taken for development from egg to filariform larva averages ten days.
- **Entrance into a new host:** Filariform larvae cast off their sheaths and enter the human body by skin penetration.
- **Migration:** On reaching the subcutaneous tissues, larvae enter into lymphatics, or small venules, pass through lymph vascular system into venous circulation, migrate via right heart into pulmonary capillaries, and enter alveolar spaces by breaking capillary walls. They further migrate to the bronchi, trachea, larynx, and crawl over epiglottis to back of pharynx and are ultimately swallowed. During migration, or after entering oesophagus, a third moulting takes place and a terminal buccal capsule is formed. The period taken for such migration is about ten days.
- **Localization and egg laying:** Growing larva settles down in small intestine, undergoes a fourth moulting and develops into adolescent worms. Provisional toothless buccal capsule formed previously is cast off during this time, and definitive buccal capsule with teeth is formed. In 3 to 4 weeks time, it attains sexual maturity. Fertilized females lay eggs in the faeces. Cycle is thus repeated. The interval, between the time of skin infection and first appearance of eggs in faeces, is about six weeks.

### *Mode of infection*

When the man walks bare-footed on faecally contaminated soil, filariform larvae penetrate directly through skin. The most common entry sites are thin skin between toes, dorsum of feet and inner side of sole. Larva can penetrate any part of skin. In gardeners and miners, skin of hands may be portal of entry. Infection through the accidental drinking of contaminated water besides transmammary and transplacental route is reported (Figure 11.8).

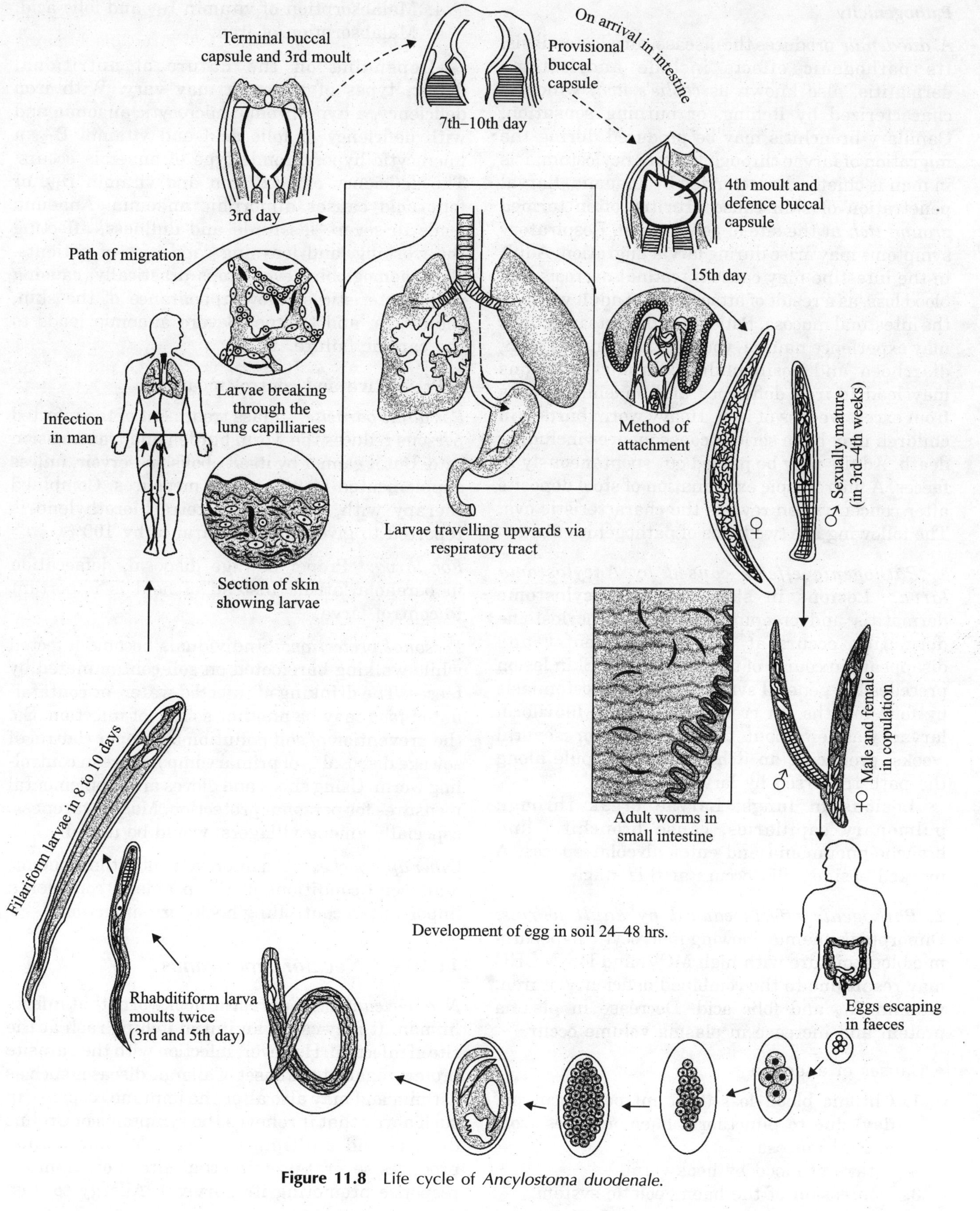


**Figure 11.8** Life cycle of *Ancylostoma duodenale*.

### *Pathogenicity*

*A. duodenale* produces the disease, ancylostomiasis. Its pathogenic effects include ancylostome dermatitis, also known as *coolie's itch,* which is characterized by itching, or burning sensation. Capillary bronchitis may be produced during the migration of larvae through lungs. Ancylostomiasis in man is chiefly characterized by anaemia. Larval penetration of skin cause pruritis, often termed *ground itch* at the site of penetration. Respiratory symptoms may arise during larval migration. Adult in the intestine may cause intestinal necrosis and blood loss, as a result of attachment of adult worm to the intestinal mucosa. Patients with acute infections may experience nausea, vomiting, abdominal pain, diarrhoea and eosinophilia. Chronic infections may lead to iron deficiency and anaemia results from excessive loss of iron. Heavy worm burden in children may have serious consequences including death. Adults may be passed out spontaneously in faeces. A microscopic examination of stool deposits after concentration reveals the characteristic ova. The following are two types of pathogenic effects:

1. *Pathogenic effects caused by Ancylostoma larva:* Lesions in skin include ancylostome dermatitis and creeping eruptions. Ancylostome dermatitis occurs at site of larval entry and disappears in course of two weeks time. Skin lesion precedes the general symptoms of ancylostomiasis by four months. In creeping eruption, filariform larvae wander about through skin for several weeks, producing an itchy reddish parpule along the path traversed by larvae

Lesions in lungs: Larvae break through pulmonary capillaries, cause bronchitis and broncho-pneumonia and enter alveolar spaces. A marked eosinophilia occurs at this stage.

2. *Pathogenic effects caused by adult worms:* Dimorphic anaemia showing macrocytic hypochromic blood picture with high MCV, and low MCHC may result due to the combined deficiency of iron, vitamin $B_{12}$ and folic acid. Decrease in plasma protein and increase in plasma volume occur.

• Causes of anaemia

1. Chronic blood loss (0.67 ml per worm per day) due to punctiform haemorrhages from jejunal mucosa.
2. Intake of blood by hookworm.
3. Depression of the haemopoietic system.
4. Malabsorption of vitamin $B_{12}$ and folic acid.
5. Malabsorption of iron.

Depending on the nature of nutritional defect, types of anaemia may vary. With iron deficiency, a hypochromic microcytic anaemia and with deficiency of folic acid and vitamin $B_{12}$, a microcytic hyperchromic type of anaemia occurs. The deficiency of both iron and vitamin $B_{12}$, or folic acid causes dimorphic anaemia. Anaemia leads to severe lassitude and dullness, affecting the working, and learning capacities of patients. The haemoglobin level drops drastically, causing a characteristic shallow appearance of the skin, conjuctiva, and tongue. Severe anaemia leads to the cardiac failure.

• Preventive and control measures

*For adult parasite*: Mass treatment of the infected persons reduces the worm burden and transmission rate. But it cannot by itself abolish reservoir, unless supplemented by sanitation measures. Combined therapy with aleopar and tetrachlorethylene is reported to give an ova clearance by 100%.

*For larva:* Proper sewage disposal, defaecation in borehole latrine instead of field is important to control larva.

*Personal protection:* Individuals become infected while walking barefooted on soil contaminated by faeces. The drinking of infected water, or contaminated food may be another source of infection. So, the prevention of soil pollution by proper control of sewage disposal is of primary importance in controlling worm. Using shoes and gloves are supplemental measures for personal protection. Mass awareness, especially among villagers, would be useful.

*Other approaches:* An improvement of the economic and social conditions of the people is considered important in controlling hookworm infection.

## 11.14.6 *Necator americanus*

*N. americanus* is a parasitic hookworm that infects human. It may cause an initial itching rash at the site of infection. However, infection with the parasite protects against the onset of allergic diseases such as asthma and may also alter the immune response in such a way that it relieves the symptoms of Crohns disease and multiple sclerosis. *N. americanus* produces several proteins that alter host immune response promoting its survival. Allergy to dust mites is reduced in infected individuals. Hookworm infection, thus, may be beneficial in the treatment of asthma. Epidemiological findings suggest that protection against asthma may be related to the intensity of parasite infection giving rise to more than 50 parasite eggs per gram of faeces. Ova are oval and transparent with a smooth, thin shell and measure 56–75 μm by 36–40 μm. They are usually passed in the 4–8 cell stage in faeces and may be embryonated.

### 11.14.7 *Angiostrongylus cantonensis*

*A. cantonensis* was first described in Taiwan in 1945 from a boy with suspected meningitis. Since then, it has been recognized throughout the Pacific areas with sporadic cases reported in other parts of the world including Central America and Cuba. It is associated with eosinophilic meningitis. *A. cantonensis* and *A. costaricensis* have now been reclassified into the genus *Parastrongylus*. *A. cantonensis* is also known as the *rat nematode*. The main cause of eosinophilic meningitis is infection with larvae. Man and rats become infected by eating infected molluscs, the giant land snail, or food which is contaminated by snails bodies. *A. cantonensis* is a long, slender worm, measuring between 17 and 25 mm long by 0.26 to 0.34 mm wide.

#### *Life cycle*

Definitive host is the rat. Adult worms inhabit pulmonary arteries. Eggs are released, and carried to lungs in the bloodstream. They break through pulmonary tract and are then swallowed. Eggs hatch in the intestine and the first stage larvae are passed in stools. First stage larvae in the rat's faeces are eaten by intermediate hosts like *Achatina, Cipangopaludina*, *Bradybaena* and also various slugs. Larvae develop within these intermediate hosts to third stage larvae. These infected molluscs are then eaten by rats where larvae can complete the life cycle. Human infection occurs by ingestion of the infective larvae in slugs, snails, or on vegetation. Larvae migrate to brain and cannot continue their growth cycle. It is the death of the larvae, which invokes inflammatory response in meninges, resulting in eosinophilic meningitis.

#### *Symptoms*

The incubation period is usually about twenty days, but can be up to 47 days. Infection in men is usually self-limiting, but can result in fatalities. Symptoms are severe headache, convulsions, vomiting, facial paralyses, parasthesia, neck stiffness and fever. It is accompanied by moderate eosinophilia in the CSF, i.e. the cell count is 100/mm$^3$–2000 mm$^3$. Eosinophilia is also common. Visual impairment, ocular pain, keratitis and retinal oedema are known.

### 11.14.8 *Angiostrongylus costaricensis*

*A. costaricensis* was first described in 1971. Human infections are most common in Costa Rica, although found in the Mexico, Central and South America. Eggs are 90 μm, oval, thin shelled and can be embryonated. The adult female measures 42 mm by 300 mm. Males measure 22 mm by 140 mm. The life cycle is similar to *A. cantonensis* with the human being an accidental host by consumption of snails and salad vegetables contaminated by infective larvae that have been shed by slugs in their mucous. Larvae of *A. costaricensus* penetrate the intestinal wall, which results in inflammatory lesions of bowel wall. Here, the life cycle ends. The common symptoms are pain and tenderness with fever, vomiting and diarrhoea. A tumour-like mass is often palpable and can mimick malignancy. The symptoms of abdominal pain, vomiting, diarrhoea and anorexia, are often mistaken for those of appendicitis. Worms can inhabit the regional lymph nodes, mesenteric and spermatic arteries causing testicular obstruction and necrosis. Sometimes, larvae and ova reach the liver and symptoms may mimic visceral larvae migrans. Eosinophilia is also found.

### 11.14.9 Anisakiasis

Anisakiasis is a collective name for infections with *Anisakis, Phokanema, Tarranova* and *Contracaecum*. Adult worms from many species belonging to family Anisakidae occur in various sea mammals, such as seals, dolphins, porpoises and whales. Larvae can cause a serious zoonosis, anisakiasis in man. Anisakiasis was first reported in the Netherlands in 1960. Since then, cases have been reported from Japan, North America, Canada, Chile and the United Kingdom with the increase in popularity of 'sushi'.

Two species of nematode have been especially responsible worldwide in causing human anisakiasis, *Anisakis simplex* and *Pseudoterranova osculates*.

The larvae of *P. osculatum* can be isolated from A. *simplex* by the possession of a caecum. *Anisakis* larvae are usually found in herring, mackerel and North American salmon. *Pseudoterranova* are found in cod, halibut, rockfish, sardine and squid. Over 100 species of fish can act as intermediate hosts. Marine mammals ingest fish and thus, the cycle is completed. Humans become infected by consuming raw, or undercooked fish, smoked fish, marinated, or salted fish which contain the third stage larvae. Human infections result from eating of raw fish. Freezing fish at –20°C for a minimum of 5 days kills larvae and thus the fish becomes suitable for consumption in dishes like sushi. Smoking the fish at temperatures above 65°C renders it suitable for consumption. Marinating, or salting the fish is not reliable to kill larvae. Adequately cleaning the fish render it safe for consumption. Human anisakiasis was virtually eliminated from the Netherlands due to mandatory freezing of herring.

#### Life cycle

Eggs are passed out in faeces of mammals. They embryonate and hatch to liberate the first stage larvae into the sea water. These larvae then become ingested by some crustaceans and moult to become L2 and L3 larvae. If fish and squid eat the infected crustaceans, third stage larvae liberate and penetrate the intestine, or muscles of that host. These larvae can encyst. Larvae reach a length of 50 mm with a diametre of 1–2 mm. The classification of Anisakids is made based on structure of digestive tract.

#### Symptoms

Larvae penetrate the intestinal wall resulting in inhabitation of the human stomach, or duodenum, but can be found in any part of the alimentary canal, or outside the gut. The end result is abdominal pain, nausea, vomiting and diarrhoea, often occurring within six hours of eating infected meal. Abdominal irritation may mimic gastric ulcer, carcinoma, appendicitis, or other conditions requiring surgery. Eosinophilic granulosus may result. Transient anisakiasis is reported in North America, which is characterized by some vomiting and distress within a few hours after ingesting the fish and then quickly subsidies. Larvae can be coughed up a few days later. Low grade eosinophilia and occult blood in the stool are common symptoms.

### 11.14.10 *Gnathostoma spinigerum*

Several species of genus *Gnathostoma* are responsible for the zoonotic infection in man. *G. spinigerum* is found in dogs, cats and several other carnivores. Human infections are reported from Japan, China, Thailand, Far East and Philippines, with man acquiring infection from eating fresh water fish. Adult females measure 25–54 mm, whereas male measures 11–25 mm. The anterior half of the worm is covered with leaf like spines. Male worms have a red tail whilst the larger females have a more curved tail. Larval worms are 4 mm long.

#### Life cycle

Life cycle involves two intermediate hosts, *Cyclops* and birds, snakes, fish and frogs, where they mature before developing into adults in definitive hosts. In definitive hosts, normally the carnivores, adult nematodes live in nodules in the stomach wall. The larvae develop inside the eggs, which are passed in faeces and then into water where they are ingested by *Cyclops*. They in turn are ingested by a number of fresh water fish, frogs, snakes and birds. Larvae pierce their gastric wall, develop into 3rd stage larvae and become encysted in muscle. Cats and dogs ingest the raw flesh of these animals and larvae excyst and mature into adults in the stomach of cats and dogs. If the infected fish, or frogs are eaten by other hosts, apart from definitive hosts, (paratenic host) such as herons, pigs and men, they do not mature, but migrate through subcutaneous tissues causing visceral and cutaneous larva migrans.

#### Symptoms

Humans are accidental hosts and after ingestion, larvae do not mature but migrate throughout the body via the intestinal wall. Symptoms include epigastric pain, vomiting and anorexia. These symptoms subside as the larvae continue their migratory path through the cutaneous tissue. The evidence of migration appears as either lesion similar to cutaneous larvae migrans, or migratory swellings accompanied by inflammation, redness, or pain. The swelling is hard and nonpitted and lasts for several days. These migratory lesions may be accompanied by pruritis and pain. There is marked eosinophilia in patients with cutaneous involvement. Ocular involvement resulting in blindness may occur in serious diseases. Eosinophilic myeloencephalitis may result from the migration of the worms along the nerve tracks. Symptoms may include pain, paralyses, seizures, coma and even death. CSF may be xanthochromic, or bloody.

### 11.14.11 *Ascaris lumbricoides*

*A. lumbricoides* eggs have been found in human coprolites from Peru dating back to 2227 BC. In the old world, *A. lumbricoides* was found in a Middle Kingdom Egyptian mummy dating from 1938 BC to 1600 BC. Extensive written records include the Egyptian medical papyri, works of Hippocrates in the 5th century BC, Chinese writings from second and third centuries BC. Life cycle including the migration of larval stages around the body was discovered only in 1922 by a Japanese paediatrician, Shimesu Koino.

Ascariasis is a disease caused by *A. lumbricoides* and is one of the most widespread parasitic infections in humans, affecting over 1.47 billion people worldwide. These worms are mainly present in communities where poverty, lack of safe drinking water and proper sanitation prevails. The parasite is also recorded in the USA where about 4 million people are infected. In Austria, ascariasis is the 3rd most prevalent helminth infestation (see Mandal, 2011).

*Ascaris* eggs are difficult to exterminate. They remain viable in the soil for as long as 15 years and are widespread in the environment. Fossilized eggs were common in archaeological sites. Unlike bacterial, or viral pathogens, eggs are easy to detect under a microscope. The relative abundance of eggs on archaeological sites can be due to their thick shell which seems to improve their preservation time. Adult female *Ascaris* lay on average 200,000 eggs per day. Roundworm is described from ancient times and then it was confused with earthworms. It is cosmopolitan, especially prevalent in tropics such as China, India and Pacific islands. The adult worm inhabits small intestine of the man.

#### *Morphology*

**Adult worm:** It resembles earthworms and is the largest intestinal nematode parasitizing man. It inhabits the human intestine and is commonly called *round worm*. The worm is found more in children than adults. When collected from the intestine, it is light brown, or pink in colour but gradually changes to white. The worm is elongated, cylindrical and tapering at both ends. The anterior end is thinner than posterior. Smooth, tough and elastic cuticle covers the body. The body is provided with four longitudinal epidermal chords, the narrow one mid-dorsal, one mid-ventral and two lateral. The mouth opens at the anterior end and possesses three finely toothed lips, one dorsal and two ventral. Thus, mouth is a triradiate aperture. Dorsal lip has 2 double sensory papillae and each latero-ventral lip has a double sensory papilla. Digestive and reproductive organs float inside the body cavity containing an irritating fluid. Irritant action is due to toxin, ascaron, or ascarase. Ascaron probably has the nature of primary albumoses (proteose). Allergic manifestations in infected individuals and laboratory workers are due to ascaron. The life span of adult worm in human host is less than a year.

**Male *Ascaris*:** It measures 15 cm with a diametre of 3–4 mm and has a curled tail with protruding spicules. Tail end is curved ventrally in form of a hook having a conical tip. The genital pore opens into cloaca from which two curved copulotory spicules protrude. The anus opens with ejaculatory duct into cloaca. Male *Ascaris* is provided with fifty pairs of pre-anal and five pairs of post-anal genital papillae.

**Female *Ascaris*:** It is longer and stouter than male and measures 25 to 40 cm in length, with a maximum diametre of 5 mm. Posterior extremities are neither curved nor pointed, but is conical and straight. The anus is sub terminal and opens directly on ventral aspect in the form of a transverse slit. Vulva opens at the junction of anterior and middle thirds of the body on the mid-ventral aspect. This section of worm is narrower and is called *vulvar waist*. The mouth has one dorsal and 2 ventral lips. Both the lips are creamy white and cuticle has fine circular striations. Adult worms appear creamy white along with fine circular striations. For diagnosis, they require to be distinguished from earthworms. Male and female *Ascaris* also differ (Table 11.1).

**Table 11.1** Differences between male and female *Ascaris*

| Male | Female |
|---|---|
| Smaller than female | Comparatively longer than male |
| Posterior end is curved | Posterior end is straight |
| Copulatory setae are present | Copulatory setae are absent |
| Vulvar waist is absent | Vulvar waist is present |
| Bears cloaca | Possesses anus |

**Eggs:** Ova can be unfertilized, fertilized, or decorticated. They show considerable variation in shape and size. Fertilized ova are easily recognized, oval in shape with a thick wall showing an irregular bumpy surface. They measure 60–75 μm by 40–50 μm. The outer covering is provided with an albuminoid coat, stained golden brown by bile. Outer wall lies directly on the top of a thick smooth shell, which is not easily distinguishable. Some have lost their albuminoid wall clear, crescentric ovum space between the ova and the egg shell at both the poles of the egg is also remarkable.

Unfertilized ova are longer and narrower than fertile ova, measuring 75–85 μm by 35–50 μm. Shell structure helps ova to withstand many chemicals and makes them ideal parasites of the intestine. In case of unfertilized egg, there is no space between the egg contents and the egg shell.

Eggs liberated by fertilized female passes out of human host with faeces. The characteristics of a fertilized egg are as follows:

1. Round, or oval in shape, measuring 60 to 75 μm by 40 to 50 μm.
2. Bile-stained and brown in colour.
3. Surrounded by a thick smooth, translucent shell with an outer albuminous coat which is thrown into rogosities, or mammillotions. This outer coat is sometimes lost (*decorticated egg*).
4. There is a clear crescentic area at each pole of a large, unsegmented ovum.
5. Fertilized eggs float in the saturated solution of common salt.
6. Unfertilized female may liberate eggs.

The characteristics of an unfertilized egg are as follows:

1. Narrower, longer and more elliptical in shape.
2. Brown in colour.
3. A thinner shell with a coating of albumin.
4. Small atrophied ovum with highly retractile granules.
5. Does not float in salt solution.

### *Life cycle*

The worm completes its life cycle in one host. No intermediate host is needed. The continuance of species is maintained by transference from one individual to another. The man is only known definitive host. Various life cycle stages are described below:

**Stage 1:** *Eggs in faeces*
Copulation takes place in host's intestine. A single female may harbour up to 27 million eggs. Fertilized eggs with unsegmented ovum are passed with faeces. They are not infective to man when freshly passed.

**Stage 2:** *Development in soil*
A rhabditiform larva develops from unsegmented ovum inside the egg shell within 40 days time, depending on atmospheric temperature and humidity. This takes place in moist soil in the presence of oxygen. A ripe egg with coiled-up embryo is infective to man. Before *hatching*, the larva undergoes a *moulting*.

**Stage 3:** *Infection by ingestion and liberation of larvae*
When ingested with contaminated food, drink, or raw vegetables, embryonated eggs pass down to duodenum where digestive juices weaken the egg shells and stimulate the enclosed larvae into activity. Splitting of eggshell occurs and rhabditiform larvae measuring 0.25 mm by 14 mm are liberated in upper part of the small intestine.

**Stage 4:** *Migration through lungs*
Larvae liberated in the small intestine do not directly develop into mature worms. Newly hatched larvae burrow their way through mucous membrane of the small intestine and are carried by portal circulation to liver. Here, they live for a period of three to four days. Finally, they pass out of liver and via right heart enter pulmonary circulation. While in lungs, they grow much bigger and increase in length from 0.2 mm to 2 mm and moult twice (first on the fifth, or sixth day and the second after the tenth day). By breaking the capillary wall, they reach lung alveoli. Time taken for such migration averages fifteen days.

**Stage 5:** *Re-entry into stomach and small intestine*
From lung alveoli, the larva crawl up bronchi and trachea aided by current caused by ciliated epithelium of respiratory tract. They are propelled into larynx and pharynx and are once again swallowed. The larva passes down the oesophagus to the stomach and localizes in upper part of the small intestine—their normal abode. Another moulting occurs between 25th and 29th days of infection.

**Stage 6:** *Sexual maturity and egg liberation*
The larva on reaching its habitat grows into adult

worm and becomes sexually maturity in about six to ten weeks. Gravid females begin to discharge eggs in stool within two months from the time of infection. The cycle is repeated again.

The migratory behavior of ascaroids in mice was categorized into the following two types:

(i) The *tracheal type*, in which the larva migrates through viscera and eventually disappears from tissues.
(ii) The *somatic type*, which consists of permanent encystment of living larva inside tissues.

Migratory pattern of *A. lumbricoides* given above represents the tracheal type. Four moultings of larva of *A. lumbricoides* occur—one in the outside, one within the egg shell, two in lungs and one in the intestine.

#### Mode of infection

Infection is effected by swallowing ripe *Ascaris* eggs (embryonated eggs) with raw vegetables cultivated on soil fertilized by infected/contaminated human excreta. Water may be contaminated and infection may occur by drinking such water. Where *soil-pollution* is common, eggs may directly be conveyed to the mouth by dirty fingers. Infection also occurs by inhalation of desiccated eggs in dust which reach pharynx and get swallowed. Eggs instead of being swallowed may hatch on moist mucous surface of upper air passage and the larva may directly penetrate into the blood stream of human host.

#### Symptoms

Small burdens of worm in the intestine may cause no symptoms. The patient may have symptoms of *pneumonitis* with cough and low grade fever during the migration of larvae through liver and lungs. This can be accompanied by wheezing, coughing and eosinophilia. In heavy worm burdens, adult worms actively migrate in the intestine resulting in intestinal blockage, vomiting and abdominal pain, but infections may also be asymptomatic. Worms can penetrate through the wall of the intestine, or into appendix, travel up the common bile duct, which may become blocked, or they may then enter gall bladder, or liver. A heavy worm burden in children may lead to severe nutritional impairments and retardation in their growth.

#### Preventive measures

The following are the preventive measures:

(i) Proper disposal of human faeces.
(ii) Treatment of parasitized individuals.
(iii) Education of children in schools on sanitary laws and *hygiene*.

### 11.14.12 *Enterobius vermicularis*

Pinworms (*E. vermicularis*) also known as threadworm, or seat worm commonly infect children and toddlers. This is a common parasitic infection in North America and Europe. Pinworms infects more than 400,000,000 people throughout the world, or approximately 10% of humans.

#### Life cycle

Human-to-human spread is favoured by close, crowded living conditions. Animals do not harbour pinworms. This is a very easily transmissible infection. Worms can grow no larger than about half an inch and are big enough to be seen by the naked eye. They are yellow-white, look like a fine piece of thread with pointed ends and most active at night. The best time to see them on anus is a few hours after bedtime. Male worms are about 2–5 mm long while female ones range from 8–13 mm. The posterior end of male is curved and bears copulatory bursa. Females possess vulva. After the male fertilizes the eggs, the female travels to anus, deposits eggs on skin near the anal opening, and usually dies. The number of eggs laid by a female varies between 5,000 and 17,000. Eggs may be fertilized, or unfertilized.

*Fertilized eggs* float in saturated solution of common salt while unfertilized eggs do not float in such solution. Eggs, transferred beneath the fingernails by scratching, are passed into mouth, from where eggs, or larvae make their way to intestine. This life cycle usually requires 15 to 43 days for completion.

They usually spread from one child to another, and children's can pick up eggs with fingers while playing. Eggs can survive on a child's skin for several hours. They can easily survive for two weeks on clothes, bedding and toys. If the eggs remain on child's hands, or toys, they can enter bodies, if the child puts finger, or toys in mouth. Eggs continue to stay in the upper part of the intestine until they hatch. After they hatch, worms move down the length of intestine, and then finally out of the anus where they lay eggs. These eggs may survive outside the body for as long as two weeks given sufficiently moist and warm conditions.

Some eggs become air borne and found in land elsewhere, but majority stay on skin of buttocks. At body temperature, eggs develop quickly and become infective in about six hours. When ingested by another person, eggs hatch in the small intestine. Juvenile worms grow into adult, sexually mature worms in about a month. These tiny worms are quite complex in that they have mouth, throat, gastrointestinal tract, and a nervous system. Male and female worms have a complex reproductive tract and reproduce sexually. A protozoan parasite *(Dientamoeba fragilis)* is transmitted in humans from the eggs of pinworms. Thus, they may occur simultaneously. The worm does not multiply in the human body and has a life span of up to three months.

#### *Causes and symptoms*

Infection can be *asymptomatic*, or result in mild gastrointestinal upsets. A common symptom is perianal (around the anus) itching. Scratching of perianal skin to relieve itching can lead to bacterial infections that result in more itching. Eventually, this cycle produces a great deal of discomfort. Children infected with pinworms often show symptoms including restlessness, irritability, and insomnia. In human females, adult pinworms can enter vagina and cause additional irritation. Since the pinworm almost stays in gastrointestinal tract, or vagina, there is usually no systemic illness. A few children develop intense nighttime itching of skin around the anus. Girls who develop vaginal pinworm may experience vaginal itching, or a vaginal discharge.

#### *Prevention*

Pinworm infections and reinfections can be diminished by the following measures:

(i) Children should wash their hands before meals and after using the restroom.
(ii) Keep children's finger-nails trimmed.
(iii) Discourage nail-biting and scratching the anal area.
(iv) Change children's underwear each day.
(v) Bathe children in the morning to reduce egg contamination.
(vi) Open bedroom and curtains during the day as eggs are sensitive to sunlight.
(vii) Change night clothes, underwear, and bedding and wash them.
(viii) General health awareness is also necessary.

### 11.14.13 Filariasis

The disease filariasis known from antiquity is caused by seven worm species. In India, mosquitoes viz. *Culex quinquefasciatus* and *Mansonia* transmit, respectively, Wuchereria *bancrofti* and *Brugia malayi*. Lymphatic filariasis was and is common along the Nile, although there are no written records. The first definitive report of lymphatic filariasis only began to appear in the 16th century.

*W. bancrofti* is responsible for 90% of lymphatic filariasis. Presently, 120 million worldwide cases of lymphatic filariasis are estimated. It largely affects areas across the broad equatorial belt (Africa, the Nile Delta, Turkey, India, the East Indies, Southeast Asia, Philippines, Oceanic Islands, Australia, and many parts of South America) and humid tropical foci in Africa, the Americas, Asia, and numerous islands in the Pacific Ocean. Mosquito vectors of *W. bancrofti* prefer human blood. Humans appear to be naturally infected with *W. bancrofti*. Reservoir host is unknown.

As per the estimation, more than 300 million individuals live in endemic zone and 6 million fresh filarial attacks occurs each year in India. The number of chronic infection which is around 15 million is also alarming in our country. *Wuchereria* infection is prevalent mainly along the south-east coast including West Bengal. In the case of *Brugia,* the districts, viz., Quilon, Alleppey, Kottayam, Ernakulam and Trichur of Kerala are endemic. Endemic *Brugia* cases have also been reported from Assam, Orissa, Madhya Pradesh and West Bengal. Both *Brugia* and *Wuchereria* may occur in same endemic area as reported from Kerala. Filariasis has also been reported from Pondicherry, Laksha Deep and Tamil Nadu. Every third individual in India faces the risk of filariasis. The disease occurs mainly in the areas where humidity and high temperature are favourable for breeding of mosquitoes.

Periodic *Brugia* and *Wuchereria bancrofti* are found almost exclusively in the man. *Brugia malayi* subperiodic strains are common in domestic cats, various wild carnivores especially *Macaca* and *Presbytis* monkeys, from which transmission to man is not unusual via diurnally active *Mansonia*. Local epidemiology is influenced by behaviour and ecology of mosquito vectors. Seven principal epidemiological zones of Bancroftian filariasis are based on regional variations of vectors. Brugian filariasis is restricted to parts of Southeast Asia and Australia (see Mandal, 2011).

***Brugian filariasis:*** *Brugia malayi* occurs only in South Asia, where its distribution and prevalence have been reduced by the control of vector *Mansonia* by the simple method of removal of host plants from their breeding places. Subperiodic strains occur in swamp-forest habitats, transmitted mainly by *M. bonneaedives* group and by *Coquillettidia crassipes* as a zoonotic cycle. Periodic strains of *B. malayi* are transmitted also by *Anopheles* and are usually not zoonotic. *B. timori* is restricted to Indonesian islands of Flores and Timor and of others nearby. It is apparently not zoonotic, and only known vector is *Anopheles barbirostris*. *B. timori* cannot apparently be transmitted by other mosquito species such as *A. subpictus* which is a vector of both malaria and *W. bancrofti* among people in that area.

### 11.14.14 *Wuchereria bancrofti*

*W. bancrofti* infects lymphatic system, cause lymphatic filariasis throughout the tropics and subtropics and is transmitted by the mosquito. There are two strains of *W. bancrofti*. Nocturnal periodic strain is widely distributed in endemic regions of Africa, India and the Far East and also in some parts of China, Korea and Japan. Microfilariae remain in their highest concentrations between 10 p.m. and 2 a.m. Sub-periodic strain is found in the Pacific region and has a microfilaraemia all the time, with the highest numbers being detected between noon and 8 p.m. It is one of the three parasites that cause lymphatic filariasis. Sub-periodic strain affects over 120 million people primarily in Africa, South America, other tropical and sub-tropical countries. If untreated, it develops into a chronic disease called *Elephantiasis*. Limited treatment modalities exist. No vaccine has been developed. Infection rates in some communities in East Africa exceed 30% of adults causing revolting swellings of the legs, or the genital system, known as elephantiasis.

The adult worm occurs in the tightly coiled nodular masses in major lymphatic ducts. The main vector is *C. quinquefasciatus,* a mosquito that is particularly common in towns and cities, breeds in organically polluted water, rests in houses and feeds during the night on its human occupants. Typical breeding sites include drains blocked with domestic refuse, accumulations of domestic waste water, inadequately covered septic tanks and pit latrines. In rural areas throughout Africa, *Anopheles gambiae* and *A. funestus* are involved in the transmission of the disease. Other anopheline species may transmit bancroftian filariasis in rural areas. In Papua New Guinea, *Mansonia* may serve as a vector. The effects of *W. bancrofti* are also found in the ancient text. Greek and Roman writers noted the similarities between enlarged limbs and cracked skin of an infected individual and those of an elephant. Since Greek and Roman times, this condition has been commonly known as elephantiasis. *W. bancrofti* was named after physician Otto Wucherer and parasitologist Joseph Bancroft, both of whom extensively studied filarial infection. The parasite is believed to be brought over to the new world by the slave trade, and then persisted throughout the areas surrounding Charleston, South Carolina until its sudden disappearance in the 1920s. It exhibits considerable sexual dimorphism.

#### *Morphology*

The adult worm is long, slender, white, threadlike and smooth with rounded ends. It has a short cephalic region, dispersed nuclei throughout its body cavity, with no nuclei on the tail tip. The male worm measures 40 mm long and 100 μm wide, and features a curved tail. In contrast, the female measures 6 cm to 10 cm long and 300 μm wide, nearly 3 times larger in diametre than the male. Females are ovoviviparous and can produce thousands of juveniles known as *microfilariae*. Microfilariae retain egg membrane as a sheath and are often considered advanced embryos. Microfilariae are 230–275 mm in length and 6 to 10 μ in thickness. When stained with Romanowsky's stains, along the central axis of microfilaria can be observed a column of granules, called the *somatic cells*, or *nuclei*.

The following are the features of microfilaria:

1. Granules are absent at the cephalic space.
2. In the anterior half of microfilaria, there is an oblique area devoid of granules, called the *nerve ring*.

3. About midway along the length, there is an anterior V-spot which represents the rudimentary excretory system.
4. Posterior V-spot represents cloaca, or anal pore.
5. Genital cells are located anterior to anal pore.
6. The internal body of Manson extends from anterior V-spot to G-cell 1.

The tail of microfilariae tapers to a delicate point and exhibits no terminal nuclei. The sheath of microfilariae stains with haematoxylin stain. Its body is gently curved, and the tail is tapered to a point. The nuclear column (cells that constitute body of microfilaria) is loosely packed. Nuclei can be visualized individually and do not extend up to the tip of the tail.

### Life cycle

*W. bancrofti* completes their life cycle in two hosts. Human serve as definitive hosts and mosquitoes as intermediate hosts. Adult parasites inhibit the lymphatics of the human host. They are viviparous. The first stage larvae known as microfilariae are present in circulation. They migrate between deep and peripheral circulation. *W. bancrofti* exhibits nocturnal periodicity. During day, it is present in deep veins and during night, it migrates to peripheral circulation. The common vectors are *Culex, Anopheles, Mansonia,* and *Aedes*. The recorded vectors of *W. bancrofti* are as follows:

*Culex quinquefasciatus, C. pipiens, C. annulirostris, Anopheles bancrofti, A. Farauti, A. gambiae, Aedes aegypti, Ae. darlingi, Ae. Kochi, Ae. Vigilax, Mansonia uniformis*

When the mosquito vector takes its blood meal, *W. bancrofti* is ingested via mosquito's proboscis into the blood stream of the human host. Infective larvae enter through wound and migrate to peripheral lymphatics, where they grow to mature male and female worms. They can live there for several years. After mating, the gravid females lay sheathed microfilariae into peripheral blood, where they can be detected 8–12 months after the initial infected bite. Mosquito acquires infection by ingestion of microfilaria in blood meal. Microfilariae lose their sheath on arrival in the stomach of mosquito due to action of gastric juices. Larvae migrate to thoracic muscles and transform into infective larvae over a period of 6–14 days. Larvae then migrate to mouth parts of the mosquito which infects the host during a blood meal. Blood stages of filariae, microfilariae, vary in times, when they are present in peripheral blood, corresponding to peak biting time of the vector. Thus, in nocturnal periodic forms, microfilaria are found in peripheral blood circulation at night; during day, they reside in deep tissues, particularly the lungs.

Larvae move through the lymphatic system to the regional lymph nodes, predominantly in legs and genital area. They develop into the adult worms over the course of a year. They reach sexual maturity in afferent lymphatic vessels. After mating, adult female worm can produce thousands of microfilariae that migrate into blood stream. A mosquito vector can bite the infected human host, ingest microfilariae, and thus repeat life cycle of *W. bancrofti*. In humans, adult *W. bancrofti* reside in lymphatic ducts and are found mostly in lymph glands of afferent lymphatic channels in the lower part of body. Microfilariae produced by female worms have a membrane "sheath". This "sheath" along with area in which worms reside, makes the identification of type of microfilariae in humans easier.

Microfilariae are found mainly in peripheral blood and can be found at peak amounts from 10 p.m. to 2 a.m. The cause of this periodicity remains unknown, but the advantages of microfilariae being in peripheral blood during these hours may ensure that the vector, the night time mosquito, will have a higher chance of transmitting them elsewhere. There are physiological changes associated with sleeping, such as lower body temperature, oxygen tension and adrenal activity and an increased carbon dioxide tension, among other physical alterations, any of which could be the signals for the rhythmic behaviour of microfilarial parasites. If the host sleeps by day and is awake at night, their periodicity is reversed. In the south Pacific, where *W. bancrofti* shows diurnal periodicity, it is known as *subperiodic*.

### Wolbacteria-endosymbionts

For almost thirty years, it is known that filarial worms contain endosymbiotic bacteria. These are found in hypodermis of male and female worms, and in oocysts, embryos, and larval stages. Presently endosymbionts are classified as genus *Wolbachia*, a genus of bacteria that are common endosymbionts of arthropods.

### *Symptoms*

Pathogenic effects are produced by living, or dead adult parasites. A light infection causes filarial fever, headache and depression. Pathogenesis is dependent on the immune system and inflammatory responses of the host. Asymptomatic phase usually consists of high microfilaraemia and individuals show no symptoms of infection. This happens due to cytokine IL-4 suppressing activity of TH1 cells in our immune system. This can occur for years until inflammatory reaction rises again. In inflammatory (acute) phase, antigens from female adult worms elicit inflammatory responses. Worms in the lymph channels disrupt the flow of lymph causing lymphadema. The individual exhibits fever, chills, skin infections, painful lymph nodes, and tender skin of lymphedematous extremity. These symptoms often lessen after 5–7 days.

Other symptoms that may occur include orchitis-inflammation of testes accompanied by painful enlargement and epididymitis, which is inflammation of spermatic cord. Obstructive (chronic) phase is marked by lymph varices, lymph scrotum, hydrocoel, chyluria (lymph in urine) and elephantiasis. Microfilariae are not normally present in this phase. A key feature of this phase is scar formation from the affected tissue areas. Other features include thickening of skin and elephantiasis, which develops gradually with the attack of lymphatic system. Elephantiasis affects men mainly in legs, arms and scrotum. In women, legs and arms are affected (Figures 11.9–11.14).

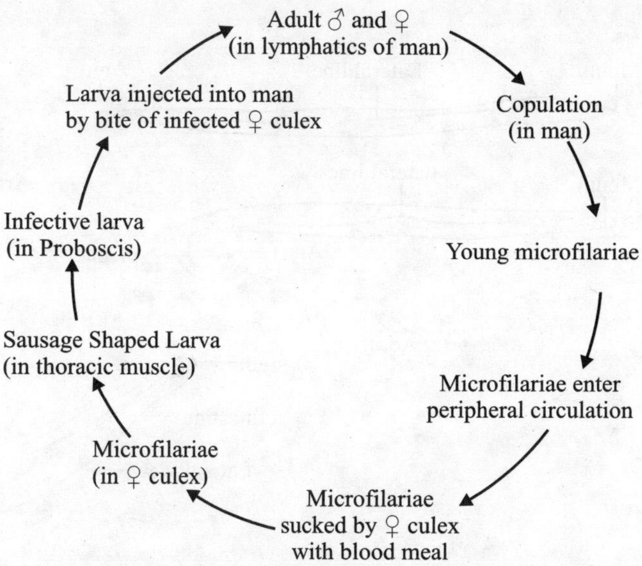


**Figure 11.9** Life cycle of *W. bancrofti*.

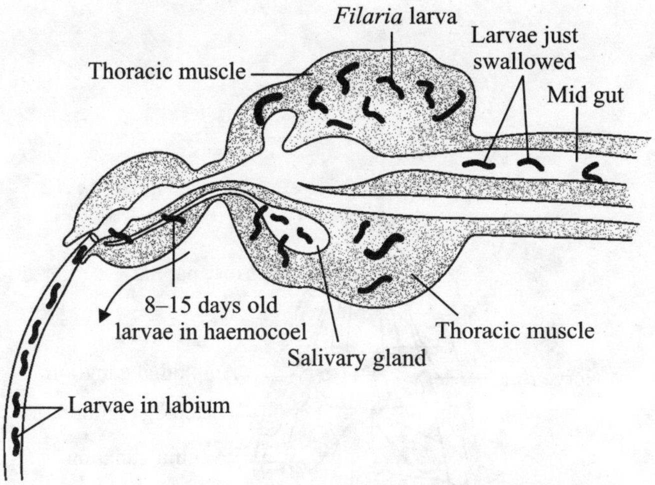


**Figure 11.10** *Wuchereria* larvae wandering in the body of the *Stegomyia* (Modified after Bahr).

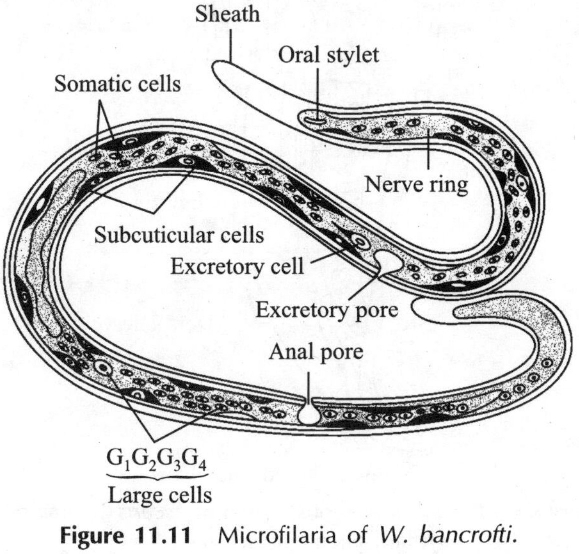


**Figure 11.11** Microfilaria of *W. bancrofti*.

## 11.14.15 *Brugia malayi*

*B. malayi* is a cause of lymphatic filariasis in South East Asia. There are two strains of *B. Malayi*.

(i) Nocturnal periodic strain which is widely distributed in Asia, the microfilariae being in their highest concentrations between 10 p.m. and 2 a.m.

(ii) Sub-periodic strain, which is found in Malaysia, Indonesia and the Philippines where humans exhibit a microfilaraemia all the time, the highest number being detected between noon and 8 p.m.

Nocturnal periodic Brugian filariasis is primarily a rural disease, being transmitted by *Anopheles* and *Mansonia*, the mosquitoes that usually bite during the night. Nocturnal sub-periodic *B. malayi*

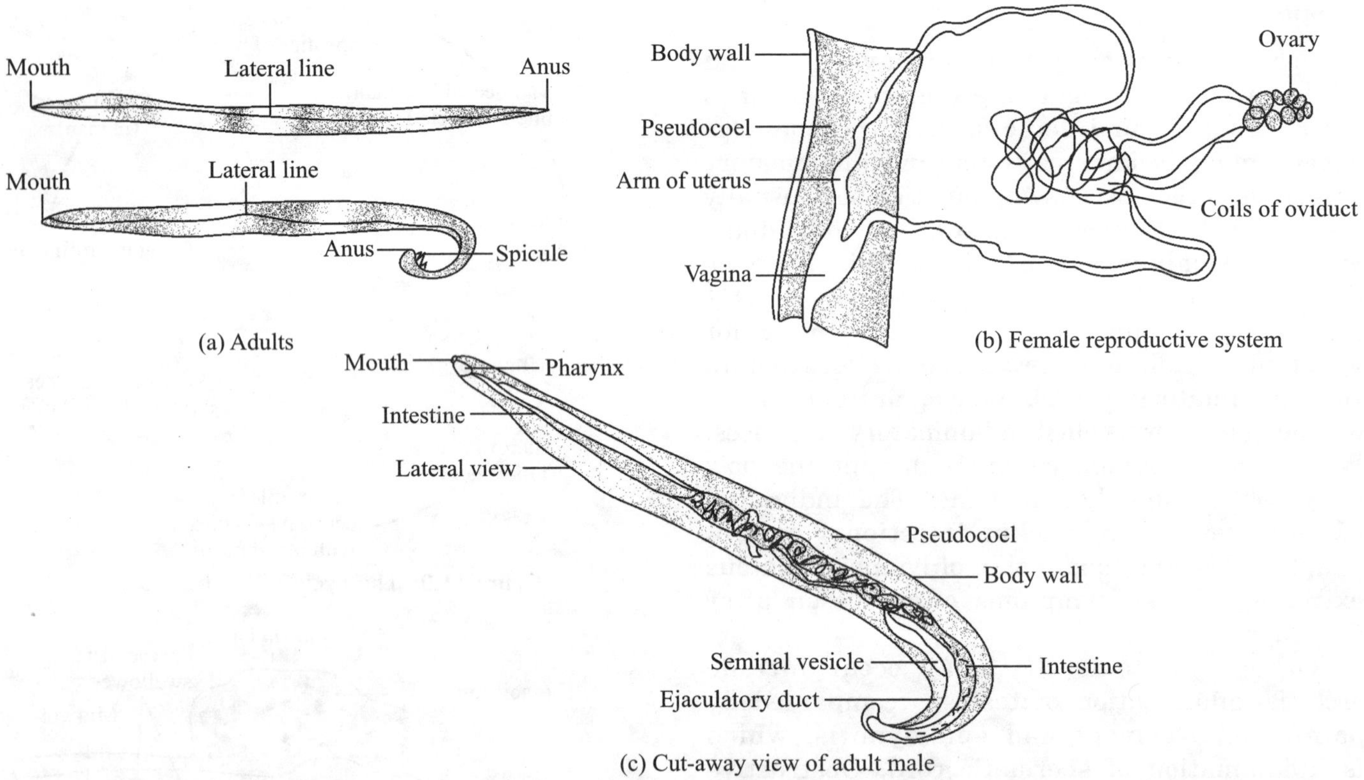


**Figure 11.12** *Ascaris lumbricoides.*

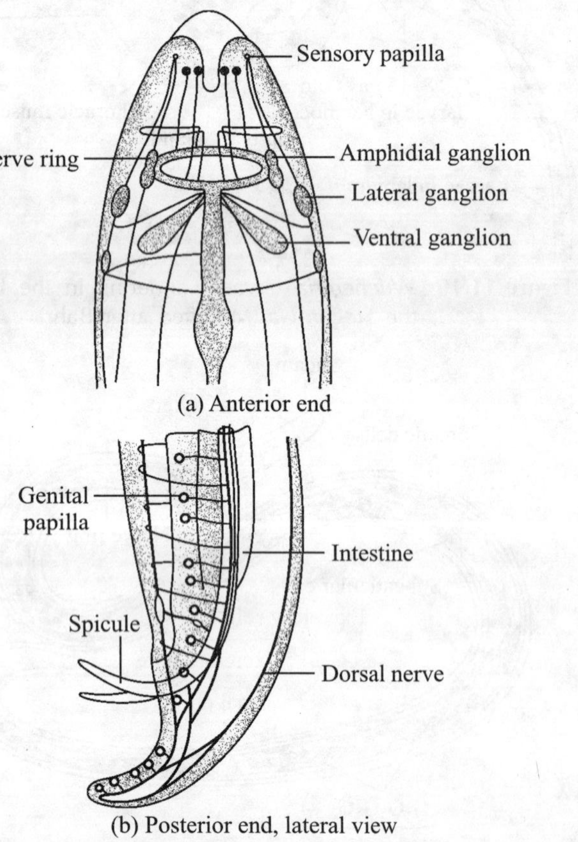


**Figure 11.13** Central nervous system of *Ascaris* (Redrawn from Hyman, after Goldschmidt).

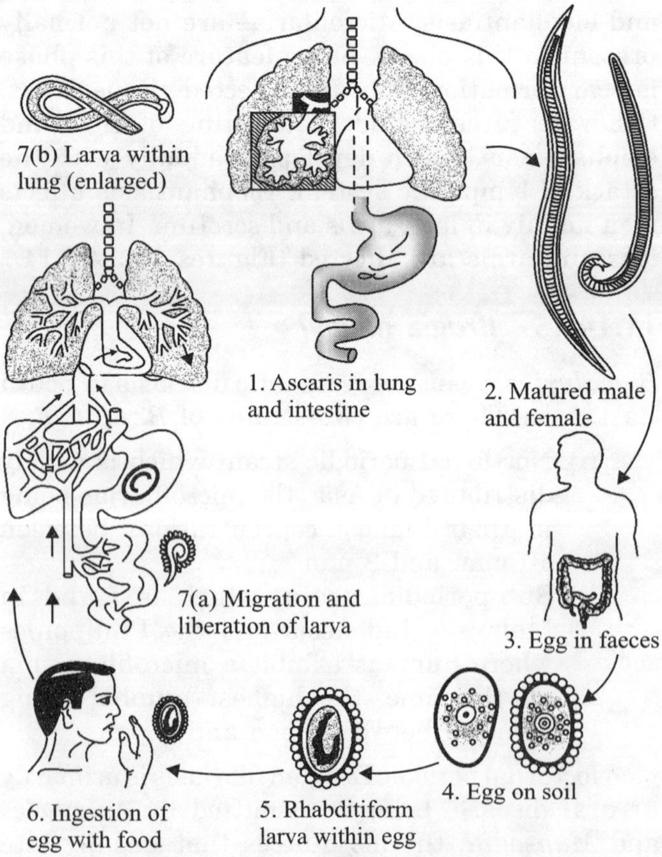


**Figure 11.14** Life cycle of *Ascaris.*

is transmitted almost exclusively by *Mansonia*. Often, different species are involved than those involved in transmitting the periodic form. *M. bonneae*, the important vector in Malaysia, breeds in swamp forest and bites during night, although sometimes in the day as well. This species like *W. bancrofti*, also parasitizes the lymph nodes and lymphatics. Adults of the two species are indistinguishable causing Malayan filariasis.

### *Morphology and Life cycle*

Adult worms are smaller than those of *W. bancrofti*. Microfilariae are 170–230 mm in length and have two terminal nuclei that are distinctly separated from other nuclei in the tail. The last terminal nucleus is small and is bound at the tip of tail. Sheath stains deep purple with haematoxylin stain. Adult worm inhabits the lymphatics. Female worms produce sheathed microfilariae which circulate in peripheral blood. Mosquito acquires infection by the ingestion of microfilaria during blood meal. Microfilaria loses its sheath on arrival in the stomach of mosquito. Larvae migrate to thoracic muscles and develop into infective larvae over a period of 6–14 days. Larvae then migrate to mouth parts of mosquito and enter the skin of the definitive host through puncture wound, when a blood meal is taken. Infective larvae enter peripheral lymphatics, where they grow to mature male and female worms.

### *Symptoms*

Clinical features are similar to those of *W. bancrofti*. However, in *B. malayi*, unlike *W. bancrofti*, genital involvement, hydrocoel and chyluria are rare. Many patients are asymptomatic. Patients may have fever. Lymphangitis and lymphadenitis develop in lower extremities. An inflammatory reaction occurs in lymphatic vessels that harbour adult worms. Oedema develops which may resolve after the first few attacks. In long-standing disease, after several episodes of lymphangitis, thickening and various changes in skin known as *elephantiasis* occur. Some patients with lymphatic filariasis do not exhibit microfilariaemia. They may have high eosinophilia, high IgE level and high antifilarial antibody titres.

## 11.14.16 *Brugia timori*

*B. timori* is found in the islands of Indonesia and exhibits a strictly nocturnal periodicity. Life cycle and disease closely resemble those of *B. malayi*. Microfilariae can be distinguished from those of *B. malayi* in that they are about 310 mm in length. Sheath satins pink with Giemsa and nuclei at the tip of tail are similar to those of B. *malayi.*

## 11.14.17 *Loa loa*

*L. loa*, known as *African eye worm*, is a filarial nematode endemic in rain forests of West and Central Africa. It is transmitted by *Chrysops*, also known as mango-, or horse- flies. Humans are the only known reservoir. It is estimated that about 2–13 million humans are infected with larvae. Adults migrate in subcutaneous tissues of men and monkeys, and eventually migrate across the eyeball under conjunctiva. They live in subcutaneous and deep connective tissues. Microfilariae are found in peripheral blood, where they can be taken by *Chrysops* (day biting fly). Adults can live in tissues for up to 17 years. Once microfilariae have been taken up by *Chrysops* during blood meal, they develop within fat body in to L3 within 10–12 days. Microfilariae, L3 reenter hosts' blood stream, when fly takes another blood meal. They reach adult stages within 4–6 months, live in subcutaneous and deep connective tissues. Microfilariae exhibit diurnal periodicity and the highest numbers being detected in the blood between 10 a.m. and 2 p.m.

### *Morphology*

Adult males are 2–3.5 cm long and females from 5–7 cm. Microfilariae measure 250 × 300 mm. They possess a sheath which stains blue–grey with Delafield's haematoxylin. Sheath does not stain with Giemsa. Tail gradually tapers to a rounded end, densely packed nuclei extend to the tip.

### *Symptoms*

Many patients appear to be asymptomatic. The migration of adult worms through subcutaneous tissues often goes unnoticed, unless passing beneath the conjunctiva of the eye. They can be seen crossing the eye and a rapid process takes approximately 15–20 minutes. Hypereosinophilia and increased antibody levels, especially IgE, are also noted. Eye worm episodes are equally common in men and women with common recurrences. There is an increased incidence with age. The common pathology associated is Calabar swellings. The inflammatory swellings result in a localized subcutaneous oedema. These swellings are due to the host's response to the worm, or its metabolic products and can be found anywhere in the body, but most commonly in extremities. These swellings last from 1–3 days, and develop rapidly, usually accompanied by localized pain, urticaria and pruritis. There is a higher frequency of calabar swellings in women with common recurrences. Serious complications such as cardiomyopathy, encephalopathy, nephropathy and pleural effusion have been recorded.

## 11.14.18 *Onchocerca volvulus*

*O. volvulus* causes onchocerciasis, the world's second leading infectious cause of blindness. The disease is transmitted from human to human through black fly. Rarely life threatening, the disease causes chronic suffering and severe disability. In Africa, it is a serious obstacle to socioeconomic development. It is often called *river blindness*. O. volvulus was first described in 1893 and relationship between worm and blindness was recognized by 1915. A treatment for the disease was elusive. Today, the worm still infects millions, and confounds health organizations. *Simulium* flies breed in fast flowing rivers, i.e. well oxygenated water, because their larvae have an obligatory aquatic stage during which they require high oxygen tension. Onchocerciasis is associated with fast flowing rivers. That is why blindness Onchocerciasis can often referred to as 'river blindness'. Fertile riverine areas are frequently abandoned for fear of disease.

### *Life cycle*

Adult lives in the human body. Male and female worms entwine in nodules in subcutaneous tissue of skin. After mating, the female releases around 1,000 microfilariae per day into the surrounding tissue. Microfilariae live for 1–2 years, moving around the body in subcutaneous tissue. The third-stage larva is the infective one. It migrates to black fly's proboscis and is able to infect another human when released during a fly's blood meal. An infected *Simulium* releases third-stage filarial larvae into the skin of human. Adult females produce millions of microfilariae. When female black flies take a blood meal, they ingest microfilariae. Microfilariae in fly host transform to L2 larvel stage. L2 larvae then molt to L3, the infective stage for humans.

Larvae develop into the adult filariae in subcutaneous tissue which are then surrounded by connective tissue and can be palpated as nodules. Some nodules contain numerous male and female worms. Adult worms can live up to 15 years. Female worms measure 33 to 50 cm in length and 0.27 to 0.40 mm in diametre, while males reach a length of 19 to 42 mm and a diametre of 0.12 to 0.2 mm. Microfilariae have a length of 220 to 360 μm, a diametre of 5 to 9 μm and they can survive for up to 2 years. Microfilariae are typically found in skin and in lymphatics of connective tissues, sometimes in peripheral blood, urine, or sputum. After ingestion, microfilariae migrate from black fly's midgut through hemocoel and invade thoracic muscles. There, they live as intra-cellular parasites and first develop into so-called *sausage stage*, subsequently into second- and third-stage infective larvae.

### *Symptoms*

One early sign of O*nchocera* is raised nodules in the areas over bony prominence. The death of microfilariae causes an inflammatory response that leads to skin rashes, lesions, intense itching and skin depigmentation. Microfilariae can enter the eye by passing along sheaths of ciliary vessels and nerves from under bulbar conjunctiva directly into cornea, via nutrient vessels into optic nerve, and via posterior perforating ciliary vessels into choroids. Dead microfilariae in the eye lead to an inflammatory immune response and eventual formation of secondary cataracts and ocular lesions. Reactions to dead microfilariae around these nodules can lead to several unpleasant conditions. In skin, there is destruction of elastic tissues and formation of redundant folds along with a loss of pigmentation. Because of this, heavy infections often lead to progressive blindness.

## 11.14.19 *Mansonella*

Members of genus *Mansonella* are filarial nematodes which rarely cause serious disease. They can be found in geographical areas where *W. bancrofti, Loa loa* and *Onchocera volvulus* also occur and therefore must be differentiated from these pathogenic microfilariae. Unlike pathogenic blood filariae, they do not exhibit periodicity as found in the blood inhabiting pathogenic filariae.

### *Life cycle*

There is a general life cycle for *Mansonella*. Microfilaria is picked up by *Culicoides* (biting midges) during a blood meal. Larvae develop inside the body of the vector and are reintroduced into the human host, when the vector takes another blood meal. They are found in sites around the human host.

## 11.14.20 *Mansonella perstans*

Microfilariae have been found in Africa and South America. This is reported as a mildly pathogenic species in men and apes. They are found in deep connective tissue and serous cavities. Adult worms live in the peritoneal, pleural and pericardial cavities and their size is comparable to pathogenic species already discussed. Microfilariae are unsheathed and are about 200 mm in length and nuclei extend to the tip of tail which is rounded.

### *Symptoms*

It is difficult to assess the disease. However, pruritis, fever and subcutaneous swellings are associated with infection. Adult worms appear to cause little, or no host reaction. Eosinophilia is common.

## 11.14.21 *M. ozzardi*

*M. ozzardi* are confined to the new world and West Indies. These are nonpathogenic filarial nematodes causing nodules in the skin of vertebrate hosts.

### *Morphology*

Adult worms are located in mesenteric tissues and their size is comparable to pathogenic species. Microfilariae are found in peripheral blood and range between 173 and 240 mm in length. Nuclei do not extend to the tip of the tail which has a pointed end. Male adult worms are almost unknown. Infections caused by *M. ozzardi* are generally symptomless. However, lymphadenopathy, arthralgia, fever and eosinophilia have been recorded. Small thin microfilaria does not have a sheath. Nuclei do not extend to the end of the tail. The tip of the tail tapers and is difficult to see. The worm stains very lightly and is found in blood and skin. Comparative accounts in these regards are given in Tables 11.2 and 11.3.

**Table 11.2** Morphology of microfilariae

| Species | Size of microfilariae | Morphology of microfilariae |
|---|---|---|
| *Wuchereria bancrofti* | 210–320 mm by 8–10 mm | Sheathed. Tail pointed and clears |
| *Brugia malayi* | 170–260 mm by 5–6 mm | Sheathed. Tail pointed with 2 nuclei |
| *Loa loa* | 230–300 mm by 6–8 mm | Sheathed. Tail blunt with nuclei |
| *Mansonella perstans* | 200 mm by 6 mm | Unsheathed. Tail blunt with nuclei |
| *M. ozzardi* | 250 mm by 6–7 mm | Unsheathed. Tail pointed and clear |

**Table 11.3** Comparison of the main human filarial nematodes

| Species | Geographic location | Periodicity |
|---|---|---|
| *Wuchereria bancrofti* | Tropics/ Subtropics | Nocturnal. |
| *W. bancrofti* | Pacific | Diurnal subperiodic |
| *Brugia malayi* | SE Asia and SW India | Nocturnal |
| *B. malayi* | Indonesia | Nocturnal subperiodic |
| *B. timori* | Indonesia | Nocturnal |
| *Loa loa* | West/Central Africa | Diurnal |
| *Mansonella perstans* | Africa/ S. America | Nonperiodic |
| *M. ozzardi* | Central and S. America | Nonperiodic |

## 11.14.22 *Dracunculus medinensis*

*Dracunculus* is found throughout the Middle East, South East Asia and Africa. The intermediate hosts (*Cyclops)* often have a very restricted distribution being found in deep wells. Control is done by filtering drinking water and killing the *Cyclops* with insecticides. Because of the very localized distribution of the intermediate host and the absence of alternative hosts, complete eradication of the parasite is a possibility, and there was a WHO campaign to eliminate the parasite by the year 2000. Since the 1970s, the global incidence has decreased from several million to less than 75,223 cases, and 73% of which were found in Sudan. This parasite is commonly called *Guinea worm*.

The adult female worm measures 120 cm in length and 2 mm in breadth. Infective larvae present within *Cyclops* lead active life inside the human body and ultimately reaches sites like legs, arms, shoulder, breast and genitalia from where they come in contact with water. When the gravid female worm comes beneath skin surface, it secrets a toxin that produces blister. Blister ultimately becomes ulcer through which the worm larvae are released into water.

Larvae present in the water are ingested by the *Cyclops*, reaches the body cavity of *Cyclops* and develop into infective third stage larvae. In rural areas, Guinea worm infection in farm workers seriously affects agricultural activities and results in socio-economic deprivation. Guinea worm infection may lead to teatanus, and sometimes unusual travelling of worm in pericardium, spinal canal and eyes causes serious concern. For prevention, *Cyclops* may be destroyed by the chemical treatment of water. Filtered, or boiled water is safe against the infection. The provision of protected piped water supply is recommended as the best preventive measure.

## REVIEW QUESTIONS

### Short Answer Questions

1. What is the typical morphological feature of nematodes that differentiates them from Platyhelminthes?
2. What are the morphological similarities and differences between nematodes and annelids?
3. Are nematodes diploblastic, or triploblastic animals?
4. What is the main evolutionary innovation presented by nematodes?
5. How does the excretory system of nematodes work?
6. How is the nervous system of nematodes organized?
7. What type of reproduction occurs in roundworm?
8. What typical feature do nematode sperm cells have?
9. What the term nematode signifies?
10. Give two morphological features of nematodes.
11. What is amphid?
12. What is phasmid?
13. What do you understand by osmoconformer?
14. What is deirid?
15. What is pseudocoelomocyte?
16. Comment on the H system of excretory system of nematodes.
17. In which sex copulatory bursa is found?
18. What do you understand by aberrant migration in nematode?
19. Name the nematode which serves as genetic specimen.
20. Name one scientist who was awarded the Nobel Prize for his work with nematode.
21. Give the scientific name of whip worm.
22. What is trichinosis?
23. Comment on rhabditiform larva.
24. Comment on filariform larva.
25. What is homogonic development?
26. What is heterogonic development?
27. What is autoinfection?
28. What is larva currens?
29. What is ancylostomiasis?
30. What the term "miners anaemia" signifies?
31. What does the name *Ancylostoma* denotes?
32. Comment on ancylostome dermatitis.
33. What is anisakiasis?
34. Name the disease caused by *Ascaris*.
35. What is vulvar waist?
36. Give the scientific name of pinworm.
37. Name the causative organism of *Bancroftian filariasis*.
38. What type of respiration is found in *Ascaris*?
39. Mention the function of amphids.
40. How male *Ascaris* can be distinguished from the female?
41. Mention the name of larval form of *Ascaris*.
42. What type of body cavity is found in *Ascaris*?
43. How many moults occur during life cycle of *Ascaris*?
44. Where the fourth moult of rhabdiform larva occurs?
45. Mention the number of lips in *Ascaris*?

46. Where would you find *Ascaris lumbricoides*?
47. What type of ectoderm is found in *Ascaris*?
48. Which type of testis is present in *Ascaris*?
49. What is the shape of sperm in *Ascaris*?
50. What is the infective stage of *Ascaris* for man?
51. Through which agency the infection of *Ascaris* takes place?
52. What is the duration of migration period in *Ascaris*?
53. What is the mode of formation of ova in *Ascaris*?
54. Mention the name of food materials of *Ascaris*.
55. What is the excretory product of *Ascaris*?
56. How many pseudocoelocytes are present in *Ascaris*?
57. Comment on *Ascaris* pneumania.

## Long Answer Questions

1. Describe the morphology and life cycle of whip worm with suitable diagram.
2. Describe the life cycle, pathogenecity and control of *Ancylostoma duodenale* with the help of diagram.
3. Describe the life cycle, pathogenecity and control of *Ascaris* lumbricoides.
4. Describe the morphology and life-cycle of pin worm.
5. Write a note on epidemiology of lymphatic filariasis.
6. Describe the importance of *Onchocera volvulus*. Add a note on its life cycle and management.
7. Represent the life cycle of *Ascaris* lumbricoides.
8. Discuss the life cycle of the *Wuchereria bancrofti* with suitable diagram.

CHAPTER 12

# Phylum Annelida

Annelida includes about 9,000 species. Most of them inhabit the aquatic environments. Some are terrestrial. They are commonly called the *segmented worms* and are widespread in the sea, fresh water and soil. More then 1,100 annelids are described. The french use of the name Annelids comes from the Latin word 'anellus', meaning a little ring. The septum limits each segment, divides it from the neighbouring segments and has a fluid-filled cavity called *coelom*. Excretory, locomotory and respiratory organs are generally repeated in each segment.

A distinctive feature is the presence of the chaetae. Chaetae (also called *setae*) are bundles of chitinous, cylindrical structures held together by the sclerotinized protein. They are produced by a micro-villar border of certain invaginated epidermal cells. Chaetae show variation, from long thin filaments (capillary chaetae) to stout multi-pronged hooks.

Earthworm, a common annelid makes soil healthy. *Tubifex* are used to feed the aquarium fishes. Leeches have medicinal property. Marine polychaetes maintain the food chain and ecological balance of the sea.

Annelids range in size. The giant earth-worm, *Michrochaetus rappi* (*Michrochaetus michrochaetus*) is the largest with an average length of 1.36 m (54 inches). The smallest one *Chaetogaster annandalai* when full grown attains 0.5 mm in length. Annelids live quietly in holes, or lead active lives. The head is followed by a long thin body of many similar segments. Body consists of a head including the mouth (prostomium) and sometimes a peristomium, along with the posterior end called the *pygidium*. The post segmental pygidium includes the zone from which new segments are proliferated during the growth.

They are coelomate, provided with simple, or complex chaetae to each body segment. The growth occurs laterally by the enlargement of the segments during the juvenile stages and through the addition of new segments. Distribution of most annelids is not well known. The 2000 IUCN Red List includes *Mesonerilla prospera*, *Phallodrilus macmasterae* under the critically endangered category and the four species, *Driloleirus americanus*, *Driloleirus macelfreshi*, *Komarekiona eatoni* and *Megascolides australis* in vulnerable category. Another 145 species including the *Hirudo medicinalis* and the Palolo Worm *Eunice viridis* are suspected of being in the need of conservation.

The Class Polychaeta (*Poly* meaning many, *Chaeta* meaning bristle), the most diverse group, are predominantly marine and mostly gonochoristic (either male, or female), some are sequential hermaphrodites (one sex first and then change to the other sex). The pelagic polychaetes swim, or drift, prey on other plankton and few groups occur in the fresh water and moist terrestrial surroundings. Reproduction takes place by the production of modified reproductive segments which may, or may not become independent of the parent worm before mating. These segments are destroyed immediately after the release of their gametes.

The oldest unequivocal fossil polychaetes, such as *Canadia* from the Cambrian belong to Phyllodocida. Several likely appearances from the Ordovician include the Serpulidae, Spionidae. By the end of the Carboniferous, most major polychaetes lineages had appeared. The exception appears to be *Scolecida* with the earliest known fossils being the dubious *Archarenicola* from the Triassic and one assignable to Paraonidae from the Cretaceous.

Oligochaetes with an indistinct head live in marine, fresh water and terrestrial habitats. They are less diverse than the polychaetes, normally hermaphrodite and possess a clitellum as adult. Clitellum looks like a bandage of skin wrapped around the animal.

Hirundinea contains well known leeches which suck blood and exhibit head to tail looping, the mode of locomotion. Except the primitive *Acanthobdella peledina*, leeches have no chaetae but have 33 body segments. They have two suckers, one of which is located at the anterior (head) end and composed of segments 1–4 and the other sucker is located at the posterior (tail) end and composed of segments 25–33 of the body. Hirundinea occurs in the fresh water, marine and terrestrial environments.

## 12.1 GENERAL CHARACTERS

1. Schizocoelous, bilaterally symmetrical.
2. Segmented.
3. Development is typically protostomous; segments arise by teloblastic growth.
4. Digestive tract is complete, usually with regional specialization.
5. Head is composed of pre-segmental prostomium and peristomium.
6. Dioecious, or hermaphroditic.
7. Trochophore larva is generally found.
8. Marine, terrestrial and fresh water species.
9. Size varies from much less than 1 mm in length to more than 3 m.
10. Neotenotrocha (smallest) is only a few hundred microns in length.
11. Earthworm (*Megascolides australis*) can reach 1 m in length.
12. *Haementeria ghilianii* (largest leech) is more than 40 cm in length.
13. A fluid-filled cavity between the outer body wall and the gut is called the *coelom*.
14. Coelom is often used as a storage area for the gametes and as a hydrostatic skeleton.
15. Coelom is generally divided by inter-segmental septa.
16. In leeches and several small polychaetes, little, or no coelomic space exists.
17. An external cuticle covers the body that is never shed, or moulted.
18. Epidermal microvilli secrete a network of fibres that are partly collagenous and contain scleroprotein.
19. Chaetae are cuticular structures and contain large amounts of chitin.
20. Beneath the epidermis and its basal lamina lies a layer of the circular muscle.
21. Circular muscle layers, a nearly continuous sheath around the body forms, except in polychaetes are present.
22. In many cases, the longitudinal muscles are present as four distinct bands.
23. Parapodia, as unjointed segmental extensions of the body wall, are found in many polychaetes.
24. Parapodia equipped with musculature usually carry the chaetae.
25. Parapodia consist of two elements, a dorsal notopodium and a ventral neuropodium.
26. Sometimes a variety of cirri and gills are found.
27. Brain or cerebral ganglion originates and usually resides in the head.
28. Brain varies in structure, with mobile, active forms having the most complex brains,
29. Sessile, or burrowing forms have simple brains with little differentiation.
30. Brain is connected to the ventral nerve cord by the circumpharyngeal connectives, which run down each side of the pharynx.
31. Ventral nerve cord is usually made up of a pair of cords that are bound together and runs along the body length.
32. Six major kinds of sensory structures are palps, antennae, eyes, statocysts, nuchal organs and lateral organs.
33. Nuchal organs are ciliated, paired, chemosensory structures, innervated from the posterior part of the brain.
34. Closed circulatory system is present in most polychaetes.
35. Coelom and the circulatory system are involved in excretion.
36. Nephridia serve as the excretory organ.
37. Gonoducts, or coelomoducts transfer gametes.

## 12.2 CLASSIFICATION OF THE ANNELIDA

The following are the important features of annelida:

1. Cylindrical, metamerically segmented, and elongated with a thin cuticle.
2. Vermiform, triploblastic and bilaterally symmetrical.

3. Body cavity is a true coelom, often divided by the internal septa.
4. The presence of chitinoid setae.
5. Often provided with the coelomoducts.
6. Body possesses the terminal mouth and anus.
7. Body possesses three separate sections, a prosotomium, a trunk and a pygidium.
6. Nervous system possesses an anterior nerve ring, ganglia and a ventral nerve chord.
7. Cerebral ganglia and double ventral nerve cord with segmentally arranged ganglia are often present.
8. True respiratory organ is absent.
9. Reproduction is normally sexual.
10. Feed a wide range of material.
11. Live in most environments.
12. Nephridia are metamerically arranged.
13. Either unisexual, or bisexual.
14. Larva, if present, of the trochophore type.
15. Length varies between less than 1 mm to 3 m.
16. Live in the sea, fresh water and damp soil.

Classification given here is based on the scheme adapted by Marshall and Williams (1972).

***Class Polychaeta***

1. Well-marked internal and external segmentation.
2. A pair of parapodia in each body segment is present.
3. Parapodia are provided with bundles of setae.
4. A distinct head and sometimes eyes are present.
5. No permanent gonads.
6. Fertilization is external.
7. Trochophore larva is found in the life cycle.
8. Chiefly marine.

***Subclass Errantia***

1. Except head and anal segments, all body segments are similar.
2. Prostomium is distinct with appendages.
3. Acicula is present.
4. Predatory and free moving.

**Example:** *Aphrodite, Polynoe, Nereis.*

***Subclass Sedentaria***

1. Sedentary.
2. Body segments and parapodia are dissimilar.
3. *Acicula* is absent.
4. Prostomium is small.
5. Burrowers, or tube dwellers.

**Example:** *Chaetopterus*, *Arenicola*, *Sabella*, *Terebella.*

***Class Archiannelida***

1. Size is small.
2. Parapodia and setae are reduced, or lacking.
3. Vascular system is simple, or lacking.
4. Nervous system is ectodermal.

**Example:** *Polygordius*, *Protodrilus*, *Dinophilus.*

***Class Oligochaeta***

1. Well-marked external and internal segmentation.
2. Chaetae are not situated on parapodia.
3. Hermaphroditic.
4. A clitellum is present at sexual maturity.
5. Eggs are deposited in cocoon.
6. No free larval stage is found in the development.

***Order Plesiopora Plesiotheca***

1. Male gonopores are present on segment following testes.
2. Spermatheca is found in region of genital segments.

**Example:** *Tubifex*, *Chaetogaster.*

***Order Plesiopora Prosotheca***

1. Male gonopores are found on segment following testes.
2. Spermatheca is present in front of genital segments.

**Example:** *Lumbricillus*, *Aspidodrillus.*

***Order Prosopora***

1. Aquatic.
2. Mesonephrostomal.
3. Male gonopores in same segment are present as the last pair of testes.

**Example:** *Lumbriculus.*

***Order Opisthopora***

1. Meganephrostomal.
2. Male gonopores are present at least one segment behind the last pair of testes.

**Example:** *Pheretima, Megascolex, Microscolex.*

### Class Hirudinea

1. Body is marked externally by the secondary rings and annuli.
2. Mouth lies in front of the posterior sucker.
3. Nephridia are segmentally arranged.
4. Hermaphroditic.
5. Testes are numerous.
6. A single pair of ovary is present.

### Order Acanthobdellida

1. Chaetae are present in five anterior segments.
2. Anterior sucker is absent.

**Example:** *Acanthobdella.*

### Order Rhynchobdellida

1. Jawless leeches with the proboscis.
2. Colourless blood is present.
2. Marine and fresh water, none are the terrestrial.

**Example:** *Placobdella*, *Pterobdella.*

### Order Gnathobdellida

1. Aquatic, or terrestrial leeches.
2. A nonreversible pharynx is present.
3. Three pairs of jaws are present.

**Example:** *Limnobdella*, *Phytobdella.*

### Order Pharyngobdellida

1. Fresh water, or amphibious leeches.
2. A nonprotrusible pharynx is present.
3. No jaws.

**Example:** *Dina*, *Erpobdella* (Figures 12.1–12.5).

## 12.3 EARTHWORM

### *Pheretima sp.*

Over thousand of earthworm species are known. *Lumbricus* is the most common in Europe and North America. *Megascolex* and *Drawida* are common in the Southern India and *Eutyphoeus* in the Gangetic plan of India. *Pheretima* is represented by nearly 500 species from South-East Asia, Japan, Sri Lanka and Australia, of which 13 species inhabit India including *P. posthuma*, *P. hetrochaeta*, *P. rubusta*, *P. elongata*, *P. compalata*, *P. planata*, and *P. marrisi*. Description of earthworm hereunder refers largely to *Pheretima posthuma*.

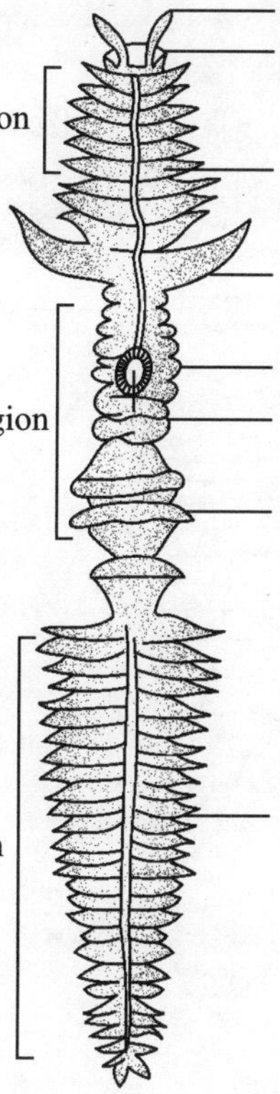


**Figure 12.1** *Chaetopterus.*

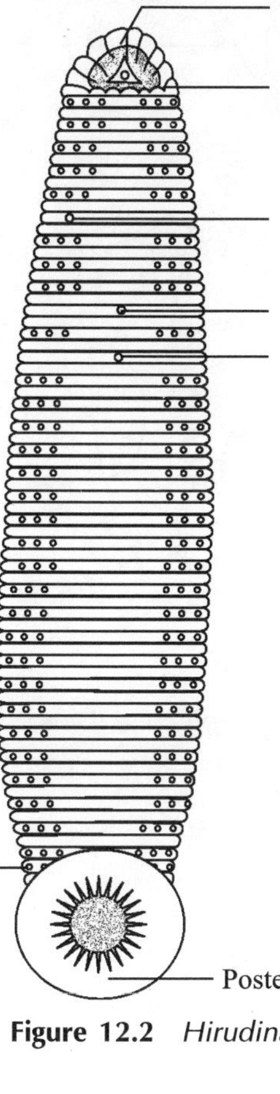


**Figure 12.2** *Hirudinaria.*

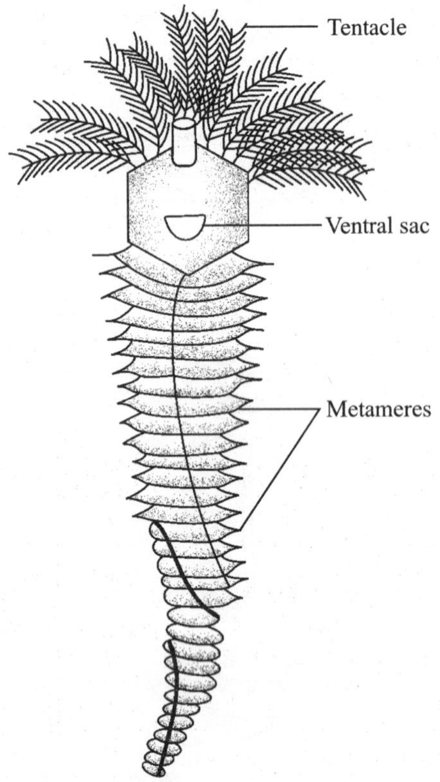


**Figure 12.3** *Sabella.*

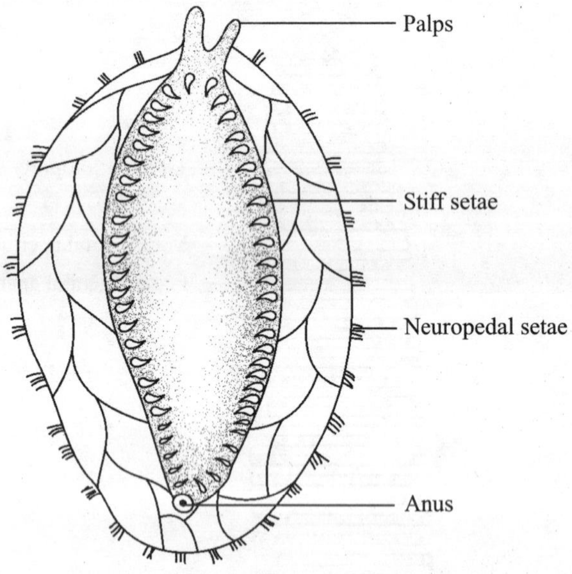


**Figure 12.4** *Aphrodite* (Dorsal view).

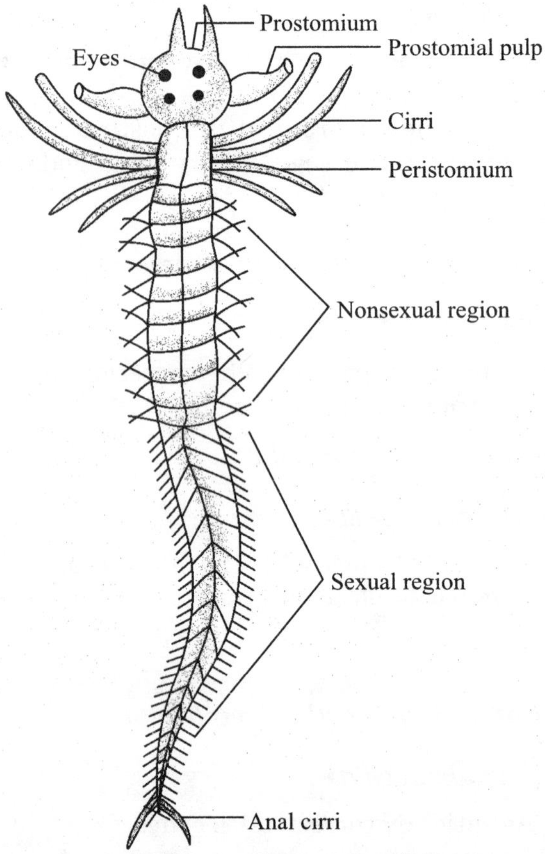


**Figure 12.5** Heteronereis stage of *Nereis.*

### *Systematic position*

Phylum : Annelida
Class : Oligochaeta
Order : Plesiopora Plesiotheca
Genus : *Pheretima*
Species : *Posthuma*

### *Habit and habitat*

These are burrowers, found in soil rich in decaying organic matters, usually in gardens, pastures and lawns, irrigated farmlands, near banks of ponds and lakes and rivers. They do not prefer to live in the sandy clay and acidic soils which are deficient in organic matters, but generally inhabit an upper layer of the earth to a depth of up to 45 cm, although go down up to 3 m, or more during the summer in search of moisture. The rainy season is the most favourable time for them to live in superficial soil, crawl on the surface in abundance and breed.

They are nocturnal in habit, come out of their burrows in search of food, reproduction and exploring fresh habitat. They are cold blooded, or poikilothermal, as their body temperature fluctuates with fluctuation in the surrounding environmental temperature. The way of excavation of their burrows is interesting. They eat on soil. The earthworm goes on eating soil progressively downwards. The organic contents of soil are digested and absorbed in the gut. The residual soil is discharged through the anus at opening of burrow as the cast. The cast shows their presence and forms heaps of small round pellets lying separately near the burrows. Natural life spans of the earthworm vary from three and a half years to ten years.

### *External features*

**Shape and size:** Body is long, elongated, cylindrical, narrow and bilaterally symmetrical. The anterior end is tapering. The posterior end is about blunt. The dorsal surface is marked by the presence of dark median line of dorsal blood vessel, which runs throughout the body length just below the skin. The presence of genital openings and papillae in the anterior body part mark the ventral surface. Size varies from species to species and individual to individual. A mature worm may attain size up to 150 mm in length and 3–5 mm in width (Figures 12.6 and 12.7).

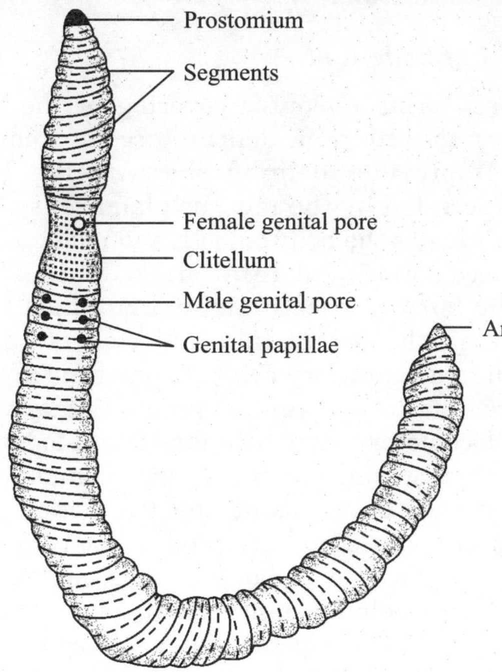


**Figure 12.6** *Pheretima posthuma.*

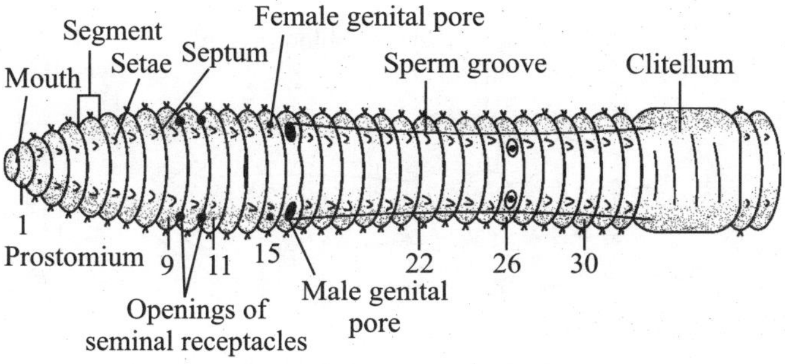


**Figure 12.7** Dorsal view of dissected earthworm.

**Colouration and segmentation:** This is slimy to touch and glistening dark brown in colour due to the porphyrin pigment in the body wall, which protects it from the effect of bright light. The dorsal surface of the body is darker than the ventral surface. The body consists of about 100–120 small ring like segments, or somites, or metameres. It exhibits true segmentation, i.e. external segmentation corresponds internal segmentation. The whole body surface is divided into many ring like segments by distinct series of annular grooves representing the external segmentation. The interior of body is divided by inter-segmental septa, or coelosepta. All body segments are alike, except the first and last segments.

**Setae:** In all body segments except the first and last, a clitellum, i.e. ring of chitinous setae lying embedded in the middle of each segments and projects backwardly. Each seta, arises from a setigerous sac of skin, is pale yellow, curved, S-shaped, rod like with a swollen middle part called the *nodulus*. Setae help in locomotion by holding the earth as they are directed backwards. Setae can be moved in any direction under the control of special muscles and extended, or withdrawn at will of the worm.

**Clitellum, or cingulum:** In mature worm, a prominent circular band of glandular tissue known as *clitellum* completely surrounds segments 14th to 16th forming a girdle around these segments. The glandular cells of clitellum secrete mucus, albumen and materials for forming cocoons, which help in fertilization as eggs are laid in them. Due to clitellum, body is divided into the following regions:

**Pre-clitellar region:** The anterior body part composes from segment 1st to 13th constitutes this region. The first segment is characteristically called the *perristomium* (peri = around + stoma = mouth) which surrounds the mouth. At anterior end of peristomiom, a small fleshly lobe called the *prostomium* (peri = around + stoma = mouth) is present. Prostomium is not a segment and remains separated from the peristomium by a groove. The prostomium is an extension of the peristomium at its dorsal side and ventral to it there is a mouth.

**Clitellar region:** The segments 14th, 15th and 16th form this region.

**Post-clitellar region:** All body segments after the clitellum represent this region. It starts from

17th segment and extends up to the last body segment.

**Body wall:** The body wall is very thin, soft, shiny and elastic and consists of the following layers:

*Cuticle:* This thin, elastic, noncellular and iridescent layer covers the body externally. It is made of two layers of collagenous proteins forming the fibres and a polysaccharide with a small quantity of gelatin. It is secreted by the columnar epithelial cells or supporting cells of epidermis and remains perforated by many minute pores through which the integumentary nephridia and epidermal glands open to outside. This layer is protective in function.

*Epidermis:* Cuticle is followed by a single layer of columnar cells forming the epidermis which consist of following four types of cells:

- *Supporting cells:* These long columnar cells constitute the major part of epidermis.
- *Gland cells:* These cells are situated between the supporting cells and are of the following two types:
  - — Mucous gland cells: Such cells are club shaped, large in numbers, secrete mucus to keep the body moist, prevent animals from desiccation and helps in locomotion by lubricating the body. Mucous secreted by these cells cements wall of burrows to keep the animals lubricated and smooth. Mucous gland cells are also known as *goblet cells*, which open at the surface by minute pores.
  - — Albumen gland cells: These are cylindrical, fewer with uniformly distributed fine granules. They secrete albumen.
- *Basal cells:* Small and undifferentiated cells which fill up the spaces between inner side of other epidermal cells. The cells later differentiate to from supporting- and gland cells and may be called the *replacing cells*.
- *Sensory cells:* Cylindrical, lie in groups having sensory hair like protoplasmic processes at their outer free ends and also known as *epidermal receptor cells*.

*Muscle:* Muscles of the body wall are present just below the epidermis. Two distinct muscle layers are outer circular and inner longitudinal muscle layer. Circular muscle layer is thin, forms a continual layer around the body and contains many porphyrin pigment granules. Longitudinal muscle layers is nearly two time thicker than the circular muscle fibres and lies in separate longitudinal bundles, which are separated together by thin connective tissue septa. Contraction of circular muscles makes the body long and narrow, while that of longitudinal muscles makes the body short and broad. Longitudinal muscle layers are followed by a very thin strip of circular muscle fibres. Musculature of the body wall consists of smooth muscle fibres.

### Coelomic epithelium

A thin layer of coelomic epithelium lines the body wall musculature internally. A membranous layer consisting of thin flattened squamous cells of mesodermal origin is found. This layer forms the outer boundary of the coelom and is called the *parietal peritoneum*, or parietal layer of the coelomic epithelium.

### Functions of body wall

Body wall forms an outer covering of the body to protect the internal, delicate organs from the injury. It provides definite shape to the body. Mucous secreted by the mucous gland cells of the epidermis keeps the body smooth and moist to help in the locomotion and respiration. Mucous also keeps the burrow smooth and moist, which helps in checking the desiccation and cementing the wall of burrow. Sensory cells of epidermis are the only receptors to receive the stimuli. Setae, which are the locomotory structure, are found lodged in it. The contraction and relaxation of muscles of the body wall bring about the locomotion with the help of setae. Thin, soft and moist body wall helps in the gaseous exchange, and simplifies the respiration. Coelomic epithelium, or peritoneal layer secretes the coelomic fluid. Cuticle checks the excess evaporation (Figure 12.8).

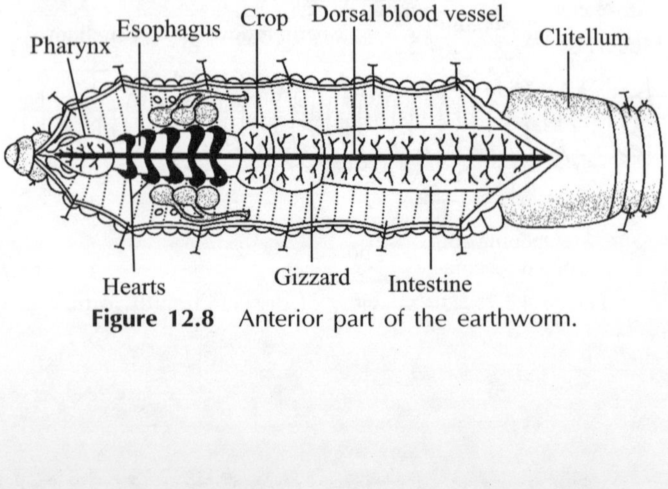


**Figure 12.8** Anterior part of the earthworm.

### *Setae, setal sac and associated structures*

Setae are found in every body segment except the first, last and clitellar segments. In the *Pheretima*, nearly 80–120 setae are arranged in a ring in the middle of each segment. Such an arrangement of setae is called *perichaetine*. In *Lumbricus*, setae occur in two pairs on eight sides of a segment. This is called *lumbricine arrangement*. Each seta is found lodged in a setal sac. Setae are composed of chitin and scleroprotein. Seta are pale yellow, curved, S-shaped, pointed, rod like structures with swollen middle part called the *nodulus*.

Two sets of muscles are found attached at the base of the setal sac, protractor and retractor muscles. Protractor muscles radiate outwards from the base of setal sac to join the circular muscles. Retractor muscles run inwards to join the inner thin strip of the circular muscle band situated just above the parietal peritoneum. These muscles protrude and withdraw the setae respectively. Setae fall out of the body wall when worn-out and are replaced by the new ones repeatedly (Figure 12.9).

### *Coelom*

Coelom is a spacious cavity between the body wall and alimentary canal and encloses all the internal, or visceral organs of the body. Hence it is also called the *perivisceral cavity*. Parietal epithelium lines this on the outer side and visceral epithelium on the inner side. The splitting of the embryonic mesoderm called the *schizocarp* derives this cavity, a true colon.

Outer lining of the colon lying below the body wall is called the *parietal layer* or *somatic layer* and that lies around the alimentary canal is called the *visceral* or *splanchnic* layer. Many cells of the splanchnic layer containing yellow granules are called the *chloragogen cells*.

Coelom is divided into small compartments due to the development of intersegmental septa, which grow from the intersegmental grooves of the body wall. Each septum is a sheet in interlacing muscle fibres and remains surrounded from both side by the peritoneum. The first four segments lack intersegmental septa and hence they have a continuous coleom. The first intersegmental septum is located between the 4th and 5th segments. It is thin and membranous. Next five septa are thick, muscular and obliquely placed between the segment 5/6, 6/7, 7/8, 8/9, 9/10.

The contraction of these cone-like septa increases pressure on coelomic fluid by which the anterior segments become turgid and are used for digging the burrow.

Intersegmental septum between the 9th and 10th is absent. Remaining septa from 11th and 12th to the posterior end are thin, membranous and transverse. The first three of them (11/12, 12/13, 13/14) are complete without apertures, thus isolating their coelomic chambers. Remaining septa from the 14/15 to the posterior end are perforated by many minute apertures. These apertures are surrounded by the unstriped muscles. Such septa are characteristic of *Pheretima*. With this perforation, the coelomic chambers of these segments communicate. Coelom is filled with the coelomic fluid, which is milky white alkaline, colourless, watery matrix of plasma and many coelomic corpuscles. Corpuscles are differentiated into four types:

- **Granulocytes:** These are numerous, large sized, spherical with a concavity having small nucleus and many fat droplets. They form

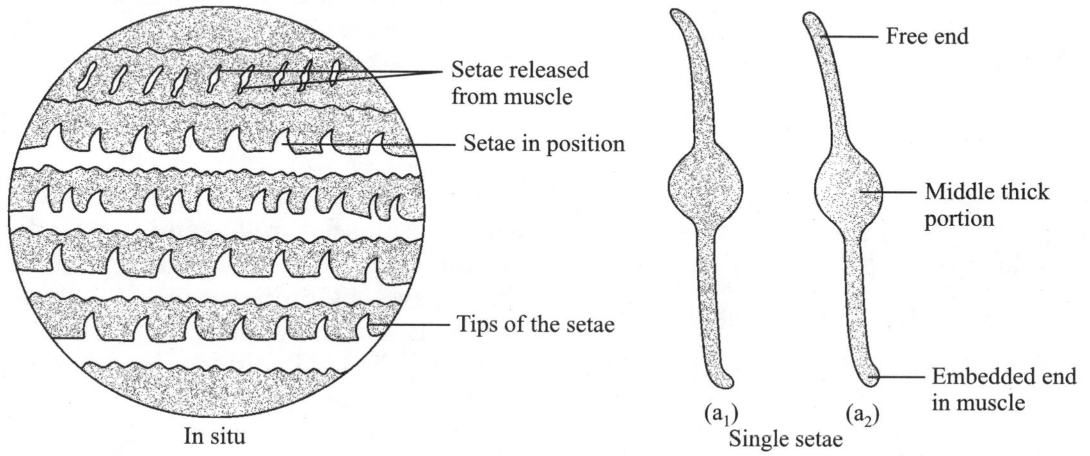


**Figure 12.9** Setae of earthworm

pseudopodia like folds to act as the phagocytes. They are probably nutritive and help in removing harmful bacteria, foreign particles and other microorganisms that happen in the coelomic fluid by the phagocytosis. These are also called the *eleocytes*.

- **Amoebocytes:** They are numerous, much smaller, having many pseudopodia like process and look like stars and nucleated. These are phagocytic, take up excretory and foreign matter and act in defense mechanism against the parasites.
- **Mucocytes:** Elongated, vase-shaped corpuscles with an expanded end like a fan and other end being narrow. The narrow end bears the nucleus. Their function is not definitely known.
- **Leucocytes:** Small flattened disc shaped nucleated cells, fewer in number, comprises nearly 10% of corpuscles with clear cytoplasm. Their functions are still uncertain. Coelomic fluid flows from the anterior to posterior on ventral side and from the posterior to anterior on dorsal side (Figure 12.10).

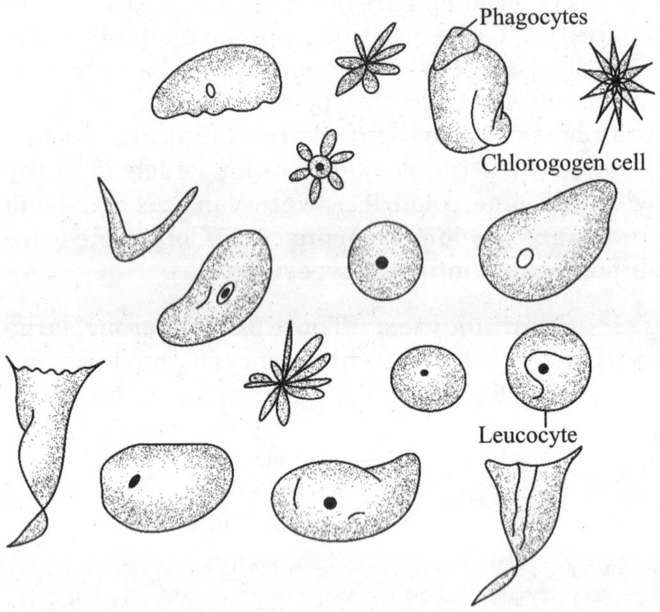


**Figure 12.10** Corpuscles in coelomic fluid of *Pheretima*.

### *Functions of coelomic fluid*

The following are the functions of coelomic fluid:

1. It oozes out through the dorsal pores and keeps the body moist, to facilitate respiration.
2. It kills the bacteria of soil, which come on the body surface of worm.
3. It helps in removing the excretory wastes from the body.
4. It helps in transportation of the materials from one part of body to the other parts.
5. It helps in destroying harmful bacteria, microorganisms and foreign materials.
6. It also helps in fixing setae into the ground during locomotion. This is achieved by making certain segments turgid and stiff by restricting coelomic fluid in them and by closing sphincter apertures of septa between them.
7. Luminosity produced by some earthworms is due to the coelomic fluid.

### *Lymph glands*

These are small, paired, glandular structures, present in the coelom on sides of the dorsal vessel from 26th segment up to last, also contain phagocytes, which ingest the foreign materials and digest them.

### *Locomotion*

No specialized locomotory organs are found, although they are very active and crawl rapidly, when out of burrows. The process of locomotion is a cumulative effect of the contraction and relaxation of both muscle layers of the body wall aided by setae and hydrostatic pressure, which is created by the coelomic fluid. Forward locomotion is brought about by an increase in hydrostatic pressure of the anterior body segments (usually first nine segments) and at the same time, the contraction of circular muscle layer begins at the anterior end and passes backwards. This causes the anterior region to extend forward and make it thinner in diameter.

Extending and thinning pass backwards as a wave of contraction at the rate of 2 to 3 cm per second. By this means, the worm is pushed forward. Anterior end now grips the substratum and setae act as hooks by their posteriorly directed points. After the wave of contraction of circular muscles passes down the front half of worm, the circular muscles relax and longitudinal muscles of the body wall of anterior segments contract in a wave beginning from the anterior end. This shortens and thickens the anterior end, causing the posterior body of worm to be dragged forward. At this stage, the segments in a state of longitudinal contraction do not move but remain anchored to the ground by the extended setae.

Contraction of the longitudinal muscles passes backwards like a wave. Again a wave of contraction of the circular muscle starts from the anterior end making it thinner and extends forward. This wave starts before the first one reaching the posterior end. Thus, locomotion is brought about by the alternate contractions of the circular and longitudinal muscles causing wave of thinning and thickening to pass backward. This involves partly a pushing of anterior end and partly a pulling of the posterior end, with setae playing only an accessory role. The worm moves at the rate of 25 cm per minute. Nervous system coordinates the activities of circular and longitudinal muscles. So contraction of any layer brings about the relaxation of other muscle layer.

Earthworms can move backward also, as they usually do during their withdrawal from the burrow and during excavation. The reversing and directing of setae brings about the backward movement. The coelomic fluid serves as a hydraulic skeleton because a decrease in its pressure results in the relaxation of muscles. The earthworms can move on a smooth and hard surface like glass by the mucous for adhesion, as setae cannot anchor the substratum.

### *Digestive system*

Digestive system consists of the alimentary canal and digestive glands.

**Alimentary canal:** The canal runs as a straight tube through the length of the body from the mouth to anus. It consists of the following regions:

**Mouth and buccal chamber:** This is a crescent aperture lying below the prostomium. Mouth leads into a short, thin-walled tube called the *buccal chamber*, or *buccal cavity*, running up to third segment. The lining of buccal cavity is folded and is surrounded by the muscular strands. The buccal cavity can be protruded through the mouth, with the help of special muscles that run backward from the buccal chamber to the body wall.

**Pharynx:** Buccal chamber is followed by the pharynx, which extends up to the fourth segment. It is a pear-shaped, broad, muscular structure and remains separated from the buccal cavity by a groove. The roof of pharynx is very thick and projects into the pharyngeal cavity as pharyngeal bulb. Its lateral wall internally from two horizontal folds, or shelves, one on each side which divides the lumen of pharynx into two chambers—the dorsal salivary chambers and ventral conducting chamber.

Roof of pharynx is lined by the ciliated epithelium, above which there are many muscles with connective tissue and blood vessels, and outside of these structures there are pharyngeal, or salivary gland. These are grouped as small, whitish, unicellular gland of the chromophill cells. These glands open into the salivary chamber of pharynx through their fine ducts. They secrete mucin for lubrication of food and a proteolytic enzyme for digestion of the proteins. Ventral conducting chamber of pharynx serves as the passage for the ingested materials.

Pharyngeal wall remains about the body wall by a thick mass of the muscular strands like the buccal chamber. The contraction and relaxation of these muscular strands compress, or dilate the pharyngeal lumen. Thus, pharynx acts as a pump during the feeding. Mouth is pressed against the bits of humus of mud. Then pharynx undergoes a series of contraction resulting into sucking of food into buccal chamber and then pumping the same backwards into oesophagus.

**Oesophagus:** Behind the pharynx, there is a short, narrow, thin walled oesophagus, which runs up to the 8th segment. The oesophageal wall is folded internally and lacks gland.

**Gizzard:** The oesophagus is followed by the gizzard, a prominent, oval, hard, thick walled and highly muscular organ lying in the 8th, or between 8th and 9th segment. It has a thick wall of circular muscles lined by the columnar cells, which is further lined by the cuticle. Gizzard grinds food into a fine state.

**Stomach:** The short, narrow, thin-walled, tubular stomach extends up to the 14th segments and follows the gizzard. Anterior and posterior opening of the stomach are sphinctered. Its thin wall is highly glandular and vascularized, but less muscular. Its internal wall is folded transversely. The glandular cells are situated in the epithelial lining of the stomach and secrete a proteolytic enzyme. Some calciferous glands are found in the epithelial lining which produces calcium and $CO_2$. Calcium probably neutralizes the contents of the alimentary canal. Calciferous glands are excretory, remove ions of calcium and carbonate as calcite into stomach from where it passes out with the mud through the anus.

**Intestine:** It is a long, wide and thin walled tube extending from the 15th segment up to the anus, with a beaded appearance due to the correspondence to septa but bulges in each segment. The intestinal lining has ciliated- and glandular cells and it folds to form the villi. One of these villi becomes larger and well developed than other to form the typhlosole. Typhlosole, thus, hangs in lumen of the intestine and runs mid-dorsally form the 26th segment up to the last except posterior 24th and 25th segments. Typhlosole divides intestine into the following three regions:

- *Pre-typhlosolar region:* The first part of intestine lying between the 15th to 26th segments constitutes pre-typhlosolar region. Here, the intestinal lining is folded to form villi which are highly vascular. Typhlosole is absent in this region. In 26th segment, two short and conical outgrowths, one on the either side are given off from the intestine. These are called the *intestinal caeca* which are extended anteriorly up to the 22nd segment with a special blood supply. Internally, the caeca have many longitudinal folds which appear as villi in transverse sections. Some workers believe that these caeca are digestive glands and secrete an amylolytic enzyme for the digestion of the starch. Usually, active digestion occurs in this region (Figure 12.11).
- *Typhlosolar region:* Second part of the intestine lying between the 26th to last segment (except the posterior 24th segment, 25th segment) is provided with an internal median fold of dorsal wall of the intestine called the *typhlosole*, which is in fact, well-developed intestinal villi. Typhlosole increases absorptive surface of intestine. The process of digestion is probably completed in this region; hence it is major site of absorption.
- *Post-typhlosolar region:* Intestine in last 24th and 25th segments have no typhlosole. It is the third region of the intestine and is called the *rectum*. It is a thin walled, vascularized, without villi and glandular cells, contain small pellets of mud, which are passed out from the anus as the faeces at opening of the burrows. These are actually castings (Figure 12.12).

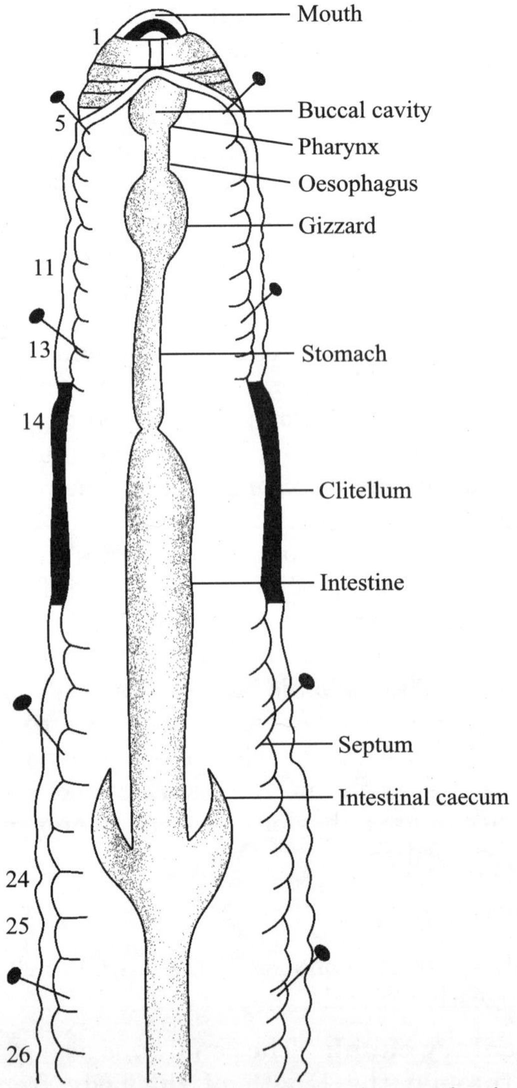


**Figure 12.11** Anterior end of *Pheretima* showing the alimentary canal.

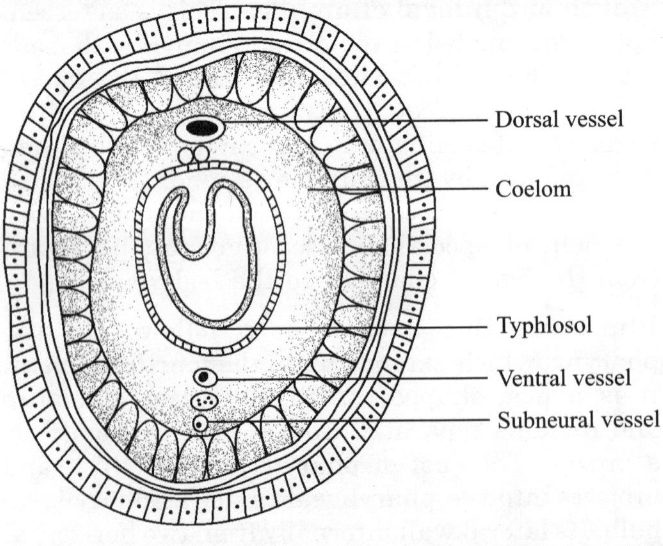


**Figure 12.12** T.S. of earthworm through typhlosar intestine.

**Anus:** It is a small circular opening and is placed at the centre of terminal end of last or the anal segment of the body.

### *Histology of alimentary canal*

The canal is made of four layers—an outer layer of visceral peritoneum which forms outermost covering layer of the alimentary canal. Most cells of this layer around stomach and intestine are modified to become the chloragogen cells or chloragocytes containing the yellow granules of chloragosomes. Next to this, two muscles layer, the outer layer of longitudinal muscle fibres and the inner layer of circular muscle fibres are found.

The arrangement of the muscle layers in the gut wall is just reverse to the body wall. Both these muscle layers are well developed around the pharynx and oesophagus, but poorly developed around the stomach and intestine. Longitudinal muscle fibres are absent in gizzard. In the entire gut wall, muscles are involuntary and unstriped. The fourth layer is the epithelial lining of columnar cells which are ciliated in roof of the pharynx. In stomach and intestine, it is mostly glandular and ciliated to form intestinal villi. In gizzard, it mostly secretes cuticle.

**Food and feeding mechanism:** Omnivorous, feed on all somewhat organic humus, decaying matters, small protozoans, nematodes, insects and other soil microorganisms. They commonly ingest soil in large quantities. During feeding, buccal chamber is everted out and food is drawn into the mouth by the sucking action of the muscular pharynx.

### *Digestive glands and digestion*

Pharyngeal, or salivary gland cells, glandular cells of the stomach, intestine and intestinal caeca are supposed to be digestive glands, which secrete the digestive enzymes for the food digestion. Various enzymes are secreted for the digestion due to their omnivorous mode of the feeding habit. Swallowed soil passes through buccal chamber into the pharynx, where it receive salivary secretion from the salivary gland cells. The salivary secretion contains mucin and a proteolytic enzyme.

Mucin lubricates the food and food passage, while proteolytic enzyme hydrolyses proteins into peptones and proteoses. Then food comes into the gizzard through oesophagus. Gizzard further grinds the food and soil due to contraction of its well-developed circular muscles into a fine state, so that the enzymes of stomach, intestine and intestinal caeca can act on it. Firstly, after reaching food in fine state into stomach, it is subjected to action of more proteolytic enzymes secreted by glandular cells of this region. This results in conversion of proteins into peptones.

Then the food reaches into the intestine and comes across the enzymes secreted by the glandular cells of the intestine and intestinal caeca. Usually, these enzymes are proteases which hydrolyze the peptones into amino acids. The lipase hydrolyzes the fat into fatty acids and glycerol. Amylases acting on the carbohydrates and converts them into the monosaccharide. Cellulose hydrolyzes the cellulose, if any and chitinase hydrolyzes chitin. The digestion occurs in the stomach and fully completed in intestine.

**Absorption:** Digested food is absorbed by the intestinal villi, more particularly by the typhlosole. Absorbed food passes into the blood stream through the well developed capillary network of the intestine from where transported to the tissues. Coelomic fluid also transports the digested food to the tissues.

**Egestion:** The undigested remains of the food are passed out through the anus as castings at opening of the burrows.

### *Chloragogen cells*

Some cells of the visceral peritoneum around stomach and intestine are modified as chloragogen cells having yellow granules, chloragosomes in them. They have vital importance in metabolism as they play a role of liver of vertebrates. These cells take up the waste from the blood and probably from the coelomic fluid and store it as yellow granules of the chloragosomes. When become filled with such granules, these cells either pass out through dorsal pores, or excretory organs, or excreta remains permanently in chloragogen cells throughout life.

These cells are the chief centre of synthesis and storage of glycogen and fat. They manufacture glycogen from fatty acids and store it as reserve food and pass it onto the coelomic fluid when needed. Thus, they are excretory as they serve for storing reserve food. They are also related with the deamination of amino acids, formation of ammonia and synthesis of urea.

**Histology of the gut wall:** The external layer derived from peritoneal epithelium is followed by longitudinal muscle fibres and circular muscle fibres. Glandular and ciliated cells are found in the internal epithelium (Figure 12.13).

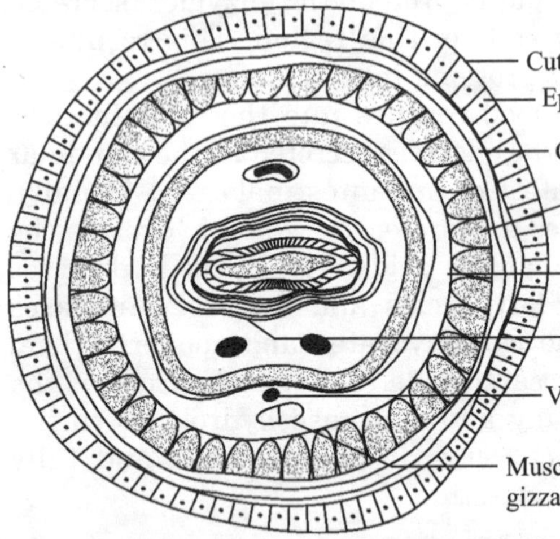


**Figure 12.13** T.S. of earthworm through gizzard.

### *Blood vascular system*

**Open and closed blood vascular systems:** Blood vascular system of *Pheretima* is closed type consisting of blood vessels, hearts, loops, capillaries and blood glands.

**Blood:** Blood is red due to the presence of haemoglobin, which is not contained in corpuscles but is found dissolved in plasma. Plasma also contains many colourless and nucleated corpuscles.

**Blood vessels:** These are of two types—collecting blood vessels and distributing blood vessels, which are closed tubes. Arrangement of blood vessels in the anterior thirteen segments is somewhat different from that behind thirteen segments. Therefore, the blood vessels can be described under the two categories.

The blood vascular system in the first thirteen segments is modified considerably and differs markedly from that of intestinal region. It consists of the following:

#### *Arrangement of blood vessels in the anterior 13 segments*

**Dorsal vessel:** A distributing vessel, which has neither dorso-intestinals, nor commissural vessels opens into it. It sends out all the collected blood from posterior regions of the body into hearts and anterior region of the gut, where it divides into three branches, distributed over the pharyngeal bulb and roof of buccal chamber. However, it supplies to stomach, gizzard, oesophagus, pharynx and other related parts.

**Ventral vessel:** This remains as distributing vessel in these segments also, but extends only up to the second segment. Ventro-intestinals are absent, hence, it does not supply to the alimentary canal in this region, However, ventro-tegumentary vessels, a pair in each segment, supply blood to the integument, nephridia, septa and reproductive organs.

**Supra-oesophageal vessel:** The shortest longitudinal vessel extending from ninth to 13th segment is situated above the stomach. It receives blood from lateral oesophageals by two pairs of anterior loops that encircle stomach in 10th and 11th segments. It sends its collected blood by the latero-oesophageal hearts in segments 12th and 13th to ventral vessel.

**Lateral oesophageals:** Sub neural vessel bifurcates in the 14th segment to form two lateral oesophageals. These vessels are closely attached to wall of the stomach form 10th to 13th segments and communicate with ring vessels. These vessels receive a pair of branches in each segment bringing blood form body wall and septum. They also collect blood from reproductive organs and nephridia.

#### *Arrangement of blood vessels in the segments behind the 13th segments*

**Dorsal vessel:** This contractile, thickest collecting vessel receives two pairs of dorsointestinal and a pair of commissural vessel.

**Ventral vessel:** This thin walled distributing vessel gives out a pair of ventrotegumentary vessel and a ventrointestinal vessel in each segment behind the 13th segment.

**Subneural vessel:** Being a collecting vessel, it collects blood from the ventral body wall.

#### *Intestinal blood plexus*

**External plexus:** Capillary network at the outer surface of intestine is known as *external plexus* which receives blood from ventral vessel through ventro-intestinal and passes it to internal plexus.

**Internal plexus:** Capillary network present between the circular muscle layer of intestine and its internal epithelial lining is known as *internal plexus*. It absorbs nutrients from gut and connects to dorsal blood vessel through the dorso-intestinals.

***Commissural vessels***

These vessels connect the dorsal and sub-neural vessels, receive blood from nephridia, body wall and reproductive organs through capillaries and then, they send it to dorsal blood vessel.

***Integumentary vessels***

Coming from the ventral vessels, they supply blood to integument for aeration and numerous capillaries of commissural vessel in each segment collect the aerated blood.

***Nephridial vessels***

Originating from ventro-tegumentary vessels, it supplies blood to nephridia.

**Hearts and anterior loops:** In the posterior segment, behind the 13th segment, dorsal and ventral blood vessels have no direct connections, but anteriorly both of these vessels are connected together by 4 pairs of pulsatile hearts which are neurogenic, i.e. the heart beat originates in the nerve cells of hearts. Hearts are contractile and encircle alimentary canal in segments 7th, 9th, 12th and 13th.

Hearts of segments 12th and 13th are joined above with the dorsal and oesophageal vessels. These *latero-oesophageal* hearts having thick muscular walls and a pair of valves at each junction with dorsal and supra-oesophageal vessel and another pair of valves at the ventral end. Other hearts of segments 7th and 9th are called *lateral hearts* which connect dorsal vessels to the ventral vessels. There are two pairs of loop-like vessels in the 10th and 11th segments, which connect supra-oesophageal with lateral oesophageals. These vessels are neither muscular nor pulsatile and are called *anterior loops*. Blood from the lateral oesophageals flows through these loops into supra-oesophageal.

**Blood vessels of gut:** On either side of the stomach, ring-like vessels, connect the supra-oesophageal with lateral-oesophageal. Through these vessels, blood flows upwards, from lateral-oesophageals into supra-oesophageal. Buccal cavity, pharynx and gizzard receive blood supply from the dorsal blood vessel directly (Figures 12.14–12.15).

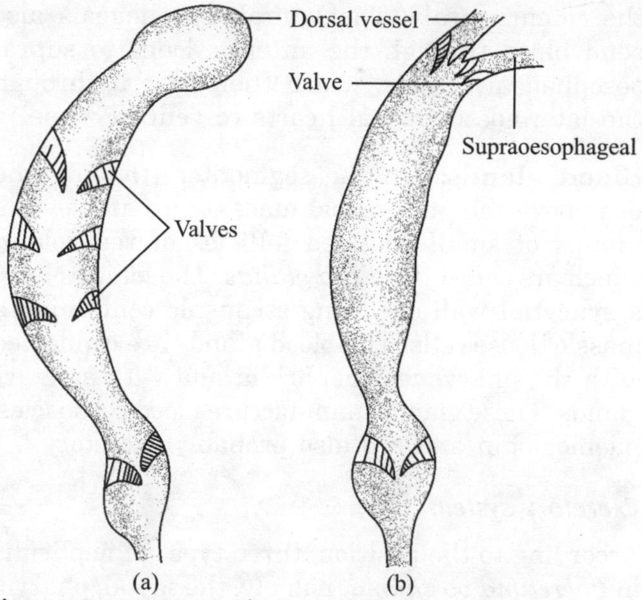


**Figure 12.14** Sectional view of (a) lateral and (b) latero-oesophageal heart.

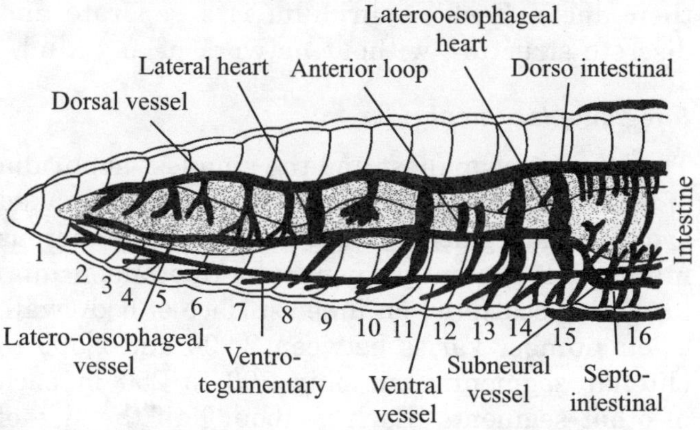


**Figure 12.15** The major vessels and hearts of *Pheretima* in the anterior segments (Lateral view).

**Circulation of blood:** Blood collected by the dorsal vessel through dorso-intestinals, blood plexuses of intestine and commissurals is given out partly to anterior alimentary canal, but mainly through the hearts to ventral vessel. In ventral vessel, the blood flows forwards to the anterior regions in front of the hearts, but major portion of blood flows backwards. This is distributed through the ventro-tegumentaries to body wall and organs in coelom and through ventro-intestinal vessels to alimentary canal.

All parts receive blood from the ventral vessel. From ventral body wall, blood is collected by the sub-neural vessel, which also receives some blood through the lateral- oesophageal vessel from the anterior region. The blood passes from the sub-neural vessel to the dorsal vessel through the commissurals. The lateral-oesophageals also send blood through the anterior loops to supra-oesophageal vessels, which then pass it through the latero-oesophageal hearts to ventral vessel.

**Blood glands:** In the segments, 4th, 5th and 6th above the pharyngeal mass, there are several groups of small rounded follicles of red colour, which are called the *blood glands*. The follicles have a syncytial wall enclosing a capsule containing a mass of loose cells. The blood glands are connected with the pharyngeal nephridia and with salivary glands. These glands manufacture blood corpuscles, haemoglobin and are also probably excretory.

### *Excretory system*

According to the position, three types of nephridia in *Pheretima posthuma*, namely, the septal, pharyngeal and integumentary are found. They differ from one another in size and in respect of opening of their ducts. Each nephridium is a separate and discrete structure without network of any kind.

### *Integumentary nephridia*

These are the smallest and the simplest nephridia of earthworms and are hardly visible to the naked eye. In each segment, except the first six, numerous minute integumentary nephridia (metanephridia) are found scattered on inner surface of body wall. Their number varies between 2,000 and 2,500 in clitellar segments, but only 200 to 250 in each of other segments. Each is about half the size of a septal nephridium and lacks nephrostome and neck. Its inner end is closed. Terminal nephridial duct has no cilia and opens out directly on the body surface through a minute nephridiopore in the body wall. These nephridia are exonephric. Each has separate duct and opens separately to exterior on the skin.

This occurs in paired tufts lying at sides of oesophagus in the 4th, 5th and 6th segments. These nephridial tufts have three pairs of ducts, one in each segment, which open into buccal cavity and the pharynx in the second and fourth segments.

### *Pharyngeal nephridia*

A pair of large bunches of these nephridia, mixed with blood glands, occurs in each of the 4, 5 and 6 segments on the dorso-lateral sides of the gut. These nephridia resemble septal nephridia in size and integumentary nephridia in structure. These lack funnel and neck parts. Terminal ducts of a bunch unite to form a common duct. Thus, three pairs of common ducts are formed. Two common ducts of the sixth segment extend forwards and open into buccal chamber in the 2nd segment. Ducts of 5th and 4th segments open into pharynx in 4th segment. Thus, the pharyngeal nephridia are enteronephric like septal nephridia. Some modern biologists hold the view that these nephridia secrete some digestive enzymes instead of serving for excretion. This contention needs confirmation.

### *Septal nephridia*

These nephridia are attached to both sides of the septa and do not open on the skin, but are attached with an elaborate system of ducts which ultimately open into intestinal lumen mid-dorsally at segmental intervals. The system of ducts and openings is segmental in arrangement with a pair of septal excretory canals on each septum which open into a pair of supra-intestinal excretory ducts lying in mid-dorsal line of the worm above dorsal wall of the gut. These ducts communicate with intestinal lumen at each intersegmental place by a narrow ductule.

The septum between segments 15 and 16 and every septum behind this one bear two semicircular rows of septal nephridia on its anterior and posterior surfaces. Of the four rows of nephridia attached to a septum, two are located on the right and two on the left sides of intestine. Each row has 20 to 25 nephridia. There are 80 to 100 septal nephridia in each segment behind the 15th, but only 40 to 50 in the 15th and the last segments and none in the anterior 14 segments. Including the septum between segments 15 and 16, each intersegmental septum bears on its posterior surface a pair of loop-like septal excretory canals.

Septal nephridia of both rows open by their terminal ducts into the septal excretory canal of their side. A pair of narrow and closely placed, parallel supra-intestinal excretory ducts extend on dorsal surface of the intestine from the 15th segment up to posterior end of the body. All septal excretory canals open into supra-intestinal excretory ducts of their sides. Supra-intestinal excretory ducts are closed behind, but near the posterior septum of each segment either one of these opens by a minute ductule into intestine.

Excretory fluid of septal nephridia is discharged, not directly to outside, but into the intestine. That is why these are called the *enteronephric nephridia*.

Septal nephridia are the typical nephridia (holonephridia). Each is divisible into four parts—nephrostome, neck, body and terminal duct (Figures 12.16 and 12.17).

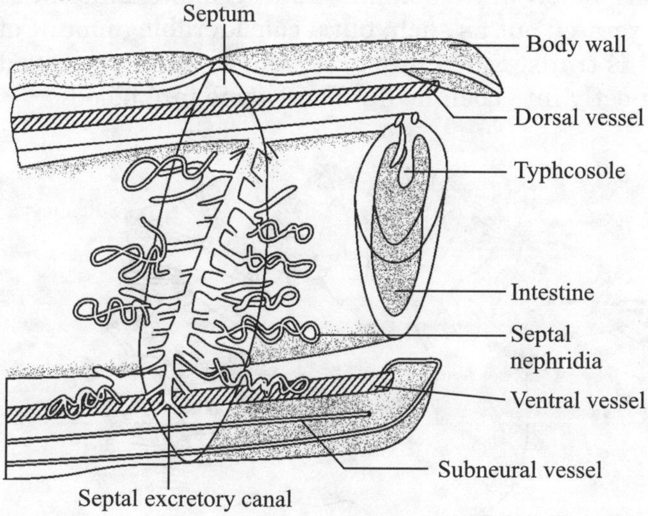


**Figure 12.16** Relationship between septal nephridia and septal excretory canal.

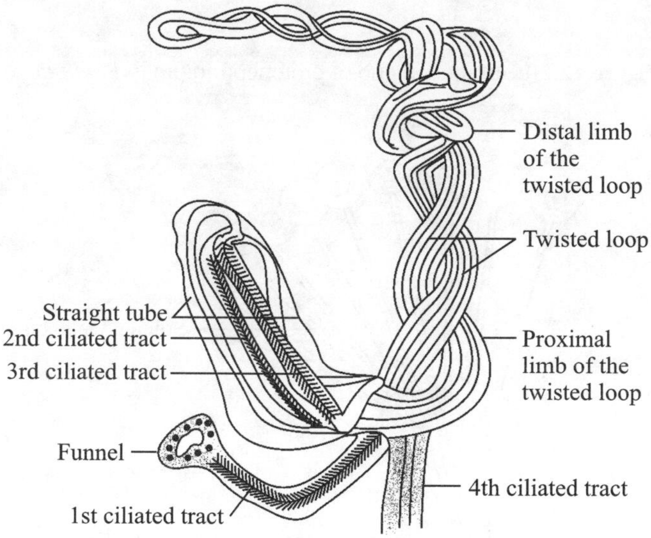


**Figure 12.17** Septal nephridium.

**Nephrostome, or nephridiostome:** This is flattened, ciliated with funnel-like proximal end and opens into coelom by an elliptical aperture. Cavity of the funnel is an intracellular tunnel through a large central cell. Margin of the aperture is lined by a row of small and ciliated marginal cells. Major part of the margin extends forwards like a hood to form its upper lip which consists of 8, or 9 larger marginal cells. The remaining part called *lower lip* consists of 4, or 5 marginal cells. All marginal cells are covered over by the coelomic epithelium.

**Neck:** Ciliated funnel leads into a short, narrow and ciliated tubule called the *neck*. This is also an intracellular tunnel through a series of cells covered over by a thin layer of the connective tissue.

**Body of nephridium:** Neck is connected to the main body of the nephridium, which is divisible into a small straight lobe and a twice larger twisted loop. The latter is formed of two limbs, the proximal and distal limb, which are twisted 9 to 134 times around one another like a rope. At the base of the nephridium, the straight lobe continues as distal limb of twisted loop, while the proximal limb is formed by the fusion of neck and terminal duct.

**Terminal duct:** The proximal limb of the twisted loop is formed at the base of nephridium by the fusion of neck and terminal duct. The latter is short, narrow and consists of the ciliated terminal part of the nephridial duct enclosed in similar connective tissue matrix. Septal nephridium has a remarkable feature as it lies wholly within the bounds of a single segment.

Enteronephric nephridial system is probably a device for conservation of water. Gut epithelium reabsorbs the water of the excretory fluid while letting the solid excretory matter passes out through anus along with faeces. Each nephridium in *Pheretima* is connected with two blood vessels, one bringing blood to it and the other taking away the blood. Septal and integumentary nephridia are supplied with blood from the ventral vessel through parietal vessel and septal branch. Blood is returned from these nephridia to general blood stream via the dorso-lateral and sub-neural vessel. Pharyngeal nephridia receive their blood supply from dorsal vessel and blood is returned from them to lateral oesophageal vessels.

Histologically, the body of nephridium consists of a connective tissue matrix into which the tubule of the neck enters and extends as a very long, intracellular, nephridial duct forming several loops. Duct extends four times in the straight lobe making two loops. In the twisted loop, it extends thrice in most part of the two limbs, but only twice in their thin apical part. In the straight lobe, two segments of nephridial duct are ciliated like tubule of the neck. Excretory structures of annelids are shown in Figures 12.18–12.21.

### *Physiology of excretion*

Earthworms excrete nitrogenous waste products of protein metabolism as mucous and urine. Mucous consisting of waste mucoproteins is mainly secreted out by mucous gland cells of the epidermis of skin and keeps the skin moist to destroy harmful bacteria and other microorganisms on body surface and to lubricate the wall of burrows. Urine is excreted by the nephridia. It is acidic and mainly contains ammonia, urea and water. Thus, earthworms are both ammonotelic and ureotelic. Traces of creatinine also occur in urine. Urea is mainly formed in chloragogen cells. Chloragogen cells deaminate excess amino acids and absorb ammonia from the blood and coelomic fluid. Some ammonia is excreted out as such, but a considerable amount of it is transformed into urea, which is then released mostly into coelmic fluid and some into blood.

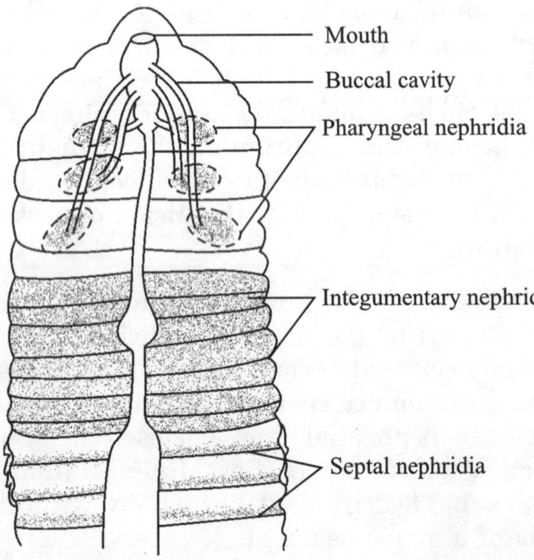


**Figure 12.18** Position of various nephridia in earthworm.

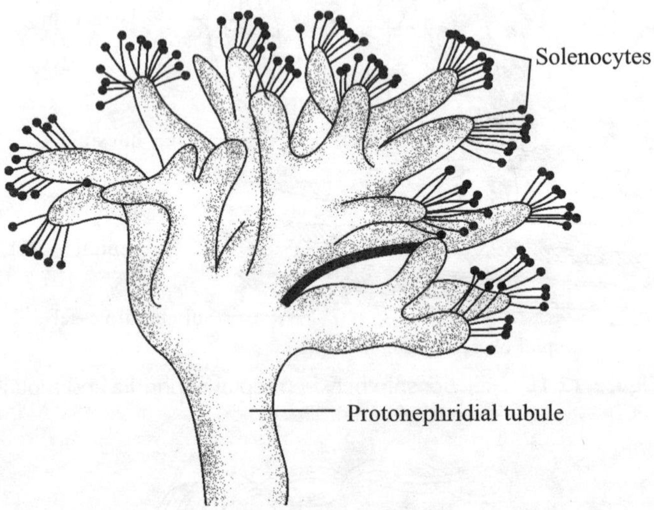


**Figure 12.20** Branched end of protonephridium in *Phyllodoce.*

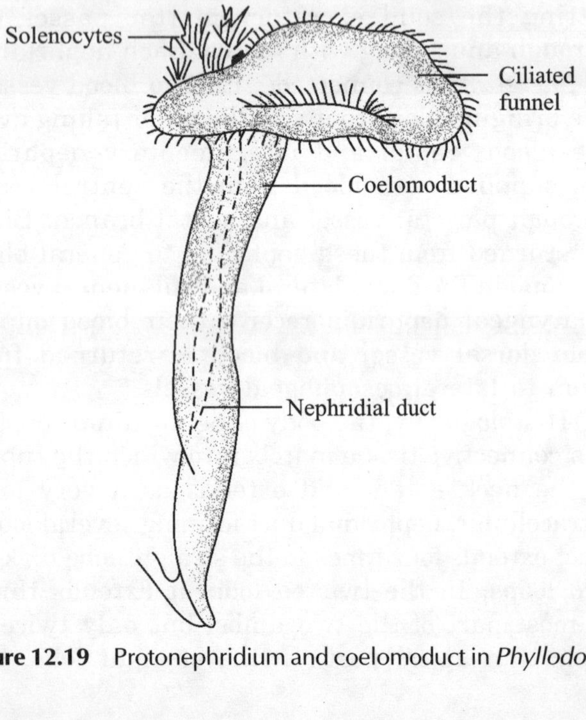


**Figure 12.19** Protonephridium and coelomoduct in *Phyllodoce.*

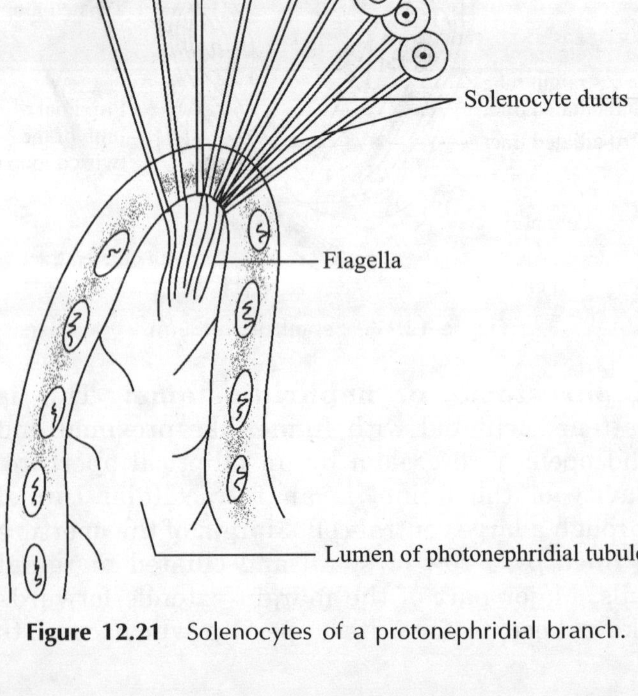


**Figure 12.21** Solenocytes of a protonephridial branch.

Nephridia extract water and excretory substances, together with some useful materials from both blood and coelomic fluid by ultrafiltration and concentrate excretory substances in filtrate by reabsorbing water and useful materials back into blood. During rainy season, the urine is dilute due to less reabsorption of water. In summers and winters, water is conserved by its increased reabsorption in the nephridia, so that the urine is hypertonic to blood and coelomic fluid. Thus, besides being excretory, the nephridia are also osmoregulatory in function. In *Pheretima* and other earthworms of warm countries, septal and pharyngeal nephridia are enteronephric. This is presumably an additional device to ensure water conservation because, if needed, more and perhaps all water can be absorbed back from the urine in intestine.

### Respiration

Earthworms respire aerobically, but lack specialized respiratory organs. Gaseous exchange, i.e. intake of $O_2$ into the blood from atmosphere in exchange of $CO_2$, therefore, occurs in skin, which is thin and richly vascularized for this purpose. Since gaseous exchange always occurs in solution, it occurs only when the skin is moist.

Mucous secreted by some glandular cells of the skin epidermis and the coelomic fluid, which constantly oozes out through the dorsal pores, keep the skin always moist. This is also the reason why earthworms always live in the moist soil. While blood is circulating through cutaneous capillaries, its haemoglobin binds the $O_2$, forming an unstable compound, the oxyhaemoglobin. When oxygenated blood circulates through various tissues, oxyhaemoglobin readily releases its oxygen due to low $O_2$ tension and high $CO_2$ tension in the tissues. In case oxygen is not available, earthworms can survive by anaerobic respiration for 6 to 30 hours. If $O_2$ is not available for longer time, or skin dried up for sometime, earthworm dies due to asphyxiation.

### Nervous system

Nervous system is well developed, concentrated and consists of central, peripheral and sympathetic, or autonomic divisions.

**Central nervous system:** This consists of nerve ring, or brain and ventral nerve cord.

**Nerve ring:** A pair of cerebral, or supra-pharyngeal ganglia is found together to form the so-called *brain*, which lies dorsally in the third segment in groove between the buccal chamber and pharynx. From brain, a pair of circum-pharyngeal connective nerve arises, one on each side. They encircle the pharynx and meet ventrally in the 4th segment by a pair of fused sub-pharyngeal ganglia. Thus, a nerve collar is formed around the pharynx which is called the *nerve ring*, or *brain ring*.

**Nerve cord:** A nerve cord originates from the sub-pharyngeal ganglia and runs on mid-ventral side beneath the ventral vessel, but above sub-neural vessel it extends up to posterior end of the body. Nerve cord is double, made of two longitudinal cords which are fused together. In each segment, from the 5th to last, nerve cord has a pair of ganglia fused together as swellings, which represents the segmental ganglia. Nerve cells are distributed all over the nerve cord and are not confined to ganglionic swellings.

Double nerve cord is solid and enclosed in a common sheath called the *perineurium* consisting of three layers. Both nerve cords are separated internally by a vertical epineurium. Ventro-lateral regions of the nerve cord contain bipolar and tripolar nerve cells all along the length and middle region have many nerve fibres with the supporting neuroglia. In upper part of nerve cord, there are four giant fibres surrounded by the epineurium. They are all tubular, filled with homogeneous plasma like matter and conduct impulses throughout nerve cord rapidly. However in the region of segmental ganglia, partition between the two nerve cords is absent.

### Peripheral nervous system

From brain, 8 to 10 pairs of nerve arise, which supply to prostomium, buccal cavity and pharynx. Two pairs of nerves arise from the circum-pharyngeal connectives, and supply to the first segment and buccal cavity. From the sub-pharyngeal ganglia, three pairs of nerve arise to supply the 2nd, 3rd and 4th segments. Nerve cord and ganglia of each segment give out three pairs of nerves which supply various parts of segments in which ganglia are situated. Thus, nerves originating from the central nervous system supply different parts of the body to constitute the peripheral nervous system.

### *Sympathetic nervous system*

Sympathetic nervous system includes nerve plexuses which are situated in wall of the alimentary canal and some other internal organs. Such plexuses are connected with peri-pharyngeal connective by the fine nerves and help in coordinating function of related organs. Nervous system, like those of higher animals, has both sensory and motor neurons. Nerves of nerve cord therefore have both sensory and motor fibres, i.e. mixed function. Stimuli from the skin are transmitted by the sensory fibres to the nerve cord and stimulus is transferred either to an association neuron first, or it may go to muscles (effectors) which cause them to contract. This circuit of stimuli, or impulses forms a simple reflex arc. Movements of circular and longitudinal muscles are coordinated, so that the contraction of one brings about the relaxation of the other. Impulses are conducted more rapidly in giant fibres resulting into sudden contraction of the whole body (Figure 12.22).

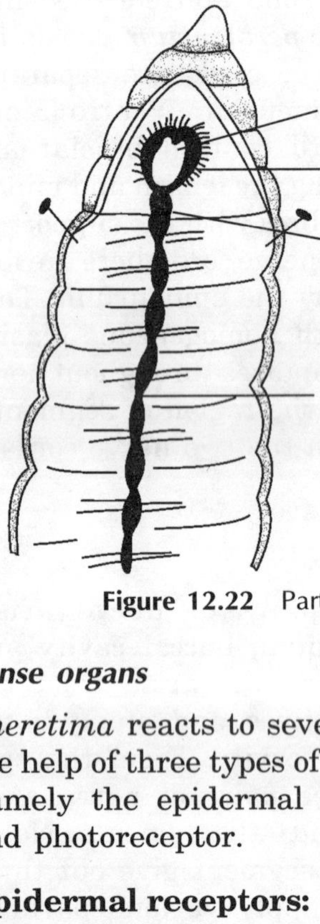


**Figure 12.22** Part of nervous system.

### *Sense organs*

*Pheretima* reacts to several external stimuli with the help of three types of sense organs, or receptors namely the epidermal receptor, buccal receptor and photoreceptor.

**Epidermal receptors:** These consist of an ovoid group of tall cells in epidermis, which causes an elevation of cuticle. These cells are separated from each other. Each cell has a nucleus near middle and it ends above in slender, hair-like processes which penetrate cuticle and project beyond it. Lower ends of these cells are supplied with nerve fibres. Epidermal receptors are found all over the epidermis but are numerous in lateral and ventral regions. They are tactile in function, i.e. tango-receptors and perceive thermal and chemical stimuli. Hence, earthworms are very sensitive to touch and vibrations transmitted through solid objects, though they cannot hear at all.

**Buccal receptors:** These are found in epithelium of buccal cavity in large numbers. They consist of groups of tall cells which project beyond epithelial cells. These cells have sensory, hair like processes and their nuclei lie below middle part of cells. These receptors sense smell, i.e. olfactor and taste food, i.e. gustatoreceptors. These cells can distinguish tastes of different vegetable food but their sense of smell is poorly developed. However, earthworms can smell different types of decaying leaves used as food (Figure 12.23).

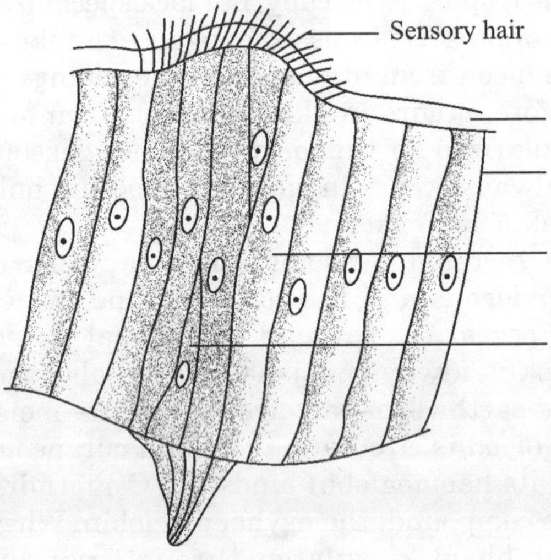


**Figure 12.23** Buccal receptor.

**Photoreceptor:** A photoreceptor is a single ovoid cell in inner part of the epidermis. It has a nucleus and cytoplasm with a network and an optic cell (phaosome, or lens) generally of curved shaped and made of a hyaline substance. One, or two nerve fibres enter this optic cell. Photoreceptors are found mostly on prostomium and the first segment. They occur in lesser number in other segments including the last. They are absent on the ventral surface. Photoreceptors (Figure 12.24) are sensitive to light and are called the little eyes, or ocelli. Actually the network of a photoreceptor is formed by branching and re-branching of nerve entering it and is called the *retinella*. Earthworms

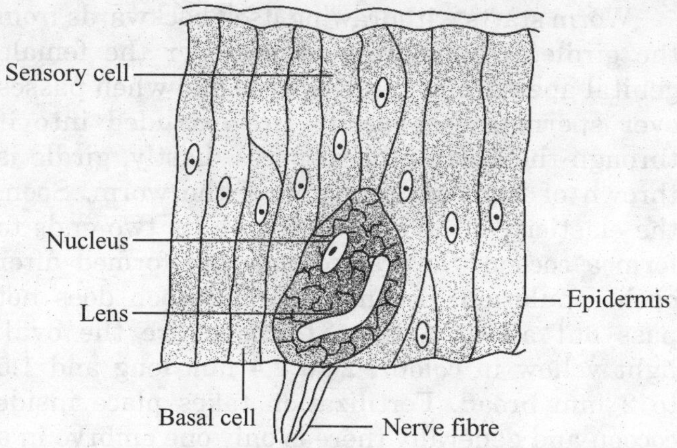

**Figure 12.24** Photoreceptor.

show a negative response to even a very weak light. Hence, they retreat into their burrows during day and come out at night, i.e. they are nocturnal in habit.

**Reproductive system:** *Pheretima* is hermaphrodite. Cross-fertilization occurs because of the relative position of male and female genital apertures. Male gametes mature first and earlier than female gametes, so that self-fertilization is prevented. *Pheretima* reproduces sexually only.

### *Male reproductive organs*

Male reproductive organs are discussed as follows:

**Testes and testis sac:** Testes are two pairs, very minute, whitish and lobed structures, situated one pair each in segment 10th and 11th, found attached with posterior surface of 9/10 and 10/11 inter segmental septa. Each testis consists of 4 to 8 finger like lobules. Each lobule contains the rounded cells in masses called the *testis sacs*. Testis sacs are situated on ventrolateral sides of the ventral nerve cord beneath the stomach. A large spermatic funnel with folded ciliated margin is present. Testis sacs remain in communication with seminal vesicles. Spermatogonia are shed into testis sacs; pass into seminal vesicles where they mature to form the spermatozoa.

**Seminal vesicles:** There are two pairs of large, white, sac like bodies, located in segments 11th and 12th into which testis sacs open by narrow ducts. Anterior seminal vesicles are smaller than posterior ones. Seminal vesicles located in 11th segment are enclosed by posterior large testis sacs, while those of 12th, the segment are exposed in coelomic cavity. Seminal vesicles are placed ventro-laterally beneath the stomach and as they develop as septal outgrowths, they are called the *septal pouches*.

**Vasa deferentia:** Each spermatic funnel leads into a thin, narrow, thread-like vas deferens. Two vasa deferentia of each side run very closely and laterally to nerve cord and in the 18th segment they join the prostatic duct.

**Prostate glands:** A pair of large, white, solid, irregularly glandular masses situated one on either side of the gut in segments 16th, 17th to 20th, or 21st is prostate glands. Each gland consists of a glandular region and a nonglandular region. From each gland emerges a prostatic duct in the 18th segment. This duct joins two vasa deferentia and these three ducts are enclosed in a common sheath to form a common prostatic spermatic duct, which opens separately through a male genital aperture on ventral side of the 18th segment.

**Accessory glands:** These are two, paired, rounded structures situated one pair each in segments 17th and 18th on ventral body wall at lateral sides of nerve cord. These glands open to outside. Two pairs of genital papillae are situated on either side of the mid-ventral line in 17th and 19th segments externally. Secretion of these glands keeps the two worms together during copulation. Spermatogonia are shed into testis sacs and pass onto seminal vesicles, where spermatozoa are formed. These spermatozoa again enter testis sac and through spermatic funnels go down the vasa deferentia and are discharged through the male genital apertures along with secretion of prostate glands.

### *Female reproductive organs*

Female reproductive organs are discussed as follows:

**Ovaries:** A pair of white, small, lobulated ovaries, one on either side of the ventral nerve cord in the 13th segment remains attached with posterior face of inter segmental septum of 12/13 segments. Each ovary possesses several finger-like processes with developing ova in a row, which gives it a beaded appearance. Ova remain in various stages of development in each ovarian lobe, being mature in distal part and remain immature in proximal part.

**Oviducts:** Below each ovary in the 13th segment, there is a small ovarian funnel with folded and ciliated margins. Each ovarian funnel leads into a short, conical and ciliated oviduct. Oviducts of both sides converge to meet below the nerve cord and open by a single median genital aperture ventrally in the 14th segment. Mature ova are released from ovaries, received by ovarian funnel, passed through the oviducts and discharged through genital aperture.

**Spermathecae:** They are four pairs in number, one pair each in segments 6th, 7th, 8th and 9th and are situated ventro-laterally. Spermatheca is flask shaped with main body, the ampulla which is continued into the narrow duct, the neck with a small diverticulum on its inner side. These are also called the *seminal receptacles* as they store spermatozoon from another worm during copulation. Spermatheca open externally by their small ducts as spermatheca pores are situated inter segmentally between segments 5/6, 6/7, 7/8 and 8/9.

**Copulation:** Copulation usually occurs in rainy season (July to October) at dawn. Two earthworms half emerge from their adjacent burrows and come closer to lie in contact to each other by their ventral surfaces with anterior ends pointing in opposite directions. In such position, male genital apertures of one worm lie opposite to spermathecal pores of the other. Areas of male genital apertures are raised into papillae and inserted into spermathecal pores. There occurs a mutual exchange of spermatozoon and prostatic fluid between the two copulating worms. Copulation lasts for about an hour, or so and then the worms get separated and recede into burrows.

**Cocoon, or ootheca formation:** This starts after copulation. Epidermis of clitellar segments, i.e. 14th, 15th and 16th segments contains three kinds of glands. Unicellular mucous glands produce mucous for copulation. Cocoon secreting glands secrete wall of cocoon and albumen glands produce albumen in which eggs are deposited in the cocoon. A membranous girdle is formed by cocoon secreting glands of clitellum. This girdle soon hardens and then albumen is deposited between the girdle and body wall.

Worm starts withdrawing itself backwards from the girdle. When girdle passes over the female genital aperture, it receives eggs and when passes over spermatheca sperms are extruded into it through the spermathecal pores. Lastly, girdle is thrown of from anterior end of the worm. Soon, the elasticity of its wall closes up its two ends to form a cocoon. Several cocoons are formed after each copulation, because spermatozoon does not pass out all at a time. Cocoons are the oval, light yellow in colour, 2 to 2.4 mm long and 1.5 to 2 mm broad. Fertilization takes place inside cocoon and generally there is only one embryo in a cocoon. Cocoons are laid (from August to October) in damp situation. Cocoon contains fertilized egg and albuminous substance secreted by glands of clitellum for nourishment of the embryo. The development occurs in cocoon without larval stage (Figures 12.25–12.26).

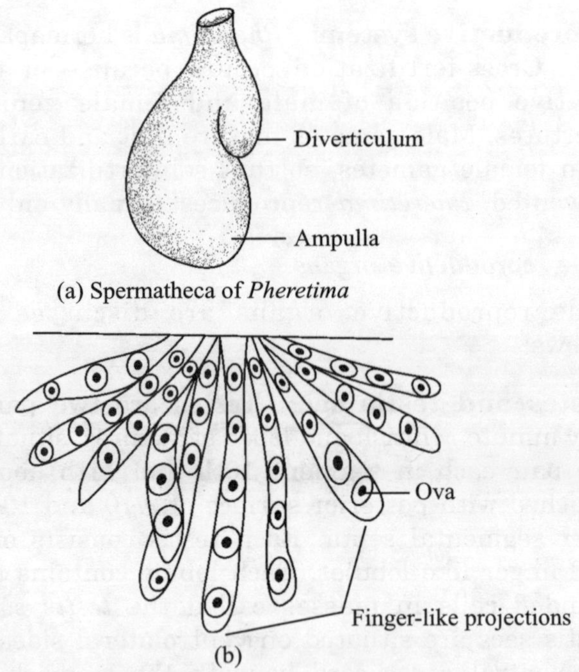


**Figure 12.25** Ovary of *Pheretima*.

## 12.4 LOCOMOTION IN *Nereis*

*Nereis* is a marine, burrowing polychaet worm. It walks and crawls by its body muscles, parapodia. Coelomic fluid form a hydrostatic skeleton.

**Parapodia:** Parapodium is a Greek word meaning side foot, i.e. foot developed from lateral side of the body. Parapodia are muscular, hollow

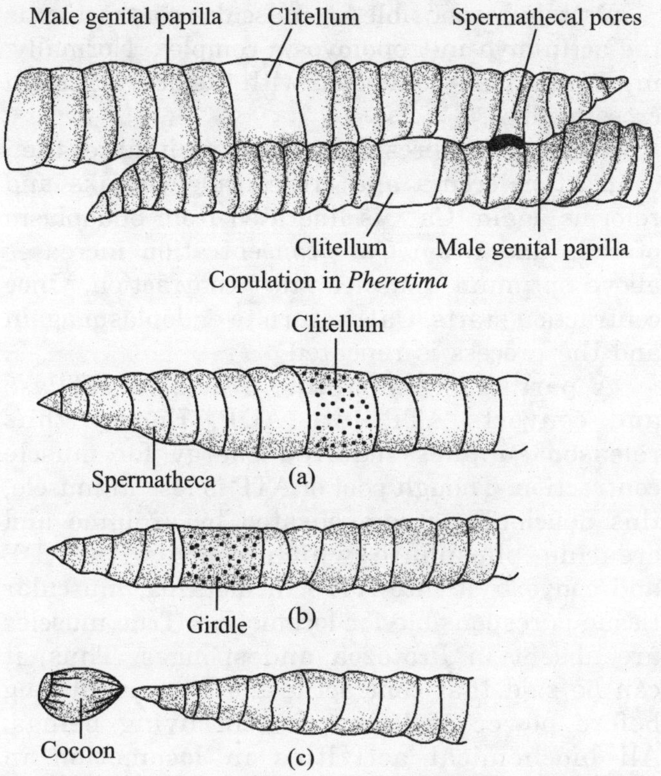


**Figure 12.26** Formation of cocoon in *Pheretima*.

outgrowths from lateral side of each segment except prostomial and anal segment. Each parapodium is latero-vertical in position, biramous, or bilobed organ, joined with body through hinge joins. Of two lobes, dorsal one is notopodium and ventral one is neuropodium. From each notopodium and neuropodium, a fleshy outgrowth arises called the *cirrus*.

Cirrus of notopodium is known as *dorsal cirrus* and that of neuropodium is known as *ventral cirrus*. From margins of myopodium and neuropodium, bunches of setae, or chaetae project out, and for this animal belongs to *polychaeta*. Setae, or chaete originate from sitigerous sacs which are situated in both podiums. These setae are attached to intrinsic parapodial muscles which act to protrude setae, or retract them, or spread them. Setae of each lobe are protected by a large chitinous rod known as *aciculum*.

**Parapodial muscles:** Parapodial muscles that take part in locomotion are oblique muscles. This originates from midventral region near nerve cord and distributed in lateral side. Some fibres are inserted at base of neuropodium. Function of muscles is to suppress pressure created in transverse axis and maintain body shape.

**Dorsal and ventral parapodial muscles:** These muscles originate from circular muscles lateral to longitudinal muscle and are inserted on various parts of the neuropodium and notopodium. These muscles contract, or protract parapodium, create up and down movement of parapodia and fix seta on surface, or raise them from surface.

**Intrinsic parapodial muscles:** These mussels are inserted on different parts of parapodia and control movement of parapodia.

**Acicular muscles:** These are paired and antagonistic, i.e. protractor and retractor group and are deeply embedded within parapodia. They control movement of setae as well as protraction and retraction of parapodia. Besides parapodial muscles, the longitudinal muscles of body play an important role in movement and locomotion of *Nereis*.

### *Mechanism of locomotion*

Locomotion in *Nereis* can be resolved into three types as:

**Slow creeping locomotion:** Parapodia take active part. Each parapodia of same segment works antagonistically. However, before this pair becomes active, adjacent posterior pair shows same slow wriggling movement. Immediately after this, wave of contraction of parapodial muscles along entire length of the body take place. Before slow creeping starts, parapodia remain well established with soil surface by their setae. When creeping starts, the acicular muscles contract, resulting in lift of the setae from the ground, or substratum.

Parapodium moves first inward, then upward and finally towards the anterior side. This movement of parapodium is known as the protraction of preparatory stroke. Parapodial protractor muscles and longitudinal muscles of the body work together for the preparatory stroke. During effective stroke, protractor and longitudinal muscles relax. As a result, the seta of parapodia become anchored again on substratum and a force is now applied by contraction of antagonistic pair rotation of parapodia. This application of force is known as retraction of effective stroke. Pulling force develops during the effective stroke drags animal forward. During this progression, parapodia of the opposite side bend in opposite direction. In this way, the distance travelled by cumulative effect of preparatory and effective stroke is known as *span of locomotion* and covered distance is known as *stride*.

**Rapid creeping:** For rapid crawling, the above mentioned muscles contract and relax in quick succession, but rapid contraction and relaxation of longitudinal muscles play most important role. This muscle contracts and relaxes very rapidly parallel to parapodial muscles. Force thus generated overcomes tension of setae-substratum attachment and pulls body forward. Contraction wave created by the longitudinal muscles passes from the posterior side to anterior side and the animal moves forward.

**Swimming:** When rapid creeping takes place in water, it is resolved into the swimming and here contraction wave passes from the posterior side to anterior side.

### *Role of fibrillar system in locomotion*

The molecular level of contraction and relaxation of micro fibrils for locomotion is known. Micro fibrils are formed by polymerization of the large protein molecules. Locomotion by any means and ways takes places through some mechanism. This can be resolved to action of contractile fibrils and can be further resolved to protein molecules. Of contractile fibrils, the role of myofibrils are known in details. Most important features of myofibril are its power of contraction. Fibres formed by protein molecule are known as *myofilaments*. Myofilaments join together to form the myofibrils which are of two types:

(i) thin and long filaments, known as *actin* and
(ii) thick and short filaments, known as *myosin*.

Protein responsible for muscular contraction is the actin-myosin-tropomyosin complex. Normally, myosin remains connected with the actin through cross bridges.

When actin fibres slide over myosin fibres then contraction occurs and cross bridge breaks and reforms again. $Ca^{++}$ is liberated from endoplasm of cells and when $Ca^{++}$ concentration increases above optimum, muscles start contraction. Once contraction starts, $Ca^{++}$ return to endoplasm again and the process is repeated.

A part of myosin protein acts as ATPase and converts ATP into ADP. Energy thus released supplies required energy for muscle contraction. Though pool of ATP is less in muscle, this deficiency is compensated by arginine and creatine phosphate which react with ADP and converts it into ATP. In metazoa, muscular tissue is responsible for locomotion. True muscles are absent in Protozoa and sponges. Thus, it can be said that contractile fibril appeared long before power of locomotion in living beings. All biochemical activities in locomotion in molecular levels are same in all animals starting from Protozoa to mammals. Microfibrills are formed of long chains of protein molecules and when they contract, they form bundles and when they relax they form mesh by cross bridging the protoplasm. Kinetodesmata of ciliates are examples of these. On the origin of micro-fibrillar system and their functional role, flagellates are regarded as most primitive, while ciliates are believed to be evolved from flagellates. Sarcodina (*Amoeba*) are advanced over other groups (Figures 12.27–12.28).

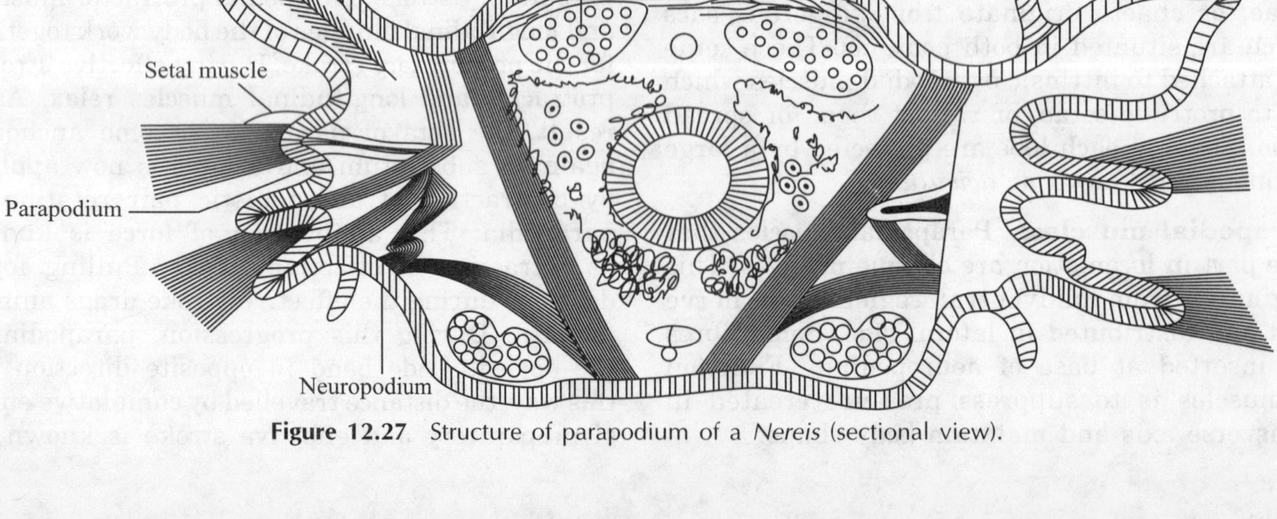


**Figure 12.27** Structure of parapodium of a *Nereis* (sectional view).

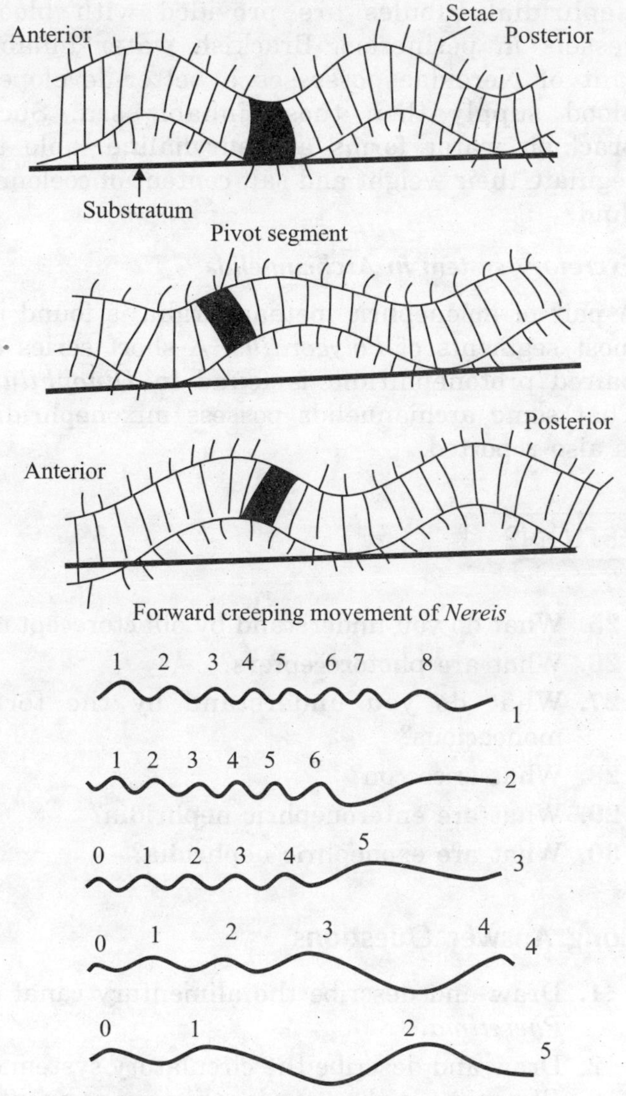


**Figure 12.28** Movement in *Nereis*.

## 12.5 NEPHRIDIA IN ANNELIDS

Nephridia are metamerically arranged, a sort of tunnel between coelom and outside, typically consists of a nephrostome, an intracellular duct penetrating septum, nephridial duct and is derived from a single cell, called the *nephroblast*. They are ectodermal in origin.

### *Types of nephridia*

Depending on developmental stages of nephridium these are:

**Provisional, or embryonic nephridia:** These are temporary structures and disappear after permanent nephridia start to develop. Embryonic head nephridia are found in head segment of many polychaetes and oligochaetes. They are branched in *Echiurus* and *Polygordius*. Embryonic trunk nephridia sometimes persist as in *Nereis*, where permanent nephridia do not develop.

**Permanent nephridia:** These are permanent excretory structures although absent in some anterior segments of most oligochaetes. Depending on size and number present in a segment, nephridia may be:

- *Meganephridia (Holonephridia):* Large sized metanephridia and in a given segment only one pair is found as in *Lumbricus*, *Chaetcogaster* and *Nereis*.
- *Micronephridia, or Meronephridia:* Very minute and are numerous in a segment as found in *Pheretima*. Nephridia may be protonephridia which are primitive nephridia of closer tubes along with their blind end projecting into coelom. Blind end is associated with solenocytes lying separately, or in group. Solenocytes are round ciliated cells connected to nephridium by a thin tube. Such nephridia are found in *Phyllodoce*, *Vanadis*.

### *Metanephridia*

Open type nephridia, typically consist of a convoluted ciliated tube and remain embedded in a protoplasmic syncytial mass. Nephrostome projects into coelomic cavity of segment anterior to one bears the nephridia through septum. Nephridiopore is located on the ventral surface near the base of ventral cirrus. Metanephridia are found in *Perinereis*, *Neanthes*.

**Nephromyxia:** Compound nephridia formed by union of nephridia and coelomoducts are called *nephromyxia*. They are of three types:

- *Protonephromyxium:* Coelomoduct is combined with a protonephridium as found in *Phyllodoce*.
- *Nephromyxium:* Coelomoduct is attached with metanephridium as found in *Hesione*.
- *Mixonephrium:* When coelomoduct is combined to a nephridium in a way that they form a single organ as found in *Arenicola*.

Nephridia may be categorized into the following types:

**Exonephric nephridia:** When nephridia open outside the body, they are called *exonephric* as in *Lumbricus.*

**Enteronephric nephridia:** When nephridia open into gut, they are called *enteronephric* as found in *Pheretima.*

### *Excretory system in Polychaeta*

Excretory organs are either protonephridia, or metanephridia. Ammonia, the chief excretory product, passes from coelomic fluid through solenocytes in pronephric species and through open nephrostome in metanephric species. Nephridial tubules are provided with blood vessels in perinereis. Brackish water inhabitant of Nereidae possesses a better-developed blood supply then those inhabit sea. Such brackish water forms are euryhaline, able to regulate their weight and salt content of coelomic fluid.

### *Excretory system in Archiannelida*

A pair of holonephric metanephridia is found in most segments of *Polygordius*. A short series of paired protonephiridia is found in *Dinophilus*. That some archiannelids possess mixonephridia is also reported.

## REVIEW QUESTIONS

### Short Answer Questions

1. Give two characters of Phylum Annelida.
2. Give two characters of Polychaeta.
3. Give two characters of Oligochaeta.
4. Give two characters of Hirudinea.
5. Give two examples of Polychaeta.
6. Give two examples of Oligochaeta.
7. Give two examples of Hirudinea.
8. Give two distinguishing features of Errantia.
9. Give two distinguishing features of Sedentaria.
10. Name the class to which *Aphrodite* belongs.
11. What are setae?
12. What is clitellum?
13. What is peristomium?
14. What are nephridiopores?
15. What do you understand by perichaetine arrangement?
16. Define chloragogen cell.
17. What is coelom?
18. What are eleocytes?
19. What are mucocytes?
20. What is typhlosole?
21. Mention the function of typhlosole.
22. What is nephrostome?
23. What are integumentary nephridia?
24. What are pharyngeal nephridia?
25. What do you understand by olfactoreceptor?
26. What are photoreceptors?
27. What do you understand by the term monoecious?
28. What is coccon?
29. What are enteronephric nephridia?
30. What are exonephric nephridia?

### Long Answer Questions

1. Draw and describe the alimentary canal of *Pheretima*.
2. Draw and describe the circulatory system of *Pheretima* in the anterior thirteen segments.
3. Draw and describe the excretory system of earthworm.
4. Draw and describe the reproductive system of earthworm.
5. Give an account of nervous system of earthworm with suitable diagram.
6. Give an account of coelom and coelomic fluid in earthworm.
7. Give an account of locomotion in earthworm.
8. Give an account of external features of earthworm with suitable diagram.
9. Give an account of locomotion in *Nereis* with diagram.
10. Write a note on nephridia in annelids with diagram.

# CHAPTER 13

# Phylum Arthropoda

Arthropods constitute a natural unit of segmented animals and contain the largest number of animals more than 3 times as many species as all the remaining phyla together. Arthropods, with paired jointed appendages and chitinous exoskeleton overlying the hypodermis, exhibit remarkable development in many aspects among the nonchordates. Each body segment, which is often specialized and distinct than annelids, bears a pair of specialized appendages. Metamerism is homonomous in the worms, but distinctly heteronomous in arthropods.

Typically, of three body divisions—head, thorax and abdomen, the head and trunk are recognizable. Joints termed sutures occur in exoskeleton and skeletal areas bounded by the sutures are called the *sclerites*. Inorganic substances, particularly the lime, are often added to chitin to provide additional strength to skeleton.

Chitinous covering, an effective barrier to further growth, increases in thickness and growth becomes possible by periodic shedding of the cuticle called the *molting*, or *ecdysis*, after which body covering becomes extremely soft and pliable. During this period, rapid increase in size occurs. Ecdysis often varies greatly in different groups. Crustaceans shed the skin periodically throughout life.

Circulatory system is an open type. The body fluid is found in the vessels throughout its course and often spilled into sinuses. A dorsal vessel, or heart lies dorsal to the digestive canal. Sinuses, or cavities called the *haemocoel* are the development of a portion of circulatory system. Central nervous system consists of a ladder like chain of ganglia and a brain, or supraoesophageal ganglion. Reproduction is mostly sexual, but parthenogenetic and paedogenetic habits are found in some groups. Polyembryony is noted in some insects. Hermaphroditism is rare. The centrolecithal egg often undergoes partial, superficial cleavage. Different larval stages are found in various groups.

## 13.1 GENERAL CHARACTERISTICS

1. Two out of every three known species are arthropods.
2. Insects constitute the most successful animal group.
3. Three important features are body segmentation, a hard exoskeleton, and jointed appendages.
4. Trilobites had pronounced segmentation.
5. Body is completely covered by the cuticle made of protein and chitin.
6. Exoskeleton is strong and relatively impermeable to water and provides protection and points of attachment for the muscles.
7. Moulting occurs in the life cycle.
8. Well-developed sense organs include eyes for vision, olfactory receptors for smell and antennae for touch and smell.
9. Most sense organs are located at the anterior end.
10. Haemolymph is propelled by the heart through the arteries into sinuses and returns to the heart through valved pores.
11. Open circulatory systems evolve convergently.
12. Most aquatic species have gills with thin, feathery extensions with an extensive surface area.
13. Terrestrial species generally have internal surfaces specialized for the gas exchange.

14. Insects have the tracheal systems, the branched air ducts leading into interior from pores in the cuticle.

## 13.2 BODY WALL

Body wall is composed of complex cuticle secreted by the epidermis. Each body segment is boxed by the sclerites, tergite and sternite, respectively. Side regions, or pleura are unsclerotized and provided with various minute, floating sclerites. Fusion, or loss of adjacent sclerites may occur.

Muscle bands are attached in point where inner surfaces of sclerites project inward as apodemes. The outermost multilayered epicuticle has a protective lipoprotein layer—the cement layer. A well-developed waxy layer is found beneath it in arachnids and insects. Waxes serve as barrier to water loss and in combination with lipoprotein layer provide protection against bacteria.

The innermost layer of epicuticle, the cuticulin layer consists primarily of proteins and usually possesses a thin, dense outer layer and a less dense inner layer. The procuticle consists of an outer exocuticle which hardens the exoskeleton. An inner endocuticle lies beneath the epicuticle which is tough, flexible and made up by protein and chitin. The cuticle is generally hard and inflexible except at the joints. Protein molecules are cross-bonded to one another by orthoquinone linkages to form the rigid sclerotized structure.

Sclerotization begins in cuticulin layer of epicuticle and progresses to procuticle. Mineralization occurs by deposition of calcium carbonate in procuticle as found in crustaceans. The epidermis contains various unicellular glands. Appendages, the outgrowths of body wall are equipped with sets of extrinsic muscles. Intrinsic muscles which are called the *articles*, or *podites* move various limb segments. Limb articles are organized into basal most groups called the *protopod* and distal most group called the *telopod*. The exoskeleton of telopodites forms a flagellum. Structures such as gills, gill cleaners and swimming paddles emerge from the protopodite.

## 13.3 SUPPORT AND LOCOMOTION

The exoskeleton provides support and maintains the body shape. Muscles arranged as short bands, extend from one body segment to next. Joints between body and limb segments are bridged by areas of very thin, flexible cuticle in which procuticle is reduced and not hardened. These thin areas are called the *arthrodial*, or *articular membranes*. Each articulation is bridged by one set of flexor muscles and opposing set of muscles is called the *extensors*.

Articular membrane may not form a complete ring of flexible material rather interrupted by points of contact called the *condyles* between hard cuticles. A dicondylic joint allows movement in one plane. The motion at a joint is limited by cuticular processes called the *locks*, which prevent overextension and over flexion. A plethora of locomotor devices are found for movement in water, air and on land. Movement through water involves swimming. Aerial locomotion is found in pterygotes and in certain spiders. Some terrestrial species engage in short-term aerial movements to escape. Fleas and some crustaceans simply jump, others jump and glide. Arthropods that move in contact with substratum under water, or on land by various forms of walking, creeping, crawling, or running are called the *pedestrian*, or *reptant*.

Many swimming crustaceans employ ventral appendages as paddles that are restricted to particular body regions. Appendages are flexed on recovery stroke. On the power stroke, limbs are held erect. In many crustaceans, less sophisticated swimming is performed by antennae. Pedestrian locomotion is highly variable.

Except a few strongly homonomous vermiform types, most arthropods are unable of lateral body undulations. Leg moves up and down as well as forward and backward. The action at each joint is coordinated by actions of all others. In some arachnids and crustaceans, certain limb joints lack extensor muscles and limbs are extended by an increase in the blood pressure. Raising and lowering of limb are done by extensor and flexor muscles. Anterior-posterior limb movements are done in two basic ways:

1. The ball-and-socket type of joint at the point of limb-body articulation typically carries out these actions in crustaceans, insects and myriapods. Promotor- and remoter muscles associated with these joints rotate the limb forward and backward.
2. Many arachnids perform multidirectional limb movements by uniplanar dicondylic joints.

Various pedestrian locomotion called the *gaits* results from many factors and is limited by biological and physical constraints such as rates of muscle contraction and necessity for coordinating leg movements. Animal maintains an appropriate distribution of legs in various phases of power and recovery strokes to support its weight. Gaits in insects and myriapods attempt to establish principles of arthropod walking, crawling and running based on metachronal model, which states that legs on each side of body move in metachronal waves from back to front and waves overlap to various degrees depending on the speed of movement. Leg movement sequences, stepping patterns and stride lengths are variable in crustaceans and arachnids

## 13.4 CIRCULATION AND GAS EXCHANGE

In the open haemocoelic system, blood is driven from the heart chamber through short vessels into haemocoel and returns to the heart via a noncoelomic pericardial sinus and perforations in heart wall called the *ostia*. Blood flows back to heart along a decreasing pressure gradient resulting from lowered pressure within pericardial sinus as the heart contracts. Variations exist in size and shape of heart, number of ostia, length and number of vessels, arrangement of haemocoelic sinuses and circulatory structures associated with the gas exchange.

Blood containing amebocytes and sometimes the clotting agents transports nutrients, wastes and gases. Colourless blood carries gases in the solution. Some forms contain hemocyanin, few contain haemoglobin. Both pigments are always dissolved in hemolymph. On land, increase in cuticular permeability facilitate gas exchange and increases the threat of water loss. Larger crustaceans possess gills, which are commonly branched, or folded, exposed, or unprotected, thin walled cuticular evaginations. In some species, gills are carried beneath the protective extensions of the exoskeleton.

Insects and arachnids possess gas exchange structures as invaginations of cuticle. External gills are placed internally, remain moist and act as humidity chambers, allowing oxygen to enter solution for uptake. Many arachnids possess book lungs containing lamellae. Insects possess tracheae, which open externally through pores called the *spiracles*. In insect tracheal systems, inner ends of tubules lie in haemocoel, or are embedded in tissues, allowing direct gas exchange between air, blood and internal organs. Some terrestrial crustaceans have pseudotracheae on abdominal appendages, which are short, branching tubes for bringing air close to blood-filled spaces.

## 13.5 EXCRETION AND OSMOREGULATION

In most adult crustaceans, only a single pair of nephridia persist. In arachnids, as many as four pairs of nephridia open at the bases of walking legs. In the arachnids; myriapods, insects and tardigrades, Malpighian tubules serve the purpose of excretion. Various nephridial types are functionally complex and efficient than open metanephridia. Materials uptake from haemocoel by the inner ends of nephridia involves passive movement in response to filtration pressure, as well as active transport.

The fluid entering the nephridium, when passes along plumbing system of nephridium, selective reabsorption, particularly of salts and nutrients occurs. The urine exiting the nephridial pores are markedly different from hemolymph. Malpighian tubules perform the same process, but they rely on assistance from gut and form 'primary urine'. Very little reabsorption of nonwaste material occurs along the length of tubule. The hindgut concentrates the urine by reabsorbing the nonwaste materials. Marine crustaceans excrete most of their nitrogenous wastes as ammonia, the remainder is excreted as urea, uric acid and amino acids. Terrestrial arachnids, myriapods and insects excrete predominantly uric acid.

## 13.6 NERVOUS SYSTEM AND SENSE ORGANS

The arthropod 'brain' (cerebral ganglia) comprises two, or three regions. The anterior ganglia form the protocerebrum and when present the deutocerebrum. The posterior most tritocerebrum usually forms circumenteric connectives around the oesophagus to a ventral suboesophageal ganglion. Crustaceans, hexapods and myriapods possess a tripartite brain composed of two anterior ganglia and the posterior ganglion. In cheliceriforms, the deutocerebrum is absent. Each of these regions gives rise to a major pair of nerves to particular head appendages. The segmental ganglia of the ventral nerve cord fuse with one another in different groups.

The exoskeleton has a major effect on the nature of sensory receptors. Most external mechanoreceptors and chemoreceptors are cuticular processes, pores, or slits, collectively called the *sensilla*. Tactile receptors are cuticular projections in the form of movable bristles. Sensilla as fine "hairs", or setae are mechanically moved by external vibrations and impart movement to underlying sensory neurons. Special thin, or hollow setae often associated with head appendages are also found. Through proprioception arthropod knows where its appendages are. Arthropod versions of these 'strain gauges' are called the *campaniform sensilla* in insects, *slit sensilla* in arachnids and *force-sensitive* organs in most crustaceans.

Three basic kinds of photoreceptors are simple ocelli, complex ocelli and compound eyes. Compound eyes, especially suitable for detecting movement, comprise from a few to many distinct photoreceptive units, called the *ommatidia* which has its own field of vision through square, or hexagonal cuticular facets. Visual fields of neighbouring ommatidia overlap to some extent. Visual acuity is affected by the degree of overlap among the fields of vision of neighbouring ommatidia—the greater the overlap, the poorer the visual acuity.

Compound eyes with many small facets produce higher-resolution images than eyes with fewer, larger facets. The crustacean-hexapod model is mention-worthy in this regard. Each ommatidium is covered by the cornea. The special epidermal cells that produce the corneal elements are corneagen cells which may later withdraw to the sides of ommatidium to form primary pigment cells. The core of each ommatidium comprises a group of crystalline cone cells, sometimes a crystalline cone stalk and a basal retinula.

An ommatidium with a four-part crystalline cone is diagnostic of crustacean-hexapod eyes and is termed as *tetrapartite ommatidium*. The crystalline cone is a hard, clear structure surrounded by primary pigment cells. The retinula is formed by 8 to 13 photosensitive retinular cells and is surrounded by secondary pigment cells, which isolate each ommatidium from its neighbours. The core of cylinder is the rhabdome made up of rhodopsin-containing microtubular folds of the cell membranes of retinular cells.

Light that enters through the facet of a particular ommatidium is directed to its rhabdome and the crystalline cone. Fixed focal length of lens makes accommodation to objects at different distances difficult. Light is shared among all rhabdomeres of a given rhabdome. The xiphosurans have ommatidia that differ in details from insect-crustacean design. The versatility of compound eyes is the result of the distal and proximal screening pigments located in cells surrounding the ommatidium core.

Distal screening pigments are located in iris cells. Proximal pigments are often located in retinular cells and secondary pigment cells. Sometimes screening pigments migrate in response to light and change their positions. In bright light, screening pigments may disperse so that nearly all the light that strikes a particular rhabdomere enter through facet of its ommatidium. The screening pigment prevents light that strikes the facets at an angle from passing through one ommatidium and into another.

Many crustacean eyes are fixed in this condition. Such appositional eyes maximize resolution. In dim light, screening pigments may concentrate usually distally, allowing light to pass through more than one ommatidium before striking rhabdomeres. The image thus formed by each ommatidium is superimposed on images formed by neighbouring ommatidia. Such super positional eyes function as efficient light-gathering structures, while sacrificing visual acuity and image formation capabilities.

## 13.7 CLASSIFICATION WITH CHARACTERS

***Characters***

1. Bilaterally symmetrical.
2. Metamerically segmented.
3. Chitinous cuticle is present.
4. Appendages are jointed.
5. Haemocoel is present.
6. Open type of circulation is present.
7. Cerebral ganglionic mass is connected to a ventral chain of ganglia.
8. Anterior photoreceptors are present.
9. Sexes are typically separate.

Scheme of classification followed here is according to Marshall and Williams (1972).

***Subphylum Onychophora***

1. A firm exoskeleton is absent.

2. Arthropod-like cuticle is present.
3. Jointed appendages are absent.
4. Numerous unjointed, clawed appendages occur in pairs.
5. Cilia line the coelomoducts.
6. Development is direct.

**Example:** *Peripatus*, *Peripatopsis*, *Peripatoides.*

### *Subphylum Tardigrada*

1. Soft skinned animals.
2. Vaguely segmented.
3. A sucking proboscis is present.
4. Mouth is surrounded by the papillae.
5. Four pairs of unjointed legs are present.
6. Organs of respiration are absent.
7. Heart and blood vessels are absent.
8. Two eye spots are the only sense organs.

**Example:** *Macrobiotus.*

### *Subphylum Pentastomida*

1. Worm like.
2. Superficial annulation is present, or absent.
3. Alimentary canal is straight, simple.
4. Heart, organs of respiration and excretion are absent.

**Example:** *Linguatula*, *Porocephalus*, *Cephalobaena.*

### *Subphylum Trilobitomorpha*

1. All are extinct.
2. Exoskeleton was hard.
3. Appendages were delicate.

**Example:** *Dalmanites*, *Phacops.*

### *Subphylum Chelicerata*

1. Body is divided into prosoma and opisthosoma.
2. Prosoma with six, or eight post-oral segments.
3. Chelicerae are pre-oral.
4. Respiration is carried out by book gills, tracheae, or cutaneous.
5. Genital opening is present on the 8th segment.
6. Deutocerebrum and protocerebrum are fused in brain.

### *Class Merostomata*

1. Branchial respiration occurs.
2. Five pairs of appendages are present in prosoma.
3. Five, or six pairs of appendages are present in metasoma.
4. Metasoma without appendages, but with a telson.

### *Subclass Xiphosura*

1. Unsegmented carapace.
2. Opisthosoma has a dorsal shield with serrated edges.
3. Chelate chelicerae are three jointed.
4. Compound eyes and ocelli are present.
5. Telson represents the several fused segments.

**Example:** *Limulus*, *Tachypleus.*

### *Subclass Eurypterida*

1. Prosoma six segmented.
2. Opisthosoma of twelve segments and a telson.
3. Chelicerae were three jointed
4. Last pair of legs was modified into paddles.
5. Labrum was absent.
6. Endostoma was present.
7. All members are extinct

**Example:** *Eurypterus.*

### *Class Arachnida*

1. Body is divided into cephalothorax and abdomen.
2. Two pairs of chelate appendages are present.
3. Four pairs of legs are present.
4. Genital opening is present on opisthosoma.
5. Respiration is carried out by lung book and/or tracheae.

**Example:** *Buthus*, *Palamnaeus.*

### *Subphylum Pycnogonida*

1. Body consists of cephalothorax and a rudimentary abdomen.
2. Four simple eyes are present.
3. First pair of appendages may be chelate.
4. Second pair of appendages is termed *palpi*.
5. Abdomen lack of appendages.
6. Cement glands are 4–7 in number in male.

**Example:** *Nymphon.*

### *Subphylum Mandibulata*

1. Body is divided into head, thorax and abdomen.
2. Chelicerae are absent.
3. One, or two pairs of antennae are present.
4. A pair of mandibles is present.

***Class Crustacea***

1. Six anterior segments are fused with acron to form a head.
2. Chelate appendages are five pairs.
3. Except antennules, all appendages are biramous.
4. Open type of vascular system.
5. Green glands, or shell glands are present.
6. Compound eyes are frequently stalked.

***Subclass Cephalocarida***

1. Trunk segments are 19 in number.
2. Anterior nine trunk segments bear appendages.
3. Both pairs of antennae are short.
4. Eyes are absent.
5. Earliest known larva is metanauplius.

**Example:** *Hutchinsoniella.*

***Subclass Branchiopoda***

1. Number of trunk segments varies.
2. Trunk appendages are foliaceous, rarely leg like.
3. A pair of caudal styles is present.
4. A cephalic carapace is often present.
5. Compound eyes are present.
6. Larva is nauplius, or metanauplius.

***Subclass Ostracoda***

1. Unsegmented, or indistinctly segmented.
2. A bivalvad carapace is present.
3. Trunk appendages are not foliaceous.
4. A pair of caudal style is present.
5. Paired eyes are present, or absent.
6. Antennae are well developed.
7. Larva is a nauplius.

**Example:** *Cypris*, *Notodromus.*

***Subclass Mystacocarida***

1. Body is divided into head, thorax and abdomen.
2. Thorax is five segmented.
3. Abdomen is six segmented.
4. A pair of caudal styles is present.
5. Only a nauplius eye is present.
6. Both sexes are separate.
7. Metanauplius is earliest known larvae.

**Example:** *Derocheilocaris.*

***Subclass Copepoda***

1. Body is divided into head, thorax and abdomen.
2. Thorax is six segmented.
3. Abdomen is variably segmented.
4. Carapace is absent.
5. Compound eye is absent.
6. Median nauplius eye is present.
7. Typically, larva is nauplius

**Example:** *Cyclops*, *Calanus.*

***Subclass Branchiura***

1. Temporarily parasitic on fish, or amphibian.
2. Thoracic appendages are biramous.
3. Abdomen unsegmented, without appendages.
4. Mouth is suctorial.
5. Eyes are compound.
6. Antennae are reduced.
7. Genital aperture opens at bases of the fourth thoracic leg.

**Example:** *Argulus.*

***Subclass Cirripedia***

1. Almost exclusively marine.
2. Sessile.
3. Free, or parasitic.
4. Imperfectly segmented.
5. Six pairs of biramous appendages are present.
6. Abdomen is limbless and rudimentary.
7. Paired compound eyes are absent in the adult.
8. Usually hermaphroditic.
9. Young passes through the nauplius and cypris stage.

**Example:** *Lepas*, *Balanus*, *Sacculina.*

***Subclass Malacostraca***

1. Telson is present.
2. Carapace is present.
3. No caudal styles.
4. Paired compound eyes are sessile, or stalked.
5. Antennules are often biramous.
6. Female gonopore lies on the 6th thoracic segment.
7. Male gonopore lies on the 8th thoracic segment.
8. Nauplius stage occurs before hatching.

**Example:** *Squilla*, *Mysis*, *Oniscus.*

***Class Pauropoda***

1. Antennae are triflagellate.
2. Eyes are absent.
3. Presence of pseudocelli.
4. Mandibles are unjointed.

5. Presence of a telson.
6. Heart and tracheal system are absent.

**Example:** *Pauropus, Eurypauropus.*

### *Class Diplopoda*

1. Antennae short, club shaped.
2. Mandibles have three podomeres.
3. Presence of gnathochilarium.
4. A single ventral gonad is present.

**Example:** *Polyzonium, Polydesmus.*

### *Class Chilopoda*

1. Presence of filiform antennae, mandibles.
2. Second maxillae do not form labium.
3. Prehensile poison claws are present.
4. Intromittent organ is absent

**Example:** *Scolopendr, Lithobius.*

### *Class Symphyla*

1. Second pair of maxillae forms a labium.
2. Out of 14 trunk segments, 12 are podous.
3. Cerci with silk glands is present.
4. Gonads are paired structures.

**Example:** *Scutigerella, Hanseniella.*

### *Class Insecta*

1. Body is divided into head, thorax and abdomen.
2. Thorax carries three pairs of legs.
3. Antennae, mandibles and maxillae are present.
4. Thorax is 3 segmented.
5. Wings are one pair, two pairs, or absent.
6. Tracheate arthropods.
7. Two genital segments in female and one in male are present.

#### *Subclass Diplura*

1. Eleven segments are found in the abdomen.
2. Segmented antennae.
3. Cerci are present.
4. Malpighian tubules are absent, or vestigial.
5. Primarily wingless.

**Example:** *Campodea.*

#### *Subclass Protura*

1. Presence of eleven abdominal segments and a telson.
2. Antennae are absent.
3. Malphigian tubules are rudimentary.
4. Primarily wingless.
5. Minute, whitish insects.

**Example:** *Acerentomum.*

#### *Subclass Collembola*

1. Six abdominal segments are present, of which three carry appendages.
2. No metamorphosis.
3. No malphigian tubules.
4. Tracheal system is absent mostly.
5. Primarily wingless.

**Example:** *Podura.*

#### *Subclass Apterygota*

1. Primarily wingless.
2. Mandibles having a single point of articulation.
3. Possess progenital abdominal appendages.

**Example:** *Lepisma.*

#### *Subclass Pterygota*

1. Adults are winged, or secondarily wingless.
2. Mandibles articulate at two points, or modified.
3. Mouthparts are variable.
4. Adult lack progenital abdominal appendage.

**Example:** *Chloeon, Aeschna, Gryllus, Musca* (Figures 13.1–13.8).

## 13.8 CLASS CRUSTACEA

The cuticle of most crustaceans becomes hardened through addition of carbonate and phosphate of lime. Members are typically aquatic though some, for example, the land crabs are terrestrial. Typically, they have two pairs of antennae. A carapace originates as a fold at the head end and grows backward as a continuous shield for protecting inner parts and obscures the external evidences of segmentation of covered parts. Sometime, the carapace encloses the whole body. Often the head and all, or part of thorax is covered by the carapace.

### *Appendages*

Appendages are typically Y-shaped and biramous. The stem of the Y, the protopodite attaches to the body wall and consists of 2 segments—the prooximal coxopodite and a distal basipodite. Distally, the basipodite bears 2 branches. One branch situated near the median line of body is the endopodite and the other is the exopodite.

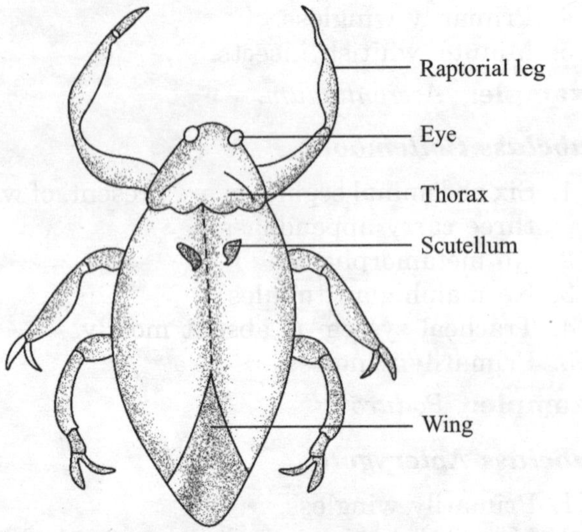


**Figure 13.1** *Belostoma.*

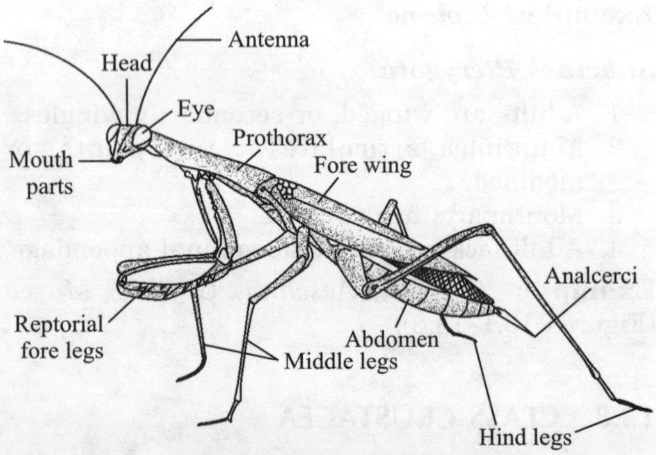


**Figure 13.2** *Mantis.*

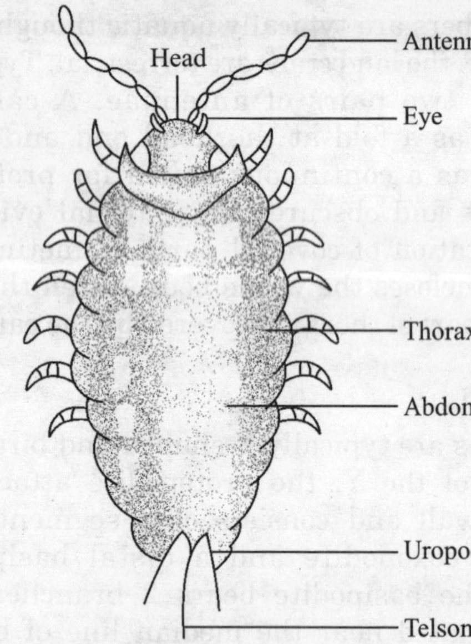


**Figure 13.3** *Oniscus.*

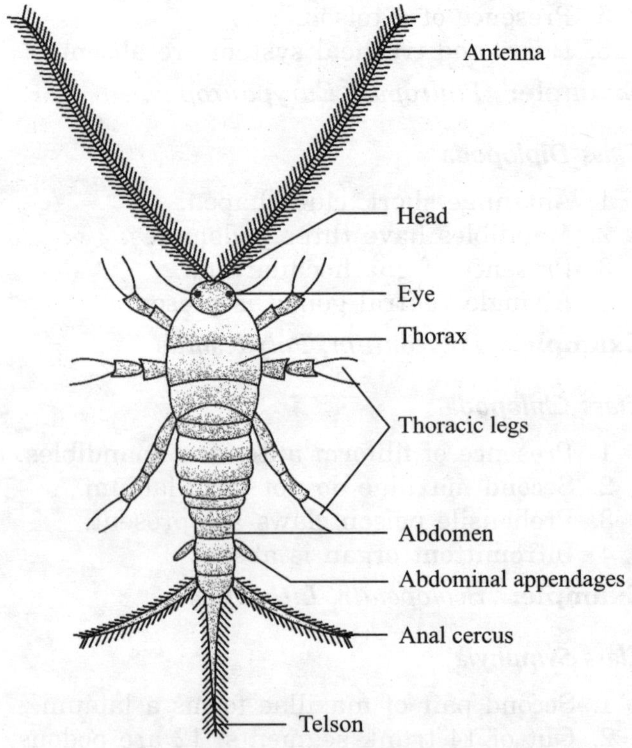


**Figure 13.4** *Lepisma.*

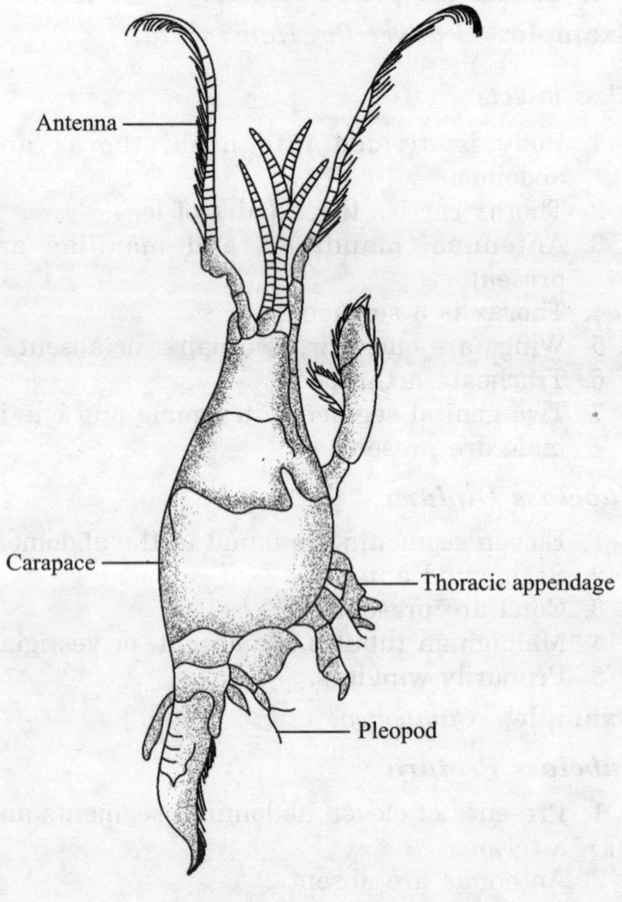


**Figure 13.5** *Hippa.*

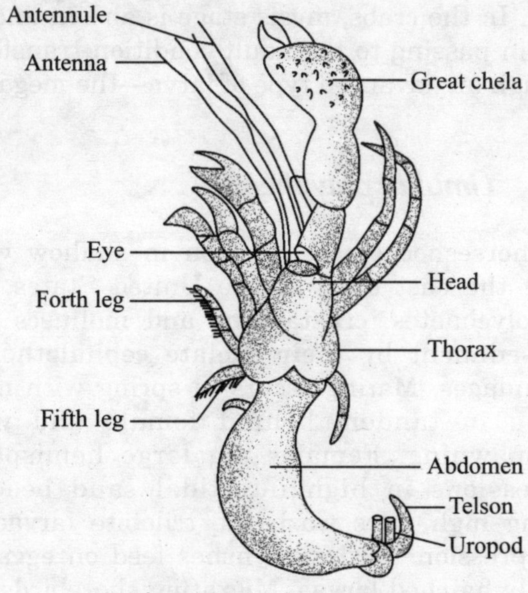


**Figure 13.6** *Eupagurus.*

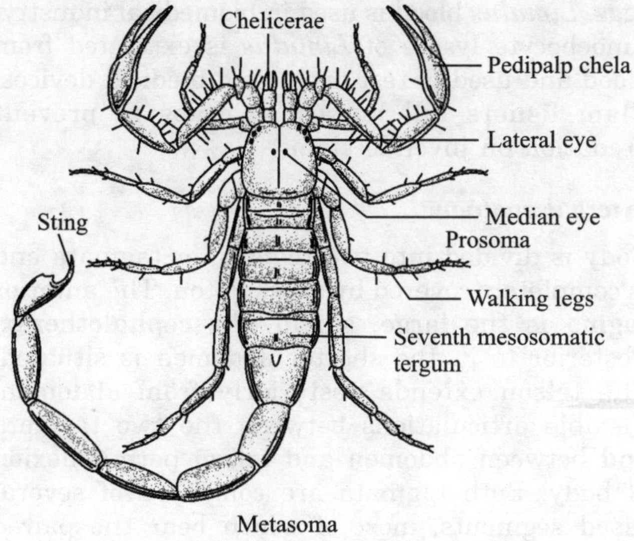


**Figure 13.7** *Buthus.*

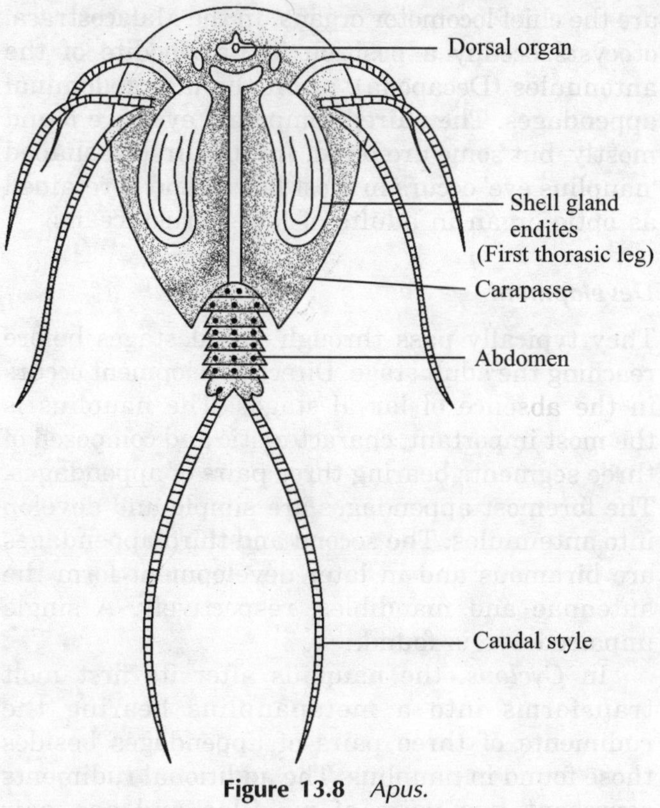


**Figure 13.8** *Apus.*

Appendages often become modified. In the walking legs of crayfish and lobster, the protopodite bears a single series of segments to form the leg. In larval lobsters, thoracic legs are distinctly biramous and in later development, the exopodite disappears.

No free larval stages exist in crayfishes and the entire development is completed within egg. In some members, a more primitive type of appendage is found. In the Phyllopoda, leaf-like feet on thorax is found though the antennae are biramous. The phyllopod appendages bear several lateral processes, called the *endites projecting* from central axis of foot segments. Body segments are variable in number. The anterior five somites fuse with prostomium to form head. This may unite with some, or the entire thorax to form cephalothorax. Abdominal somite also varies in number.

### *Respiratory and excretory organs*

The gill located in a gill chamber, which is formed by the carapace wall. The body wall carries out respiration usually. The entire body surfaces sometimes functions for respiration. Excretion is carried out by green glands and shell glands which open at the bases of second antennae and second maxillae, respectively. These organs occur together in larvae, but in adult one, or the other fails to develop. They are similar in fundamental structure and each consists of a terminal vesicle, which communicates with an external pore by a slender, coiled tubule. The chitin lined digestive canal is largely ectodermal. The stomodaeum includes pharynx and stomach. Much of the intestine is the proctodaeum. The mesenteron is a relatively short region into which a paired digestive gland, the hepatopancreas empties.

### *Sensory organs*

Antennae and antennules are tactile organs, innervated from the brain. Sometimes antennae are the chief locomotor organs. In the Malacostraca, otocysts occupy a position in protopodite of the antennules (Decapoda), or on the last abdominal appendages. The paired compound eyes are found mostly, but some are blind. An unpaired X-shaped 'nauplius eye' occurs in most larvae and is retained as optic organ in adults of lower crustaceans.

### Development

They typically pass through larval stages before reaching the adult stage. Direct development occurs in the absence of larval stages. The nauplius is the most important, characteristic and composed of three segments bearing three pairs of appendages. The foremost appendages are simple and develop into antennules. The second and third appendages are biramous and in later development form the antennae and mandibles, respectively. A single unpaired eye is found.

In *Cyclops*, the nauplius after its first molt transforms into a metanauplius bearing the rudiments of three pairs of appendages besides those found in nauplius. The additional rudiments represent two pairs of maxillae and one pair of maxillipeds. A pair of rudimentary thoracic appendages is present in metanauplius. In each of three successive molts, Cyclops larva acquires an additional pair of stump like rudiments of thoracic legs. Thus at the end of larval period, four pairs of thoracic rudiments are found. A final larval moult results in adult.

In the parasitic copepods, nauplius and metanauplius stages are passed within egg membranes. The larva at hatching is termed *copepodid*, which bears similarity to general organization of free-living copepod. The zoea is usually the first larval stage and characteristic of Malacostraca. It may be preceded by a nauplius. It consists of a cephalothorax bearing biramous appendages and an abdomen without appendages. Head bears a pair of lateral compound eyes.

The zoea passes through several moults to reach a mysis stage, in which even the posterior thoracic appendages are biramous and are used in swimming. In later development, the exopodite of these biramous thoracic appendages disappears leaving unbranched walking legs of the adult. In the Brachyura, thoracic appendages of the Zoea become walking legs later and zoea develop into free-swimming larva, which never acquire biramous condition, the characteristic of mysis larva. In the crabs, mysis stage is absent and the zoea in passing to the adult condition transforms through an advanced type of larva—the megalops.

## 13.9 *Limulus polyphemus*

The horseshoe crabs, common in shallow water along the east coast of the United States, feed on polychaetes, crustaceans and molluscs from the sediment by their chelate cephalothoracic appendages. Mating occurs in spring with males riding in tandem behind female and waits for spawning. Females dig large hemispheric depressions in high intertidal sand beaches. Spring high tides wash the trilobite larvae out of depression. Birds and fishes feed on eggs and freshly hatched larvae. Migrating shorebirds may consume over 500 metric tons of horseshoe crab eggs. *Limulus* blood is used in biomedical industry. Amoebocyte lysate of *Limulus* is extracted from blood and used to test drugs and medical devices. Clam fishers kill horseshoe crabs to prevent predation on juvenile clams.

### External anatomy

Body is divided into two regions, or tagmata and is completely covered by exoskeleton. The anterior tagma is the large, shield like cephalothorax. Posterior to it, the shorter abdomen is situated. The telson extends posteriorly from abdomen. Flexible articulations between the two tagmata and between abdomen and telson permit flexion of body. Both tagmata are composed of several fused segments, most of which bear the paired appendages (Figure 13.9).

### Cephalothorax

The dorsal exoskeleton of cephalothorax is strongly convex and shaped like a horse's hoof with smoothly curved anterior margin and a posterior recess. Three longitudinal ridges on dorsal surface of carapace are found. One situated on the midline is the median ridge. Others found in the lateral side are lateral ridges. Three ridges possess small teeth. The carapace bears a pair of lateral compound eyes and two anterior median ocelli, or simple eyes. Ocelli are located on either side of a tooth at the anterior end of median ridge. Concave ventral surface of cephalothorax bears six pairs of appendages. The next four appendages are

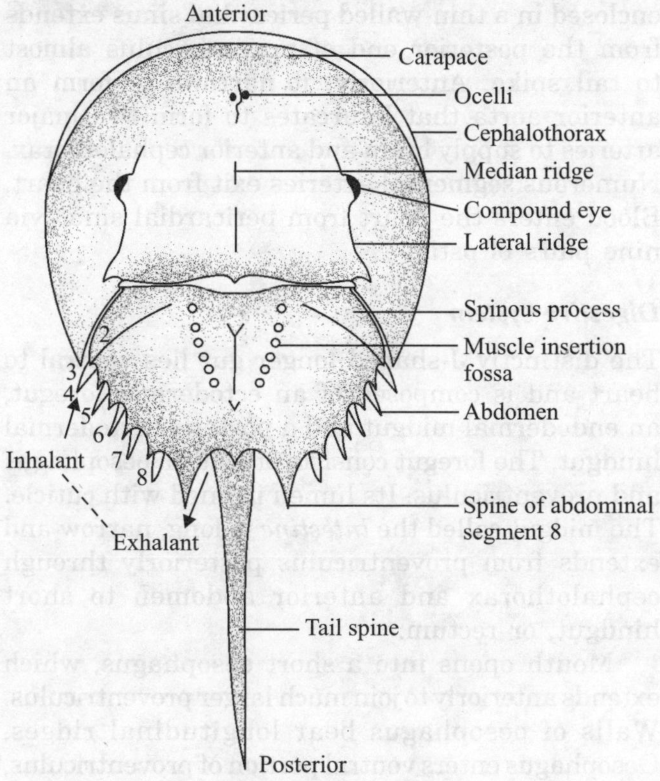


**Figure 13.9** *Limulus.*

chelate walking legs, although the first of them is sometimes called the *pedipalp*.

In females, pedipalp is almost identical to other walking legs, but in mature males, it is modified and is not chelate rather provided with a terminal bulbous swelling and a hook for holding female during mating. Pedipalps are followed by three pairs of unspecialized walking legs. Pedipalps and walking legs are morphologically similar and each is composed of six articles namely the coxa, trochanter, femur, patella, tibia and tarsus. Coxa bears a spiny gnathobase on its medial surface and attaches to body proximally. Walking legs are prehensile with chelae formed of tibia and tarsus. Tarsus, a movable finger closes against immovable finger extending from tibia.

The sixth cephalothoracic appendage is pushers and differ from the more anterior limbs. The fifth article of pusher bears four petal-like spines, which flare apart when pushed against sand. The gnathobase of this leg bears a heavily cuticularized set of teeth. *Limulus* has a 'mandible' at the posterior end of food groove. Laterally, the coxa of the pusher leg bears a narrow flabellum. Perhaps it is sensory. The flabellum articulates with the coxa. The sixth article bears a pair of short spines. In juveniles, one of these spines is long and biarticulate.

Right and left rows of five gnathobases lie beside ventral midline and form lateral borders of food groove. Mouth is funnel-like opening between coxae of second pair of legs. It is directed posteriorly facing anterior end of food groove and is provided by a large upper lip. When feeding, the crab uses its chelae to capture small prey and place them in food groove. Feeding and walking motions of legs move coxae, causing gnathobases to crush and grind food in groove. Spines move food inexorably toward mouth. The gut is J-shaped. The excretion is carried out by four pairs of saccate nephridia, known as *coxal glands* in cephalothorax. The four nephridia on each side drain by a single duct to nephridiopore located at the base of the fifth appendage.

### *Abdomen*

The abdomen is concave ventrally, composed of nine segments, and bears seven pairs of appendages. The concavity is due to branchial chamber containing book gills. The first abdominal appendages, the chilariae are small, oval and composed of only one article on posterior cephalothorax. The second abdominal appendages are two broad lamellae, which are fused to form genital operculum and protect the remaining abdominal appendages. Two gonopores are located on hidden posterior face of genital operculum near its articulation with body.

Each limb of genital operculum is biramous. Two limbs of each pair are fused along midline. Each consists of a basal protopod from which arise two distal rami, an endopod and an exopod. The large protopod attaches to body. The wide, short, uniarticulate exopod articulates with protopod along most of its distal lateral margin. A slender and biarticulate endopod attaches on medial corner of distal margin. The next five pairs of delicate, flap like appendages bear book gills. Appendages are broad, bears numerous flat and cuticular gill lamellae on posterior surface of protopod. Lamellae resemble the pages of a book and are ventilated by movements of their respective appendages. Blood circulates through the interior of lamellae. Gases are exchanged over thin permeable cuticle.

Dorsally, the abdomen bears a median longitudinal row of three teeth. A large spinous process is situated laterally on each side of the anterior end of the abdomen. A row of six movable spines are found along each lateral margin of abdomen. These and other features are external manifestations of abdominal segmentation. Abdomen consists of nine fused segments. Segment 1 is vestigial segment of the chilariae. Segment 2 bears spinous process and genital operculum. Segments 3–8 bear the book gills (absent on Segment 8), dorsal muscle depressions and lateral spines. Segment 9 has no appendages, or lateral spines. Tail spine extends posteriorly from abdomen. Anus is a ventral opening in soft flexible articular membrane connecting the abdomen and telson (Figure 13.10).

### Internal anatomy

Digestive caecum composed of tiny branching tubes which are connected with midgut. In some areas, tubules of digestive caecum may be accompanied with white deposits of calcium phosphate. Calcium phosphate is removed from blood and released into lumen of these tubules to be discarded in faeces. In adults, tubules of gonad, like those of digestive caecum are intermingled with those of caecum and the two occupy roughly the same areas of cephalothorax. A large oval, pale area on midline in anterior cephalothorax is the proventriculus, which is surrounded on three sides by the digestive caeca. Two oblique rows of circular white masses extend posterolaterally, one on each side from proventriculus. These muscles operate cephalothoracic appendages.

### Haemal system

The haemocoel is reduced to small spaces with coxal glands and gonads. The long tubular heart, enclosed in a thin-walled pericardial sinus extends from the posterior end of proventriculus almost to tail spike. Anteriorly, it narrows to form an anterior aorta that bifurcates to form two major arteries to supply brain and anterior cephalothorax. Numerous segmental arteries exit from the heart. Blood enters the heart from pericardial sinus via nine pairs of ostia.

### Digestive system

The distinctly J-shaped longer gut lies ventral to heart and is composed of an ectodermal foregut, an endodermal midgut and a posterior ectodermal hindgut. The foregut consists of mouth, oesophagus and proventriculus. Its lumen is lined with cuticle. The midgut called the *intestine* is long, narrow and extends from proventriculus posteriorly through cephalothorax and anterior abdomen to short hindgut, or rectum.

Mouth opens into a short oesophagus, which extends anteriorly to join much larger proventriculus. Walls of oesophagus bear longitudinal ridges. Oesophagus enters ventral portion of proventriculus, also called the *crop*, whose walls are not ridged, rather wrinkled. Proventriculus forms the bend in 'J' of the gut. Oesophagus is the short arm of the 'J' and midgut begins the long arm.

The posterior dorsal end of proventriculus narrows and extends into the lumen of anterior midgut. Midgut walls are thinner than proventriculus. Two pairs of hepatic ducts from digestive caeca open into anterior midgut. The anterior hepatic ducts opens ventro-laterally just posterior to anterior end of midgut. The posterior hepatic ducts open laterally and are a little farther posterior. Both

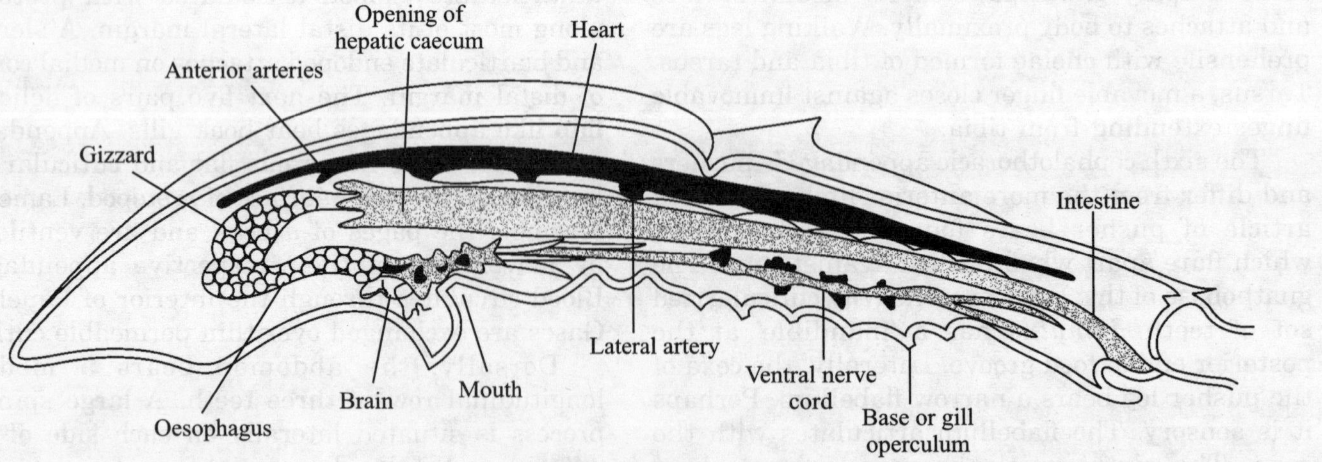


**Figure 13.10** Sagittal section of the *Limulus*.

pairs of ducts join anterior fourth of midgut. The lumina of the digestive caeca tubules are continuous with that of midgut. The midgut is the secretory and absorptive region. The brush border of microvilli on midgut epithelium produces colours. At about the level of the third movable abdominal spine, the midgut joins rectum. The rectum is short and empties ventrally via the anus. Anus opens at the base of the tail spine.

### *Nervous system*

The nervous system is well developed including central nervous system and peripheral nerves. The double ventral nerve cord is present. The nerve cord of cephalothorax remains hidden beneath a white sheet of connective tissue. Many fused ganglia, connectives and commissures form an oval circumesophageal nerve ring around oesophagus. The nerve ring lies on the floor of cephalothorax.

Many nerves exit the ring to supply the appendages, sense organs, muscles and organs of cephalothorax. The nerve ring consists of dorsal brain and paired ganglia of cephalothoracic and anterior abdominal appendages. Six pairs of cephalothoracic appendages, chilaria and genital operculum are innervated from ganglia of nerve ring. Two longitudinal ventral nerve cords extend posteriorly from nerve ring to abdomen.

Five pairs of segmental ganglia are the swellings along nerve cord in the abdomen. From these arise nerves to supply five pairs of book gills. The last of these five pairs of ganglia is also the end of nerve cord and from it arise nerves to supply the fifth book gill and many other posterior structures including tail spine. This terminus is located at the level of the fourth and fifth book gills.

### *Excretory system*

Four pairs of saccate nephridia, also called *coxal* glands are modified metanephridia which are coated on each side and share a common duct to a common nephridiopore on the fifth cephalothoracic appendage. Coxal glands are brick red in adults and lie at the bases of cephalothoracic appendages 2–5 (Figure 13.11).

### *Reproductive system*

In adults, the paired gonads either testes, or ovaries are composed of highly branched, arborescent tubules extending throughout haemocoel, mostly in association with digestive caeca. Each gonad empties via a gonoduct to a genital aperture on genital operculum. The spermatozoa are flagellated. Fertilization is external and development produces a trilobite larva.

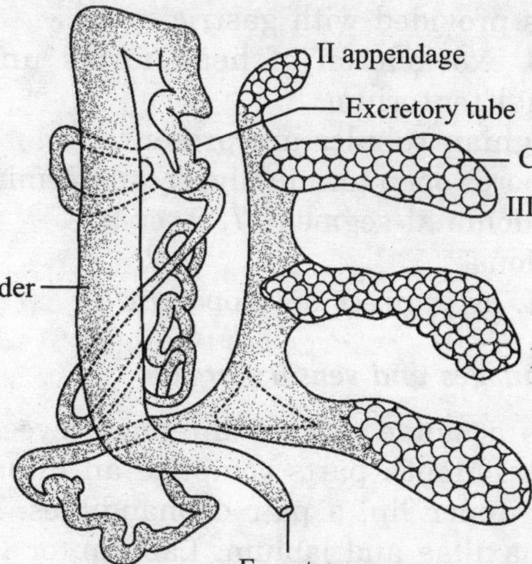


**Figure 13.11** Excretory organ of *Limulus*.

## 13.10 INSECTA

Insects are air-breathing arthropods with distinct head, thorax and abdomen, except in some larvae and in some modified adults. They possess a single pair of antennae, 3 pairs of thoracic legs and usually 1, or 2 pairs of wings in the adult stage. Reproductive organs open near the caudal extremity of the body. Insects are adapted to the greatest variety of conditions and are successful in outnumbering all other animals both in species and individuals. They also occupy a position of economic importance.

### *Characters*

1. Body composed of 19 true somites comprising a head (5 somites), thorax (3 somites) and abdomen (11 somites).
2. Head bears compound eyes; antennae, clypeo-labrum, mandibles, maxillae and labium.
3. Legs are uniramous, present on three thoracic segments of adults.
4. Legs composed of coxa, trochanter, femur, tibia, tarsus, post-tarsus; tarsus often subdivided; post-tarsus typically clawed.
5. Gas exchange occurs by spiracles and tracheae.

6. Gut is provided with gastric caeca.
7. Fused exoskeleton of head forms unique internal tentorium.
8. Malpighian tubules are present.
9. Gonopores open terminally, or sub terminally on abdominal segment 7, 8, or 9.
11. Dioecious.
12. Direct, or indirect development.

### *Head appendages and sensory organs*

Head bears a single pair of antennae, eyes and mouth parts. Mouth parts comprise an unpaired labrum, or upper lip, a pair of mandibles, hypo pharynx, maxillae and labium. Labium, or lower lip, is the second pair of maxillae of which at least the basal portions are fused along median line. Both the maxillae and labium are composed of several sclerites and bear palpi called the *maxillary* and *labial palpi*, respectively.

All the head appendages are articulated with immovable parts, forming the head capsule. Mouth parts are modified in form and function. In some, they are suited for holding and chewing food. Many insects take only liquid food and in them mouth parts are modified to form a sucking tube. The significant suctorial modifications are found in Hemiptera, Lepidoptera, Diptera (Figures 13.12 and 13.13) and Hymenoptera. The jointed beak of Hemiptera consists of a labium partially covered above at its base by labrum.

The long, coiled proboscis of Lepidoptera is formed by parts of maxillae, while labrum, mandibles and labium are reduced, or lacking. Female mosquito has a piercing type of mouth parts. Labrum and epipharynx are fused with hypopharynx to form food channel. Linear mandibles and maxillae are used in puncturing the skin of victim. Labium forms a sheath for other mouth parts. Both the suctorial and mandibulate functions are carried out by mouth parts of honeybee.

Mandibles are used for cutting, crushing and other purposes. Maxillae and labial palpi are folded to form a sheath within which an elongated portion of labium serves as a tongue. Eyes, antennae, mandibles, maxillae and labium are borne on distinct segments and an additional pair of appendages starts to form in the embryo but never becomes functional in an adult. Probably head form by fusion of six original somites, only five of which have retained their appendages, or their rudiments in adult.

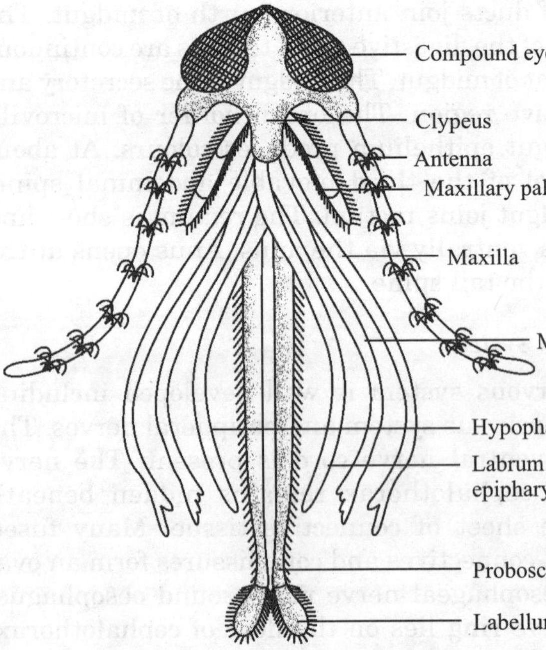

**Figure 13.12** Mouth parts of female *culex.*

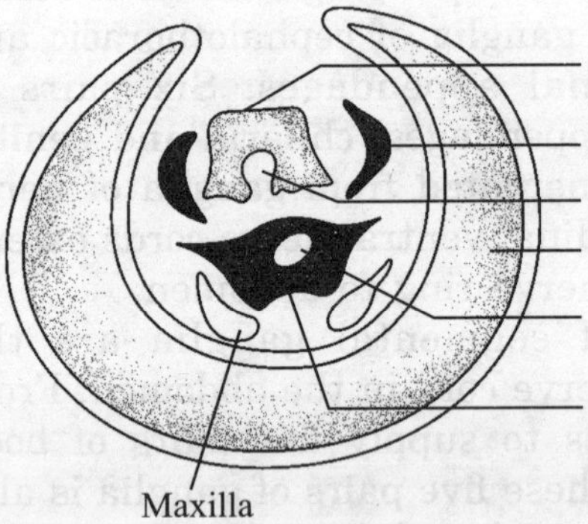

**Figure 13.13** T.S. of mouth parts.

A number of sclerites are fused to form head capsule. Often, sutures separating the sclerites are visible. In the Orthoptera, epicranial suture is the most constant which originates at the margin of occipital opening and extends as a median suture over the dorsal surface of the head. At its ventral extremity, epicranial suture bifurcates to form an inverted Y. Between the arms of the Y, an unpaired sclerite called the *front exists*, which mostly bears median ocellus.

An additional unpaired sclerite ventral to front is called the *clypeus* which on its ventral border is articulated with labrum. Several paired sclerites of head, including lateral surfaces and parts dorsal to arms of epicranial suture, constitute epicranium. Vertex is just dorsal to front and lies between compound eyes and usually bearspaired ocelli. Genae and postgenae form lateral portions of epicranium. A chitinous supporting structure called the *tentorium* usually consists of a central plate from which two, or three pairs of arms pass to exoskeleton of head.

Both simple and compound eyes are found. Compound eyes occur on sides of head in most adult insects. All holometabola have simple eyes in larval stage. Antennae are tactile. In some insects at least seven different types of microscopic sensory organs are found. Most of them are probably tactile, auditory and olfactory. Hairs of graduated lengths are found on feathery antennae of mosquitoes (Figure 13.14) and moths. The auditory organ is the tympanal organ consisting of a drum like membrane for reception of sound waves. In the common grasshoppers, tympanal organ is located on each side of the first abdominal somite, while in katydids a similar organ lies on second joint of each front leg. The end organs of taste and smell are usually located on epipharynx, hypopharynx and labial palpi, but they are not restricted to any given organ. Tactile organs and sensory structures of unknown functions occur as modifications of body covering over most of surface of an insect.

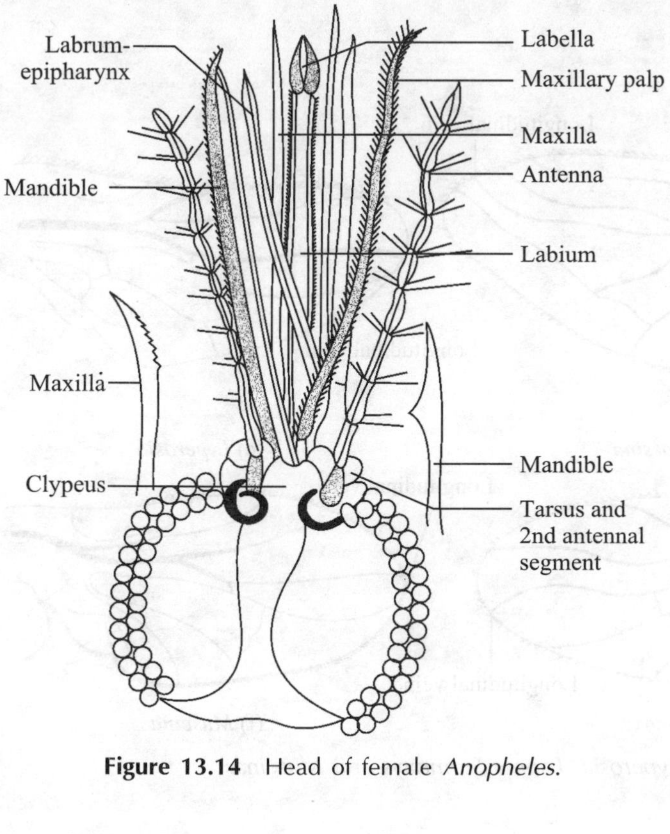


**Figure 13.14** Head of female *Anopheles*.

### *Thorax*

Thorax bears legs and wings in adult and composed of three firmly united segments—the prothorax, mesothorax and metathorax. In many insects, the last two bear wings. Each somite is composed of several sclerite groups, which are recognized as comprising parts of tergum (dorsal wall), sternum (ventral wall) and pleura (lateral walls).

### *Legs*

Each leg is articulated to thorax wall partly by small articular sclerites in the region of articulation of sternum and pleuron. Five divisions in each leg are coxa, trochanter, femur, tibia and tarsus. In most insects, each region is a single segment except the tarsus, which commonly has five segments. Numerous modifications in the form are found in legs.

### *Wings*

Generally, mesothorax and metathorax bears a pair of wings, but sometimes only one pair of wings are borne on mesothorax. Rudiments of the second pair of wings are found as halteres on metathorax. Modifications of the primitively membranous wings occur in some cases. In beetles, mesothoracic wings are thickened into horny structures, the elytra for protection of metathoracic wings and dorsal surface of the body. In some bugs, bases of front wings (hemelytra) are horny while tips are membranous. In Orthoptera, the entire front wings are thickened and are called the *tegmina*. Typically, wings develop as saclike folds of the body wall. In fully developed wings, this saclike form is obscured because two walls of sac become so closely applied that they appear as a single delicate membrane.

A framework for supporting this membrane is composed of hollow tubes, or veins which originate as modified tracheae. In early development, tracheae are accompanied by nerves and blood-filled outgrowths of body cavity. Wing venation in most groups shows constancy among members of the same species. Between the genera and larger groups variation in details of arrangement of veins exist. Probably wings of all insects and their principal individual veins are homologous.

Comparisons of numerous wings and arrangement of tracheae before the wings reach their functional form have led to a hypothetical plan of primitive venation from which all wing patterns are derivable, though all of homologies are not observable.

In the plan of primitive venation, eight principal veins are present and most of these may be branched. Beginning with anterior margin, when wings are at right angles to the body, these veins are costa (C), subcosta (Sc), radius (R), media (M), cubitus (Cu) and anal (A) and jugal (J). Changes from hypothetical type occur either by addition of new veins, or through reduction in the number of veins through atrophy, or through coalescence of two, or more veins. Cross-veins frequently connect two of the longitudinal veins.

At the base of wing, there are small articulatory sclerites including three axillary sclerites (Ax), humeral plate (HP) and median plate (mp). Longitudinal veins commonly fork and branches are indicated by subscripts. Radius often has a main anterior branch, $R_1$ and a main posterior branch, the radial sector (Rs), that is further subdivide into $R_2$, $R_3$, $R_4$ and $R_5$. When two, or more veins are fused together to form a single vein, the names of veins are joined by a plus sign ($Sc + R_1 + Rs$). Some longitudinal veins are joined by cross veins that are usually called humeral (h), radial (r), sectorial (s), radio-medial (r-m), medial (m), mediocubital (m-cu) and cubitoanal (cu-a) according to longitudinal veins involved (Figures 13.15 and 13.16).

Membranous areas enclosed by veins are formally named according to longitudinal vein just anterior to the cell. For example, the veins $M_2$ and $M_3$ define a cell, cell $M_2$. This may be divided by a median cross-vein to form a proximal cell, first $M_2$ a distal cell, second $M_3$. For taxonomy, cells may be called *basal*, *marginal,* or *submarginal*, etc., or may be given special names such as dorsal, depending on group. Veins surround a closed cell, whereas an open cell has one side bordering wing margin. The wing area anterior to the anal vein and its branches is called the *remigium*. The anterior edge close to base of wing is the humeral region.

### *Abdomen and its appendages*

Abdomen bears variable number of segments in various insects. In embryo, abdomen consists normally of eleven segments, but later adjacent segments may coalesce. The wall of each segment is composed of a tergum. Sternum is united by a pair of pleural membranes. Segments are typically without appendages and approximately similar except near caudal extremity. In the Thysanura, rudimentary abdominal limbs occur. First seven abdominal segments are usually lost during early

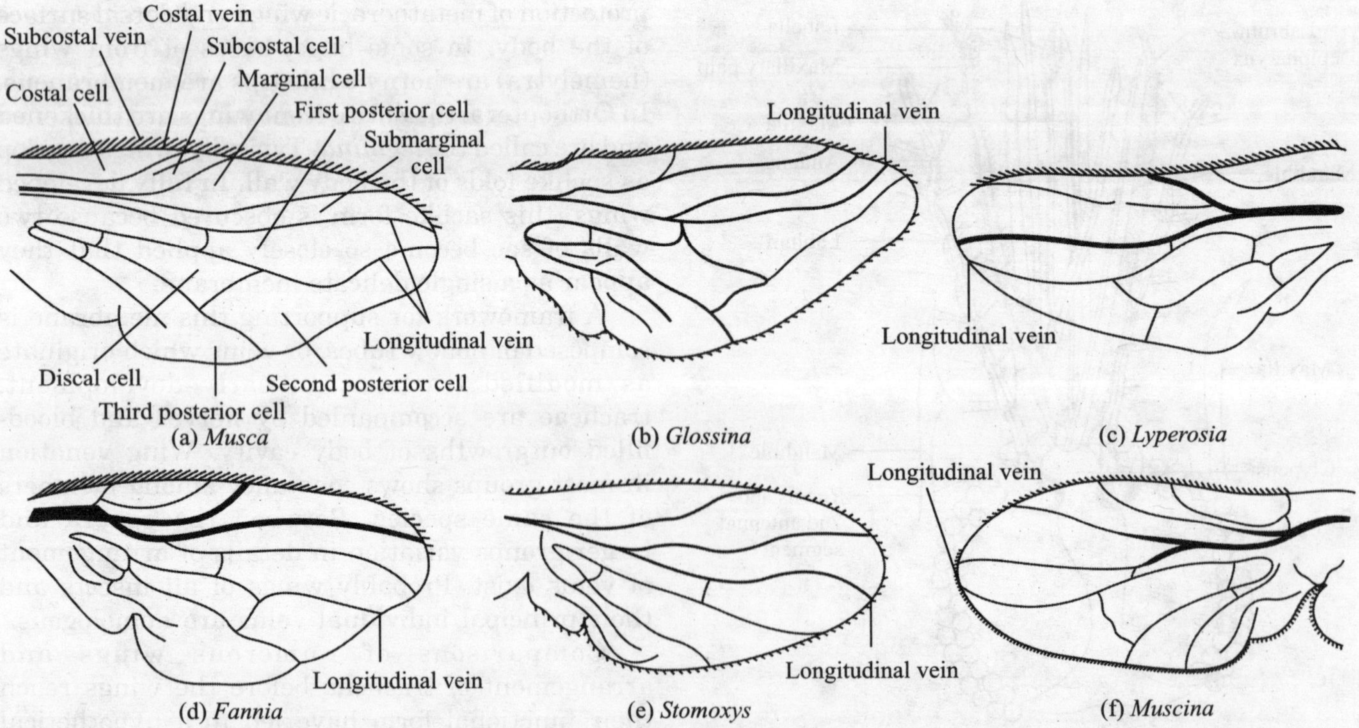


**Figure 13.15** Wings of *Musca, Glossina, Lyperosia, Fannia Stomoxys* and *Muscina.*

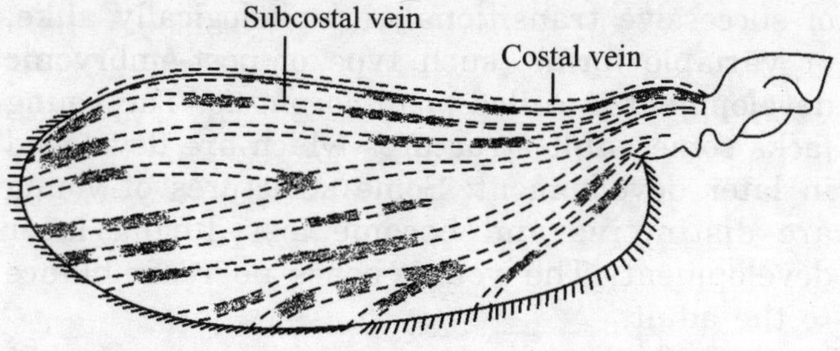


**Figure 13.16** Wing of an anopheline mosquito.

embryonic life; while last two, or three pairs frequently persist to form genitalia—the genital claspers of males and the ovipositors of females. A true ovipositor consists of three pairs of valves called *gonopophyses*.

These three pairs of gonopophyses represent paired appendages of the 8th, 9th and 10th abdominal somites. In Hymenoptera, ovipositor is modified as stinging organ and has poison glands associated with it. Inner pair of gonopophyses in males is modified as intromittent organ, while other 2 pairs of gonopophyses are modified as clasping organs to function in copulation. The eleventh abdominal somite in many insects bears a pair of caudal appendages known as the *cerci*.

### *Respiratory system*

This is a system of air tubes called the *tracheae* and *tracheoles*, which carries air to all parts of the body and is in communication with outside air through openings termed the *spiracles*. Of 10 pairs of spiracles, 2 pairs are found on thorax and 8 pairs on abdomen. Opening and closing of spiracles is under control of insect. From each spiracle, a short trachea generally leads to a main tracheal trunk of which there is one on each side of body. Branches from these two main trunks enter into the most minute parts of the body (Figure 13.17).

Tracheae arise as invaginations of body wall and infolded chitinous covering of body continues within tracheae as a chitinous internal lining of tubes. Within tracheae, chitin is not disposed in a uniform layer rather assumes the form of a coiled spiral thread lining the inner wall of tracheal vessel. Small tubes lack the spiral chitinous threads, form the most minute subdivisions of tracheal system called the *tracheoles*. Tracheae in some insects may become modified as enlarged sacs, which serve as air reservoirs.

In Collembola and in many aquatic larvae, there are no specialized organs for respiration.

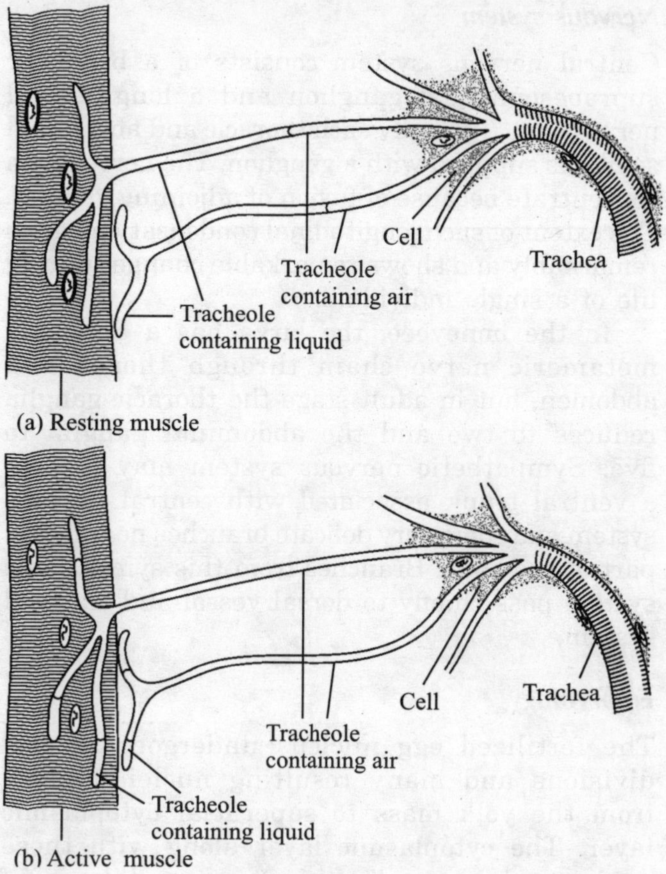


**Figure 13.17** Regulation of tracheal respiration. In active muscle the osmotic pressure is higher and the fluid is withdrawn from the tracheoles, thus allowing air to get to the muscles more easily. (Redrawn after Wigglesworth).

Respiration is performed directly through skin. Gills occur in many aquatic nymphs and larvae. Two distinct types of gills are tracheal and blood gills. In the former, the lateral or caudal evaginations of the body wall are furnished with numerous tracheae, which are continuous with vessels of tracheal system. A portion of digestive tract serve as a respiratory organ in rectal tracheal gills of dragonfly nymphs.

Blood gills are usually threadlike evaginations from body wall of aquatic insects. Spaces within these gills are in direct communication with fluid-filled body cavity. Through delicate walls of gills, a respiratory exchange occurs without requiring presence of tracheal tubes. Few insects, particularly larvae live under conditions, where oxygen is absent as in some larvae living in very deep water, the anaerobic respiration is carried on.

### Nervous system

Central nervous system consists of a brain, or supraoesophageal ganglion and a longitudinal nerve chain. Typically each thoracic and abdominal somite is supplied with a ganglion. The nerve chain concentrate because of fusion of adjoining ganglia. The extent of such longitudinal condensation varies remarkably and shows remarkable changes during life of a single individual.

In the honeybee, the larva has a distinctly metameric nerve chain through thorax and abdomen, but in adult stage the thoracic ganglia reduces to two and the abdominal ganglia to five. Sympathetic nervous system may include a ventral trunk associated with central nervous system and some very delicate branches near dorsal part of the body. Branches from this sympathetic system pass chiefly to dorsal vessel and tracheal system.

### Embryology

The fertilized egg nucleus undergoes several divisions and many resulting nuclei migrate from the yolk mass to superficial cytoplasmic layer. The cytoplasmic layer along with these nuclei undergoes cleavage forming a layer of cells surrounding a central yolk mass—the blastula and cell layer is frequently called the *blastoderm*. Nuclei which remain within yolk called the *yolk cells* become surrounded by cytoplasm.

Blastoderm becomes thickened in one region and forms germ band from which ventral surface of embryo develops. The course of this development follows two different paths. These are known as the *overgrown* and *invaginated* types of development. In the development of most of the Orthoptera, Trichoptera, Diptera, Lepidoptera and Hymenoptera, the overgrown, or superficial type of germ band occurs. The invaginated, or immersed type of germ band is found in some Odonata, Coleoptera, Thysanoptera and Hemiptera. Transitional conditions are found in some Coleoptera and Orthoptera.

### Types of metamorphosis

Most insects pass through conspicuous changes in form between the time of egg laying and the time it reaches full maturity. These changes are collectively called *metamorphosis*. Whenever the development of an embryo into adult takes place through a series of successive transitional, morphologically alike, or variable forms, such type of post-embryonic development is called *metamorphosis*. The young lacks some adult structures which are developed in later development. Some structures of young are distinctive and become lost during later development. The young bears no resemblance to the adult.

Several types of metamorphosis are recognizable. Insects which undergo a complete metamorphosis involve profound changes in form and have an inactive pupal stage are called *holometabolous*, or *holometabola*. Egg, larva, pupa and imago are all distinguishable in holometabola. Organs of one stage in development are not necessarily carried over directly into following stage. Many larval organs disappear due to phagocytic, or chemical action and new organs of adult are formed through histogenesis. In holometabola, wings and legs of adult do not develop externally on larva but develop internally as imaginal buds which emerge and become free later in development.

The heterometabola transform without a true pupal period. Stages between the egg and imago closely resemble the general bodily structure of imago and are termed *nymphs*. The absence of functional wings usually differentiates nymph from adult, but external wing pads from which wings later develop are characteristic of many nymphs. Thysanura and Collembola are wingless and throughout life retain essentially the forms they have at hatching. There are some inconspicuous changes and many are inclined to refer to insects of these two orders as the ametabola.

In both types of metamorphosis, insect undergoes radical changes in its internal organization before reaching adult stage. Even organs which are found alike in larval and adult stages do not pass over directly from one stage to next, but through histolysis the tissues of one stage disappear and entirely new tissues are formed through histogenesis.

In the holometabola, entirely new structures such as wings and legs become functional for the first time in adult form and more profound internal changes accompany metamorphosis. During larval and pupal stages, leg rudiments, wings and head appendages appear as internal buds. These imaginal discs remain throughout their early development enclosed within internal sacs. The hypoderm of body wall invaginates to form these sacs and rudiment of appendage. These sacs lies within a cavity entirely surrounded by hypoderm. Only in later development sacs open and allow developing appendages to extend freely beyond body surface.

### *Metamorphic changes in insects*

Insects provide an interesting example of morphogenetic processes in post-embryonic developmental stages. These changes are caused by massive destruction of larval tissues and differentiation of adult tissues from embryonic reserve cells called the *imaginal tissues*. Insect metamorphosis involves the balance between a hormone necessary for continued larval growth and hormone capable of stimulating adult development. Metamorphosis involves casting off cuticle several times. Moulting is called the *ecdysis* while sloughed off skin is called the *exuviae*. Interval between the ecdysis is called *stadium*, or *intermoult period* and form assumed during a particular stadium is termed *instar*.

Moulting initiates by separation of underlying epidermis from cuticle by a process is known as *apolysis*. Gap between cuticle and epidermis is filled with a fluid called *moulting fluid* containing several hydrolytic enzymes that start digesting the inner part of overlying cuticle. Simultaneously, the epidermis secrets a new cuticle, which is soft and extensible. Due to resorption of cuticle by moulting fluid, it becomes brittle and finally shed away from insects. Moulting facilitates body growth and brings about metamorphosis.

### *Importance of insects*

Honeybees, silk worms and various insects which pollinate flowers are economically important. As pests, hemipterans occupy a place of extreme significance, with chinch bug as a menace to grain farming. Hundreds of species of scale insects attack trees and shrubs. Numerous disease-carrying flies and mosquitoes, hessian fly destroys growing grain. Warbles and bots which attack domestic animals demand attention.

Gipsy moth in its ravages on shade trees, coddling moth in its molestation of apple industry and numerous cutworms are examples of Lepidoptera. Extensive damage to cotton through boll weevil and injury by wood-boring and leaf-eating insects are due to coleopterans. Predaceous insects such as coccinelid beetles, dragon flies and neuropterans aid materially in checking damage.

## 13.11 *Periplaneta americana*

*P. americana* is the widespread, nocturnal, negatively phototactic insect. It prefers dark warm, moist habitats, reaches 44 mm in length, dark reddish-brown with a pale peripheral band around the pronotum. Adults have large wings, although they do not fly. Nymphs are wingless. It is acutely sensitive to vibrations and is one of the world's fastest running insects and is found near, or in human habitations. Such insects are called *synanthropic*.

### *Major synanthropic cockroaches*

Seven species of cockroaches have the synanthropic associations with humans and are spread throughout world by human migrations and commerce. They are now serious pests and are thought to have originated in Africa. Many other species are loosely associated with humans and their activities. The American cockroach, *P. americana*, is an exotic that was introduced to North America from Africa in the early 17th century. This is the second most abundant cockroach pest in the United States.

*P. fuliginosa* is an introduced peri-domestic species. *P. brunnea* is an African native spread by human travel and commerce and has a pale border around the pronotum, but it lacks the pale areas on forewings characteristic of *P. australasiae*. It is common in the southeastern United States and reaches 37 mm in length. Both sexes are winged and can fly. *P. australasiae* is almost as serious a domestic. *Blatta orientalis* probably originated in North Africa and is now a common pest in the southern, mid-western and northwestern United States. The German cockroach, *B. germanica* originated in Africa and travelled first to Europe and from there to the North America. It is the most abundant domiciliary cockroach in the United States and is an important pest.

### *External anatomy*

*Periplaneta* are elongate, oval in outline and strongly dorso-ventrally depressed, or flattened. Body is divided into head, thorax and abdomen. Head is inconspicuous in dorsal view. Pronotum of thorax is very large. The remainder of thorax and abdomen are hidden by two pairs of wings. Six pairs of similar, strong, spiny legs are present.

### Head

Head is flattened dorsoventrally, more, or less pear-shaped with mouthparts extending posteriorly from narrow end of the pear. Typical unspecialized mouthparts are found, which are adapted for biting and chewing. Head consists of a complex of fused sclerites, which constitute epicranium, or head capsule. Head capsule is divided into several regions and is associated with unfused sclerites and appendages.

This sclerotized capsule encloses mouthpart muscles, brain, anterior most region of gut and provides sites of attachment for mouthparts and antennae. Head can be retracted beneath prothorax. Antennae arise on the face immediately ventral to eyes. The epicranial suture marks the junction of three fused sclerites that form face. This suture is usually visible as a faint, yellow, Y-shaped line. Two arms of the 'Y' begin at the antennal sockets and converge between two eyes (Figure 13.18).

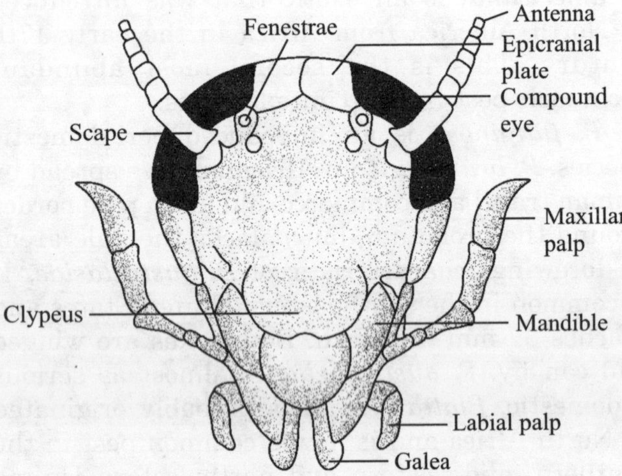


**Figure 13.18** Dorsal view of head of cockroach.

From the convergence, the single stem of the 'Y' extends dorsally between the eyes over top of the head. It divides the top of head into a pair of sclerites, epicranial plates. The part of capsule between the two arms of the 'Y' is the front of capsule. The front is an unpaired sclerite.

Two white circular areas between the antennae and compound eyes are the fenestrae. Two additional pale circles the frons, the ventral and a little medial to fenestrae, mark the sites of muscle insertions inside the capsule. The vertex occupies dorsum of capsule between the two eyes. It is divided into two by the stem of epicranial suture and includes two epicranial plates. The immovable clypeus is an unpaired sclerite on face; lies ventral to frons and the two are joined along a transverse, indistinguishable suture. A third unpaired sclerite, the labrum is attached along a transverse, movable articulation with ventral edge of clypeus. Ventrally slender, transverse cervical sclerites are embedded in soft integument of cervix.

### Antenna

Anterior most head appendages are two antennae which arise from a socket ventral to compound eyes. The first article of antenna is scape and the second is pedicel. From the pedicel arises a long sensory flagellum of 75–90 articles. Antennae are equipped with mechanoreceptors.

### Preoral cavity

Mouthparts surround and enclose a chamber, the preoral cavity ventrally, from which mouth opens. The preoral cavity precedes mouth and is not regarded to be part of gut, although mechanical and chemical digestion begins there. Five types of mouthparts recognized are

(i) biting and chewing mouthparts
(ii) chewing and lapping mouthparts
(iii) piercing and sucking mouthparts
(iv) sponging mouthparts
(v) siphoning mouthparts

Mouthparts of cockroach is biting and chewing type. This is a primitive mouthparts from which other types of mouthparts have evolved. It is responsible for biting, chewing and swallowing of plant and animal tissues and consists of a labrum, a pair of mandibles and a pair of first maxillae, a pair of second maxillae and a hypo-pharynx.

### Mandible

Two mandibles lie beside mouth and form the lateral walls of preoral cavity. They are derived from a pair of segmental appendages. The cutting edge of mandible consists of a lateral shearing incisor with three, or four sharp denticles and a more medial grinding molar. This is largely hidden and is the dark, heavily sclerotized. The mandible articulates with head capsule by two balls and socket-like condyles—one anterior and one posterior. The anterior condyle lies at dorsolateral corner of clypeus on the face.

The posterior condyle articulates with ventral corner of gena and lies on the side of head capsule. Its plane of motion is transverse. Movement in this plane brings two mandibles together on midline. There is a soft cuticular area in the base of mandible called the *submolar region*. The remaining part of mandible is hard. Teeth of left mandible lie dorsal to teeth of right mandible. Both mandibles work like a saw to cut food into pieces. Mandibles are jaws and works by four muscles.

### *Maxilla*

Two maxillae are appendages lying beside the preoral cavity, which consists of large 5–articulate palp. Each maxilla consists of a proximal cardo that articulate with head capsule and a distal stipes. The cardo is short, transverse, partly sclerotized article extending laterally from head capsule. It remains hidden by larger stipes. It is elongated and articulates with cardo proximally, beside the ventral edge of eye. Distally three processes arise from the stipes—the maxillary palp, lateral galea and lacinia.

The maxillary palp consists of five articles. The palp is sensory with chemo- and mechano-recepotors, which evaluate the suitability of food. From the distal end of stipes arise a lightly sclerotized, lateral galea and median, sclerotized, bladelike lacinia, which bears denticles distally. The lacinia assists mandible in cutting food into smaller particles. The lacinia is completely hidden by soft bulbous galea. The galea acts as a sheath to cover lacinia and because of it, the lacinia is not visible until the galea is moved aside (Figure 13.19).

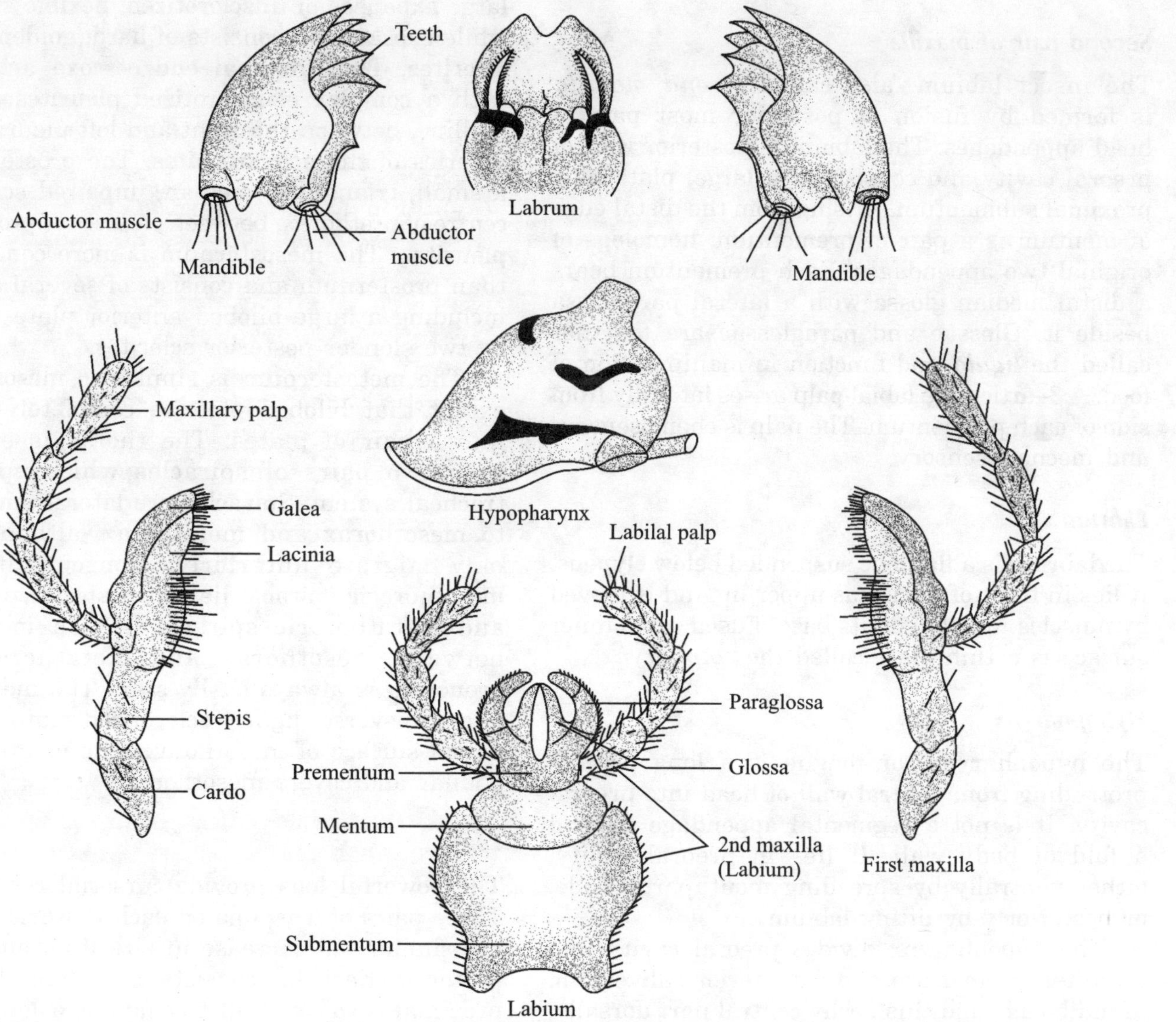


**Figure 13.19** Mouth parts of cockroach.

### First pair of maxillae

Each of first pair maxilla consists of two parts—the protopodite and endite. The endite consists of endopodite and exopodite. The protopodite consists of cardo and stipes, which is at right angle to each other. The outer distal end of stipes bears five-jointed exopodite called the *maxillary palp* and inner distal end bears an endopodite, which consists of an inner hard, pointed lacinia and outer soft galea. When mouthparts are not in work, two galea completely enclose laciniae. The basal tip of cardo bears a condyle for articulating with exoskeleton of head. An inner groove separates the cardo from stipes. A membrane from middle of stipes extends up to the head. First pair of maxillae is responsible for holding food and is also used for cleaning antennae, palps and front legs.

### Second pair of maxilla

The insect labium, also called *second maxilla* is formed by fusion of posterior most pair of head appendages. The labium is posterior wall of preoral cavity and consists of a large, plate like, proximal submentum. Arising from the distal edge of mentum is a pair of prementum, homologs of original two appendages. Each prementum bears a distal median glossa with a lateral paraglossa beside it. Glossae and paraglossae are together called the *ligula* and function in manipulation of food. A 3–articulate labial palp arises laterally from side of each prementum. The palp is chemosensory and mechanosensory.

### Labrum

The labrum is a flat lobe suspended below clypeus. It lies in front of mouth as upper lip and is moved by muscles inserted on its base. Fused to its inner surface is a thin plate called the *epipharynx*.

### Hypopharynx

The hypopharynx, or tongue is a long process protruding from ventral wall of head into preoral cavity. It is not a segmental appendage instead a fold of body wall. It lies in preoral cavity, either ventrally by spreading mouthparts aside, or posteriorly by lifting labium.

The hypopharynx divides preoral cavity into an anterior cibarium and a posterior salivarium. Mandibles bound this fleshy central part dorsally, ventrally by labium and laterally by maxillae. A membrane from inner border of labium is continuous with its ventral side. The salivary duct opens in the middle of ventral side and near the base. Hypopharynx is also termed the *tongue,* or *lingua*.

### Thorax

The thorax is the middle tagma, which is specialized for locomotion, bears three pairs of legs and two pairs of wings and houses the muscles. It consists of prothorax, mesothorax and metathorax. In ventral view the prothorax, mesothorax and metathorax are all visible. Each thoracic segment has complement of sclerites consisting of dorsal tergite, lateral pleurites and ventral sternites.

The ventral surface of thorax consists of large expanses of unsclerotized, flexible sternum, while exoskeleton consists of hard, golden brown, sclerites. The proximal end of coxa articulates with a complex of sclerotized pleurites. On the midline, between the right and left pleurites, lies a series of thoracic sternites. The prosternum is a small, triangular, median, unpaired sclerite in centre of prothorax, between clusters of prothoracic pleurites. The mesosternum is more complicated than prosternum and consists of several sclerites including a large bilobed anterior plate followed by two slender posterior sclerites.

The metasternum is similar to mesosternum except that bilobed plate is completely divided into a pair of plates. The thorax is equipped with two pairs of spiracles which open into tracheal system. Spiracles are lateral and belong to mesothorax and metathorax, although they may migrate anteriorly. Consequently, the mesothoracic spiracle lies on posterior prothorax and metathoracic spiracle lies on membrane between mesothorax and metathorax. The pronotum is always easily seen. The mesonotum is a transverse, lightly sclerotized plate covering dorsal surface of mesothorax. The metanotum is similar and covers mesothorax.

### Legs

The powerful legs provide cursorial competency. Three pairs of legs, one on each thoracic segment are similar but increase in size from anterior to posterior. Each leg consists of a large, flattened, proximal coxa, a small trochanter, a long femur, a tibia and a long tarsus. Femur and tibia bear strong spines. The tarsus is a series of 5–articulate tarsomeres of tarsomeres 1–4 each bear a posterior pad-like adhesive pulvillus. Tarsomere 5 ends with a pair of tarsal claws beside a pad-like arolium.

The distal arrangement of claws and arolium is sometimes called the *pretarsus*. The arolium is an adaptation for clinging to smooth surfaces. When flexed, the femur fits neatly into a recess in the surface of coxa. Articulations between coxa and trochanter and between femur and tibia are dicondylic joints, which restrict movement to a single plane. Coxa-body articulation is monocondylic and permits movement in a variety of planes as in the tibia–tarsus articulation.

### Wings

Meso-thorax and metathorax of adults of both sexes bear a pair of large wings. Anterior mesothoracic wings, or forewings are obvious in dorsal view, whereas equally large metathoracic wings are almost entirely hidden by the forewings. Parchment-like forewings protect the more delicate hind wings, which are membranous and fold in pleats when at rest under forewings. Folded wings completely cover the dorsal surface except for the head and pronotum

### Abdomen

Abdomen is the largest of three tagmata, which contains most of viscera. The abdomen consists of 11 segments of which the 11th is reduced and fused with 10th, so that a maximum of ten segments are distinct and visible. The first seven segments called the *pregenital*, or *visceral segments* are similar, unspecialized, visible and countable. The segment 9 in males and 8–9 in females are genital segments which are modified for reproductive functions. The segment 10 (and vestigial 11) are post genital and are modified. Dorsally each segment is covered by a sclerotized tergite and is covered ventrally by a sclerotized sternite. Consecutive sternites articulate with each other by unsclerotized articular membranes. Laterally, tergites and sternites are connected by unsclerotized, flexible pleura.

Most abdominal segments lack appendages, but segment 10 bears a pair of large, segmented cerci. Embryologically, cerci arise as appendages of 11th segment but in adults they are attached to segment 10 since segment 11 fuse with segment 10. The cercus does not contain intrinsic muscles. Extrinsic muscles extend from cercus to abdomen. Cerci bear vibration-sensitive hair, perhaps to ground vibrations including sound. Eight pairs of spiracles are present in abdomen, which are smaller than thoracic spiracles. Abdominal spiracles are located laterally on anterior-dorsal corner of pleura of the first eight abdominal segments. Abdominal spiracle 1 is located on the lateral edge of tergite 1 rather than at corner (Figure 13.20).

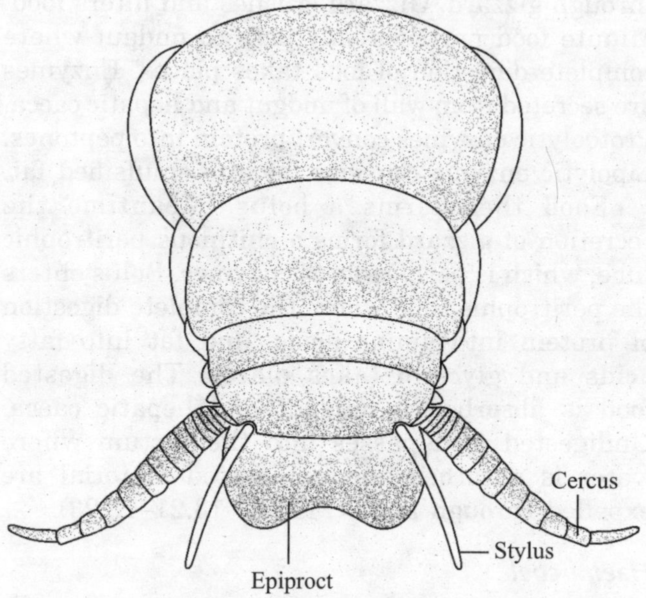


**Figure 13.20** Ventral view of the posterior abdomen of male cockroach.

### Alimentary system

The alimentary system includes alimentary canal comprising foregut, midgut and hindgut and digestive glands. The foregut includes mouth, pharynx, oesophagus, crop and gizzard. The midgut called the *mesenteron* is undivided. Near the junction of foregut and midgut, there is hepatic caeca. The junction is also marked by eight hollow slender tubes and numerous Malpighian tubules. The hindgut is divisible into ileum, colon, rectum and anus. The salivary glands, the inner lining of midgut and haepatic caeca altogether constitute the digestive glands.

Cockroach is omnivorous, macrophagous and all kinds of organic matter constitute the food of cockroach. Even it eats dead insects and dead bodies of its own kind. Maxillae pick up food particles, which are transferred to mandible. The food is cut into further particles and chewed by mandibular teeth. For this habit, mouthparts of cockroach are called the biting and chewing type.

Chewed food is pushed into preoral cavity by maxillae, mandibles and labium. Food is mixed with saliva coming from salivary glands. The saliva mixed chewed food passes to mouth form preoral cavity. The saliva contains amylase, which converts carbohydrate into glucose, which is absorbed in crop. In the crop, some digestion occurs because enzymes form midgut enters crop through gizzard. Gizzard crushes and filters food. Minute food particles only pass to midgut where complete digestion of food takes places. Enzymes are secreted from wall of midgut and hepatic caeca. Proteolytic enzymes convert protein into peptones. Lipolytic enzyme converts fat into emulsified fat.

Food then forms a bolus. Meantime the secretion of gizzard forms a chitinous peritrophic tube, which is permeable to enzymes. Bolus enters the peritrophic tube where the complete digestion of protein into amino acids, and fat into fatty acids and glycerol takes places. The digested food is absorbed in midgut and hepatic caeca. Undigested food passes into the rectum where water is absorbed and undigested material are expelled through anus (Figures 13.21–13.23).

### *Haemocoel*

In the embryonic stage, a large perivisceral coelom is found. But as the growth proceeds, blood vessels enlarge, swell into irregular spaces, finally they completely obliterate perivisceral coelom and replace it by such irregular spaces called the *sinuses*, or *lacunae*. Sinuses join to form body cavity which is filled with blood and contains all visceral organs lie. Body cavity itself is a part of circulatory system and is called the *haemocoel*, i.e. blood filled cavity.

Haemocoel extends from anterior end of body to posterior end, lacks epithelium of a true coelom and is divided into three bigger sinuses by two horizontal septa. Septa are muscular membranous structures. One dorsal diaphragm and other ventral diaphragm are present. Sinuses are upper dorsal sinus, or pericardial sinus enclosing heart. The middle perivisceral sinus lodges various visceral organs and ventral perineural sinus, or sternal sinus encloses ventral nerve cord. Both the diaphragms are perforated. As a result three sinuses remain in communication with each other.

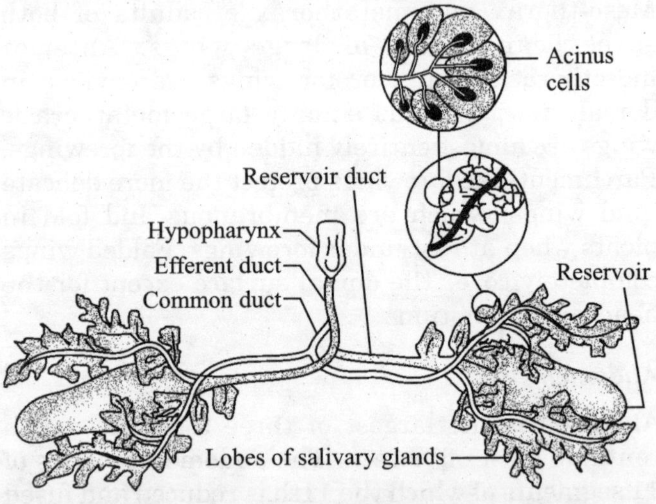


**Figure 13.21** Salivary apparatus of cockroach.

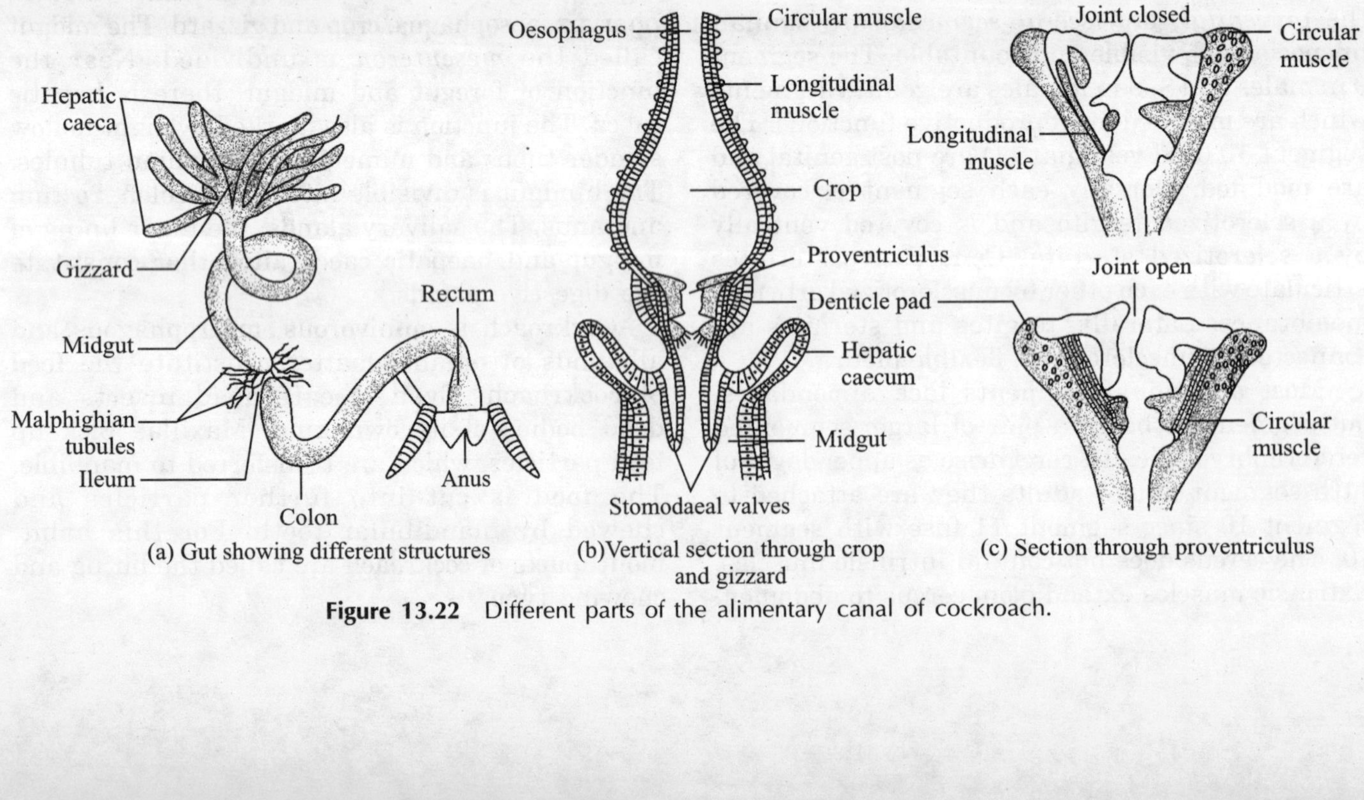


**Figure 13.22** Different parts of the alimentary canal of cockroach.

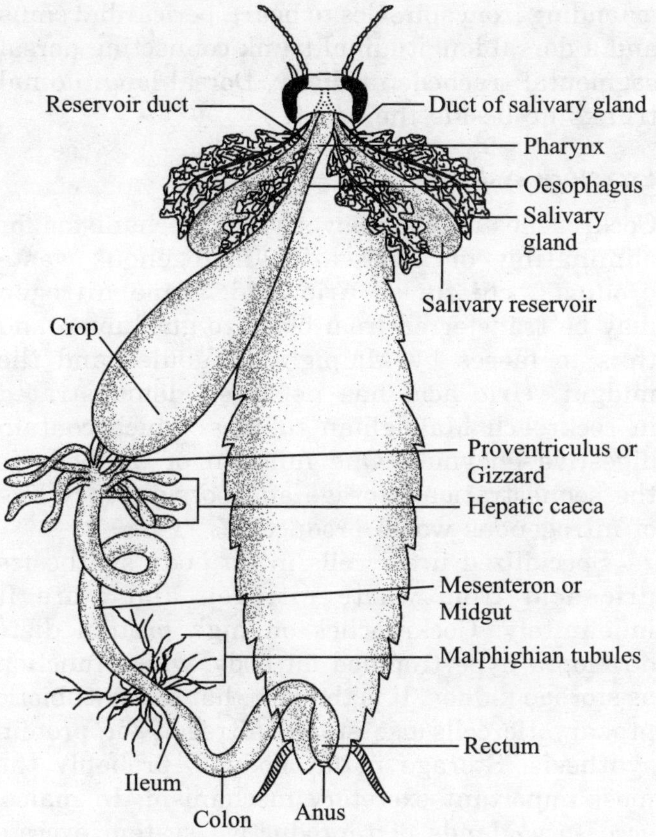


**Figure 13.23** Digestive system of cockroach.

Attached to dorsal diaphragm there is a series of paired alary muscles. They are triangular and their pointed outer ends are inserted into terga.

### *Haemal system*

Haemal system consists of haemocoel, blood and heart. Perforated horizontal membranes are called the *diaphragms*. A shallow dorsal pericardial sinus, a spacious perivisceral sinus and a small ventral perineural sinus divide the haemocoel. Together the sinuses form haemocoel. Heart surrounded by pericardial sinus lies immediately under tergites. Diaphragm is the sheet of connective tissue with longitudinal tergal muscles and separates pericardial sinus from the much larger perivisceral sinus.

These two sinuses are part of haemocoel. Heart is a longitudinal middorsal tube extending along the length of body in pericardial sinus and resting on dorsal diaphragm. Blood is pumped anteriorly during systole by contractions of circular muscles of heart wall. Blood leaves heart, enters cephalic haemocoel and passes posteriorly through three sinuses of haemocoel, eventually passing through perforations of the diaphragms to enter pericardial sinus.

During diastole, contractions of radiating alary muscles dilate the heart and draw blood into its lumen through ostia. Heart is now refilled with blood and ready to enter another systole. Blood vascular system is poorly developed and it is of open, or lacunar type because blood vessels open not into capillaries, but into spaces. Blood comes in contact directly with tissues, hence the rate of circulation is low.

### *Heart*

Heart is an elongated tube with muscular wall, lying mid-dorsally beneath terga of thorax and abdomen consisting of thirteen segmentally arranged funnel-shaped chambers. At lateral side of each chamber, there is a pair of ostia, one on each side of the valves, which allow blood to enter only heart guarding the ostia. Heart is made of a single layer of cells having striated muscles. Cells are lined both the outside and inside by a delicate membrane.

First chamber of heart is continued anteriorly into an anterior aorta, which opens into haemocoel. Alary muscles are twelve pairs whose alternate contractions and relaxations cause dorsal diaphragm to raise and fall and results blood to flow from perivisceral to pericardial sinus, and finally into heart as it beat. Muscular wall of heart contracts in a wave from behind to forward. Blood is forced into the anterior aorta from where it reenters haemocoel and goes slowly to organs and appendages including wings. An accessory pulstile vesicle at the base of each antenna also pumps blood (Figure 13.24).

### *Haemolymph, or blood*

Haemolymph is the circulatory media with colourless plasma and many corpuscles called the *haemocytes*. No respiratory pigment is found. Haemocytes does not help in transport of respiratory gases but transport various nutrients from one part of body to other and carry nitrogenous waste substances from tissues to organ of excretion. Plasma contains about 70% water and many organic molecules such as free amino acids, uric acid, proteins, sugars and fats.

Haemocytes are of three types, i.e. pro-haemocytes, transitional haemocytes and large haemocytes. According to some workers, only two

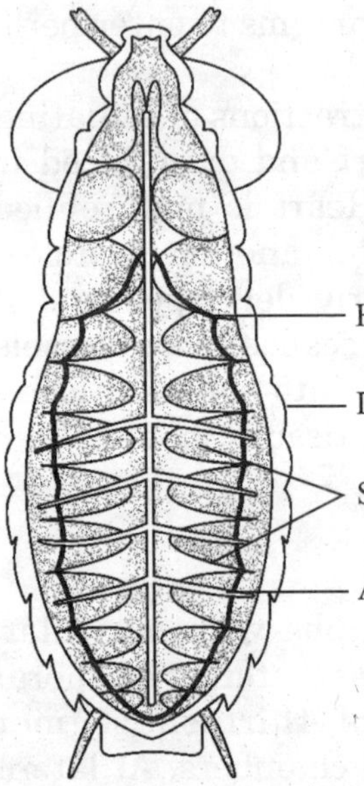


**Figure 13.24** Circulatory system of a cockroach (Dorsal view).

types of haemocytes are found—plasmatocytes and coagulocytes, or cytocytes. Plasmatocytes are polymorphic and constitute nearly 60–95% of total haemocytes. Hametocytes are phagocytic in nature which help in coagulation and for healing in wounds.

**Course of circulation of blood:** Wave of contraction in heart starts from posterior to anterior end and forces haemolymph in anterior aorta to pass into haemocoel of head and then to ventral and perivisceral sinuses then to dorsal, or pericardial sinus and finally to heart. Complete circulation of haemolymph through body takes nearly 30 to 60 minutes.

### *Respiratory system*

The system consists of ten pairs of spiracles which open into a complex system of tracheae to deliver oxygen to tissues. Removal of tergites exposes the network of branching, tubular, white, or silvery tracheae. Tracheae arise at spiracles in pleura near postero lateral corners of tergites and join with other tracheae to form a single network of air tubes for delivery of oxygen to tissues. Two pairs of spiracles are present in thorax and eight pairs in abdomen. Spiracles are absent in head.

On each side, respiratory system includes three longitudinal trunks and segmental tracheae to three-haemocoel sinuses and their viscera. Portion of system consists, on each side, of a lateral longitudinal trunk extending from spiracle to spiracle, a series of dorsal segmental tracheae extending from spiracles to heart, pericardial sinus and a dorsal longitudinal trunk connecting dorsal segmental tracheae medially. Dorsal longitudinal trunks lie beside the heart.

### *Excretory system*

Cockroaches may employ several mechanisms for eliminating, or inactivating nitrogenous waste products, chiefly as uric acid. Some nitrogen may be transferred from blood to gut lumen and thus to faeces by Malpighian tubules and the midgut. Uric acid has not been demonstrated in cockroach Malpighian tubules, which contain digestive enzymes. One function of fat body is the sequestration, storage and possibly recycling of nitrogenous waste products.

Specialized urate cells in fat body synthesize uric acid from waste nitrogen and store it indefinitely. Cockroaches on high protein diets develop a hypertrophied fat body, which function as storage kidney. It is thought that endosymbiotic prokaryotic cells use stored nitrogen for protein synthesis. Storage in fat body is probably the most important excretory mechanism. In males, accessory glands of reproductive system excrete nitrogen. Some tubules of this gland absorb and store uric acid, which is then deposited in spermatophore and eliminated during copulation.

### *Reproductive system*

**Male:** Male reproductive system consists of paired testes, each draining by a sperm duct to a common ejaculatory duct, which opens via the male gonopore in genital pouch. Associated with confluence of two sperm ducts are two clusters of secretory accessory glands. Near the confluence, sperm ducts are expanded to form seminal vesicles. Testes are located dorsolaterally in segments 4 and 5. They are embedded in white fat body. Prior to copulation, a spermatophore is formed in ejaculatory duct when secretions of accessory glands enclose and harden around a mass of sperm from seminal vesicles.

During copulation, the phallomeres open female's genital pouch and hold male's genitalia in correct position. Ejaculatory duct on ventral phallomere is held adjacent to opening of spermathecal papilla and spermatophore is released. Spermatophore is glued in place by secretions of male's phallic gland. In about 24 hours, or less, sperm exit the spermatophore and are stored in female's seminal receptacle.

Male accessory gland is by far the most conspicuous of male's internal genitalia. It secretes the covering of spermatophore and is a mass of worm-like diverticula in posterior haemocoel under tergites 7–8 beside anterior end of posterior colon and anterior rectum. Two sperm ducts enter gland and seminal vesicles are enclosed in mass of tubules. Thick ejaculatory duct, formed by union of sperm ducts, exits ventrally from centre mass of accessory glands and extends deep into ventral body wall and then posteriorly to gonopore on dorsal surface of ventral phallomere. Most of phallic gland is ventral to nerve cord. It is deeply ventral in posterior abdominal cavity, ventral even to ventral diaphragm and is a little to right of midline (Figures 13.25 and 13.26).

**Female:** Female internal genitalia consist chiefly of a pair of ovaries, two colleterial glands with separate ducts, seminal receptacle and oviducts. Each ovary is a bundle of tapering tubes, the ovarioles, which empty into an oviduct. Lateral oviducts, one from each of the two ovaries, join to

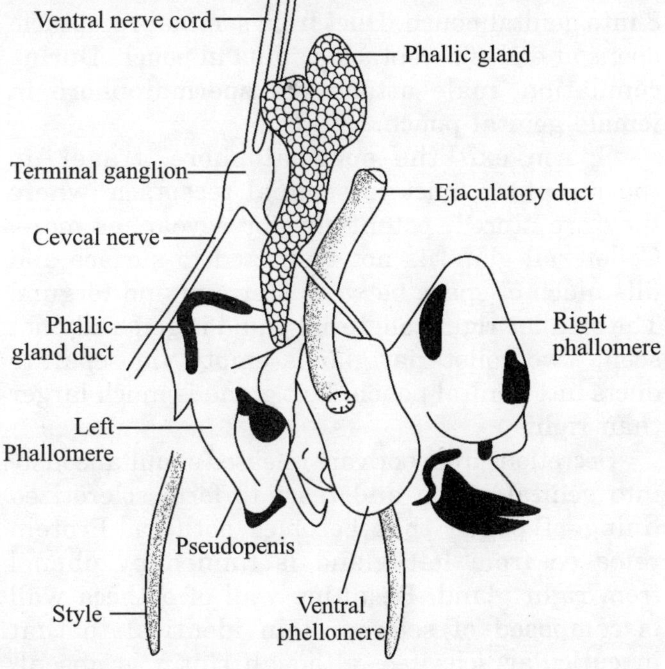


**Figure 13.26** Dissection of cockroach showing reproductive system.

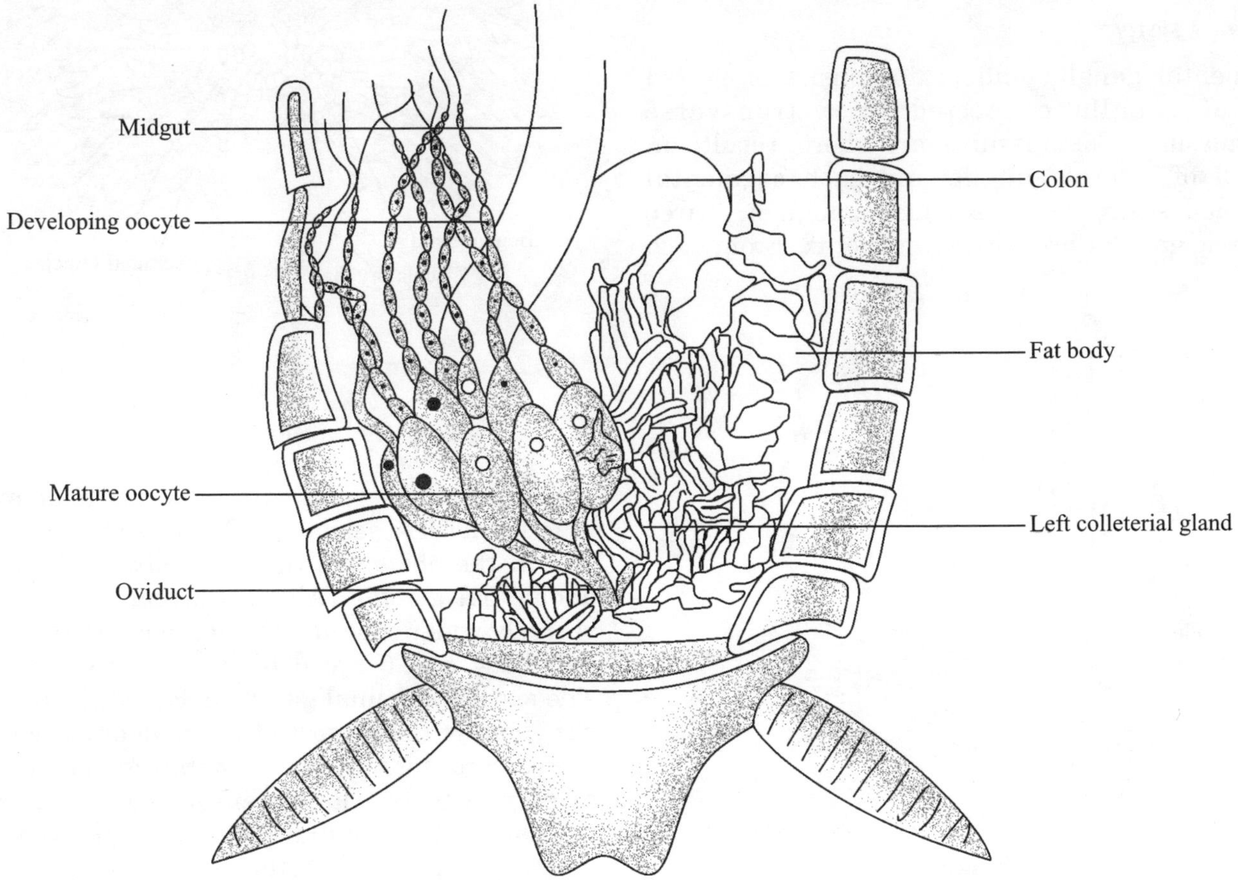


**Figure 13.25** Posterior haemocoel and genital pouch of cockroach.

form a common oviduct that opens through sternite 8 into genital pouch. Duct from seminal receptacle opens at end of receptacle papilla in pouch. During copulation, male attaches a spermatophore in female genital pouch.

Sperm exit the spermatophore, travel up the receptacle duct to seminal receptacle where they are stored, potentially for a year, or more. Colleterial gland is not restricted to surface and fills much of space between sternum and tergum. The smaller right colleterial gland is generally not seen. Two colleterial glands empty via separate ducts into genital pouch. Left gland is much larger than right.

Secretions from both are released simultaneously into genital pouch and react to form sclerotized lining of pouch that becomes ootheca. Protein released from left gland is tanned by phenol from right gland. Resulting wall of ootheca wall is composed of scleroprotein identical to that in cuticular sclerites, although chitin is absent. Eggs are deposited into this lining as they leave ovarioles and lining then hardens (tans) to become ootheca (Figures 13.27–13.29).

### *Nervous system*

Segmental ganglia and brain consist of paired lateral ganglia connected by a transverse commissure. Commissures are short, results in fusion of lateral ganglia and each segmental ganglion seems to be a single ganglion. Three thoracic ganglia are located in thorax. From each

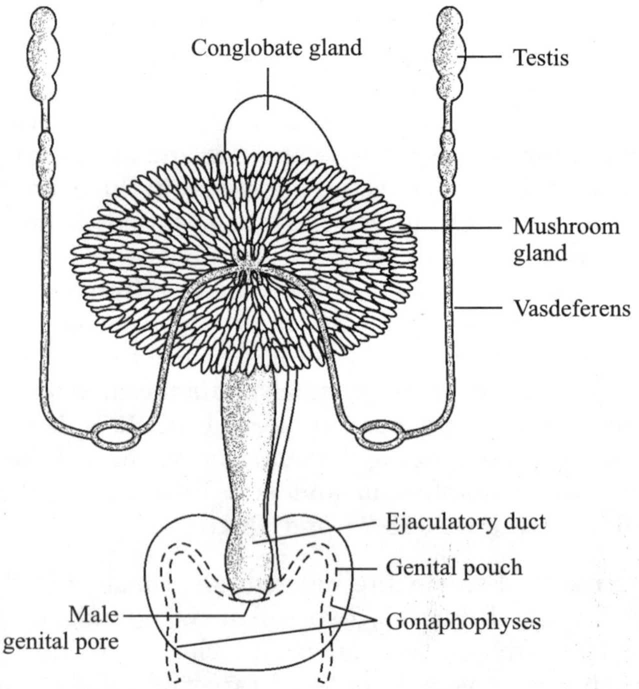


**Figure 13.28** Male reproductive system of cockroach.

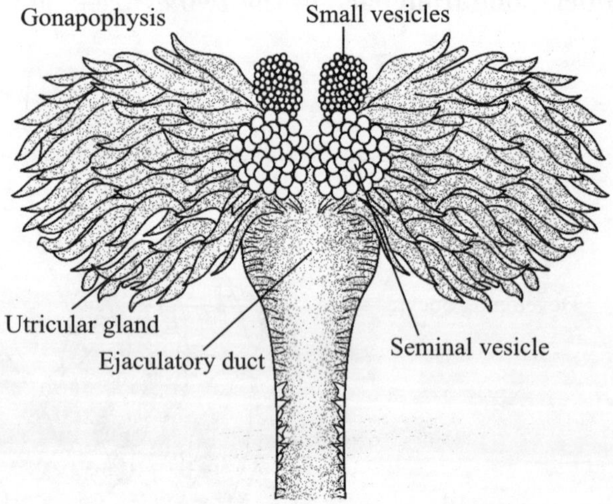


**Figure 13.29** Utricular gland of cockroach.

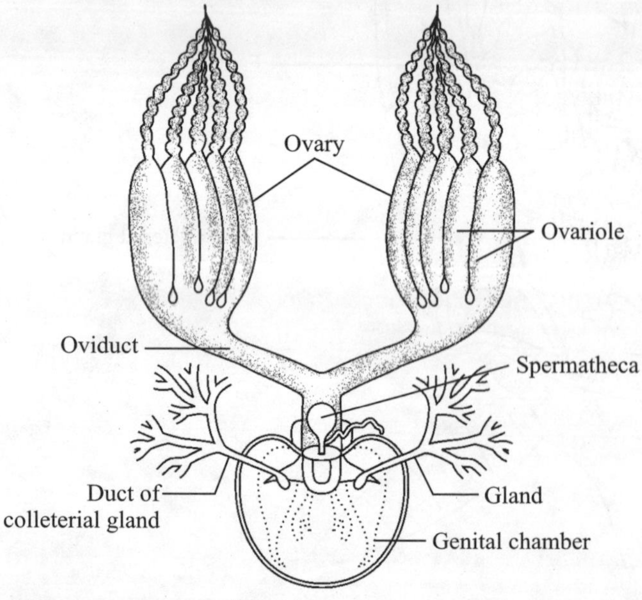


**Figure 13.27** Female reproductive system of cockroach.

of them extend several pairs of nerves to abundant muscles of these segments. These are large ganglia.

In the abdomen, there are six much smaller abdominal ganglia. Abdominal ganglia (1–3) are scarcely wider than double nerve cord. Abdominal ganglia 4 and 5 are usually readily discernable. The sixth abdominal ganglion, called the *terminal ganglion*, is the largest of the abdominal ganglia. Nerve cord ends at the terminal ganglion which gives rise to two large sensory cercal nerves to the cerci. The terminal ganglion is the coalesced ganglia of several posterior segmental ganglia. (Figures 13.30 and 13.31)

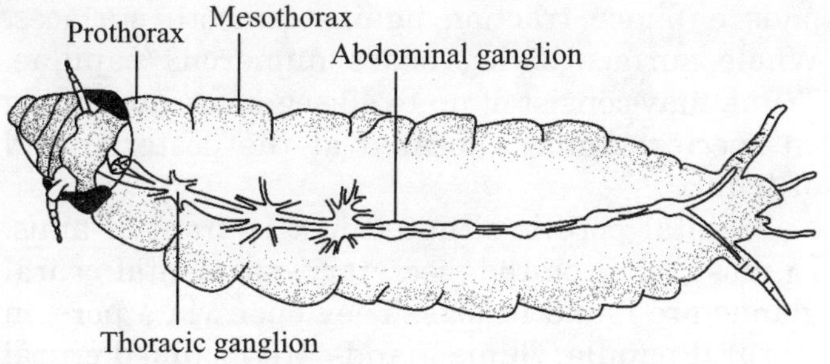


**Figure 13.30** Dissection of cockroach showing nervous system.

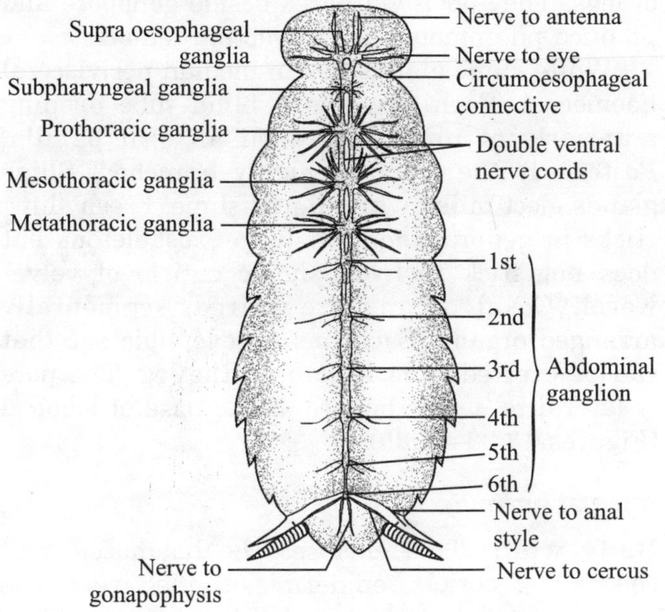


**Figure 13.31** Nervous system of cockroach.

## 13.12 *Peripatus*

*Peripatus* belongs to the group Onychophora which are also known as *velvet worms*. They are wormlike, terrestrial, inhabit particularly the humid localities of the tropical, or temperate southern hemisphere. Body is caterpillar-like, soft, elongated and about cylindrical. Body surface is dry, hydrophobic and velvety. *Peripatus* lives in damp forests under leaves, stones, in the bark of tree and crevices of rocks. No species is found in the north of tropic of cancer. Distribution of *Peripatus* is discontinuous and it occurs in widely separated countries like the Central America, South America, South Africa, India, and Australia with no forms in intervening tracts.

According to Barnes, onychophora consists of two families. The family Peripatidae are more, or less equatorial in distribution, extending no further north than the Himalayas, Caribbean Sea and Central Mexico. The family Peripatopsidae is restricted to the southern hemisphere. Thus, the South African species are closely related to those found in Chile than to those in equatorial Africa. Distribution of onychophoran is accounted for past geological connection of the American and African Landmasses. Peripatopsidae, with about 50 known genera, is found in temperate regions of the southern hemisphere including the Chile, New Zealand, Australia Tasmania and South Africa. Peripatopsids are blue, grey, green, or sometimes brown in colour.

Trunk contains up to 25 pairs of legs. Gonopore lies between the last pair of legs. Peripatidae, with about 10 known genera, is recorded in the Caribbean islands, Central America and southern Mexico, tropical South America, Malaysia and the Congo Basin of Central Africa. Peripatids are usually reddish brown. Trunk possesses 22–43 pairs of legs. Gonopore is found between the penultimate pair of legs. The family includes the *Peripatus.*

### *External anatomy*

**Papillae:** The large and small papillae arranged in closely spaced rings and cover body along with appendages. Papillae are provided with tiny scales and sensilla with various functions, chiefly chemoreceptive and mechanoreceptive. The larger papillae with a terminal chitinous sensory bristle are mechanoreceptive. Bristles are absent smaller papillae. Sensilla of the antennae and lips are chemoreceptive.

**Tagmata:** Body is divided into a small anterior head and a much larger posterior trunk, but demarcation between the two is indistinct. Both the head and trunk have paired segmental appendages. Body is not segmented except segmentally arranged lobopods. Inconspicuous external nephridiopores, coxal organs and crural papillae are also segmentally arranged.

**Head:** The three-segmented head bears the paired antennae, mandibles and oral papillae. The large median opening on ventral surface of head is the preoral aperture. It opens into preoral cavity, which precedes true mouth. True mouth opens from the inside wall of preoral cavity into pharynx. Preoral cavity is not the part of gut. Mandibles protrude into preoral cavity and salivary glands empty here. Preoral aperture is a distinct chamber and is surrounded by a circular outer lip and a semicircular inner lip. Lips lie close to preoral aperture. Scales of inner lip completely surround the aperture (Figure 13.32).

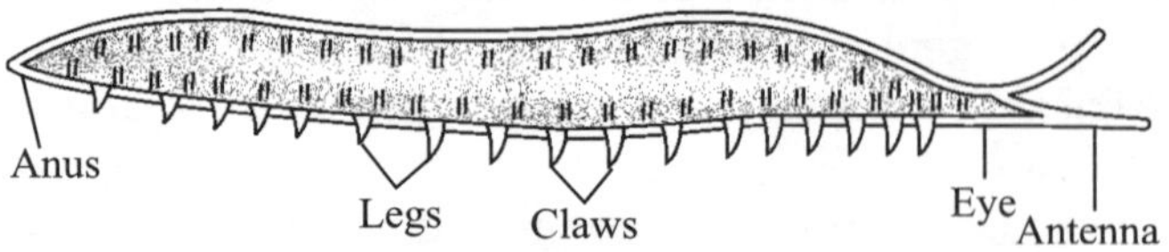


**Figure 13.32** *Peripatus.*

Anterior labrum constitutes the roof of preoral cavity. Muscular, median labrum is the dorsal anterior wall of preoral cavity and is provided with finely toothed mid-ventral ridge. Two sensory antennae lie dorsolaterally at the anterior end of head. They are conspicuously annulated. Annuli bear chemoreceptive and mechanoreceptive sensory papillae.

A pair of mandibles is located laterally in walls of the preoral cavity. Each mandible consists of a base from which emerge two sclerotized toothed jaws. Two jaws are located posteriorly to labrum and extend from sides of cavity into lumen. Mandible is equipped with a complex musculature. At the time of prey capture, the preoral aperture is opened and mandibles are extended to lacerate prey. Mandibles move independently of each other in longitudinal planes from anterior to posterior, but do not oppose each other in transverse plane.

Saliva from a pair of salivary glands is released into preoral cavity, subsequently injected into lacerated prey. Salivary enzymes digest prey externally and resulting liquid is pumped to midgut by pharynx. Oral papillae are the posterior most pair of head appendages. They lie laterally beside preoral aperture. Distal end of each papilla bears opening of duct from one of two large slime glands. Oral papillae are sensory in nature. A tiny eye, looks like a translucent white bead embedded in the integument, is found dorsally at the base of each antenna. A retina and optic ganglion are internal.

**Trunk:** This is devoid exoskeleton coverings. Its skin is drawn into several ridges. Trunk bears up to 43 pairs of similar legs and are found at regular intervals. Leg is hollow and cone-shaped. The tip of leg bears a flat pad and two claws. Claws are used to grasp the substratum during locomotion on rough surfaces and spinous pads enhance traction against smooth surfaces. Whole surface of leg bears numerous papillae. Trunk may consist of up to 43 segments depending on species. Anus is located at the posterior end of trunk.

Genital pore is situated just in front of anus. In most taxa, paired, secretory, segmental crural glands are found in legs. They open via a pore on a crural papilla. Slime glands are modified crural glands. Two anterior accessory genital glands of males are as modified crural glands of last pair of legs. They open via pores beside gonopore and produce pheromones for attracting females.

White slime glands lies in median perivisceral haemocoel. Each gland is a blind tube opening anteriorly at an aperture on an oral papilla. Posteriorly the gland is highly branched. Slime glands eject milky white liquid slime. Fresh slime sticks in general tenaciously to exoskeletons but does not stick to hydrophobic cuticle of velvet worm. Coxal organs are paired, segmentally arranged organs. Each has an eversible sac that can be everted from a pore on the leg. The pore is lateral to nephridiopore at the base of lobopod (Figures 13.33–13.35).

### *Internal anatomy*

**Body wall:** This encloses the haemocoel and consists of cuticle, epidermis, connective tissue dermis and muscle layers. Epidermis is made up of a single layer of cells and situated immediately beneath the cuticle. Dry, hydrophobic, flexible, chitinous cuticle is outermost and is nearly 1 mm in thickness. Hypodermis bears numerous short papillae, each containing a fine seta. Various pigments for colouration are found in hypodermis. Dermis is collagenous.

**Muscles:** Beneath hypodermis three sheets of muscle complete body wall. These are circular muscle, diagonal muscle and longitudinal muscle. Dorsoventrally arranged muscles separate body cavity into two lateral and one median cavity. Besides muscles of body wall, some muscles remain connected with mouth, pharynx, jaws and the alimentary canal.

**Haemocoel:** Body cavity is a haemocoel and is called *mixocoel*, which extends as four compartments—one central, two laterals and one pericardial. A diaphragm separates dorsal pericardial sinus from ventral perivisceral

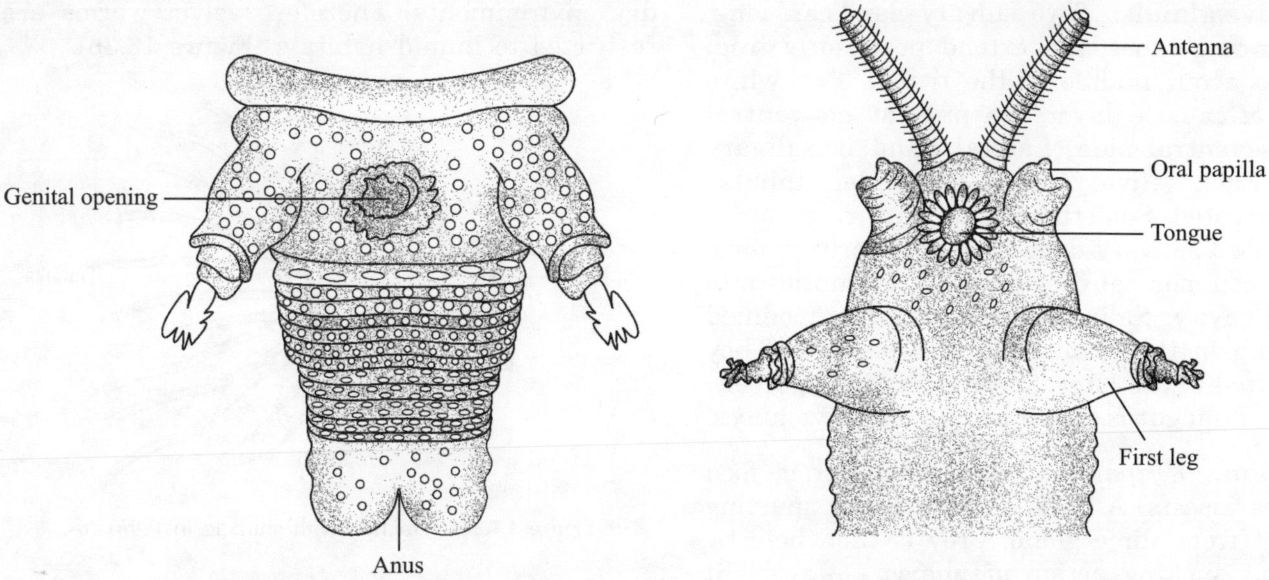


(a) Ventral view of posterior end (b) Ventral view of head

**Figure 13.33** Morphology of *Peripatus.*

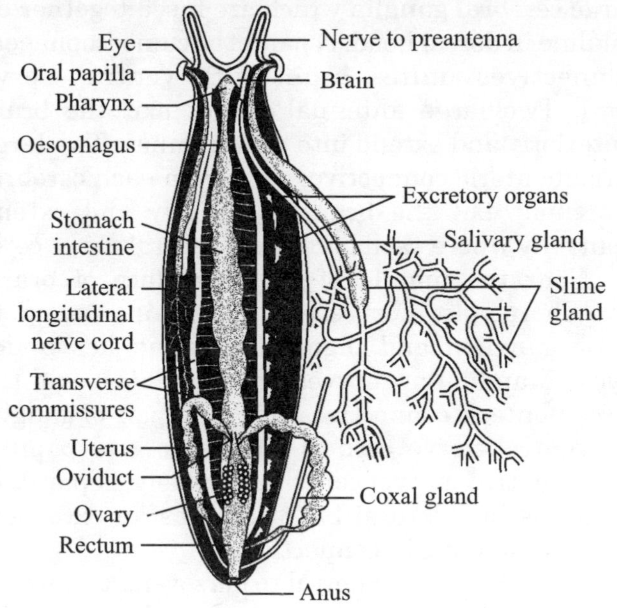


**Figure 13.34** Anatomical structure of *Peripatus.*

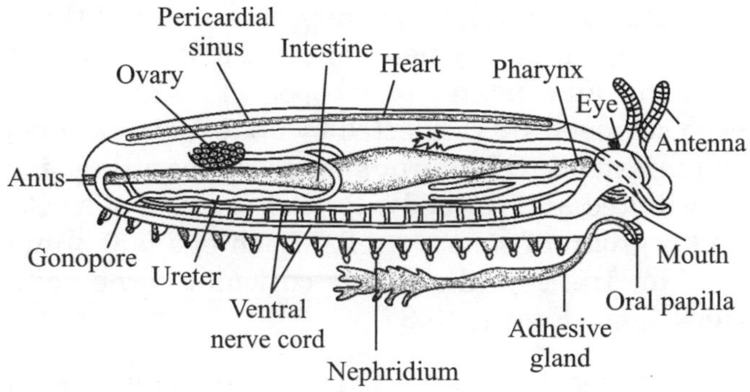


**Figure 13.35** Internal structure of *Peripatus.*

haemocoel. Lateral haemocoels extend into lobopods. Median perivisceral haemocoel contains the gut, slime glands and reproductive system. Lateral perivisceral haemocoels contain salivary glands, ventral nerve cords and nephridia.

Salivary glands are lateral to nerve cords. Fluid-filled haemocoel is a hydrostatic skeleton, which allows the circular and longitudinal body wall muscles to antagonize each other in a locomotory system. Longitudinal fibres of inner muscle layer lie dorsal to diaphragm and pericardial sinus. Heart probably appears as a collapsed tube. It extends from metamere bearing first pair of legs to metamere immediately before last. Heart is enclosed by a pericardial sinus.

**Digestive system:** The alimentary canal begins with mouth which is enclosed by oral lobes. Mouth leads into a pharynx, cavity of which is lined by chitin. Movement of pharynx produces sucking action. Circular muscles are found in pharyngeal wall. Pharynx extends posteriorly and opens to chitinous oesophagous having weak muscular wall. Oesophagus leads into midgut. Midgut sometimes called stomach. Intestine, or stomach-intestine, is a long, thick, irregular extension covering most of the length of trunk. Midgut epithelium is secretory and absorptive and final digestion and absorption occurs here. Hindgut, or rectum, is short, sinuously curved and narrow. It extends from posterior midgut to terminal anus.

**Digestive glands:** Two salivary glands are long, blind-ending tubes and extend posteriorly from head to about middle of the trunk. Two white bands of muscle layer are present on ventral to dorsoventral side. Lateral band is salivary gland. Each salivary gland is slender, tubular and elongated. Sometimes a salivary reservoir is found. Two salivary ducts join anteriorly to form a short common salivary duct which empties into preoral cavity. Salivary glands are the modified nephridia in the segment of oral papillae. They are located in the lateral perivisceral haemocoels. Lining of midgut is also produces digestive juices.

**Nutrition:** *Peripatus* is insectivorous and its food includes insects. A prey is captured by spurting of slime from slime gland. Prey is then held by lips and sucking action of pharynx. Jaws slit open the prey. Saliva is poured over the prey for digesting soft parts of prey. Partly digested material is passed posteriorly. Various juices from lining of midgut digest the food further. Digested materials are absorbed by absorptive cells in the midgut lining.

**Vascular system:** Heart is a dorsal, muscular, segmentally ostiate tube. Ostia are segmentally arranged and their number depends on the number of segments. Heart pumps blood from and to haemocoel. It extends from head to level of gonopore; remains enclosed in dorsal pericardial sinus and adheres to dorsal diaphragm. Heart is probably closed posteriorly, but opens anteriorly into head haemocoel. Arteries and aortae are probably absent. The flow of blood occurs through pericardial sinus, ostia, heart, perivisceral haemocoel and back to pericardial sinus. Blood contains haemocytes and nephrocytes and is colourless.

**Respiratory system:** This system is a network of trachea. It consists of abundant tufts of very fine (1–3 μm in diameter), unbranched tracheae extending from spiracles to tissues. Tracheal tubes are usually delicate, distally branched and possess faint transverse striations. Each metamere contains about 75 spiracles. Midgut, brain and muscles are well supplied by tracheae. Tiny spiracles are scattered over body surface and there may be as many as 75 per segment. Each spiracle opens into an atrium from which tracheae radiate to tissues. Spiracles are the most abundant on dorsal surface and cannot be closed rendering animal susceptible to desiccation in dry environments. Therefore velvet worms are restricted to humid habitats (Figure 13.36).

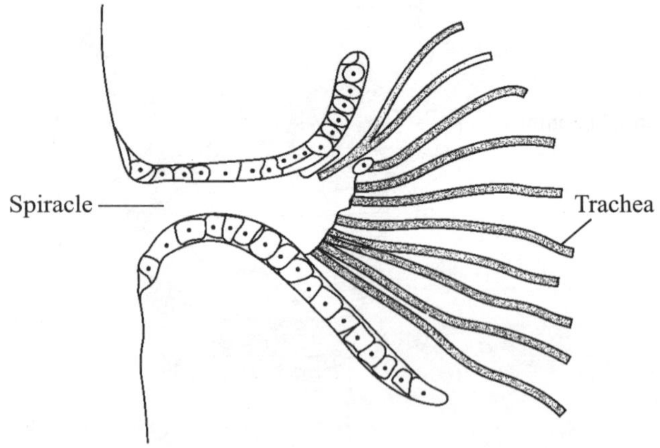


**Figure 13.36** Trachea with spiracle in *Peripatus*.

**Nervous system:** This system consists of brain and a pair of longitudinal nerve cords, peripheral nerves and sense organs. Brain consists of two large cerebral ganglia which are fused together on midline in dorsal head. A pair of circumoesophageal connectives unites brain with ventral nerve cord. Two large antennal nerves exit the brain anteriorly and extend into the antennae. Two large circumenteric connectives, one from each cerebral ganglion, exit the brain posteriorly and extend ventrally where each becomes a ventral nerve cord.

Near the middle of dorsal surface of brain, two short optic nerves exit the brain, extend to optic ganglion and pigmented retina lies under eyes. Mandibular nerves to mandibles exit the circumenteric connectives. The first pair of ganglia on ventral nerve cord is similar to oral papilla. Two ventral nerve cords are widely separated, each lies in a lateral haemocoel, as an extension of a circumenteric connective.

Nerve cords are ventral to dorsoventral muscles and is visible through this translucent sheet of muscles. Ganglia are inconspicuous obscure swellings. Each trunk segment has a pair of these obscure ganglia. Segmental nerves arise from each ganglion and extend to lobopod. Two ganglia of each segment are connected by up to 10 transverse commissures. Numerous sensory papillae are distributed all over body. Papilla consists of a seta and a group of sensory cells. Each eye is 0.3 mm in diametre, cup-like with a cuticular cornea and thick lens (Figure 13.37).

**Excretory system:** This system consists of nephridia, midgut lining and pericardial cells. A

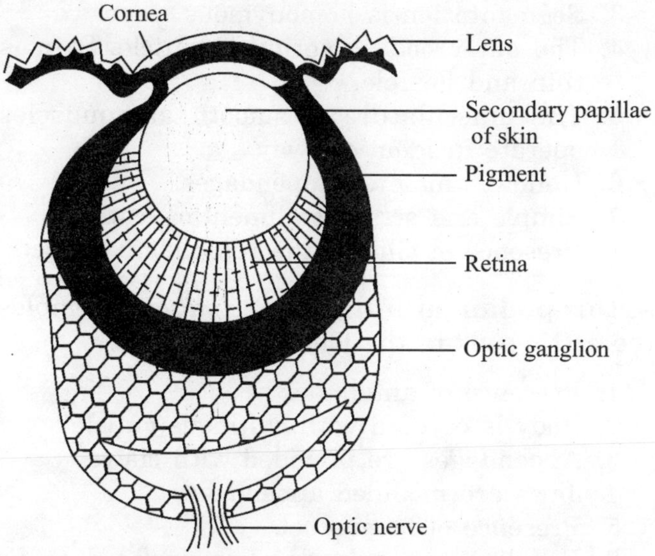


**Figure 13.37** L.S. of eye of *Peripatus*.

typical nephridium consists of an end sac and a tubule, derived from a metanephridium. Nephridia are situated in lateral perivisceral haemocoels immersed in blood, which is ultrafiltered across their walls to form primary urine.

Primary urine is modified by selective absorption and secretion as it passes through tubule to become final urine. Distal part of tubule may be expanded to form a bladder for storage of final urine. Nephridium opens through a nephridiopore. Nephridia of segments bearing legs 1–3 are reduced and smaller than all others. Nephridia of segments bearing legs 4 and 5 are much larger than other nephridia, lack bladder and have nephridiopores distally on leg. Certain cells in midgut lining collect uric acid crystals as excretory products. Pericardial cells serve as nephrocytes to remove excretory products from blood (Figures 13.38 and 13.39).

### *Reproductive system*

Both sexes are separate. Males are usually smaller than females. Sexual dimorphism is minimal except that females are larger than males and have more legs. They are gonochoric. Fertilization is internal. Sperm are transferred in spermatophores directly or indirectly. Reproductive systems of male and female are similar in consisting of paired gonads, paired ducts leaving the gonads to coalesce to form a single common duct that opens at the posterior gonopore.

Accessory genital glands are present in both sexes. Reproductive systems are found in median perivisceral haemocoel. Male reproductive organs

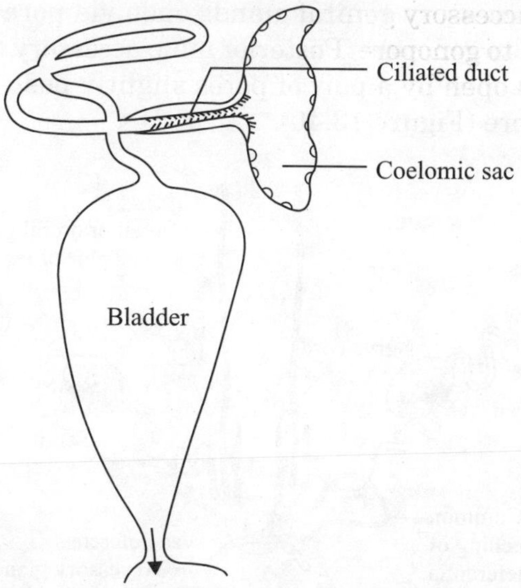


**Figure 13.38** Nephridium for the leg of *Peripatus*.

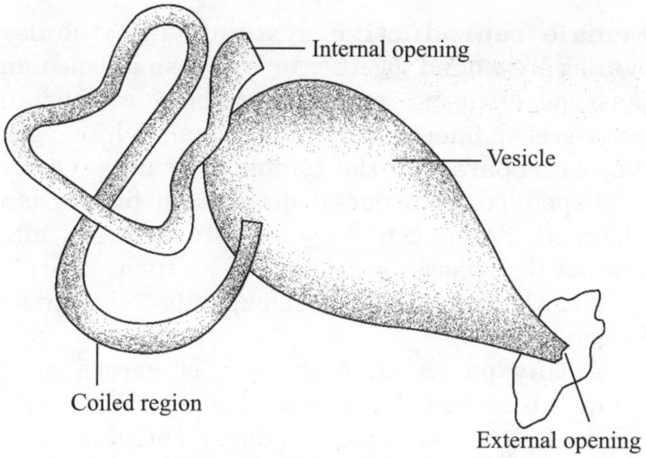


**Figure 13.39** Excretory organ of *Peripatus*.

are testes, vas deferens and vesicular seminalis. Females usually have a seminal receptacle and males have a seminal vesicle, both for storage of sperm, either allochthonous, or autochthonous, respectively. Species may be oviparous, or viviparous. Viviparous females, matrotrophic, or lecithotrophic, gestate embryos in the uterus for about a year.

**Male reproductive system:** This system consists of a pair of testes, a long convoluted vas efferens. Two vasa efferentia join posteriorly to form a single short vas deferens on left side of median haemocoel. Distally vas deferens becomes muscular, ejaculatory duct and opens through gonopore. Right vas efferens join left vas efferens to produce vas deferens. Ejaculatory duct which lies on left reach median gonopore. Two pairs of accessory genital glands, anterior and posterior are associated with male system and open via pores near gonopore. Anterior male accessory genital glands open via pores lying lateral to gonopore. Posterior male accessory genital glands open by a pair of pores slightly posterior to gonopore (Figure 13.40).

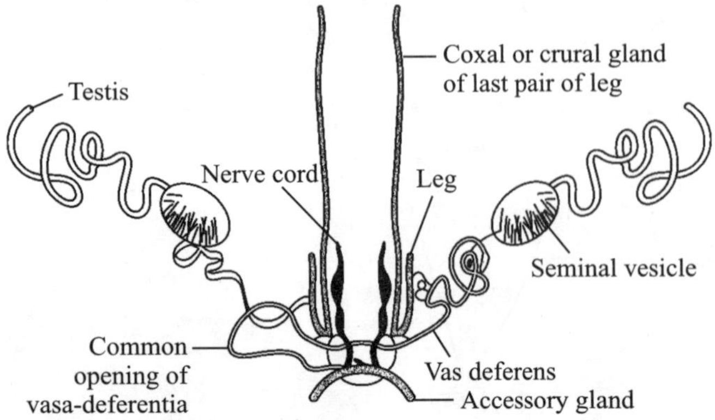


**Figure 13.40** Male reproductive organ.

**Female reproductive system:** Two tubular ovaries are wound together to form a single median mass near the posterior end of body in median perivisceral haemocoel. Small, oval, white eggs may be apparent in the combined ovaries. Ovary is suspended from dorsal diaphragm by ovarian ligament. From each ovary arises a slender white oviduct that passes anteriorly, then turns sharply posteriorly and immediately expands to become a large uterus.

A tiny pouch off each oviduct serves as a seminal receptacle for storage of sperm. Posteriorly uteri pass ventral to rectum, dorsoventral muscles and nerve cords before joining to form very short vagina, which opens at gonopore. Vagina is ventral to the rectum and hidden by it. Gonopore, also ventral to rectum and vagina, is on ventral midline between the last pair of legs, or between penultimate pair of legs.

## 13.13 AFFINITIES

The genus *Peripatus* was established in 1826 by Guilding who first obtained specimens of it from St. Vincent in the Antilles. *Peripatus* exhibits affinities with Annelida, Arthropoda and Mollusca as well as possesses peculiarities of its own.

**Annelidan affinities:** *Peripatus* resembles the annelids in the following features:

1. Vermiform body with truncated extremities.
2. True head is absent.
3. Segmentation is homonymous.
4. The outer body covering is cuticle. Skin is thin and flexible.
5. The musculature is smooth and muscles operate in identical way.
6. Hollow, unjointed appendages.
7. Simple and straight alimentary canal.
8. Presence of cilia in the reproductive tracts.

**Arthropodan affinities:** *Peripatus* resembles the arthropods in the following features:

1. Presence of antennae.
2. Body is covered with chitinous cuticle.
3. Appendages are provided with claws.
4. Jaws are modified appendages.
5. Presence of haemocoel.
6. Peculiar salivary glands, supposed to be modified nephridia.
7. Presence of lateral ostia as doors between heart and pericardial sinus.
8. Presence of tracheae as respiratory organs.

**Molluscan affinities:** *Peripatus* exhibits the following molluscan characteristics:

1. Slug like appearance of body.
2. Ladder like nervous system as found in Polyplacophora and Prosobranchia.

**Onychophoran affinities:** The following characteristic features are peculiar to Onychophora:

1. No, or indistinct segmentation.
2. Skin is not like that of arthropod.
3. Antennae are not homologous to that of arthropods.
4. Presence of slime and coxal glands.
5. Arrangement of tracheae is not arthropod like.
6. Spiracles in each segment.
7. Two ventral nerve cord are widely separated and without true ganglia.
8. Less complicated eye.

### *Taxonomic position*

Manton and other contemporary zoologists included onychophorans within Arthropoda as a subphylum, or Class. But the absence of exoskeleton and jointed limbs and primary annelidan characters in Onychophora present serious problems. It is likely that onychophorans are neither annelids nor arthropods but possess distinctive characteristics of their own. Onychophora are nowadays considered as a separate Phylum, not an evolutionary link between Annelida and Arthropoda, but a distinctive group of segment animals.

## 13.14 EXCRETORY SYSTEM OF PRAWN

Excretory system consists of the following:

- a pair of green gland, or antennary gland
- a pair of lateral excretory ducts and
- a renal sac.

### 13.14.1 Antennary Gland, or Green Gland

This gland lie at the coxal segment of antenna (it lies at the base of antenna) and hence called the *antennary gland*. It is pea, or gram seed shaped and generally opaque white. In *Astacus*, this gland appears green and hence called the *green gland*. Convention goes that all antennary glands are *green glands*. The antennary gland has three parts, namely, (a) End sac, (b) Labyrinth, or glandular plexus and (c) Urinary bladder.

***End sac***

End sac is a bean-shaped structure situated at the anterior side of labyrinth with blood lacuna inside. Lacuna joins with labyrinth through a single aperture. The end sac is a double walled structure. The outer layer is found of lacuna riched connective tissue and the inner wall is composed of excretory epithelial cells. The cytoplasm of these cells is granular with a large nucleus inside. Radially arranged septa prolongate from the wall to cavity of the end sac.

***Labyrinth, or glandular plexus***

This is a complicated structure formed by closely coiling and recoiling of glandular excretory ducts. It communicates with end sac by a single opening anteriorly and posteriorly by many opening into the urinary bladder. The glandular coiled and recoiled duct represents labyrinth, or glandular plexus. A single layer of glandular excretory endothelial cells lines the duct. The connective tissue lining the duct joins with one another to form plexus. This connective tissue is blood riched lacunae (Figures 13.41 and 13.42).

***Urinary bladder***

This is a thin walled sac, lies in the inner side of end sac and its outer wall remains against

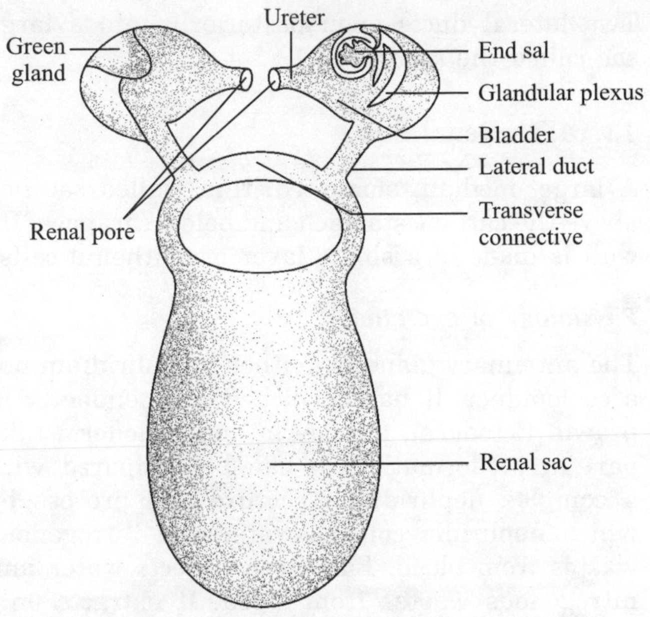


**Figure 13.41** Excretory organ of Prawn.

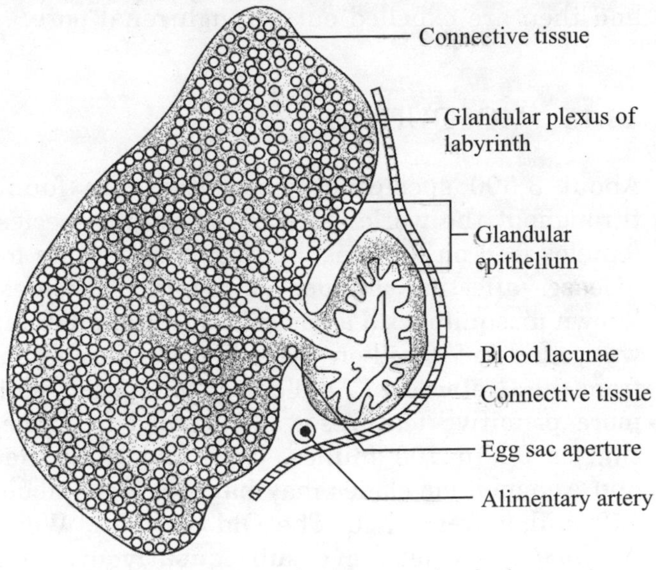


**Figure 13.42** T.S. of antennary gland of prawn.

the wall of end sac. Its inner wall forms a short duct called the *excretory duct*, or *ureter*, which opens outside through the pore called the *renal pore* which is situated on a papilla on inner side of coxal joint. The wall of bladder consists of a single layer of epithelial cells.

### 13.14.2 Lateral Excretory Ducts

A duct runs posteriorly and laterally from bladder of each antennary gland is called the *lateral excretory duct*. A transverse duct called the *transverse connective* connects both lateral ducts.

Two lateral ducts open posteriorly into a large sac called the *renal sac*.

### 13.14.3 Renal Sac

A large, median, elongated, thin walled sac lies above the cardiac stomach and below carapace. Its wall is made of a single layer of epithelial cells.

#### *Physiology of excretion*

The antennary gland is neither a nephridium nor a coelomduct. It has developed independently in prawn. Ectoderm, mesoderm and endoderm take part in its formation. It can be compared with a complex nephridial structure and process by which nephridia collects water and nitrogenous wastes from blood. Labyrinth collects water and nitrogenous wastes from blood. It extracts only ammonia from blood and other excretory products such as uric acid, etc. which restored in bladder and then are expelled out through renal pore.

## 13.15 MOSQUITOES

About 3,500 species of mosquitoes are found throughout the world. In some mosquito species, females feed on humans, and serve as vectors for diseases affecting millions of people. The oldest known mosquito with a basically modern anatomy was found in 79 million year old Canadian amber from the Cretaceous. An older sister species with more primitive features was recorded in amber that is 90 to 100 million years old. Culicinae and Anophelinae clades may have diverged about 150 million years ago. The Old and New World *Anopheles* species have subsequently diverged about 95 million years ago.

#### *Morphology of adult*

Mosquitoes are small with a clearly demarcated body and very long slender legs. Body is divided into head, thorax and abdomen. Head is specialized for sensory information and feeding. Head bears eyes and pair of long, many-segmented antennae. Antennae detects host's odours and odours of breeding sites. Auditory receptors in antennae detect whine of female. An elongated, forwardly-projecting proboscis is used for feeding. Maxillary palps of males are longer than their proboscis. In female, maxillary palps are much shorter but proboscis is elongated. Proboscis is used to collect blood. Compound eyes are separated from one another. New ommatidia are added in semicircular rows at the rear of eye. In the first phase of growth, individual ommatidia is square shaped, but later they become hexagonal. Thorax, abdomen and wings are often covered with scales. Differential colouration and pattern of scales helps in distinguishing species. Large wings are folded over segmented abdomen. Abdomen appears brown-black and slender but turns a bright red and swells following feeding.

Antennae of male are plumose. In females these are pilose. Male Anopheline palps are long and clubbed. Male Culicine palps are long but not clubbed. Female Anopheline palps are long. Female Culicine palps are short. At rest, Anophelines usually position their abdomens at an angle to surface. Culicines maintain their abdomen in a parallel position. Mosquitoes go through 4 stages in their life cycle—egg, larva, pupa and adult. Adult females lay their eggs in water. Mouthparts of female are adapted for piercing the skin of animals. Male typically feed on nectar and plant juices. Female needs blood meal for maturation of eggs. Females live longer than a month in captivity; most do not live more than 1–2 weeks in nature. Their life span depends on temperature, humidity, and their ability to get a blood meal while avoiding host defenses. Length of adult varies, but is rarely greater than 16 mm and weight up to 2.5 mg.

#### *Alimentary canal and feeding in mosquito*

Alimentary canal is composed of fore-, mid- and hindgut.

**Foregut**

1. Lined by chitinous material.
2. Subdivided into buccal cavity, pharynx and oesophagus.
3. Buccal cavity is formed by apposition of hypopharynx and labrum epipharynx. Hypopharynx forms the floor and labrum epipharynx forms the roof of buccal cavity.
4. Pharynx is a large, flask-shaped, muscular organ which acts as a suction pump.
5. Oesophagus joins the posterior end of pharynx with the anterior end of midgut.
6. Oesophagus is composed mainly of a single layer flattened epithelial cells. Below this layer, there is a thin chitinous intima, provided

with a set of fine spicules (cuticular teeth) at the anterior end projecting into lumen, the primary function of which is to rupture the blood corpuscles.

7. Oesophagus has 2 small dorsal and 1 large ventral diverticula, which contain air bubbles in the normal condition.
8. Large ventral diverticulum may have some role in buoyancy of the mosquito during long flight. It extends up to the abdomen. When the mosquito imbibes fruit juice, it first enters into diverticula, then into midgut. Excess amount may be accumulated in diverticula for future use.
9. Proventriculus is an invagination of the posterior end of oesophagus into commencement of midgut. It forms a marked annular thickening and acts as a valve regulating the entry as well as exit of blood meal.
10. Two salivary glands are present at level of first pair of legs in the thorax.

**Midgut**

1. Midgut, or mesenteron is an elongated, more, or less straight tube and consists of a single layer of epithelial cells with 2 portions.
2. First portion is tubular and lies for the most part of its length in the thorax, which is termed *cardia*.
3. At the outlet from the thorax, or a little behind this, it dilates to form an oval chamber about 3 times the breadth of thoracic portion. It terminates from the hindgut by a constriction.

**Hindgut**

1. Narrow tube, composed of ileum, colon and rectum.
2. Spherical rectum is provided with 6 rectal glands, or rectal papillae.
3. Hindgut wall is simple, composed of a single layer of regular cubical cells with a underlined chitinous intima.

**Peritrophic membrane**

1. Peritrophic membrane forms an inner tube extending from proventriculus to anus, which separates the content of gut from the epithelium.
2. It hangs down from the proventriculus like a tubular curtain.
3. Thin peritrophic membrane is composed of chitin alone with deposition of the mucous substance. Chitin is secreted from special type of large columnar cells situated at the junction of proventriculus. Mucous is secreted from cells of mesentron.
4. It is a sheet without aperture and therefore, the digestive fluid from the cells and digested products from food, must pass through it like an osmotic membrane.

***Feeding***

Mosquitoes are suctorial feeders. Piercing organ of adult female consists of 6 stylets—labrum-epipharynx, hypopharynx, paired mandibles and paired maxillae. Labium ensheaths piercing stylets, but does not penetrate the feeding puncture. Suction is provided by a cibarial pump. Mosquitoes force the slender piercing organ into skin, leaving the blunt labium at surface. Blood is taken directly into the food canal. Normal food of both sexes is nectar and plant juices. Female requires vertebrate blood for maturation of eggs.

A female mosquito sits on a host and presses its labella against the skin. Labella acts as a guide for piercing mandibles and maxillae, which are sunk into the flesh. Ensheathing labium bends back to allow the needles to go in. Labrum-epipharynx and hypopharynx together act as a tube forming a food canal, through which blood is sucked up from the wound. Suction is caused by pharynx by which blood comes into mouth. Thus mouthparts are for piercing and sucking. Mosquitoes have 3 oesophageal food reservoirs in addition to stomach. Reservoirs are used for storage of food, such as plant juices, but not for blood, which passes directly to stomach.

In host location, carbon dioxide and organic substances produced by the host, humidity, and optical recognition play important roles. In *Aedes* search for a host takes place in 2 phases. First, the mosquito exhibits a nonspecific searching behaviour until perception of host stimulants, and then it follows a targeted approach. Most species are crepuscular feeders. During heat of the day, mosquito rests in a cool place and wait for evenings. Asian tiger mosquitoes fly and feed during daytime. Prior to and during blood feeding, they inject saliva into bodies of their host. Saliva serves as an anticoagulant. Without it, female mosquito's proboscis would quickly become clogged with blood clots. Females hunt their blood host by detecting carbon dioxide from a distance. *Toxorhynchites* never drink blood.

### *Saliva*

Saliva contains a mixture of secreted proteins and negatively affects vascular constriction, blood clotting, platelet aggregation, immunity and creates inflammation. Hematophagous arthropod saliva contains at least 1 anticlotting, 1 antiplatelet and 1 vasodilatory substance. It also contains enzymes that aid in sugar feeding and antimicrobial agents to control bacterial growth in sugar meal. It contains fewer than 20 dominant proteins. Feeding mosquitoes modulate the immune response of the host (Figure 13.43).

### *Egg development and blood digestion*

After taking a blood meal, proteolytic enzymes hydrolyze blood proteins into free amino acids. In *A. stephensi*, trypsin activity is restricted entirely in posterior midgut lumen. No trypsin activity occurs before blood meal. Aminopeptidase is active in anterior- and posterior midgut regions. Aminopeptidase in anterior midgut is maintained at a constant low level, showing no significant variation with time after feeding. Alpha-glucosidase is active in anterior and posterior midguts. After blood meal ingestion, proteases become active only in posterior midgut. Trypsin is secreted into posterior midgut lumen without activation in posterior midgut epithelium. Aminopeptidase activity is also luminal in posterior midgut, but cellular aminopeptidases are required for peptide processing in both anterior- and posterior midguts. Alpha-glucosidase activity is elevated in posterior midgut after feeding in response to blood, whereas activity in anterior midgut is consistent with a nectar-processing role for this midgut region.

### *Mosquito borne disease*

The principal mosquito borne diseases are malaria, filariasis, viral diseases such as yellow fever, dengue fever and chikungunya, Japanese encephalitis. Mosquitoes are estimated to transmit disease to more than 700 million people annually in Africa, South America, Central America, Mexico and much of Asia with millions of resulting deaths. At least 2 million people are estimated to die annually because of malaria alone.

### *Control*

Many methods are used for mosquito control. Depending on situation, source reduction, biocontrol, larviciding, or adulticiding may be used to manage mosquito population. These techniques are accomplished by habitat modification, such as removing stagnant water and other breeding areas, pesticide such as DDT and natural predators.

Dragonfly nymph eats mosquitoes at all stages of development. Some cyclopoid copepods are predators on first instar larvae, killing up to 40

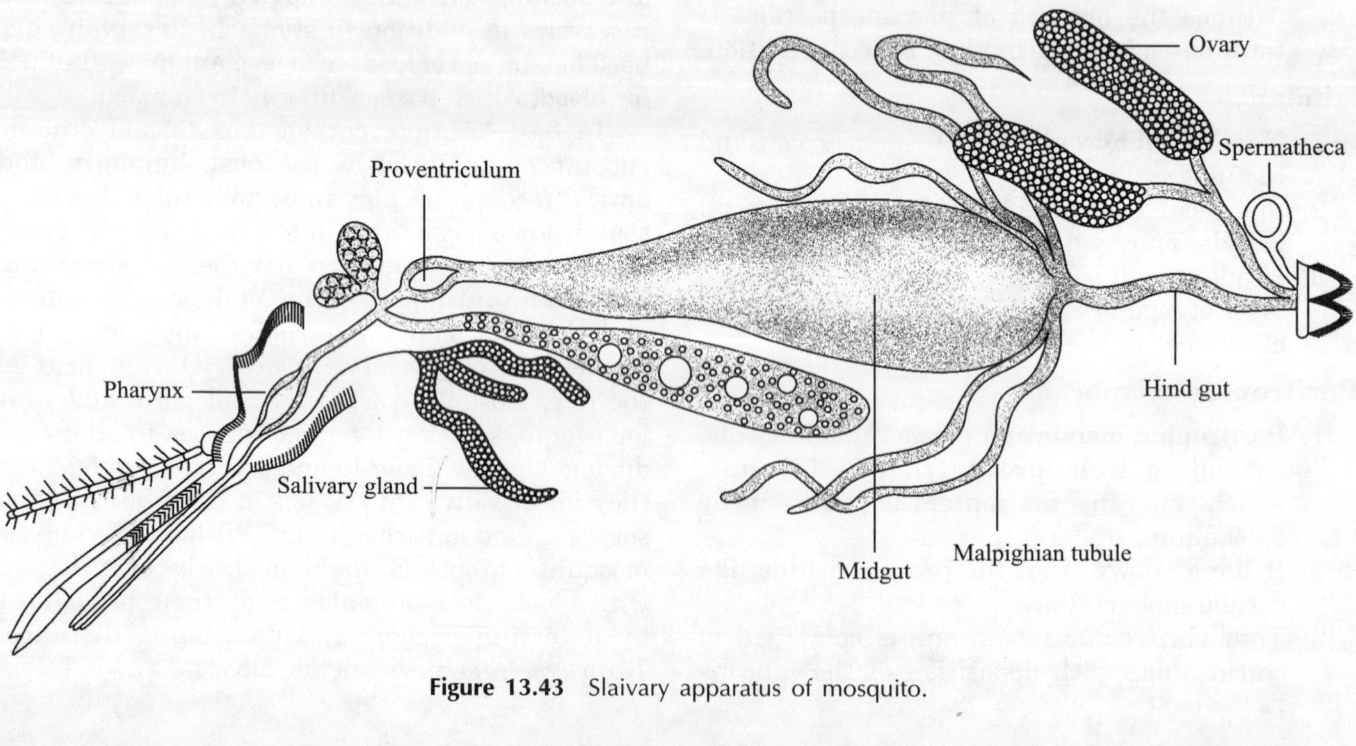


**Figure 13.43** Slaivary apparatus of mosquito.

*Aedes* larvae per day. Larval *Toxorhynchites* are known as *natural predators* of other Culicidae During its entire development, a *Toxorhynchites* larva consumes an equivalent of 5,000 larvae of the first instar (L1), or 300 fourth instar larvae (L4). Fish such as bass, bluegill, piranha, catfish, fathead minnows, goldfish, guppies and killifish consume mosquito larvae. *Bacillus thuringiensis israelensis* is used to control mosquito as a biological agent.

## 13.16 *Aedes*

Many species are native to tropical and subtropical regions and are active the entire year long. This is a domestic mosquito. They have adapted to cooler regions. In temperate regions, they hibernate over winter. Eggs from strains in temperate zones are more cold tolerant than ones from warmer regions. Long distances travel is carried out primarily through sea routes. Transport through personal vehicles, delivery trucks and trains are also important.

They are recognized by a marking in the form of a lyre on thorax, by patterns of black and silvery scales on abdomen and thorax. They lack spiracular bristles. Although female palpi are short, male palpi are longer than proboscis. They also lack pulvilli. Eighth abdominal segments of female in some cases remain hidden within the 7th segments. Legs have black and white rings along their length. Wings are generally covered with black scales. They breed in marshes and other wetland areas and are distinguished from Anopheline mosquitoes by black and slivery scales on abdomen and thorax. It may feed at any time, but bites humans between a few hours after dawn until an hour after sunset. Preferred breeding sites are stagnant water, but the dangerous areas are toilet tanks. They particularly lay eggs in water containers that have the correct amounts of specific fatty acids associated with bacteria involved in degradation of leaves and other organic matter in water.

### 13.16.1 *Aedes aegypti*

*Aedes aegypti*, the yellow fever mosquito, originated from Africa is now found worldwide. *Ae. Aegypti* spread dengue, chikungunya and yellow fever. Head has a distinct double white median line. Palps are black but white at the tip. First and second pair of legs is provided with white bands on the tarsi. Abdomen is dark with white bands on the bases of segments. Eggs are elongated, blackish with a highly chitinized shell. In larva, the antenna is bare except a single short hair on the shaft near the middle and a few hairs at the apex. Larva is also provided with long tracheal gills and short, broad siphon tube. In pupa, the respiratory trumpets are shorter and broader tin comparison to *Anopheles*.

#### *Life cycle*

Female lay eggs on damp areas such as mud, detritus, clay and rock. Eggs are very robust and survive desiccation and other environmental pressures. They generally float on the water surface. Eggs hatch depending on environmental cues generally within two days. Larva has a stout, barrel shaped siphon with one pair of subventral tufts. There are 3 pairs of setae on ventral brush. Large setae are absent on abdominal segments. Larva exhibits characteristic worm-like movement. Larval life generally lasts for 4 days. Larva transforms into pupa. Sunlight is not essential for development.

**Disease:** They are vectors of *Bancroftian filariasis* and arboviruses such as yellow fever and dengue. Yellow fever and dengue haemorrhagic fever are serious viral infections. Dengue is now the most important mosquito borne virus, with increasing global infection.

#### *Control and treatment*

Effective control is achieved by repellents and fine screening, or netting. Insecticides treatment also serves to reduce their population, but increased problems are encountered with Culicines, because they also feed during daytime. CDC traveller's page on preventing dengue fever suggests for using mosquito repellents that contain DEET (N, N-diethylmetatoluamide) (between 20% to 30% concentratione). Spray permethrin, or DEET repellents on clothing, as mosquitos may bite through thin clothing, may be an effective control measure. Use mosquito net is also necessary. Elimination of breeding places has also been emphasized in this connection.

#### *Genomics*

Effort in sequencing DNA was intended to provide new avenues for research into insecticides and possible genetic modification to prevent the spread of virus. *Aedes aegypti* was the second mosquito species to have its genome sequenced in full (the first is *Anopheles gambiae*). Published data included the 1.38 billion base pairs containing an estimated 15,419 protein encoding genes. Sequence indicates that the species diverged from *Drosophila melanogaster* about 250 million years ago, and that *A. gambiae* and this species diverged about 150 million years ago.

## 13.17 *Culex quinquefasciatus*

*Culex quinquefasciatus* (*Culex fatigans*) is the most common mosquito in our country. It is found everywhere in India up to 1930 metre altitude above sea level. It is a domestic species and remains alive up to 210 days. They can fly to a distance of 5 km from their breeding places. *C. fatigans* is a night biter. The life cycle is completed within 10–14 days. Breeding sites of *Culex fatigans* are ground pools, ditches, open drains, cisterns, any domestic collection of water, with high organic pollution. They take shelter in cowshed, in dark corners of dwelling houses; can hide behind dumped clothes and cupboards and also under fallen leaves in gardens. It has been observed to take rest by day on clothes and furnishings in houses as well as on walls and roofs. Adult is generally reluctant to feed in cage till 3, or 4 days have elapsed after its emergence. Mating takes place in captivity. It is most active at dusk and during first part of night. This species is unable to withstand wide variations of temperature and humidity. Gill observed that with a relative humidity of less than 40% *Culex* would not feed; between 40–48% *Culex* fed but if infected died within 5 days.

These mosquitoes are terrible pests of man. They are moderate sized mosquitoes. Male palpi are longer than proboscis and curved upwards at extremities. Female palpi are quite short. *Culex* is distinguished from all other mosquitoes by possession of distinct pulvilli, absence of colouration. Thorax, abdomen, legs and wings are often covered with brown-black scales giving a generally dark appearance. Abdomen may occasionally possesses white scales arranged in segments. *Culex* breeds mainly in aquatic habitats, often in areas containing large quantities of organic waste.

The following are the identifying characters of *C. fatigaus*

- Absence of a distinct pale band in the middle of proboscis
- Abdomen is provided with yellowish basal band
- Pleurae are pale brown and venter of abdomen is pale in colour.
- It is a medium sized, brown species

**Habitat:** Adults take shelter in cowsheds, dark corners of dwelling houses, behind dumped cloths and cupboards and also under fallen leaves in gardens. They take rest by day on cloths and furnishings in houses as well as on walls and roofs.

**Breeding behaviour:** *Culex* breeds in almost any kind of water. When it breeds in prolific numbers it chooses water contaminated with sewage, for example, cesspools, septic tanks, sewage farms, etc. They also breed in ground pools, cement pith drains. Maturation of ovum inside ovary takes at least 72 hours following blood meal. The earliest time eggs are deposited is at the end of four days after emergence. Oviposition takes place at night.

### *Life cycle*

Female lay dark brown, ovoid eggs in characteristic clumps of approximately 300 eggs. These eggs are often found in organic waste deposits, or polluted waters. Egg raft is boat-shaped and contain upwards of 200 eggs. Eggs are dark in colour, floats with the narrow end upwards. Broad end, which corresponds to the head of larva lies on the surface of water. Egg stage lasts for 24–36 hours. Larvae have a long and narrow siphon tube with more than one pair of subventral tufts. Antenna has a tuft of hairs at the junction of proximal two-thirds and distal third. Larvae undergo four moults and take 7 to 10 days to become mature and ultimately transform into pupa. Pupa takes about 2 days to become adults. Average life span of adults is two and a half- to three months.

**Disease:** *Aedes aegypti* is the first proved vector of yellow fever.

### *Control*

Prevention of accumulation of sewage contaminated water within a distance of 3 miles from the outer perimeter of locality is necessary. A properly planned drainage system should be installed. In ponds, ditches and other collections of stagnant water, pyrethrum extract should be added to the mixture of kerosene and crude oil and should be sprayed by a knapsack sprayer. Defective septic tank should be corrected. In industrial places, foul water may be allowed to flow into a marsh situated at a distance. Waste $H_2SO_4$ and other acids may be used for destroying the larvae.

Spraying the pyrethrum and kerosene in the house, or in cowsheds is perhaps the only effective way. Walls white washed with an emulsion of DDT is also effective. *Culex* serves as vector of *Bancroftian filariasis* and *Japanese encephalitis* throughout Africa. *Culex* prefers to bite at night and breed in an organic refuse. Improving sanitation and removing static water sources from the affected area can control *Culex*. Effective control can be achieved by repellents and fine screening, or netting. Insecticides reduce *Culex* population, but as Culicine mosquitoes also feed during the daytime, use of insecticide are not so effective. Biological control by releasing *carnivorous* fish, or other agents and genetic control through sterile male production and other forms of genetic manipulation were also attempted. Mosquito breeding place management is also necessary in this connection. Poor environmental sanitation creates favourable ecological conditions for mosquitoes. It is spreading with the spreads of indiscriminate urbanization. Chemical control is same as that of other vector insects.

## 13.18 *Anopheles*

*Anopheles* is characterized by the following features:

1. Dark and pale black scale are arranged on their wings.
2. Palps are of equal length to the proboscis, which appear terminally clubbed in males.
3. Free edge of scutellum is semilunar.
4. Abdomen never uniformly invested by broad overlapping scales.
5. Generally rest with head, thorax and abdomen being in a line.

*Anopheles* includes over 300 species and subspecies all over the world. Common malarial mosquitoes are *A. maculates, A. culicifacies* and *A. annularis*. They display many interesting behavioural adaptations in relation to ecology, feeding and breeding. They are commonly found in agricultural fields, bushes, forests, dirty places, cattle sheds and in unoccupied disused rooms. Species which rests in houses, cattle sheds and breed not very far from their habitation are referred to as *domestic species*. Wild species are found in jungle. Species which are intermediate in character enter houses to feed, and take shelter during the day in the forest. With certain exceptions they rest with proboscis, head and abdomen nearly in a straight line. Generally, they rest in cool dark places, viz., shady parts of jungle and convenient places in house and cattle sheds such as dark corners, underside of furniture and inner side of doors and windows. Some wild species found to take rest in vertical surfaces such as the bases of tree trunks, or banks of small streams. Some other hang beneath horizontal surfaces, such as provided by large leaves and some enter cervices.

They tend to move down wind, but there is no tendency for them to be carried far in this way. Species vary greatly in their range of flight. Males do not fly so far as females as the food of male are usually available in sufficient quantities in the vicinity of breeding places. Female travel longer distance in search of blood meal, generally within a quarter mile from their breeding sites, although exceptionally distances of several kilometres have been recorded. In addition to active dispersal, they frequently travel longer distance by trains, steamers and other vehicles that is termed *passive dispersal*. They cannot fly against a steady wind current, but are carried over by a strong breeze for longer distances. Some colours are said to attract them, while others do not. Dark, or navy blue is said to be the most favourite colour and yellow the least. Bright light do not attract *Anopheles*, the exception being *A. hycanus* and *A. annularis*. They generally avoid bright light. However, they do not like darkness. Possibly they prefer the dusk.

**Feeding habits:** Most species are nightbiters but some may bite in dull light. Females are bloodsucker. Males live on plant, or fruit juices. Weight of blood taken at a meal is about 1.0 mg. in *A. minimus* (Laurel, 1934). At the time of feeding, they hold their body at an angle from a given surface, the head is rotated to bring the ventral surface uppermost and contact with the food materials is made by maxillae and submentum. Particular odour of particular animal species (man, cow) attracts particular species and they only feed when environmental factors are favourable. They not only feed on human blood, their consumption of other animal blood is also reported. Some species are proved to have a distinct preference to cattle blood over human blood. This condition is termed *zoophilism*, better studied in Europe. This behaviour of Anopheles establishes a new dimension of control measure (Van Thiel et a1. 1939). Researches are being carried on for protection against mosquito bite through the presence of cattle which attract *Anopheles* and reduce the biting number on man. Interval between feeds mostly differs with species as well as temperature. Warmth stimulates biting and favours digestion. In hot weather in Assam, *A. vagus* digests blood in about 24 hours and *A. minimus* in 48 hours.

**Breeding behaviour:** They breed in a wide variety of places, most commonly in natural water such as in ponds, rice-fields, streams, rivers, nullahs, drains, tree-holes, pools and rarely breed in pots and other domestic containers. Algae, plankton and various microorganisms are consumed by larvae. Necessity of larval food appears to be a very important deciding factor in connection with their choice of breeding place. As soon as the adult mosquito emerges out of pupa, it rests for a while on pupal skin in order to dry its wings. Then it flies to a place of safety and mates. They have some peculiar habits. It is said that a mosquito hatching out of a particular breeding place returns to the same place for laying eggs and goes again to the same house for feeding.

### *Life cycle*

Holometabolous type of development is found. It passes through complete, or indirect metamorphosis. Life cycle stages include egg, its subsequent transformation to larva, pupa and finally the adult.

**Egg:** Female deposits eggs on water surface. It is difficult to detect eggs in water as they are laid separately and they are very tiny objects and appear like small specks of dust. Number of eggs laid at one time varies upon the species as in average 80 in *A. stephensi* and 118 in *A. minimus*. Eggs float on the surface of water, because floats contain air which gives buoyancy to egg mass. Frill surrounds the upper surface, characterised by striations and corrugations (Patton and Crag, 1984). Posterior end is narrower than anterior. Eggs increases in size due to absorption of water immediately after laying. Eggs are susceptible to desiccation upto 12 hours. Hatching normally occurs within 48 hours, although lower temperature delays hatching. Intraovarian life of eggs is varied, in Calcutta it is on average 4 days, while in Punjab it is 6 days.

**Larva:** Larva also called *wriggler emerges* by a spiral cut. This is usually an eruciform maggot without true legs. This stage is the longest in pre-adult life. It is not difficult to detect larvae in nature. They can be seen very easily with naked eye. First stage larva is very small; egg breaker is situated on posterior part of head. Second stage larva is relatively larger. Egg breaker is absent in this stage. Third stage is comparatively longer. Larva attains maximum size during fourth stage. Larva floats parallel to the surface of water and is characterised by its swift movement, absence of siphon-tube and respiratory opening on 8th abdominal segment. Palmate hairs arranged in pairs are present on many abdominal segments and sometimes on metathorax. Body of larva is divided into head, thorax and abdomen. Head is rather oval in shape. Two large feeding brushes help to carry the food into mouth. Being always near the water surface, it feeds on substances found on surface. Hence, it is said to be a top feeder. On either side of head there are antennae with antennal hairs. Thorax is divided into pro-, meso- and meta-thorax. The last portion is abdomen consisting of nine segments. Larva is a voracious eater and feeds freely on alga, or other organic matter present in water and even on dead insects. It does not feed for an hour before pupation. Under suitable environmental condition, this stage lasts for 8–9 days but may be extended to a month, or more.

**Pupa:** Pupa is the third stage of mosquito's life. Generally, pupation takes place at night. Comma-shaped pupa does not feed and is distinguished from others by hairs on abdomen, presence of more, or less sessile respiratory trumpet that is truncated at the end with a large opening. Accessory paddle hairs are also notable. Pupal stage generally lasts for 36–48 hours. When about to become an adult, pupa comes to rest, straightens out its abdomen, its skin splits and adult mosquito emerges.

**Adult:** Generally, adult emerges in evening, or at night, although emergence during day is recorded in cold climates. Head consists of proboscis, one pair of palps and one pair of antennae, which are together called the *head appendages*. Suctorial apparatus differ in male and female. In male, mouth parts are not well-developed, but developed in female which sucks blood. Antenna carries whorls of long hair in male, but is simple in female. This can be seen with naked eye. In adult, scutellum is evenly rounded, wing stem of median (M) fork-straight; legs are very long and slender with no distinct tibial bristles and no pulvilli. Abdomen is scaleless. Remarkable characteristics are pair of spotted wing and long female palpi which is about the same length as the proboscis (see Mandal, 2011).

**Life span:** Period from oviposition to final emergence varies with species, climatic condition, etc. In India, life cycle is generally completed within 9 days, while in Egypt it is 7 days for *A. gambiae*. It is reported that egg of *A. minimus* lasts for two days. Life span of females is generally greater than males. Life span of female of *A. culicifacies* varies from 8–34 days while that of the male of 8 days (vide Mandal and Nandi 1989).

**Disease:** *Anopheles* spreads malaria and filariaisis.

### *Control measures and management*

Antianopheline measures are widely used in antimalaria work. These measures may be grouped under the following headings:

**Elimination of breeding places:** It is an approach towards 'habitat management' that involves modification of aquatic habitats. It includes proper drainage of swampy areas, marshes, pools and ditches with stagnant water and cleaning of bushes and shrubs to make habitat unsuitable for the development of mosquito larvae.

**Chemical larvicides:** Certain chemicals such as DDT, DDD and benzene hexachloride have proved effective against mosquito larvae and pupae. But extensive uses of residual insecticides in public health and agriculture have created the problem of disturbance of ecological balance, environmental pollution and toxic hazards. Diesel, kerosene and mosquito larvicidal oil are also used.

**Genetical control:** Attempts to control anophelines by mass production of male and its sterilization by chemosterilants have been confined to *A. albimanus*. An initial small-scale field experiment in a 12 sq km area in the region of Lake Apastepeque in EL Salvador was carried out in 1972 using ENT-61585. This produced 99.8% sterility in males and 96.3% in females.

**Biological control:** Some fishes like *Haplochilus* and *Gambusia affinis* and *Poecilia reticulata eat* larvae and in certain conditions constitute a simple and fairly effective biological measure. It also includes the use of microbes that attacks mosquito larvae. Malaria, for its spread, always required a third agent, the transmitter, mosquitoes. It is, therefore, primarily an environmental and socioeconomic problem. Mass awareness, especially among the villagers, regarding modifications and management of aquatic habitat where mosquitoes breed, is essential to control the mosquitoes, preferably with less dependence on insecticide.

## GENERAL NOTES

## 13.19 RESPIRATION IN *Palaemon*

Respiration organs are bronchiostegites, or gill covers, epipodites and gills.

**Branchiostegites:** Each gill chamber is enclosed between bronchiostegite, or carapace on outer side and thoracic wall on the inner side. Gill chambers open on all sides except dorsally. Inner linings of branchiostegites are constantly bathed in fresh water, thus, forming the respiratory surfaces, which absorb oxygen dissolved in water and give out $CO_2$.

**Epipodites:** There pairs, or simple, leaf like and highly vascular epipodites are the outgrowths of coxal podomeres of 3 maxillipedes.

**Gills:** Eight gills lying inside each gill chamber, but only seven of them are exposed at once in the removal of gill cover as the eight lies hidden beneath the dorsal part of second gill.

The following test discusses the types of gills:

(i) **Podobranch, or foot gill:** When a gill is attached to the coxa of an appendage it is called *podobranch*. One podobranch remains attached to the coxa of each second maxilliped.

(ii) **Arthrobranch, or joint gill:** When a gill is attached to the arthroidal membrane connecting the appendage to the thorax it is called an *arthrobranch*. There are two arthrobranchs.

(iii) **Pleurobranch, or side gill:** When a gill is attached to the lateral wall of the segment having the limbs, it is called a *pleurobranch*. There are five pleurobranchs attached to the lateral walls of the thoracic segments bearing five walking legs.

**Structure of gill:** Gills are more, or less crescentic in shape. Each gill is larger than one in front of it. Dorsal end of each gill projects into antero-dorsal depression of gill chamber while its ventral end lies close to lower margin of gill cover. Each gill is attached in the middle of its length to thoracic wall by a small connection called the *gill root*. Nerves and blood vessels enter, or leave the gills through the gill roots. Here, gills are phyllobranch, i.e. each gill consists of two rows of leaf like rhomboidal gill plates arranged like the pages of book and lie at right angles to long axis, or the base of gill. Gill plates are largest in middle, but become gradually smaller towards two ends of the gill. Between two rows of gill plates runs a deep median longitudinal grove, which opens into gill chamber at both dorsal end and ventral end of gills [Figure 13.44(a)–(c)].

**Mechanisms of respiration:** Scaphognathite of each maxilla vibrates constantly causing a current of water in gill chamber. Fresh water enters gill chamber from behind in the form of a current along posterior and ventral margins of gill cover. Water then flows over gill and epipodites, and

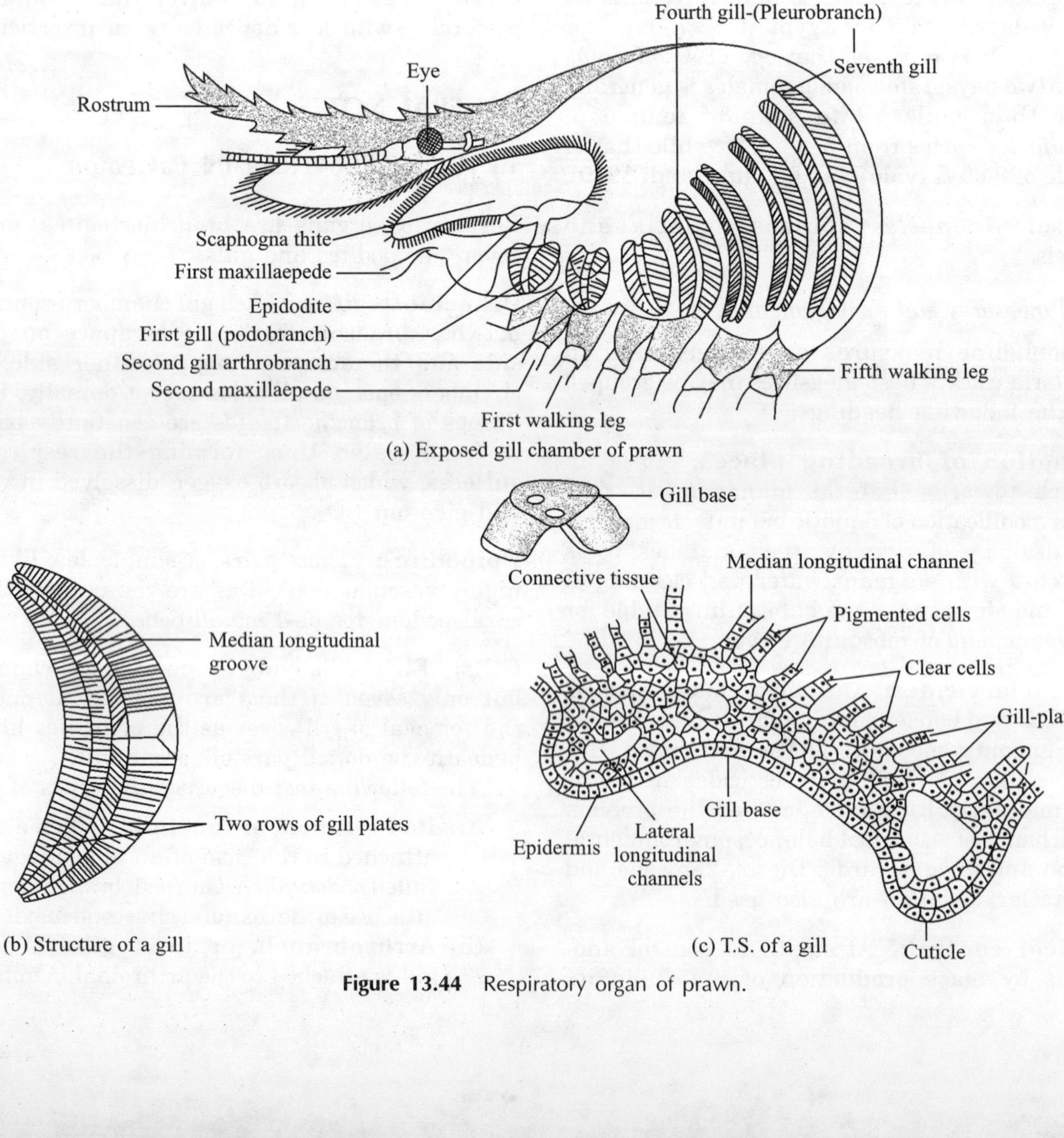


**Figure 13.44** Respiratory organ of prawn.

reaches an outer-dorsal depression of gill chamber, from where it is expelled out at the anterior end by the baling action of the scaphognathite. Gills and epipodites are richly supplied with blood and exchange of gases takes place on the gills surface and epipodites. Oxygen dissolved in water is taken in by blood and $CO_2$ diffuses out in water. Action of scaphognathite is supplemented by movements of the exopodites of maxillipeds.

## 13.20 RESPIRATION IN SCORPION

Four pairs of book lungs are situated on the ventro-lateral sides in the third, fourth, fifth and sixth mesosomal, or preabdominal segments. Book lung consists of a compressed sac-like cavity formed by invagination of the body wall and lined with a thin layer of cuticle, the opening of which is called *stigma*. Cavity of book lung is divisible into smaller proximal part called the *atrial chamber* which opens to outside by stigma, while the larger distal part is the pulmonary chamber. Each pulmonary chamber contains about 150 vertical folds, or leaf like lamellae. Lamellae are attached to inner posterior side of chamber and are arranged like pages of a book. Lamella is a hollow structure made of two thin layers of cuticle united at their edges. Outer side of lamellae bears ridges and bristles which keep adjacent lamellae apart and a little air space (inter-lamellar space) are left between them for the flow of air. Roof of atrial chamber is perforated by many linear slit like openings, set parallel with another. Atrial chamber communicates with inter-lamellar spaces containing air through these openings.

Blood flows continuously in spaces inside lamellae while inter-lamellar spaces are filled with air, so that exchange of gases takes place through thin walls of lamellae, which separate the blood contained in lamellar cavities from air in inter-lamellar spaces.

**Blood supply:** Impure blood from ventral sinus is sent to each book lung by a diverticulum from where it enters lamellae at their bases. Blood is aerated, or purified in lamellae of each book-lung and is collected into a pulmonary vein which travels dorsally and opens into pericardium.

**Mechanisms of respiration:** When the dorso-ventral muscles are contracted the pulmonary chamber is compressed and the air in between the lamellae is forced out into atrial chamber, then to outside through stigma due to contraction of atrial muscles. When the atrial muscles relax, book lung resume their normal shape and fresh air enters through stigma first into the atrial chamber and then into inter-lamellar spaces. At this time, dorso-ventral muscles relax and pulmonary chamber resumes its normal shape. Gaseous exchange takes place through the thin membranous wall of lamellae and blood becomes oxygenated and $CO_2$ is expelled out into the air.

## 13.21 LARVAL FORMS OF CRUSTACEANS

A larva is the immature form of an animal, which is unlike its parents and must pass through some sort of metamorphosis before assuming the characteristics of the adult. The primary purpose of the larval stage, or stages is to increase the possibility for the distribution, or dissemination of the species. Development of crustacean is either direct, or indirect. Development is direct in case of large eggs with enough yolk and indirect in case of small eggs with a little yolk. A few important larval forms are described as follows:

### *Nauplius larva*

1. Egg shaped, unsegmented.
2. Broad anterior end bears a median eye, large labrum and three paired appendages.
3. Median eye, or nauplius eye is made usually of three, but sometimes four ocelli.
4. Ocelli are pigmented cups with no lens and are innervated by protocerebrum.
5. Median eye may degenerate, or persists in the adult.
6. First pair of unjointed appendages is located in the head region.
7. Two groups of appendages are biramous and are present in the trunk region.
8. A stomodaeum with mouth, proctodaeum with anus, and a mid gut are also present.

**Comment:** In all crustaceans, nauplius is the first larval stage. In Cladocera, nauplius appears inside eggs. In Ostracoda nauplius, pair of bivalved shells and uniramous appendages are present. Alimentary canal is absent in *Sacculina* nauplius. In Cirripedia, a number of spines are present. In Isopoda, nauplius is maggot like.

### *Metanauplius larva*

1. Oval body is divided into cephalothorax and rudiments of abdomen.
2. Presence of a four pair of appendages in addition to the original three pairs of nauplius.
3. Two truncated processes at the posterior end are present.

This larval stage is found in *Triops*, *Penaeus*.

### *Cypris larva*

1. Body is enclosed in a bivalve shell.
2. A median and two compound eyes are present.
3. Anterior antennae are four jointed.
4. Posterior antennae are absent.
5. Six pairs of biramous, thoracic appendages are present.
6. Abdomen ends in a caudal fork.

This larva is found in *Sacculina* and barnacles.

### *Protozoea larva*

1. First antenna is four jointed.
2. Second antenna with three jointed endopodite and four jointed exopodite.
3. Mandibles are palpless, toothed and masticatory in function.
4. Maxillipeds are biramous.
5. Incompletly segmented abdomen.
6. Impression of paired eyes are visible.

This larva is found in *Penaeus*.

### *Zoea larva*

1. Body is divided into cephalothorax and abdomen.
2. Stalked compound eyes are prominent.
3. A dorsal, a frontal and two lateral spines are present in the head.
4. Six segments including a telson are present in the head.
5. Cephalothoracic appendages are seven pairs.

This larva is found in *Penaeus* and *crabs*.

### *Mysis larva*

1. Body is divisible into cephalothorax and six segmented abdomen with a telson.
2. Thoracic appendages are biramous.
3. A pair of stalked compound eye is present.

This larva is found in *Penaeus*.

### *Metazoea larva*

1. Body is divisible into cephalothorax and abdomen.
2. Well-developed thoracic limbs.
3. Biramous pleopods and a pair of telson are present.

This larva is found in *Corystes*.

### *Megalopa larva*

1. Well-developed cephalothorax.
2. Eyes are distinctly visible.
3. Five pairs of walking legs are present.

This larva is found in crabs.

### *Kentrogen larva*

1. Body sac-like, elongated.
2. Undifferentiated mass of cells is present inside the body.

This larva is found in *Sacculina*.

### *Epicaridian larva*

1. Two pairs of prominent antennae are present.
2. A pair of frontal processes and a pair of mandible are present.
3. Six pairs of limbs are present in the thorax.
4. Six pairs of biramous pleopods are present.

This is found in parasitic isopods.

### *Ericthus larva*

1. Body is divisible into cephalothorax and abdomen.
2. Three pairs of reduced anterior thoracic limbs are present.

This larva is found in stomatopods.

### *Alima larva*

1. Body parts are degenerated.
2. Body is similar to that of zoea larva.

This larva is found in *Squilla*.

### *Phyllosoma larva*

1. Transparent body with a round head and flattened leaf-like thorax.
2. Abdomen is short and narrow.

This larva is found in *Scyllarus* and *Palinurus*.

### *Significance of crustacean larva*

Primitive crustaceans pass only through nauplius stage. More evolved crustaceans begin with nauplius, but pass through several larval stages all of which resemble primitive adult crustaceans, i.e. cypris, mysis, etc. Appearance of appendages, eye, segmentation and other structural organizations advocates a distinct line of evolution of crustaceans. Thus, in spite of various limitations, it can be stated from life histories of crustaceans that an individual in the course of development passes through stages which resemble larval forms which appeared during its evolution. This generalization first made by Haeckel, through studies of developmental changes in various invertebrates and vertebrates, is known as *biogenetic law*, or *recapitulation theory*. One could scarcely take Haeckel's recapitulation theory seriously without eliminating larval adaptations from direct line of evolutionary descent. Only in those forms that show a direct development could ontogeny possibly recapitulate phylogeny (Figures 13.45–13–53).

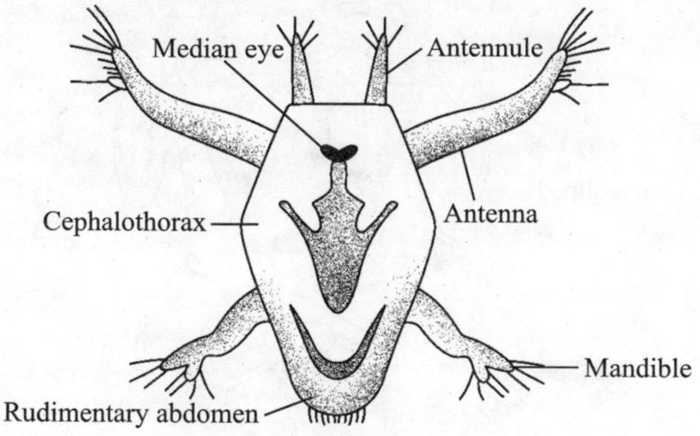


**Figure 13.45** Nauplius larva.

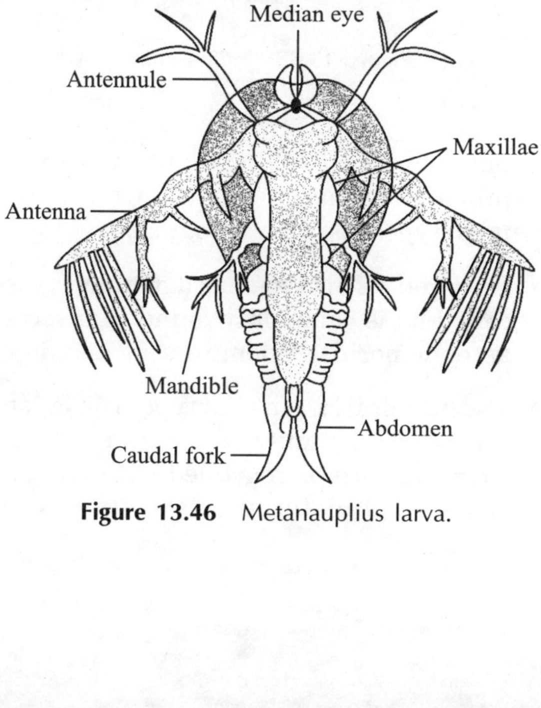


**Figure 13.46** Metanauplius larva.

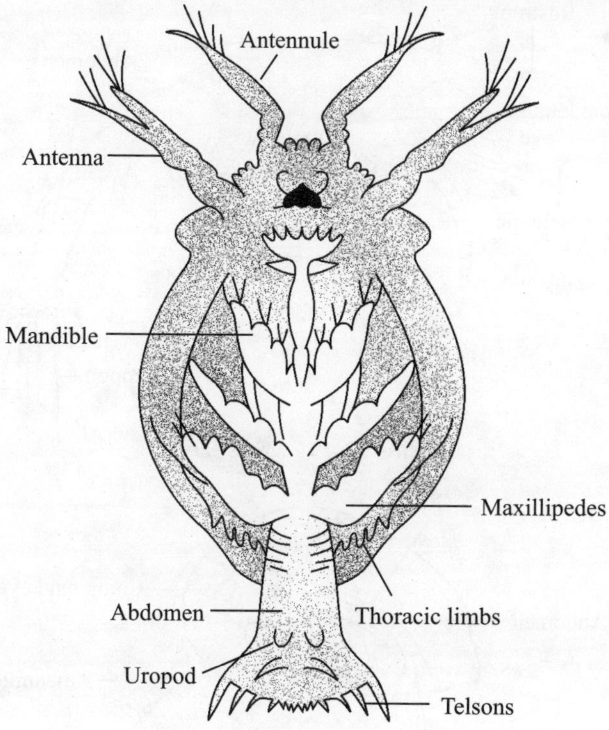


**Figure 13.47** Protozoea larva.

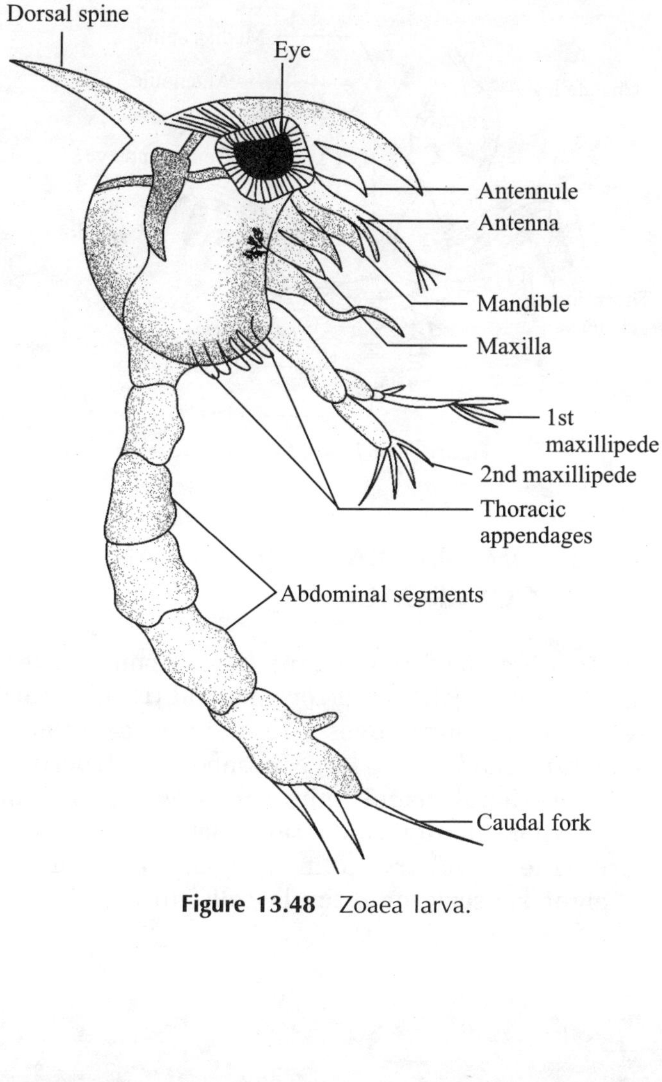


**Figure 13.48** Zoaea larva.

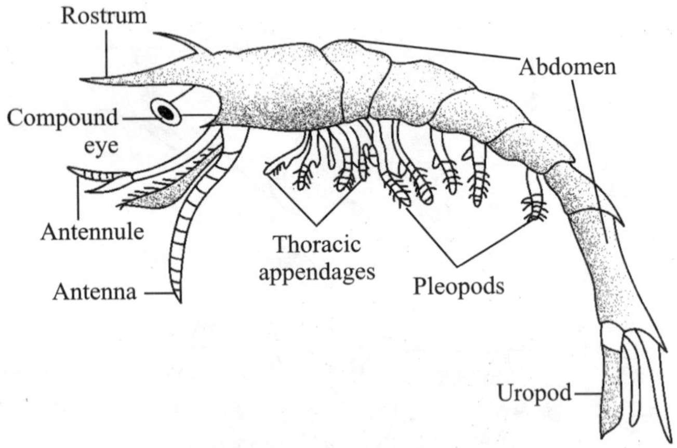


Figure 13.49 Mysis larva.

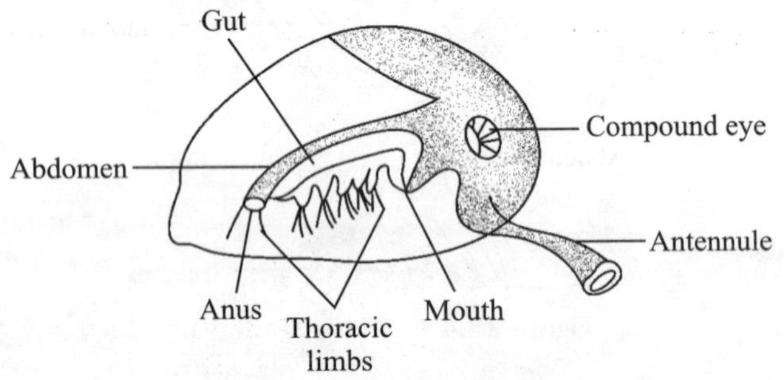


Figure 13.50 Cypris larva.

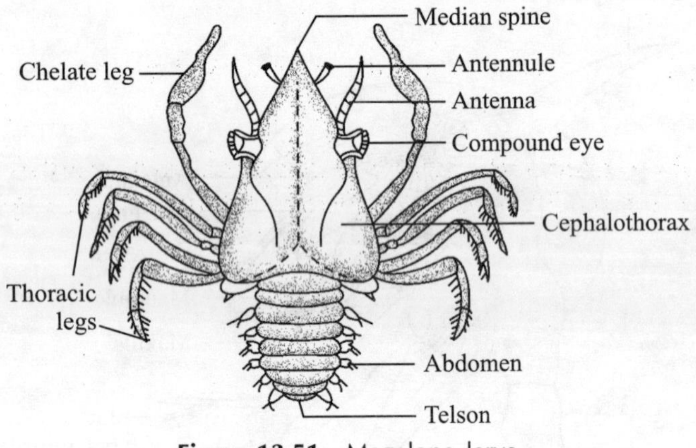


Figure 13.51 Megalopa larva.

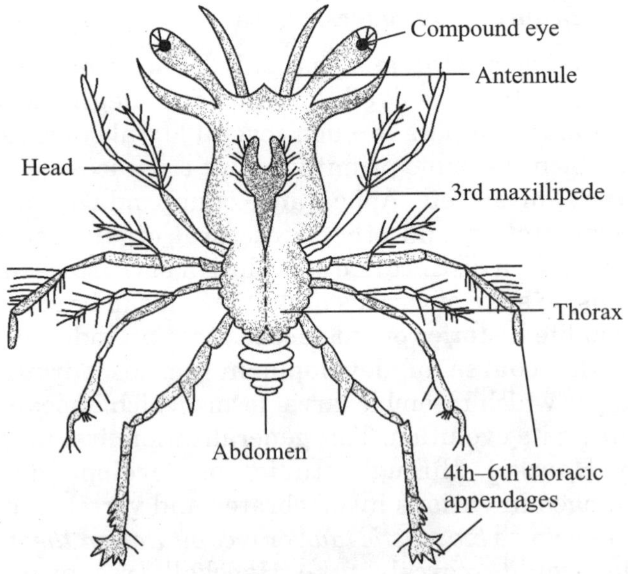


Figure 13.52 Phyllosoma larva.

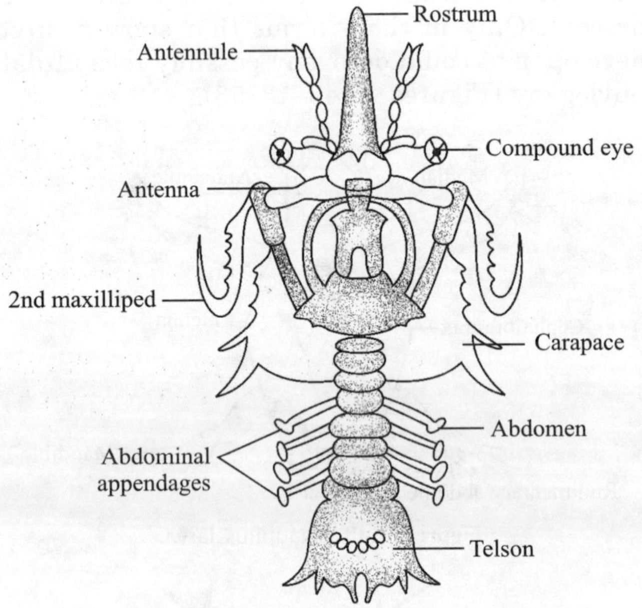


Figure 13.53 Alima larva.

## 13.22 OMMATIDIA IN PRAWN AND COCKROACH

Sense organs of prawn are (i) compound eyes, (ii) stotocysts, (iii) tactile organs and (iv) olfactory setae. Compound eyes are paired, composite structure made of a large number of structural and functional visual units called *ommatidia*, or *ocelli* lying radially. Ommatidia include two lenses, corneal lens and crystalline cone and a set of six, or eight sensory, or retinula cells, in addition to basement membrane. Compound eyes in insects show profound variations in structure. At least four major types may be recognized:

- **Eucone:** This type has a distinct crystalline cone and is commonly found in insects such as ephemerids, Odonata and Lepidoptera.
- **Pseudo cone:** In this type true crystalline cone is absent and is found in most Diptera. Cone cells are being filled with transparent semi-liquid material.

- **Acone:** This is found in Dermoptera, most Hemiptera and Coleoptera. The cone cells are elongated, transparent and serve as corneagen layer.
- **Exocone:** This type is found in some Coleoptera with a peculiar feature having deep cuticular in growth of cells beneath the cornea proper, in addition to the cone cells.

#### *Structure of ommatidium in Prawn*

- **Cornea:** Outermost transparent layer is divided into a large number of facets. Facets are thickened in the centre and behave like a lens and sheds off at the time of maturity.
- **Corneagen cells:** Each facet is followed by modified epidermal cells called the *corneagen cells*.
- **Cone cells, or vitrellae:** These are group of four much elongated cells situated beneath corneagen cells. These cells secrete and enclose a transparent and refractile crystalline cone which works like a second lens. Cornea, corneagen cells and cone cells together constitute dioptical region, whose function is to focus the light rays on the inner sensitive region.
- **Retinula cells:** Cone cells are followed by a group of seven elongated cells forming the proximal part of the axis of an ommatidium. They are elongated and provided with distally placed dilations having nuclei.
- **Rhobdome, or optic rod:** Elongated, spindle shaped and transversly striated body which is secreted and fully enclosed by the retinula cells.
- **Basement membrane:** Innermost layer that marks the internal boundary of ommatidia. Ommatidia are innervated by optic nerve fibres, coming from optic ganglia, through the fenestrae in the basement membrane.

  Retinula cells and the rhabdome constitute the receptor region. Its function is to receive the light rays focused by the dioptical region.

#### *Pigment sheath*

Each ommatidium is cut off from its neighbouring ommatidia by a sheath of dark pigment formed by the surrounding amoeboid chromatophores, which are arranged in two groups. Proximal group surrounding the rhabdome forms the retinal pigment sheath and distal group surrounding the crystalline cone forms iris pigment sheath. Amoeboid pigment cells take up different positions according to changes in intensity of light.

#### *Formation of apposition, or mosaic image*

In bright light, during day time, pigment cells spread in such a way that they completely separate optically one ommatidium from the adjacent ommatidia. Rays of light, which strike the cornea obliquely, are absorbed by the pigment cells, therefore they cannot produce a visual effect. Those rays of light that pass directly through the centre of cornea, travel through ommatidium and reach rhabdome to form an image of a part of an object. These small parts, placed together like the parts in a mosaic, form an image of entire object. This is known as *mosaic visions* in which rays are received simultaneously by distinctly separate visual elements, i.e. ommatidium and the image is made up of several compounds placed in juxtaposition. Such image is called an *apposition image*.

Sharpness of image depends upon the number of ommatidia involved and degree of their isolation from one another. A large number of ommatidia completely isolated from one another produce sharper image. An eye adapted for this type of image formation functions best at short distances only. It is therefore most of the arthropods are short sighted and usually night blind.

#### *Formation of superposition image*

In dim light, pigment cells migrate towards distal and basal parts of ommatidia and neighbouring ommatidia work in unison. In this condition, even the oblique rays of light are capable of forming a point of image, after passing through a number of ommatidia. As a result, an overlapping of adjacent points of image takes place and, thus, a continuous image called the *superposition image* is formed. The vision is not distinct, but the animal gets some sort of idea of its surrounding objects, especially of their movements. In insects such as fire flies and some moths, eyes are permanently set in the way that they are adapted for vision in the night light but they are day blind. It is probable that *Palaemon* can adjust its eyes to form both types of images according to intensity of light available (Figures 13.54 and 13.55).

***Structure of ommatidium in cockroach***

An ommatidium has a biconvex lens, or cornea, which is formed by the thickened and transparent cuticle. These form the facets. Below the lens, epidermis forms two clear corneagen cells, or lenticular cells, which secrete the lens. Below the corneagen cells, there is a transparent crystalline cone, which functions as a second lens. It is surrounded by four specialized cells. Vitrellae secrete crystalline cone and taper downwards. All these together form the focusing, or dioptical region. Below the cone and in contact with it is a spindle shaped refractive body, the rhabdome surrounded by seven photoreceptor retinular cells, or retinulae. Retinular cells secrete the rhabdome, which is made of seven rhabdomeres, one for each retinular cell. Rhabdome and retinulae constitute the receptor region and below it, there is a basement membrane. Each retinular cell joins a nerve fibre at its base and the fibres enter the optic nerves. Surrounding each ommatidium and separating it from its neighbours, in many insects there are heavily pigmented cells arranged in two groups, an iris pigment sheath around the cone and a retinal pigment sheath around the rhabdome and retinular cells. Retinal pigment sheath is absent in some insects. Only mosaic type of image formation occurs in cockroaches.

**Conclusion:** According to Villee and others, the term compound eye tends to imply that each ommatidium forms a separate image, but this is untrue. Seven retinula cells function as a single photoreceptor unit and together transmit a single signal. Image formed by the entire eye, although sometimes called the *mosaic* depends upon all of the signals transmitted by ommatidia and is really not different from that of other types of eyes. Compound eyes function differently in bright light than in weak light. In bright light, screening pigment is extended between ommatidium and each retinula responds only to light received by its facet and crystalline. Under these conditions, the eye is believed to be especially effective in detecting movement, as slight changes in the position of the moving object will stimulate different ommatidia.

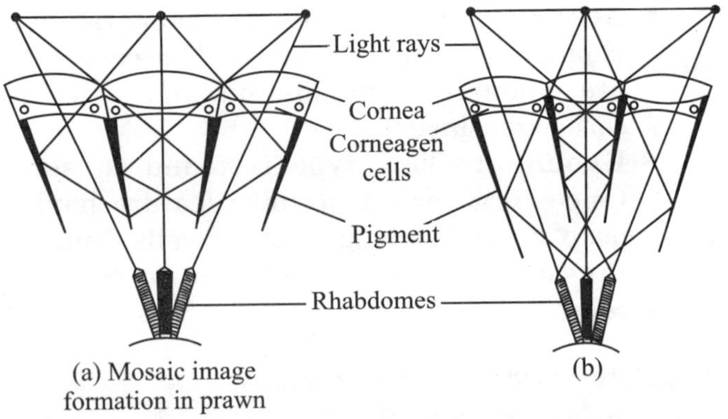


**Figure 13.54** Superposition image.

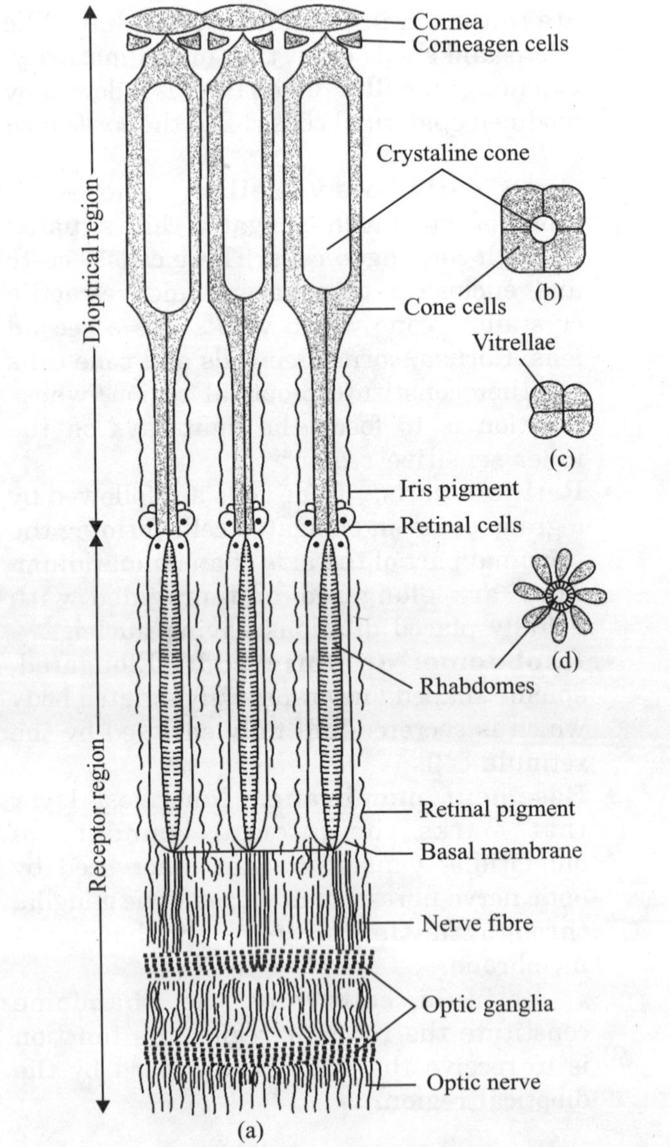


**Figure 13.55** Structure of ommatidia in prawn (a) Three ommatidia in L.S, (b) T.S. of an ommatidium through cone cells, (c) T.S. through basal ends of cone-cells and (d) T.S. through retinal cells and rhabdome

# REVIEW QUESTIONS

## Short Answer Questions

1. To which order the *Periplaneta* belongs?
2. How the male cockroach can be identified from the female?
3. What type of mouthparts is found in the cockroach?
4. Mention the number of segments by which the head capsules of cockroach is formed.
5. What type of body cavity is found in cockroach?
6. Which are the excretory organs in cockroach?
7. Name the respiratory organ of cockroach.
8. How many segments are present in the thorax of cockroach?
9. What is the number of spiracles in cockroach?
10. From where the absorption of the digested food takes place in cockroach.
11. Which is called tentorium of cockroach?
12. Through which organ the oxygen is carried to various tissues of the cockroach.
13. What is clypeus?
14. Where the stomodaeal valve of cockroach is situated?
15. What is elytra?
16. How many segments form the abdomen of cockroach?
17. Which segment bears anal style in male cockroach?
18. Mention the number of chamber in the heart of a cockroach.
19. How many ovarioles are present in the ovary of cockroach?
20. What is the function of colleterial gland?
21. In which form the nitrogenous wastes in cockroach are excreted?
22. In which sex of cockroach anal cerci are present?
23. What is corpora allata?
24. What is tergum?
25. What is vertex?
26. How many ommatidia are present in a single compound eye of *Periplaneta*?
27. How many spermthecae are found in cockroach?
28. How many hepatic caecae are found in cockroach?
29. Name the blood cells present in cockroach.
30. Which type of vision is produced in cockroach?
31. What is the smallest basic unit of a compound eye in cockroach?
32. Mention the various forms of crustacean larva?
33. Mention one character of blood of *Peripatus*?
34. Which glands are considered as excretory organ of *Peripatus*?
35. In which animal protozoea and zoea larva is found?
36. Comment on the parasitic castration.
37. Comment on the slime gland.
38. What is the significance of the term *Arthropoda*?
39. Give four characters of arthropods.
40. Give four characters of insects.
41. Give four characters of Acarina.
42. Give two differences between hard tick and soft tick.
43. Mention two differences between tick and mite.
44. Mention two differences between *Anopheles* and *Culex*.
45. Mention four characters of *Aedes*.
46. Define Diptera.
47. Define Siphonoptera.
48. Define Coleoptera.
49. Give the scientific name of sand fly.
50. Name the disease transmitted by *Simulium*.
51. Give the scientific name of house fly.
52. Name the diseases transmitted by *Culex*.
53. Give scientific name of two malaria vector.
54. What characters of arthropod distinguish it from all other animals?
55. Distinguish between complete and incomplete metamorphosis in insects.
56. Distinguish between open- and closed-circulatory systems
57. State two characters of Crustacea.
58. State a character of Apterygota.
59. What do you understand by mosaic vision?

## Long Answer Questions

1. Describe the life cycle of *Culex* with a suitable diagram. Add a note on its control.
2. Describe the life history of *Aedes* and add a note on its role in disease transmission.
3. Write a note on synanthropic insects.
4. Give an account of the structure of *Limulus*.
5. Discuss the affinities and systematic position of *Peripatus*.
6. Describe the anatomical peculiarities of *Peripatus*.
7. Describe the larval forms of crustaceans.
8. Give an account of respiration in cockroach with suitable diagrams.
9. Describe the mouth parts of cockroach with labelled diagrams.
10. Describe the excretory organ of Prawn.
11. Describe the structure of compound eye with suitable diagram.
12. Describe the circulatory system of cockroach with suitable diagram.
13. Describe the reproductive system of cockroach with suitable diagram.
14. Mention ten characteristics features of Insecta.
15. Mention ten characteristics features of Arthropoda.

CHAPTER 14

# Phylum Mollusca

The Mollusca ranks as second successful phylum, after the Arthropoda in terms of species numbers. Of about 110,000 known molluscan species, most are marine. They occupy a vast range of habitats, both aquatic and terrestrial, from the arctic seas to tropical streams, and from the valleys to mountainsides 7,000 m high. A few live in the deserts. Some are parasitic.

They range in size from almost microscopic to the largest of all invertebrates. The giant squid weighs 270 kg, measures up to 12 m in length with tentacles of about 50 m long. Generally, they secrete a long lasting beautiful shell and are believed to evolve from a flatworm like ancestor during the Precambrian, about 650 million years ago. Different classes of molluscs were predominant in the past.

Ammonites, a group of cephalopods was extremely abundant for millions of years before becoming extinct. The Nautiloid cephalopods were once very successful, but now only one species, the *Nautilus* exists. Humans have been deliberately culturing molluscs as food for a long time and eat many sorts of molluscs. Mollusc shells are used for decorations, or as a substance to carve into cameos and buttons. The oyster produces pearls. Slugs and snails are serious pests of crops in some places. Wooden ships and wharves can be destroyed by burrowing bivalves like shipworm.

## 14.1 GENERAL CHARACTERS

1. Bilaterally symmetrical, or secondarily asymmetrical.
2. Unsegmented and coelomate.
3. Coelom is limited to a small space.
4. Principal body cavity is a haemocoel.
5. Viscera concentrate dorsally as the 'visceral mass'.
6. Body is covered by thick epidermal skin and mantle.
7. Heart is composed of separate ventricle and atria.
8. Presence of large, well-defined muscular foot.
9. Buccal region is provided with radula.
10. Gut shows regional specialization including digestive caeca.
11. Large, complex metanephridia are present.
12. Most are the marine, some inhabit fresh water, or land.
13. Soft-bodied animal.
14. Generally, they are protected by a hard shell of calcium carbonate.
15. Some members have reduced, or lost their shells completely during evolution.
16. The mantle secretes the shell and creates a water-filled chamber, the mantle cavity with gills, anus and excretory pores.
17. Many feed by a strap like rasping organ, the radula, to scrape up food.
18. Most have separate sexes with gonads located in the visceral mass.
19. Many snails are hermaphrodites.
20. Life cycles of many molluscs involve a ciliated larva, the trochophore.
21. Cephalopods are the only molluscs with a closed circulatory system.
22. Shelled cephalopods called the *ammonites* were the dominant predators of the seas.

## 14.2 GENERAL FEATURES

Modern molluscs show various adaptability. Same basic anatomical characteristics are found in all, or most of them. The body is divided into the head-foot and the visceral lump. The head-foot is a muscular organ covered by cilia and rich in mucous cells. It tapers to tail at one end and bears head in the front end. The head bears mouth, eyes and tentacles. The last two may be reduced, or even absent. In shelled species, the head-foot can be drawn into the shell. Rest of the body is the visceral mass, which is entirely nonmuscular and contains the digestive- and reproductive organs. It includes the gonads, kidney, heart and digestive diverticulum.

Pallium remains attached to the dorsal surface of the visceral mass. Mantle hangs freely down the sides of pallium. The space between the mantle and visceral mass is called the *mantle cavity*, or *pallial cavity*, which generally contains gills, or ctenidia. A water current, generated by beating cilia, enters mantle cavity at the sides, passes over gills and departs centrally. Near the head, just behind mouth, there is a pair, or more of ganglia and a nerve ring from which two nerve chords arise that reach out through the body. Molluscs are truly coelomic.

### *Body wall*

Body wall comprises the cuticle, epidermis and muscles and composed of amino acids and sclerotized proteins (conchin) which is usually a single layer of cuboidal to columnar cells. Epidermis and outer muscle layer are separated by a basement membrane and sometimes by a dermis. Body wall usually includes an outer circular layer, a middle diagonal layer and an inner longitudinal layer of smooth muscles.

### *Mantle and mantle cavity*

Mantle forms dorsal body wall. Mantle cavity may be the shallow groove, or one, or two large chambers through which water is passed by ciliary, or muscular action. This cavity houses the gills, or ctenidia, receives the faecal material from the anus and products of the excretory- and reproductive systems. Mantle cavity of *Chiton* is a pair of long pallial grooves lying along the sides of foot. Water enters through the grooves from the front and sides, passing medially over ctenidia and then posteriorly between ctenidia and foot. After passing through the gonopores and nephridiopores, water exits the back end of the grooves and carries away the faecal material from the anus.

### *Shell*

Except Aplacophora, all molluscs have solid, calcareous shells. Shell glands produce shell in the mantle. In the Aplacophora, aragonite spicules, or plates are formed extracellularly. Shells vary in shape and size, but they adhere to the basic construction plan of calcium carbonate which is laid down in layers. Shells are covered by a thin organic surface coating, the periostracum which is composed of a type of conchin that is similar to that found in the epidermal cuticle. Calcium layers are generally of two types, an outer chalky prismatic portion and an inner pearly lamellar, or nacreous layer. The latter layer is lost in some groups. Both layers incorporate conchin in various ways to help bind the calcareous crystals together. Shells are often composed of varying numerous of calcareous sublayers (Figure 14.1).

### *Locomotion*

Aplacophorans lack a well-developed foot. Locomotion is primarily performed by slow ciliary gliding movements through the substratum. Chaetodermomorphans are mostly infaunal burrowers. Neomeniomorphans are mostly symbiotic on cnidarians. Except these two groups, most molluscs possess a distinct foot. The foot forms a flat, ventral, creeping sole. Sole is ciliated and imbued with numerous gland cells that produce a mucous trail. In gastropods, large pedal gland secretes substantial amounts of slime, especially in terrestrial species that glide on relatively dry surfaces. A few molluscs may move largely by ciliary propulsion. Most molluscs move primarily by waves of muscular contractions that sweep along the foot.

### *Feeding*

Two basic methods of feeding—herbivory, or predation (macrophagy) and suspension feeding are found in molluscs. Radula is usually a ribbon of recurved chitinous teeth, which may be simple, serrate, or pectinate. Radula removes food particles for ingestion. Radula of some primitive molluscs is believed to have originated in the early stage of evolution. In aplacophorans, the teeth, when

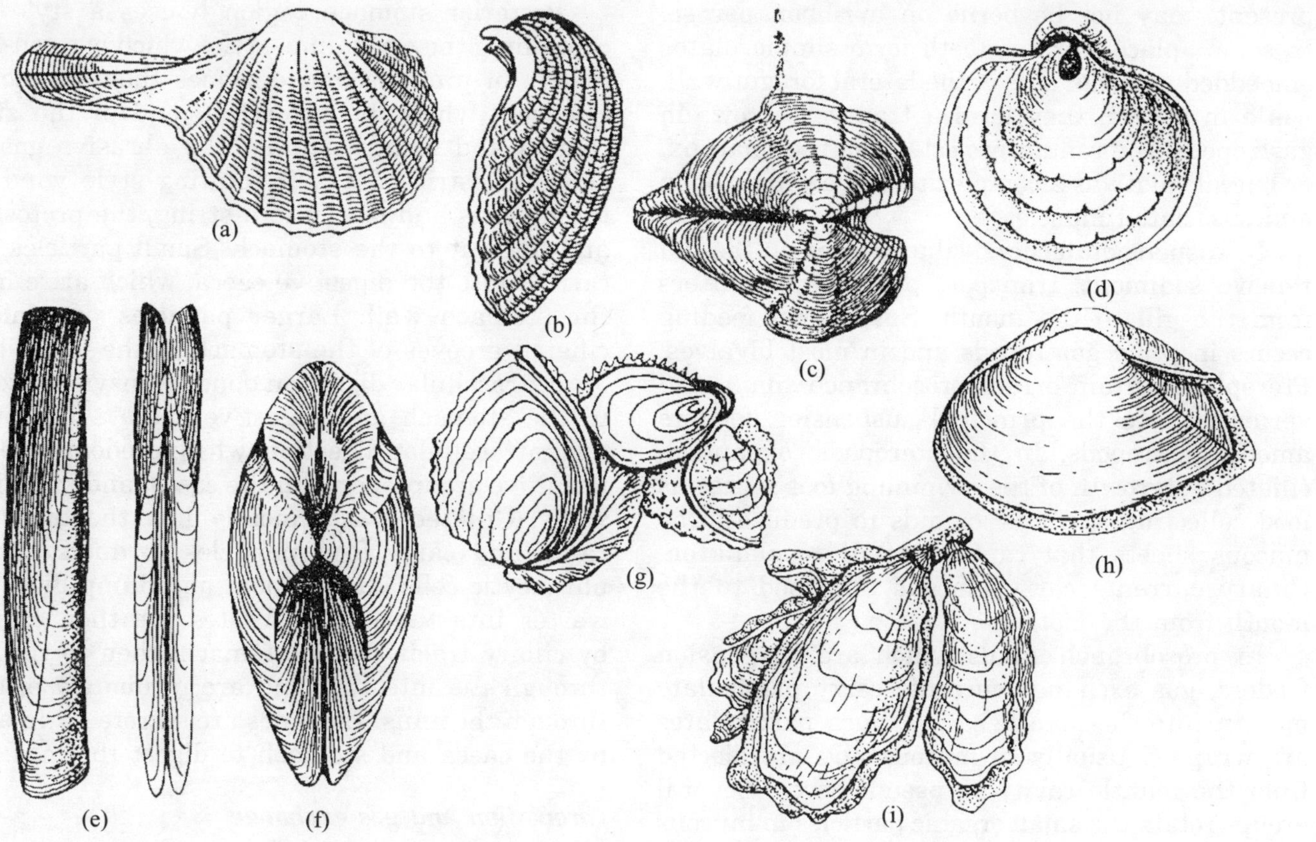


**Figure 14.1** Diversity of shells of lamellibranchs: (a) *Cardiomya* (b) *Pecten* (c) *Xylophaga* (d) *Anomia* (e) *Ensis* (f) *Leda* (g) *Chama* (h) *Corbula* (i) *Ostrea*. (Modified from Turner, in Shrock and Twenhofel).

## Mussels: The Potentially Useful Bio-monitors

**—Fatik Baran Mandal**

The mussels accumulate both essential and non-essential metals in high concentration along with other contaminants and are used as bio-monitors. Mussels belong to the Phylum Mollusca and Class Bivavia. Many properties of mussels are known. Now, it has acquired another importance for its role as bio-monitor, i.e. it can judge the water quality. Judging the water quality through living organisms (bio-monitors) have significance to the ecologists as well as to the agencies which are involved with water quality control.

Mussels, the bivalve molluscs are very interesting. They remove the suspended soil particles, algal blooms and metal ions like Cu, Zn, and Ni through their gills. Their mantle, kidney, foot and hepatopancreas are also anticipated sites of metal uptake.

The mussel can accumulate both essential and non-essential metals in high concentration along with contaminants like hydrophobic organic contaminants, polyaromatic hydrocarbons, metallothionein, and organochlorines.

The accumulation of contaminants/pollutants from the water is referred as bioconcentration which indicate the cumulative build up of nondegradable substances. The ability to absorb contaminants/pollutants/toxins makes mussel a potentially useful animal for monitoring water quality.

In the United States, *Dreissena polymorpha* is used to reduce suspected matter loads, toxins and organic pollutants. *Mytilus edulis*, another mussel has been traditionally used in the marine sector for monitoring work. The mussel, *Elliptio complanata* activity filter, concentrates, and retains faecal coliform bacteria.

Use of mussel in water monitoring practice has been discussed in a July–September 2008 issue of *Aquaculture Asia Magazine*.

*Article source*: http://www.bukias.com/articles/424767_mussels-the-potentially-useful-bio-monitors.html

present, may not be borne on a ribbon per se. In some aplacophorans, teeth form simple plates embedded in either side of the lateral foregut wall, while in others, they form a transverse row. In gastropods, the radula projects from the pharynx, or buccal cavity as a complex tooth-bearing ribbon and associated muscles.

In suspension feeders, cilia present on the gill remove sediment, transport particulate matters from the gill to the mouth. Suspension feeding occurs in some gastropods and in most bivalves. Pteropods, certain errant prosobranch snails and vermetids are the principal suspension feeders among gastropods. In the pteropods, expanded, ciliated outgrowth of the swimming foot serves as food collecting surfaces, or aids to produce large mucous sheets that capture small zooplankton. Ciliary currents carry mucous and food to the mouth from the foot.

In prosobranch snails which are suspension feeders, gas exchange currents carry particulate matter, into the mantle cavity. Such particulates are wrapped usually in mucous and are ejected from the mantle cavity as pseudofaeces. Several groups retain the small organic particles in mucous and carry them to mouth by ciliary currents. Radula in all suspension-feeding gastropods is reduced, serve only to pull mucous and food into the mouth. Suspension-feeding gastropods are usually sedentary and rely on generating water currents to bring the food.

### *Digestion*

Molluscs possess complete guts. Mouth leads inward to a buccal cavity, which contains radula. Oesophagus, a straight tube connects the foregut and stomach. Various glands are found in the gut region, including enzyme producing gland and the salivary glands. In herbivorous species, a muscular gizzard grinds up tough vegetable matters. Stomach bears one, or more ducts that lead to large glandular digestive caeca. Several sets of digestive caeca may be present.

Intestine leaves the stomach and ends in the anus. Once the food enters the buccal cavity, it is carried by the mucous strings into the oesophagus and then to stomach. In cephalopods, chunks of food are swallowed by muscular action of the oesophagus and are stored in the stomach. In many molluscs, the stomach wall bears a chitinous gastric shield and a ciliated ridged sorting area.

Posterior stomach region houses a style sac containing the crystalline style, which is a rod-like matrix of proteins and enzymes that are slowly released, when the projecting end of the style rotates and grinds against the abrasive gastric shield. Gastric cilia and rotating style wind up the mucous and food into a string, the protostyle and draw it to the stomach. Small particles are carried into the digestive caeca, which arise from the stomach wall. Larger particles pass along ciliated grooves of the stomach to the intestine.

Extracellular digestion done by enzymes occurs in the stomach and digestive caeca. Absorption and intracellular digestion, which predominates in primitive groups, occur in the caecal and intestinal walls. Ciliated tracts mostly line the digestive caeca and carry food particles to ducts, where phagocytic cells engulf them and dump digestive wastes into the ducts. Wastes are then carried by ciliary tracts to the stomach, then to the gut through the intestine and are ultimately ejected through the anus. Enzymes are secreted primarily by the caeca and stomach to digest the food.

### *Circulation and gas exchange*

The main body cavity is a reduced haemocoel. Blood contains various cells including amebocytes is called *haemolymph*. It picks up the digestive products and delivers these products throughout the body. It carries the copper-containing respiratory pigment haemocyanin, sometimes the haemoglobin and myoglobin. The heart lies dorsally within the pericardial chamber and comprises a pair of atria and a single ventricle.

In *Nautilus,* there are two pairs of atria. Atria receive the efferent ctenidial vessels, drawing the oxygenated haemolymph from each ctenidium and pass it into the muscular ventricle. Anterior artery branches and eventually opens into various sinuses within which the tissues are bathed in oxygenated haemolymph, return drainage through the sinuses and funnels the haemolymph. Most molluscs have true gills. When ctenidia are lost, they rely on secondarily derived gills, or on gas exchange across the mantle.

### *Excretion and osmoregulation*

Generally, paired tubular metanephridia often called *kidneys* are found. Nephridia are absent in aplacophorans. Three, six, or seven pairs of metanephridia occur in monoplacophorans, two pairs in the nautiloids and a single pair in other molluscs. Nephrostome opens in the pericardial coelom via a reno pericardial duct. The nephridiopore discharges the waste products into the mantle cavity.

In species with six pairs of nephridia, the first four pairs are associated with large pharyngeal pouches and the last two pairs may drain the pericardium. Typically, pericardial fluids pass through the nephrostome and nephridium, where selective resorption occurs until the final urine is ready to pass out through the nephridiopore. Pericardial sac and heart wall act as selective barriers between the open nephrostome and haemolymph in the surrounding haemocoel and in the heart. Nephridia are large, sac-like and their walls are much folded.

#### *Nervous system*

In the aplacophorans, monoplacophorans and polyplacophorans, ganglia are poorly developed. A simple nerve ring surrounds the oesophagus, often with small cerebral ganglia on either side. Each cerebral ganglion sends small nerves to buccal region and gives rise to the pedal nerve cords and visceral nerve cords. Other molluscs possess better developed three pairs of large ganglia. Two pairs, the cerebral and pleural ganglia, lie dorsal, or lateral to the oesophagus and one pair, the pedal ganglia lies ventral to the gut.

In cephalopods, bivalves and advanced gastropods, the cerebral and pleural ganglia are generally fused. From the cerebral ganglia, peripheral nerves innervate the tentacles, eyes, statocysts, general head surface and the buccal ganglia. Pleural ganglia give rise to the visceral cords, supply peripheral nerves to the viscera and mantle. Visceral cords join a pair of parietal ganglia and from there passes on to end in paired visceral ganglia. Parietal ganglia innervate the gills and osphradium. Visceral ganglia serve the organs of the visceral mass. Pedal ganglia give rise to a pair of pedal nerve cords that extend posteriorly and provide nerves to muscles of foot.

## 14.3 CLASSIFICATION WITH CHARACTERS

Two distinguishing characteristics are the presence of muscular foot for locomotion and presence of a protective calcareous shell.

#### *Characteristics*

1. Bilaterally symmetrical.
2. Soft bodied.
3. Body has more than two cell layers, tissues and organs.
4. Body is highly variable in form.
5. Possess dorsal, or lateral shells of protein and calcareous spicules.
6. Body is unsegmented with unjointed appendages.
7. A calcareous shell is present.
8. A mantle encloses the visceral mass.
9. Muscular foot is usually ventral in position.
10. Body cavity is haemocoel.
11. Feed a wide range of materials.
12. Live in most environments.
13. Radula bears minute horny teeth.
14. Heart is dorsal in position, usually lies in a pericardial sac.
15. Auricles 2, or one, ventricle 1, an anterior aorta and often other blood vessels are remarkable.
16. Respiration occurs by gills, or lung.
17. A pair of metanephridia is present.
18. A circum-oesophagal ring, ganglia and paired nerve chords are found in the nervous system.
19. The paired cerebral, pleural, pedal and visceral ganglia are joined by the connective nerves.
20. Both sexes are normally separate, some are hermaphrodite.
21. Fertilization can be external, or internal.
22. Development is direct, or with metamorphosis through the veliger larva.

Classification followed here is based on the scheme of Morton and Yonge (1964), which is accepted by Marshall and William (1972).

#### *Class Monoplacophora*

1. Thin shell is composed of periostracum, prismatic layer and a nacreous layer.
2. Five pairs of gills are present laterally to the foot.
3. Presence of preoral cephalic tentacles.
4. Circular foot is situated centrally.

**Example:** *Neopilina.*

#### *Class Amphineura*

1. Elongated and bilaterally symmetrical.
2. Mouth and anus are terminal.
3. Longitudinal, pallial and pedal cords are present with cross anastomoses.

***Subclass Polyplacophora***

1. Flattened with a broad, ventral foot.
2. Mantle bears eight transverse shell plates.
3. Ctenidia occur in numerous pairs.

**Example:** *Chiton.*

***Subclass Aplacophora***

1. Worm like.
2. Prominent foot is absent.
3. Spicules are present.

***Order Chaetodermomorpha***

1. Ventral foot groove is absent.
2. Cloaca contains the ctenidia.

**Example:** *Chaetoderma.*

***Order Neomeniomorpha***

1. Longitudinal ventral groove represents the foot.
2. Cloacal wall often contains a series of folds.
3. Ctenidial structures are absent.

**Example:** *Neomenia, Proneomenia.*

***Class Scaphopoda***

1. Burrowing, marine.
2. Shell open at both ends.

**Example:** *Dentaliun.*

***Class Gastropoda***

1. Head is well developed.
2. Foot is muscular and broad.
3. Visceropallium is enclosed in a shell of one piece.

***Subclass Prosobranchia***

1. Head bears a single pair of tentacles.
2. Shell is closed by an operculum.
3. Presence of two ctenidia.

***Order Archaeogastropoda***

1. Osphradium is poorly developed.
2. Two kidneys are present.

**Example:** *Patella, Haliotis*

***Order Neritacea***

1. Right renal organ is diminished, except for its duct.

**Example:** *Nerita, Theodoxus.*

***Order Mesogastropoda***

1. Osphradium is well developed.
2. Right renal organs remain as a genital duct.

**Example:** *Strombus, Cypraea.*

***Order Neogastropoda***

1. Osphradium is large.
2. Proboscis is eversible.

**Example:** *Murex, Terebra.*

***Subclass Opisthobranchia***

1. Reduction, or loss of shell.
2. Respiratory structures are secondary gills, never ctenidia.

***Order Cephalaspidea***

1. Moderately well-developed shell and mantle cavity are present.
2. Lateral parapodia are prominent.

**Example:** *Bulla, Acteon.*

***Order Anaspidea***

1. Shell is usually reduced and internal.
2. Mantle cavity is small.
3. Parapodia are prominent.

**Example:** *Aplysia, Akera.*

***Order Thecosmeta***

1. Parapodial fins are present.
2. Mantle cavity is well developed.

**Example:** *Limacina, Cavolina.*

***Order Gymnosomata***

1. Naked.
2. Small, ventral, parapodial fins are present.
3. Mantle cavity is absent.

**Example:** *Cliopsis, Pneumoderma.*

***Order Notaspidea***

1. Shell is external, or reduced and internal.
2. Usually slug-like and flattened.

**Example:** *Pleurobranchus.*

***Order Acochlidiacea***

1. Visceral sac is naked.
2. Presence of spicules.

**Example:** *Acochlidium.*

*Order Sacoglossa*

1. Buccal complex is suctorial.

**Example:** *Berthelinia.*

*Order Acoela*

1. Shell, mantle cavity and ctenidia are lost.
2. Respiratory branchiae are present in rows.

**Example:** *Doris.*

***Subclass Pulmonata***

1. Shell is spiral, sometimes lost.
2. Ctenidia are absent.

*Order Basommatophora*

1. Single pair of tentacles is present.
2. Eyes are present at the base of tentacles.

**Example:** *Planorbis, Physa.*

*Order Stylommatophora*

1. Two pairs of invaginable tentacles are present.
2. The posterior pair of tentacles bears eyes.

**Example:** *Arion, Limax.*

*Class Bivalvia*

1. Laterally compressed.
2. Possess a shell with two valves.
3. The head is rudimentary.
4. Radula is absent.
5. Foot is compressed.

***Subclass Protobranchia***

1. Gill filaments are not folded.
2. Foot bears flattened ventral surface.

**Example:** *Nucula, Solemya.*

***Subclass Lamellibranchia***

1. Gill filaments are folded.

*Order Taxodonta*

1. Gill filaments are free.
2. Hinge teeth are numerous.

**Example:** *Arca.*

*Order Anisomyaria*

1. Gills are usually filibranch.
2. No siphon.

**Example:** *Mytilus, Anomia.*

*Order Heteradonta*

1. Gills are eulamellibranch.
2. Heterodont type of hinge dentition is found.

**Example:** *Donax, Tridaena.*

*Order Schizodonta*

1. Hinge dentition.
2. Lack byssus structure in adults.

**Example:** *Unio.*

*Order Adapedonta*

1. Siphons are long and united.
2. Teeth are often absent.

**Example:** *Teredo, Pholas.*

*Order Anomalodesmata*

1. Mantle edges are fused.
2. Outer ctenidial lamella is reduced.

**Example:** *Lyonsia, Pandora.*

***Subclass Septibranchia***

1. Gills as a muscular septum are present.
2. Mantle edges are mostly free.

**Example:** *Poromya, Cuspidaria.*

*Class Cephalopoda*

1. Head is well developed.
2. A pair of highly developed eyes is present.
3. Sense organs are highly developed.

***Subclass Nautiloidea***

1. Shell is external and has many chambered.
2. Two pairs of ctenidia, osphradia, renal organs and auricles are present.
3. Ink gland is absent.

**Example:** *Nautilus.*

***Subclass Ammonoidea***

1. Shell is usually spiral and septate.
2. Sutures are complex.

**Example:** *Ceratites.*

***Subclass Coleoidea***

1. Shells are internal, or absent.
2. One pair of ctenidia, osphradia, auricles and renal organs is present.
3. Ink sac and duct are present.

### *Order Belemnoidea*

1. Shell is internal, chambered.
2. Posterior solid rostrum is present.

**Example:** *Belemnites.*

### *Order Sepioidea*

1. Eight arms, two tentacles and an internal shell are present
2. Proostracum is absent.

**Example:** *Sepia, Spirula.*

### *Order Teuthoidea*

1. Prostracum is elongated.
2. Eight arms, two tentacles and an internal shell are present

**Example:** *Loligo.*

### *Order Octopoda*

1. Arms are eight in number and are longer than body.
2. Internal shell is absent.

**Example:** *Octopus.*

### *Order Vampyromorpha*

1. Eight long arms remain united.

**Example:** *Vampyroteuthis* (Figures 14.2–14.5).

## 14.4 *Pila*

*Pila globosa*, or apple snail, one of the largest fresh water molluscs, is found in the ponds, pools, tanks, lakes, marshes, rice fields, streams and rivers, where a large quantity of vegetation such as *Vallisneria, Pistia* exists as their food. They are amphibious, i.e. adapted for life in the

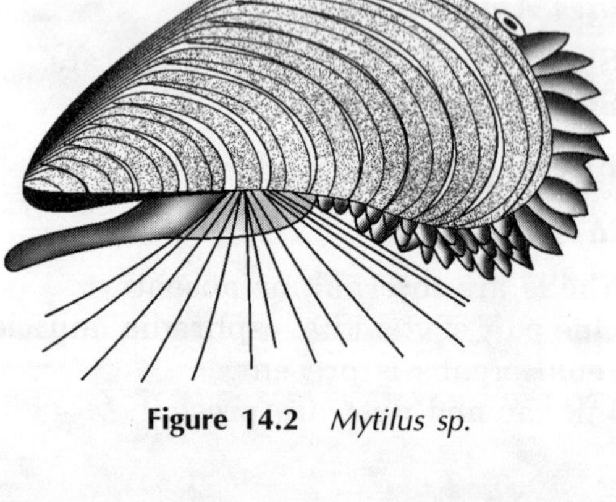

**Figure 14.2** *Mytilus sp.*

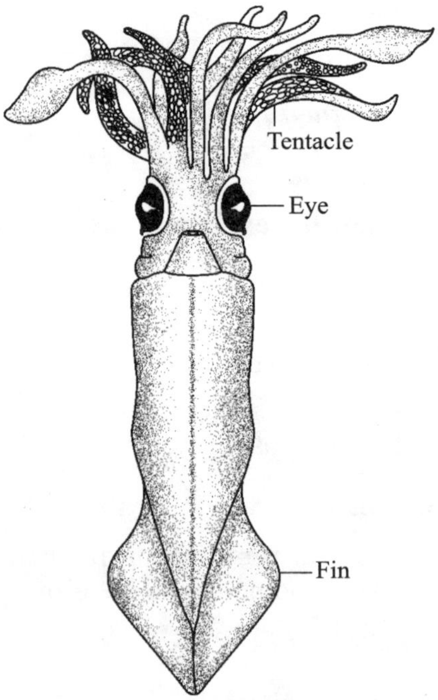


**Figure 14.3** *Sepia sp.*

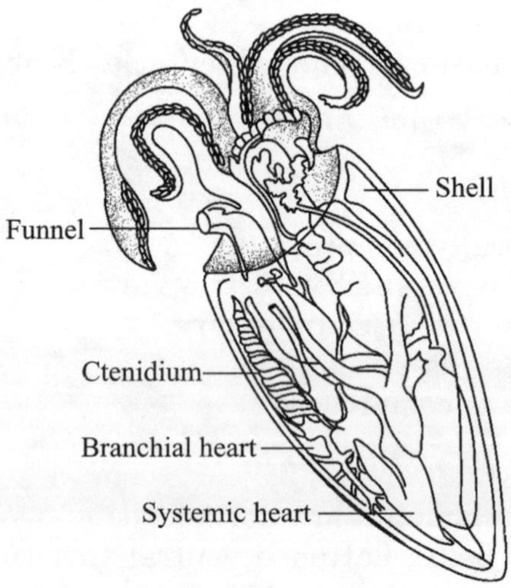


**Figure 14.4** *Loligo sp.*

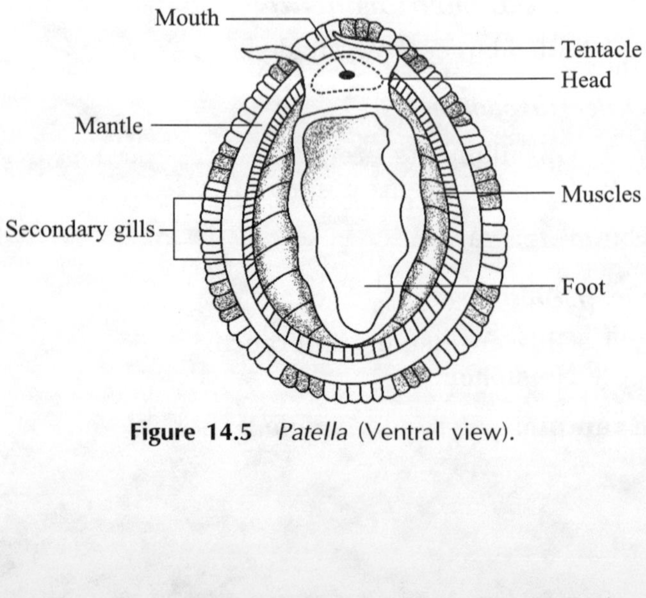


**Figure 14.5** *Patella* (Ventral view).

water and land, creep very slowly by the ventral muscular foot, covering about 5 cm per minute. During rainy season, *Pila* comes out of the ponds and travel long terrestrial routes. It overcomes the long periods of drought in a dormant condition buried in mud. This dormant state is called the *aestivation*, or *summer sleep*.

### External features

**Shell:** Shell is a univalve, coiled around a central axis in a right-handed spiral. Top of the shell is called the *apex* which is formed first and growth of the shell takes place from it. The apex contains the smallest and the oldest whorl. A spire below the apex consists of successively larger whorls followed by the penultimate whorl and largest whorl of the body whorl, which encloses most of the body. Lines between the whorls are called *sutures*.

Internally, all whorls are communicated with each other. Such a shell is called the *unilocular*. The body whorl has a large mouth, the margin of which is called the *peristome*, through which head and foot protrude. In ventral view with peristome facing to the observer, the mouth lies to right of the columella and the shell is spiral clockwise. Such coiling is called the *right handed*, or *dextral spiral*. The outer margin of the mouth is called the *inner*, or *columella lip*.

In the centre of shell, a vertical axis, or columella runs, around which the whorls are coiled. The hollow columella opens to exterior through the umbilicus. Shells with an umbilicus are called the *umbilicate*, or the *perforate*. Lines of growth of shell are visible; some appear as ridges called the *varices*. The shell varies in colour from the yellowish to brown, or even blackish.

**Operculum:** Fitting into mouth of the shell, there is a calcareous operculum. Its outer surface shows several rings of growth around a nucleus. The boss is cream coloured and remains surrounded by a groove. The operculum is secreted by the glandular cells of foot. The shell consists of an outermost pigmented layer, the periostracum that is made of horny organic conchiolin. Below this, a prismatic layer made of crystalline calcareous plates runs vertically. The innermost nacreous layer consists of crystalline calcareous plates which run longitudinally. The mantle secretes the shell.

**Body:** This consists of head, foot and visceral mass. In an expanded animal, the head and foot come out of the shell mouth, but the visceral mass lies inside the shell whorls. A columellar muscle arising from the foot is inserted in the columella. It attaches body to the shell, withdraws the animal inside and closes the operculum.

**Head:** A distinct head extends into a snout. The head bears the first pair of small tentacles, or labial plaps in the front. Behind them, there is a second pair of long tentacles. Tentacles are hollow, able to extend and contract. A pair of eyes borne on stalks, or ommatophores is present behind the tentacles.

**Foot:** A muscular foot is present below the head. Its lower surface is gray and flattened, triangular with apex pointing backwards. When the foot is withdrawn, the operculum closes mouth of the shell. A pedal mucous gland, which forms the slime during the locomotion, is found in the foot. The head and foot together constitute the head-foot-complex, which is connected to the visceral mass by an inconspicuous neck.

**Visceral mass:** A visceral mass containing the main organs is found above the head-foot-complex. This complex fills all whorls of shell, exhibits torsion and is spirally coiled. It is soft and grey to dark brown in colour.

**Mantle:** The mantle, also called *pallium* covers the visceral mass, forms a hood over the animal, when it is withdrawn. The mantle edge is thick, contains shell glands which secretes the shell. There is a supra-marginal groove above the thickened edge. Two fleshy lobes, called the *nuchal lobes*, or *pseudopodia* are joined on either side of the head, are found in mantle. The left pseudopodium is long and tubular. The respiratory siphon serves in aerial respiration and respiratory current enters through it. The right pseudopodium is less developed and not a regular tube. Respiratory current passes out through it.

### Mantle cavity, pallial complex and coelom

In the anterior part, a large space between the mantle and body, the mantle cavity has been shifted to the front by torsion. It encloses several organs, collectively known as the *pallial complex* and head can be withdrawn to it. A prominent ridge, or epitaenia runs backwards up to the end of mantle cavity near the right pseudopodium.

It divides the mantle cavity into a right branchial cavity and a left pulmonary sac.

In the branchial cavity, gill, or ctenidium, rectum, anus, genital aperture and anterior chamber of kidney (a reddish mass near posterior end of epitaenia) are found. Near the left pseudopodium, typical molluscan sense organ, the fleshy osphradium is found. The coelom includes the reduced to unpaired cavities of the pericardium, kidney and gonad. Renal and pericardial cavities communicate, but cavity of gonad is unconnected. Visceral organs are found in the haemocoel.

### *Digestive system*

Alimentary canal is distinguishable into three regions, foregut, or stomodaeum including buccal mass and oesophagus, midgut, or mesenteron consisting of stomach and intestine and hindgut, or proctodaeum comprising rectum. Midgut alone is lined by the endoderm. Foregut includes the mouth, buccal mass and oesophagus.

**Mouth:** This is a narrow vertical slit situated at the end of the snout. True lip is absent, but the plicate edges serve as the secondary lips.

**Buccal mass:** Mouth leads into a large cavity of buccal mass, or pharynx with thick walls and several sets of muscles. Anterior part of the cavity is called the vestibule. Two jaws hang from the roof of buccal mass behind the vestibule. Jaws bear the muscles and their anterior edges bear teeth like projections for cutting the vegetable food.

**Buccal cavity:** Behind jaws, a large buccal cavity exhibits a large elevation called the *odontophore* on its floor. The odontophore has protractor and retractor muscles, two pairs of cartilages, a pair of triangular superior cartilages which project into the buccal cavity and a pair of large S-shaped lateral cartilages. The front part of odontophore has a furrowed, subradular organ for cutting food.

**Radula:** Above and behind the odontophore, there is a bag like radular sac containing the transverse rows of cells called the *odontoblasts* which is believed to be a diverticulum of buccal cavity. Radula is present inside the sac. Radula is made of transverse rows of the horny teeth. Each row has seven teeth, two marginal, one lateral on each side, a central, or rachidian in the middle, thus giving a formula of 2, 1, 1, 1, 2 (Figure 14.6).

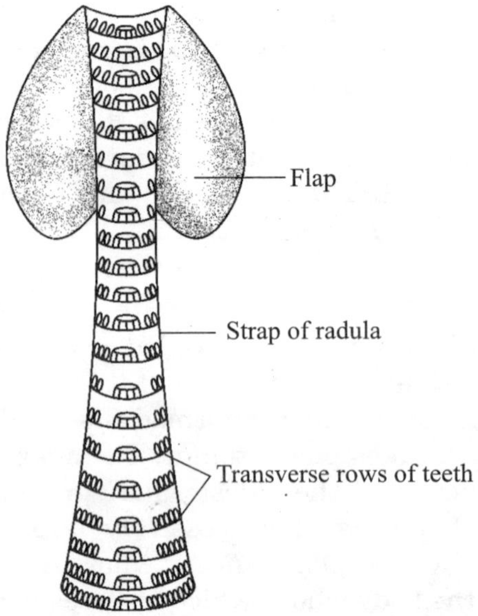

**Figure 14.6** Radula of *Pila*.

Radula moves forward and backward on odontophore, for rasping food particles. Such movements are called the *chain saw* movements. Teeth are made of chitin, which is reinforced by hardened protein. They have sharp cutting projections, which act like a file and rasp vegetable food. Teeth of radula are worn off in front and new teeth are formed all time by odontoblasts. On the roof of buccal cavity, there is a pair of grooved buccal glands.

**Oesophagus:** Buccal mass leads into long, narrow oesophagus. From near origin of oesophagus, a pair of round whitish oesophageal pouches, which are prolongations of oesophagus and probably secrete digestive enzymes, arises by short ducts and lie below the salivary glands. These pouches serve for temporary storage of food and digestion begins in them. Some extra cellular digestion takes place in stomach.

**Midgut:** This includes stomach and intestine.

**Stomach:** This begins on the left side, just below the pericardium and runs backwards as a blind pouch on postero-lateral sides of the main whorl of visceral mass. It is a rectangular sac of dark red colour, having a broad, U-shaped, internal cavity of rose red colour. Stomach is differentiated into cardiac- and pyloric chamber. Oesophagus opens into rounded cardiac chamber with longitudinal folds on its inner surface. A short bag like caecum without crystalline style arises from the pyloric chamber, which is tubular with transverse folds on its inner surface. The duct of digestive gland opens into stomach at junctions of its two chambers (Figure 14.7).

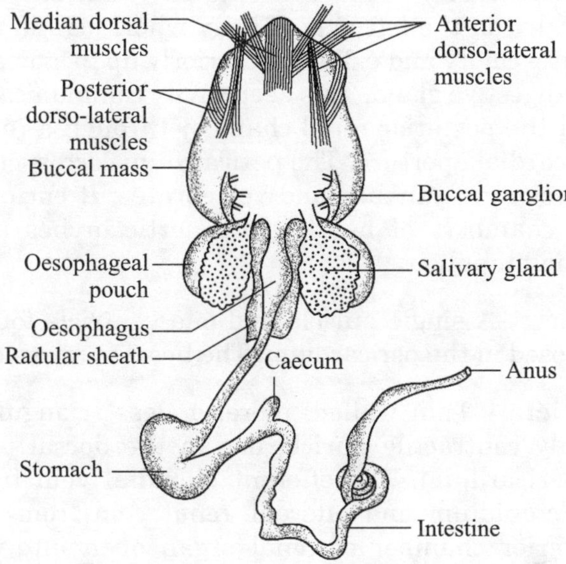


**Figure 14.7** Digestive system of *Pila*.

**Intestine:** This arises from the pyloric chamber, runs along its anterior edge, along the digestive gland beneath the posterior renal chamber and then turns upwards and backwards in visceral mass, where it forms the 2½, or 3 coils, between gonad in front and digestive gland behind, before joining rectum.

**Hindgut:** Rectum, the terminal part of alimentary canal is a thick walled tube that enters mantle cavity and passes downwards to open by an anus on the right of head.

**Salivary gland:** Two salivary glands, lying one on each side of the posterior limit of the buccal mass, partially cover the oesophagus. Surface and margins of each gland are greatly cut up, giving it the appearance of a branched type of gland. The duct of each gland, begins near its internal anterior corner and immediately enters the muscles of buccal mass and opens into the buccal cavity. The salivary secretion contains the mucous and enzymes for digesting the starch. The mucous lubricates the radula and helps in food transport.

**Digestive gland:** This gland, often called the *liver*, or *hepatopancreas*, is the triangular plate, or cone with a convex outer and more, or less flattened inner surface. The cone is spirally coiled from the tip inwards and downwards following the whorls of shell. When fresh, glands are brownish to dirty green in colour and soft. Two main ducts arise from two main lobes of digestive gland unite just before reaching the stomach to open into it, by a common aperture.

The digestive gland is made up of fine tubules, which unite with one another by the connective tissue to form the larger tubules. The terminal part of each tubule is glandular, called the *alveolus* and rest of the tubule is ciliated. The alveoli contain secretory, resorptive- and calcareous cells, or lime cells. Secretory cells produce a brown liquid, containing an enzyme, which dissolves cellulose of plants in stomach converting it into the pulp. Resorptive cells produce a proteolytic enzyme, which brings about the intracellular digestion of cellulose pulp. Calcareous cells store phosphate of lime.

**Food, feeding and digestion:** Food includes the succulent aquatic plants such as *Vallisneria*. Jaws and the odontophore cut *Pistia*. Radula moves forwards and backwards, making food into small particles. Food is masticated inside the buccal cavity. Salivary glands pour their secretion by their ducts into buccal cavity, where it mixes with food. It helps in digesting the starch. In stomach, the secretion of digestive gland digests food. Cellulose is digested inside the resorptive cells only. Thus, both extracellular and intracellular digestion occurs. Stomach is the site of extracellular digestion and digestive gland is the site of intracellular digestion and absorption. Absorption of digested food takes place mainly in the digestive gland and partly in intestine.

**Ctenidium, or gill of *Pila*:** Ctenidium, or gill is situated in dorso-lateral wall of the branchial chamber of the mouth cavity. Gill is composed of a long series of thin, triangular leaflets, or lamellae, lying parallel to each other, which are attached to the mantle wall by their broad bases, but have their apices hanging free in branchial chamber. The line of attachment of lamellae with the wall of mantle constitutes the ctenidial axis, which is provided with an afferent blood vessel and an efferent blood vessel.

Gill lamellae are the longest in size in middle and gradually become smaller towards two ends. Such gill is known as the *monopectinate gill*. Each lamella bears the transverse ridges, or pleats on both its anterior- and posterior surfaces. Pleats are low ridges which gradually decrease in size from base of the lamella to its apex. Each ridge contains branches of blood vessels. Each lamella has a smaller right side, which receives the blood from an afferent vessel and is called the *afferent side* and a longer left side from which the blood goes into the efferent vessel is called the *efferent side*. Histologically, each lamella appears to be a hollow cavity, lined on either side by epithelium containing the nonciliated columnar cells and a few glandular cells. A thin basement membrane follows the epithelial layer, below which are found the connective tissue cells, having scattered nuclei and oblique muscle fibres.

**Mechanism of respiration:** In aquatic respiration, a current of water enters the left nuchal lobe and first comes in contact with osphradium, which tests the nature of water. It enters the mantle cavity and passes over the epitaenia into branchial chamber to bathe ctenidium and then the current passes out through the right nuchal lobe. Ctenidium takes in $O_2$ from the water current and gives out $CO_2$, which diffuses into water (Figure 14.8).

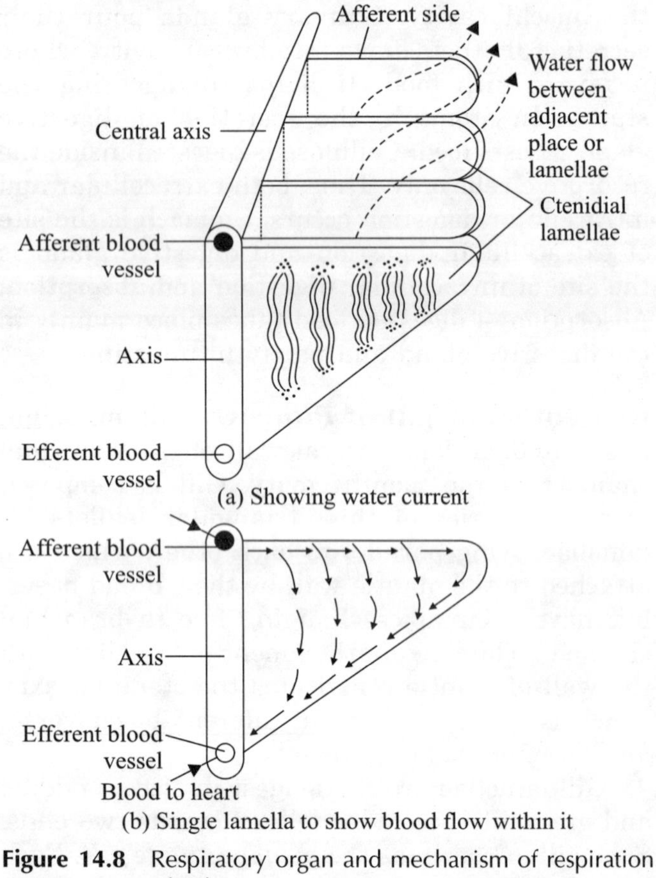


**Figure 14.8** Respiratory organ and mechanism of respiration of *Pila*.

### *Blood vascular system*

This system is open type, very much complicated, consists of the pericardium, heart, arteries, sinuses and veins.

**Pericardium:** This thin walled ovoid sac lies dorsally on the left side of body whorl behind the mantle cavity and extends anteriorly up to stomach and digestive gland. This deep cavity communicates with the posterior renal chamber through a reno-pericardial aperture. The pericardium corresponds to coelom of annelids and vertebrates. It encloses two chambers of heart, main aortic arches and aortic ampulla.

**Heart:** A single auricle and one ventricle found enclosed in the pericardium. The heart is myogenic.

**Auricle:** Thin walled, more, or less triangular, highly contractile auricle lies in the dorsal part of pericardium. The efferent ctenidial vein from the ctenidium and efferent renal vein from the posterior chamber of renal organ open into the apex of auricle, while pulmonary vein from the pulmonary sac opens at a slightly lower level at its anterior end. The auricle communicates with ventricle by an aperture, the auriculo-ventricular opening, having two semi-lunar valves. The valves are so arranged as to allow blood from auricle to flow into ventricle, but not in reverse direction.

**Ventricle:** Ventricle is ovoidal in shape, with thick spongy wall in the form of a meshwork, which greatly reduces the cavity of ventricle. Aortic trunk arises from the lower end of ventricle and divides immediately into two branches, the cephalic aorta and visceral aorta. Opening between the ventricle and aortic trunk is guarded by two semi-lunar valves, which prevent blood flow from back into the ventricle (Figure 14.9).

**Arteries:** From the ventricle arises an aorta, or aorta trunk, which divides into two branches, a cephalic aorta and a visceral aorta. Cephalic aorta has a bulbous outgrowth called the *aortic ampulla*, which helps in blood circulation and controls blood pressure. Opening of the aortic ampulla into cephalic aorta is devoid of valves. Cephalic aorta sends arteries to head and buccal mass. Visceral aorta forms the arteries and sends them to visceral mass.

**Cephalic aorta:** This aorta immediately beyond aortic ampulla gives out three arteries, first

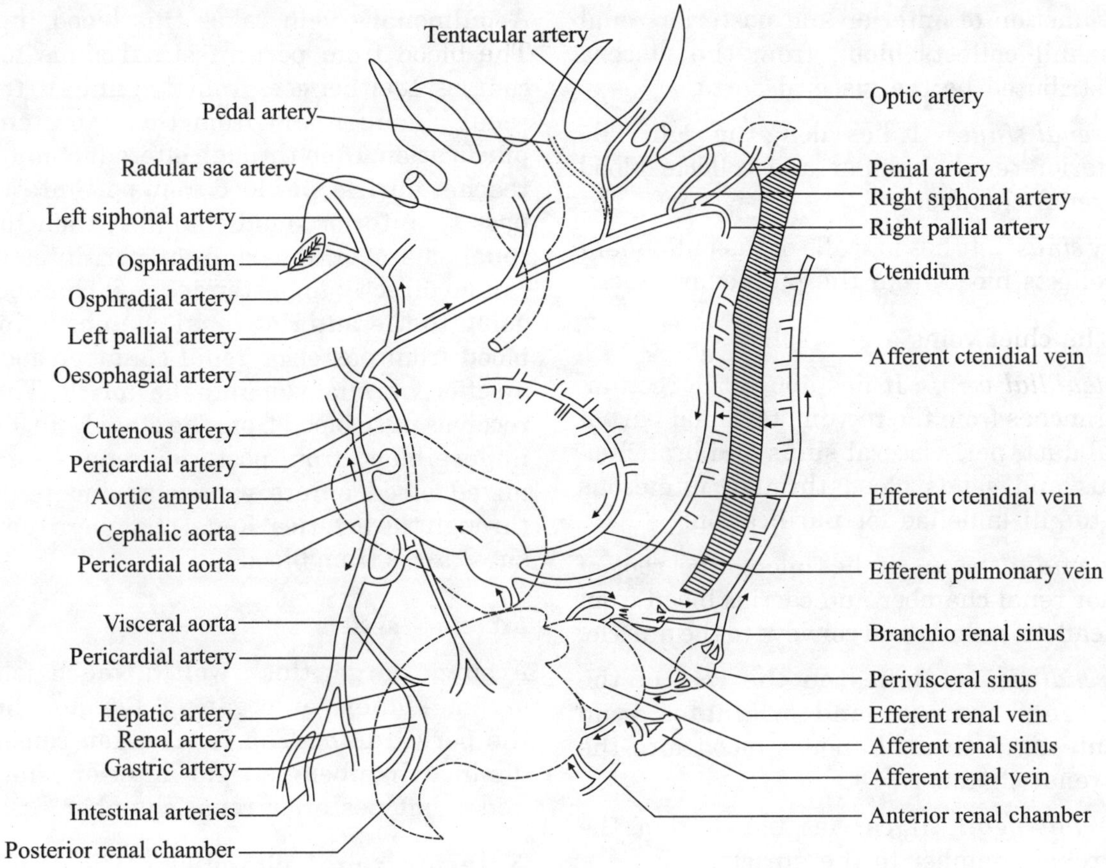


**Figure 14.9** *Pila:* Heart and blood-vascular system.

supplying to skin by the cutaneous artery, second supplying to oesophagus by the oesophageal artery and the third, being stout and thick supplies to left side of mantle (organs like left nuchal lobe and osphradium) by the left pallial artery. Cephalic aorta on its inner side gives out an artery supplying to pericardium, renal chamber and to a portion of genital organs—pericardial artery.

The main trunk of cephalic aorta then runs ahead and crosses over the oesophagus, to reach to its right side. It now gives off numerous small branches to oesophageal area and large branch to right side, which finally gives out a right pallial artery. A right siphonal artery supplies to the right part of mantle, supplying to right nuchal lobe and a penial artery supplies to the copulatory organ. The main trunk of cephalic aorta also branches to give a radular sac artery, supplying to the radular sac, an optic artery, supplying to eyestalks and the eyes, a tentacular artery, supplying to tentacles and pedal arteries, supplying to the foot.

**Visceral aorta:** This aorta runs into visceral mass, supplies to its different organs by giving off many branches. These branches are—a pericardial artery to pericardium skin and digestive gland, a gastric artery to stomach, many small intestinal arteries to roof of the posterior renal chamber, a hepatic artery to the digestive gland and gonad, several small arteries to tip of genital duct and then finally the visceral aorta terminates in rectal wall.

**Sinuses:** Blood after being distributed to various body parts through arteries and their branches, passes into small lacunae, which in turn unite to form large sinuses. Four chief sinuses are the anterior peri-visceral sinus, anterior peri-intestinal sinus, branchio-renal sinus and pulmonary sinus.

*Anterior peri-visceral sinus:* It lies above the foot and beneath the floor of pallial cavity surrounding anterior part of the alimentary canal and collects blood supplied by the cephalic aorta, except the pericardial artery and some branches of visceral aorta.

*Anterior peri-intestinal sinus:* It lies on columellar axis next to intestine and on terminal part of genital duct. It runs along the coils of intestine up to the junction of anterior and posterior renal chambers and collects blood from the viscera mainly distributed by the visceral aorta.

*Branchio-renal sinus:* It lies along the right side of the anterior renal chamber and collects blood from the renal chambers.

*Pulmonary sinus:* It lies in walls of the pulmonary sac and collects blood from the pulmonary sac.

**Veins:** The chief veins are as follows:

*Afferent ctenidial vein:* It lies above the rectum, receives branches from the rectum, terminal part of the genital duct, peri-visceral sinus and branchio-renal sinus and sends blood through numerous branches to gill lamellae for purification.

*Efferent ctenidial vein:* It lies along the roof of the anterior renal chamber and carries blood from ctenidial leaflets, mantle and conveys to the auricle.

*Afferent renal vein:* It lies on the roof of the posterior renal chamber and originates from the peri-intestinal sinus. It pours blood into the posterior renal chamber.

*Efferent renal vein:* It carries blood from the posterior renal chamber to the auricle.

*Pulmonary vein:* It collects blood from walls of the pulmonary sac and opens into the auricle.

**Blood:** Blood of the gastropods contains a respiratory pigment called the *haemocyanin*, a compound of copper and protein, which remains dissolved in plasma and gives a faintly blue colour to the blood. Phagocytic stellate amoebocytes present in plasma remove wastes substances and carry on intracellular digestion.

**Course of circulation:** All body parts are supplied with blood from the ventricle through cephalic and visceral aortae. Cephalic aorta carries blood to head and its associated structures, mantle, buccal mass, oesophagus, copulatory organ and columellar muscle. Visceral aorta supplies the entire visceral mass. Blood is collected from various parts into the peri-visceral and peri-intestinal sinuses, from which blood passes either into ctenidium, pulmonary sac, or into renal organ.

During aquatic respiration, blood from peri-visceral sinus pass to ctenidium for purification. The efferent ctenidial vein then takes this blood into the auricle. During aerial respiration, blood from per-visceral sinus passes to lung for purification. A pulmonary vein takes this blood into auricle. The blood from peri-intestinal sinus follows two courses. It either goes from this sinus to the anterior renal chamber and from there to ctenidium for purification, then through efferent ctenidial vein, it reaches auricle, or blood from peri-intestinal sinus pass to anterior renal chamber, then to posterior renal chamber (blood from peri-intestinal sinus can go directly to posterior renal chamber without going to the anterior renal chamber). In any case blood from posterior renal chamber goes through an efferent renal vein into the auricle. Thus, auricle receives pure blood from the ctenidium, or lung and impure blood from posterior renal chamber. This mixed blood enters ventricles and is distributed through the arteries. Renal chambers remove waste substances from blood.

### *Excretory system*

A single large, thick walled, much folded renal organ, kidney, or organ of Bojanus lies behind the pericardium. The renal organ consists of two distinct chambers, a right anterior renal chamber and a left posterior renal chamber.

**Anterior renal chamber:** The anterior renal chamber is a reddish organ, somewhat ovoidal in outline, situated in front of pericardium and posterior renal chamber. It projects into mantle cavity and opens into it in a deep crypt through an elongated opening lying to right of epitaenia. Upper surface of the chamber has shallow grooves corresponding to internal lamellae, which hang downwards from the roof into cavity of the chamber. Lamellae on the roof are arranged on the either side of a median longitudinal axis. Lamellae on floor are arranged on the either side of a median axis, the afferent renal sinus, which is the branch of the peri-intestinal sinus. It breaks up into numerous branches to supply the lamellae on both sides (Figure 14.10).

**Posterior renal chamber:** The posterior renal chamber is a broad, somewhat hook shaped, brownish to dusky grey in colour. This is situated behind the anterior renal chamber, between rectum on left and pericardium and digestive gland on the right side. It contains a large cavity in which a genital duct and few coils of intestine lie. The roof of chamber is richly supplied with blood vessels, resulting from a repeated division of afferent and efferent renal vessels of this chamber. The posterior

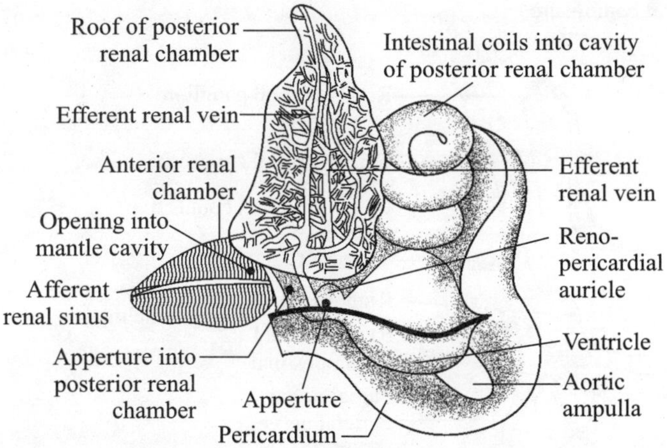


**Figure 14.10** Excretory organ of *Pila*.

renal chamber communicates with pericardium by a reno-pericardial aperture at one end and with anterior renal chamber through an aperture at other end. The floor of posterior renal chamber consists of a thin lining of tissue, which separates its cavity from organs, which project into it.

**Physiology of excretion:** Both the renal chambers excrete nitrogenous waste products. Owing to a single external opening, excretory products from posterior chamber are collected and emptied into anterior chamber, from where they are discharged into pallial cavity, through external renal opening. The kidney is a coelomoduct and communicates at one end with coelom (pericardium) and at other end with exterior (mantle cavity). The kidney removes nitrogenous waste from the blood. The waste is discharged into the mantle cavity. The excretory matter contains mostly the ammonia and ammonium compounds like urea and uric acid. For conserving water, ammonia is converted into relatively insoluble uric acid, which is striking in *Pila*. During its aquatic phase, it excretes ammonium compounds, but during its terrestrial phase, it excretes uric acid.

### *Nervous system*

The nervous system consists of paired and unpaired ganglia with their commissure and connectives. The commissure connects the similar ganglia, while connectives connect two dissimilar, or different ganglia. The paired ganglia are cerebral, buccal, pleural, pedal and visceral, while unpaired ganglia are supraintestinal and infraintestinal. These ganglia with their commissure and connectives are described as follows:

**Cerebral ganglia:** Two triangular cerebral ganglia located one on each side above the buccal mass are connected to each other by a thick cerebral commissure, which runs transversely above buccal mass and by a thin labial commissure lying below buccal mass. Each cerebral ganglion is further connected with buccal ganglion of its side, through a very slender cerebrobuccal connective. Thick, band-shaped, cerebropleural and cerebropedal connectives connect each cerebral ganglion with corresponding pleural and pedal ganglia. Each cerebral ganglion gives off several nerves supplying anteriorly the skin of snout, tentacle and buccal mass and posteriorly the tentacle, eye and statocyst.

**Buccal ganglia:** At the junction of buccal mass and oesophagus, there are two buccal ganglia which are connected to each other by a transverse buccal commissure, also connected to the cerebral ganglia by a cerebrobuccal connective. On each side, the connectives lie above the buccal mass. Nerves from each buccal ganglion supply the buccal mass, radular sac, salivary glands, oesophagus and oesophageal pouches (Figure 14.11).

**Pleuro-pedal ganglionic mass:** Pleural and pedal ganglia of each side join to form a pleuropedal ganglionic mass, which is situated below the buccal mass. In this ganglionic mass, pleural ganglion is placed towards the outer side and pedal ganglion to the inner side. The pleuropedal ganglionic mass is connected to cerebral ganglion of its side by a cerebropleural connective and cerebropedal connective. Two pedal ganglia are connected to each other by two pedal commissures, lying closely parallel to each other. The right pleuropedal mass has an infraintestinal, or a sub intestinal ganglion also fused with it. A slender loop like infraintestinal nerve behind the pedal commissure connects pleural ganglia of both sides. A statocyst is connected by a band of connective tissue of each pedal ganglion.

**Supraintestinal ganglion:** This is a slightly swollen, about fusiform, lies in a sinus about a quarter of an inch behind the pleuropedal mass of left side. It is connected with pleuropedal mass by a stout connective called the *zygoneury*. It gives off on inner side a thin supraintestinal nerve, which runs anteriorly above the intestine to right side to join right pleural ganglion. The supraintestinal ganglion also sends off posteriorly a

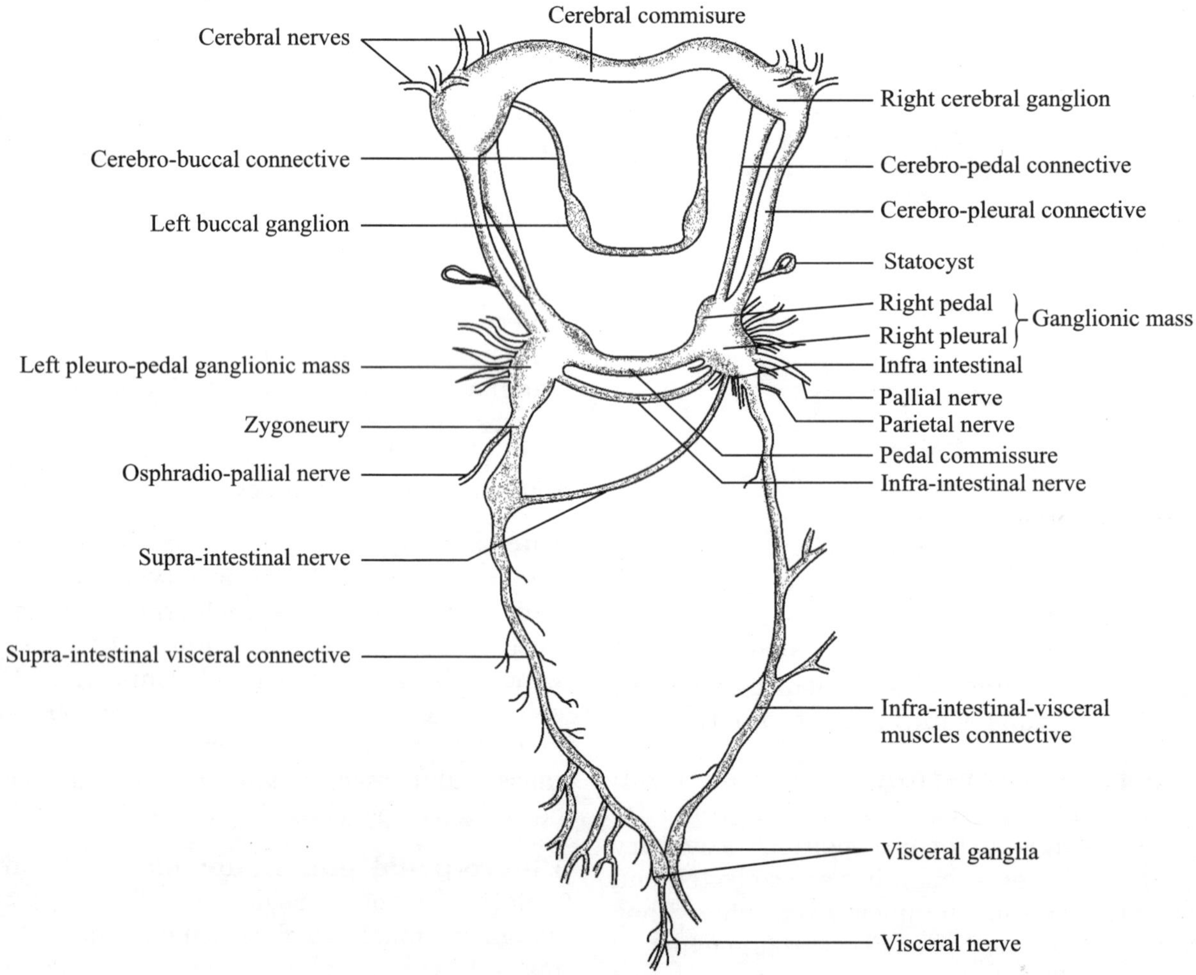


**Figure 14.11** Nervous system of *Pila.*

branch, the left visceral connective, which connects with visceral ganglion.

**Visceral ganglion:** Two spindle shaped ganglionic masses fuse to form the visceral ganglion which lies near the base of visceral mass, close to the anterior lobe of digestive gland and to the right of pericardium. Visceral ganglion is connected with supraintestinal, or left visceral connective. It is further connected with fused right pleural and infraintestinal ganglion through infraintestinal, or right visceral connective. Nerves from cerebral ganglia go to head, tentacles and eyes. Buccal ganglia send nerves to buccal mass.

Nerves from pedal ganglia innervate foot and those from pleural ganglia reach mantle, ctenidium and siphons. From visceral ganglion, nerves reach to intestine, kidney and gonads. These nerves constitute the peripheral nerves. Nervous system shows two characteristics; firstly most of ganglia except visceral are concentrated near buccal mass; secondly, visceral loop is twisted into a figure of 8 due to torsion. This twisted condition of nervous system is a primitive feature, because in most gastropods, a secondary bilateral symmetry is shown by ganglia and connectives.

**Sense organs:** There are a single osphradium, statocysts, paired eyes and tentacles.

*Osphradium:* The oval osphradium with 22 to 28 fleshy leaflets (arranged on the sides of a central axis) hangs from mantle, near the left pseudopodium. It is a chemoreceptor and tests the current of water which enters mantle cavity, through left pseudopodium. It selects the food taken in. The evolution of gastropod osphradium parallels that of the ctenidia in primitive forms. An osphradium is found for each ctenidium in prosobranchs, which have one ctenidium and a single osphradium. Osphradium disappears in those gastropods which have lost ctenidia, or have reduced mantle cavity, or are pelagic (Figure 14.12).

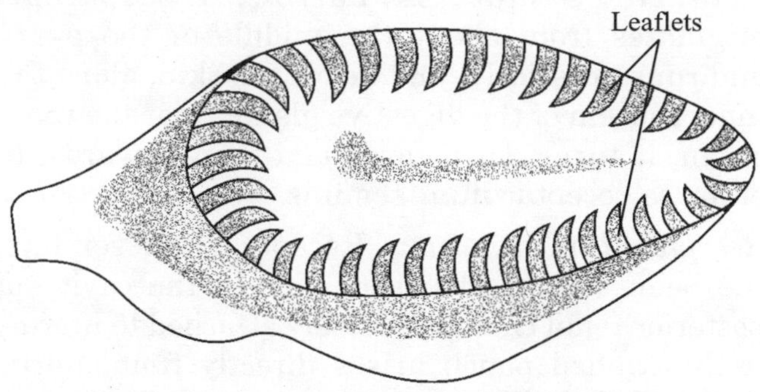


**Figure 14.12** Osphradium of *Pila globosa*.

**Statocysts:** Near each pedal ganglion, a statocyst lies in a depression in the foot. Lined by the epithelial cells, this round capsule remains surrounded by connective tissue. In the cavity of capsule, small calcareous statoconia are found. Statocysts receive the nerves from pedal- and cerebral ganglia, serve as organs of equilibrium to regulate the position of snail (Figure 14.13).

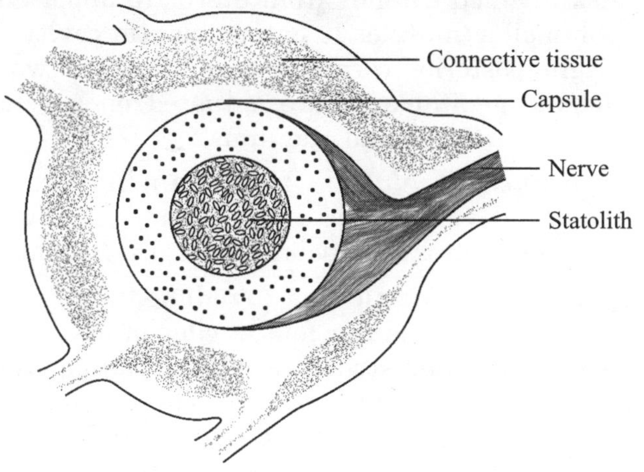


**Figure 14.13** Statocyst of *Pila globosa*.

**Eyes:** A pair of eyes, each borne on an ommatophore, is oval capsules with retina which is made of pigmented sensory cells. It continues in front as a thin non pigmented transparent cornea. The overlying epidermis is transparent and in the interior of capsule, a clear ovoidal lens surrounds a dense vitreous body. Optic nerve innervates retinal cells. Eyes are sensory to light.

*Tentacles:* Tentacles are sensory and contain both the tactile- and chemoreceptor cells and probably the gustatory cells. The first pair of tentacle is olfactory. Foot are liberally supplied with the nerves.

***Reproductive system***

Both the sexes are separate, i.e. dioecious and there is sexual dimorphism. Shell of male is smaller in size and less swollen than female. Copulatory organ is well developed in the male, but this is rudimentary in female.

**Male reproductive organs:** Male reproductive organs consist of testis with its fine vasa efferentia, vas deferens with vesicular seminalis and terminal glandular part of vas deferens, penis with its sheath and hypobranchial glands.

**Testis:** Testis is a flat, plate-like, whitish structure, triangular in outline, situated in upper part of the first 2–3 whorls of shell. It lies closely attached to digestive gland, along its upper- and inner-columellar edge and is separated from shell by a thin cutaneous membrane. The cream coloured testis is easily distinguished from the digestive gland, which is brownish, or dirty green. Minute ducts, the vasa efferentia lead downwards from different parts of testis and may unite with one another before opening into the vas deferens.

*Vas deferens:* The vas deferens arises from posterior end of testis. It consists of three distinct parts, the proximal thin tubular portion leading from testis, vesicular seminalis, and thick glandular portion which opens into mantle cavity near anal opening. The vas deferens starts from posterior end of testis and runs immediately beneath skin, along inner, or columellar edge up to posterior renal chamber. It then turns to left. On reaching the level of pericardium, it turns upwards and to right to open into the vesicular seminalis on its ventral side.

The vesicular seminalis lies to right of pericardium, immediately below the line of junction of anterior and posterior renal chambers. It is slightly curved and has a flask-shaped appearance, with its posterior blind and broadly rounded end. The vesicular seminalis opens on left side into terminals glandular part of vasa deferens. In the mantle cavity, the vas deferens lies closely attached to left side of rectum and ends in a prominent claw-shaped structure, the genital papilla having the male genital aperture, a little behind anus.

*Penis sheath and penis:* The edge of mantle bears on its inner surface a thick glandular flap of yellowish colour. The flap is attached on its right side, but is free on its left. Its edges are slightly rolled in to form a spout like sheath, the penis sheath. The penis is a long and stout flagellar structure, about half an inch long, arising from attached right side of the mantle flap. It is a slightly curved structure lying within its sheath. It is swollen at its point of attachment, but gradually tapers to free tip bearing a deep groove, all along its length, on its inner surface. Penis is capable of extension.

*Hypobranchial gland:* At the base of penis sheath, there is an oval hypobranchial gland. It consists of cells containing the small basal nuclei. The surface of glandular area is pleated. There is no duct and secretion of gland cells are apparently poured directly on surface. Spermatozoa are of two kinds, the eupyrene-sperms and oligopyrene-sperms. Eupyrene-sperms are hair-like having an elongated spirally twisted nucleus with a small conical acrosome in front and mitochondrial middle pieces behind, followed by end piece, as a long vibrating tail. Anterior and posterior limits of the middle pieces are marked by the proximal ring shaped centrosome and distal, granular centrosome, respectively. Axial filament springs up from the proximal centrosome.

In the region of middle piece, axial filament is unsheathed by the mitochondrial material. In end piece, the filament is naked. These sperms move actively forward in a spiral course. The sperm measures about 25 μ in length and 1.2 μ in breadth and they can fertilize eggs. Oligopyrene-sperms on the other hand have a very sluggish and serpent likes movement. Acrosome is poorly developed. The nucleus is elongated, broad and curved, but not spirally coiled. The middle piece is short and the number of axial filaments varies from 4–8. These sperms measure about 32.5 μ in length and 3 μ in width and they cannot fertilize the eggs. These probably have some secondary function.

**Female reproductive organs:** Female reproductive organs consist of ovary with numerous, minute ducts, main oviduct, receptaculum seminis, uterus, vagina and hypobranchial gland.

*Ovary:* Ovary lies in same position as testis, but it is not so extensive. It occupies the upper and inner surfaces on the first 2–2½ whorls and is covered over by a thin, but thick skin coat. This is much branched structure of light colour, which becomes darker in fully mature individuals. Branches of ovary consist of single layered acini, which are more, or less flask shaped, with their closed rounded ends which are directed outwards. Elongated necks of flasks descend to meet those of adjacent acini to form minute ducts, which in their turn open into main oviduct.

*Oviduct:* Oviduct is narrow, transparent, originates from about the middle of the ovary and runs anteriorly, just below the skin, along the inner margin of the digestive gland. Near the renal organ, it turns downwards and then upwards, to enter to receptaculum seminis.

*Receptaculum seminis:* Receptaculum seminis, the bean shaped structure, lies in the cavity of posterior renal chamber, closely attached to uterus. A thin-walled pouch arises directly from uterus wall is called the *pouch of receptaculum.*

*Uterus:* Uterus is the large, pear shaped structure, deep yellow in colour, lies inside the body whorl, below the intestine and right of renal chambers. The apex of uterus points forwards and is continued as vagina. Its basal portion is broad, rounded and is connected on its outer side with the receptaculum seminis.

*Vagina:* Vagina is white, or cream coloured, band like structure, found immediately beneath the skin which extends from uterus to upper end of columellar muscle. It enters mantle cavity at its right posterior corner and continues forward to female genital aperture, situated on a small papilla, a little behind the anus.

*Hypobranchial gland:* Hypobranchial gland is poorly developed. A rudimentary glandular thickening in area of hypobranchial gland is found.

***Copulation:*** Female has a rudimentary penis, lying beneath glandular fold at edge of mantle. It is a thin flagellar structure with a rudimentary groove along it inner surface. Three flagellum is about a quarter of an inch in length and has nearly same thickness throughout except the tip, where it is slightly pointed. There is no trace of folding of penis sheath. Copulation occurs either in water, or on land and lasts for 3 hours. Male and female come together, facing each other. Penis is expanded and is attached to genital papilla by its base. Penis and its sheath are inserted into mantle cavity of the female. The tip of the penis is put into female genital aperture. The spermatozoa are transferred through the vagina into receptaculum seminis.

### *Fertilization and development*

Eggs are fertilized in uterus. Oviposition starts a day, or two later. Fertilized eggs are laid in masses of 200 to 800 in moist earth, near ponds and lakes. Mollusca passes through two larval stages—trochosphere larva and veliger larva. The development of trochosphere is same as in polychaete annelids. A free-swimming trochosphere is found only in some primitive gastropods. In all others, trochosphere stage is reduced and passed within the egg. Veliger is a modified trochosphere and represents a more advanced stage of development. Its organs show greater degree of development than that of trochosphere larva. It has a ciliated apical organ on head, a curved gut, larval nephridia and a ciliated preoral prototroch. But it has organs which are not found in trochosphere. The prototroch develops into a swimming organ, the velum, which is a bilobed circlet with strong cilia. It is formed as an outward extension of the prototroch.

The velum causes a current, bringing food into the mouth and serves as an organ of locomotion for the free-swimming veliger. On dorsal side of the veliger, an embryonic shell gland, secretes a shell. Shell soon loses its simple form to assume a spiral shape, due to unequal growth. The larval retractor muscles are formed. On ventral side lies a foot between the mouth and the anus.

On dorsal side, between the velum and anus, ectoderm forms a mantle. Mesoderm forms two bands, which do not segment, but break up into cells, some of which form muscles. Torsion takes place during veliger stage. Shell and visceral mass are rotated through 180 degrees in relation to head and foot., but the coiling of visceral hump usually precedes torsion. Torsion may be very rapid, taking only a few minutes, or it may be gradual process taking up to several days. In later stage, veliger swim by its velum and creep by its foot. Gradually, the velum becomes smaller. In fresh water and terrestrial mollusca, there is no free-swimming larva. Both trochosphere and veliger stages are passed within the egg shell and a tiny snail hatches from the egg shell.

## 14.5 RESPIRATORY SYSTEM IN *Unio*

### *Ctenidia*

A single pair of elongated, eulamellibranch type of ctenidia, one on each side of foot is found. Each ctenidium is made up of two gill plates, or demibranchs, or laminae, an outer- and an inner gill plate, which are derived by folding of a single ctenidium. Ctenidia divide mantle cavity on each side into a large ventral inhalant chamber, or infrabranchial chamber and a smaller dorsal exhalant chamber, or suprabranchial chamber. Each gill plate, or lamina is formed of two similar flaps, or lamellae, joined to each other except dorsally. Thus, two lamellae of a gill plate from a narrow, long bag. Lamellae are made of several vertically parallel gill filaments, or branchial filaments.

Gill filaments are elongated. They pass downwards, then pass upwards like a 'V', so that each gill filament forms a descending and ascending limb. Adjacent gill filaments are joined by the fusion of tissues forming inter-filamentar junctions. Thus, gill filaments and interfilamentar junctions form two lamellae of a gill plate. In the inter-filamentar junction, there are holes known as *ostia*, which connect ventral inhalant chamber of mantle cavity with water tubes in laminae. Gill filaments appear as vertical lines and their interfilamentar junctions appear as horizontal striations on the lamella.

Gill filaments are covered with various kinds of cilia. Two chitinous rods support each gill filament. On the sides of filaments, there are lateral cilia and on distal surface, which first encounters inhalant current, there are frontal cilia. Bordering frontal cilia on both sides there are long cilia, called the *fronto-lateral* cilia. Between two lamellae of a gill plate, a space divided by the vertical bars of vascular tissue form the interlamellar junctions, which contain blood vessels.

The interlamellar junctions between the two lamellae divide the space into distinct compartments, called the *water tubes*, which are closed all round except dorsally, where they open into a suprabranchial chamber of mantle cavity. A ventral food groove runs longitudinally at the lower edge of each inner gill plate. Two dorsal food grooves on each side at the base of a ctenidium, one between mantle and outer lamella of outer gill plate, and another between inner lamella of outer gill plate and outer lamella of inner gill plate are found.

**Course of water current:** The constant beating of cilia of gill filaments produces a continuous current of water to enter inhalant siphon, lying posteriorly and ventrally, from where it pass to mantle cavity. Lateral cilia pass water current in and through ostia water enters the water tubes of gill plates. Then, it passes to suprabranchial chamber and passes out through exhalant siphon situated posteriorly and dorsally. The latero-frontal cilia form a flexible comb bordering the ostia. This comb forms a sieve to prevent large particles from entering the ostia. Current transports food and oxygen to ctenidia. Outgoing current carries away products of excretion and faeces, in addition to $CO_2$.

**Physiology of respiration:** When water passes through water tubes in gill, gaseous exchange takes places. $CO_2$ from blood is diffused out in water. The dissolved $O_2$ from water is diffused in blood. Thus, deoxygenated blood becomes oxygenated, which is carried to heart for distribution to various parts of the body. A comparative account of respiratory organ of *Pila* and *Unio* is given in Table 14.1.

**Table 14.1** Comparison of respiratory organ (gill) of *Pila* and *Unio*

| *Pila* | *Unio* |
|---|---|
| Aquatic and aerial respiration occur. | Only aquatic respiration occurs. |
| Single ctenidium lies on the right side of the branchial chamber. | Ctenidia are paired, lying on both sides of the foot. |
| Ctenidium is made of a large number of gill plates, or lamellae. | Ctenidium is made of two gill laminae. Each lamina is made of two gill lamellae; total being four lamellae in a ctenidium. |
| Arrangement of gill lamella is *monopectinate*. | Arrangement is eulamellibranch. |
| Gill filaments forming a lamella are covered with the ciliated epithelium at their free ends and their basal parts are made of glandular and nonciliated cells. | Many vertically placed gill filaments form a lamella. These are covered by the different types of cilia all along their outer border, supported internally by the chitinous rods and filled with the blood. |
| Suprabranchial chamber is not found. | Water tubes of each gill lamina open dorsally into a suprabranchial chamber. |

## 14.6 TORSION IN GASTROPODA

Torsion is a fundamental feature of gastropods. This suggests their departure from primitive molluscan plan. It is a process in larval gastropods, whereby viscero-pallium is rotated anticlockwise through 180° from its initial position on head-foot complex. As a result the mantle cavity and anus are found at the anterior end of the body. The gut, nervous system and blood vascular system are twisted correspondingly. During embryonic development, mantle cavity forms first at the posterior end of body, but shifts to an anterior position later on.

**Mechanism:** The original probable mechanism which brought about the torsion was contraction of larval retractor muscles, the entire rotation being completed in a few minutes. In most gastropods in which torsion has been carefully studied, larval retractor muscles account for the 90° of rotation, different growth for rest. Even in very early veliger larva, mesodermal bands develop asymmetrically. Right band is distinctly larger than left and can be distinguished as five large mesodermal cells.

As these cells elongate to form muscle cells, they gradually displace visceral hump to left side causing asymmetry. These cells converge on the right side in posterior region of larval shell and are inserted into the anterior end of body as larval retractor muscles. There are no related muscle cells on the left side. When the larval muscle cells attain contractile power, the process of torsion begins. If muscle contraction accounts for only 90° of rotation, the process usually lasts for a few hours. At the end of this first stage, the mantle cavity, which originally lay ventrally and posteriorly, comes to the right side and the foot projects on left. Stage 2 of torsion is usually longer in duration, relying on differential growth (Figure 14.14).

In general, the later the onset of torsion, the shorter the duration of stage 2, because of beginning of torsion in more advanced stage of development. It is possible to distinguish five ways in which torsion can be brought about 180° rotation by muscle contraction alone. This was probably the original mechanism and is known only for *Acmaea*. Rotation at 180° is achieved in two stages, an initial 90° movement by the larval retractor muscle contraction, followed by a slower rotation through the remaining 90° by differential growth processes. This is probably the most common mechanism of present day forms, like *Haliotis, Patella*. 180° rotation is achieved by differential growth processes alone,

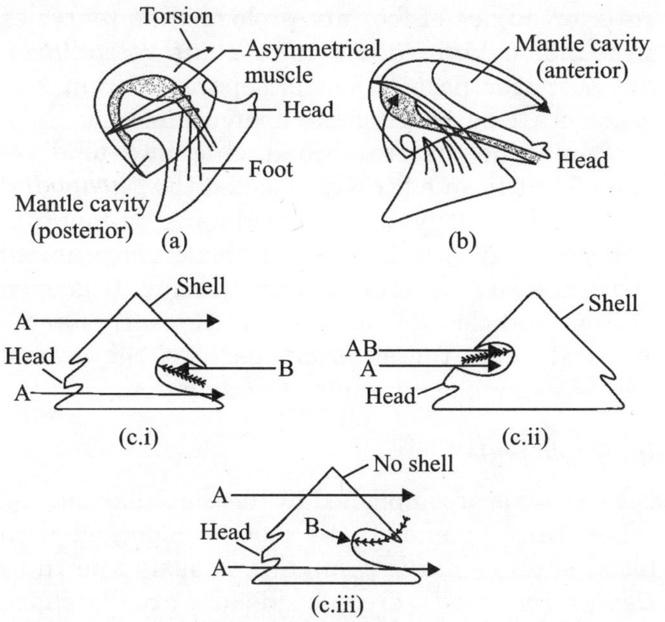


**Figure 14.14** Torsion in gastropods. Limpet veliger (advanced stage): (a) before torsion, (b) after torsion. Arrows indicate direction of twisting . (c) diagrams of water currents (A. as the animal moves forward, and B. respiratory current) in (i) prosobranch without torsion, (ii) prosobranch with torsion, (iii) opisthobranch without shell and torsion.

e.g. *Vivipara*. Torsion is achieved by differential growth processes. Change in the position of anus is being halted at a site appropriate to adult state as in *Aplysia.* Torsion is no longer recognizable as a movement of visceropallium, the organs being in post-torsional, position from their first appearances as in *Adalaria*.

### *Effects of torsion*

Torsion has some effects on structural organization of body. The mantle cavity which was posteriorly placed during larval stages opens anteriorly after torsion. The anus, ctenidia and renal apertures which were posteriorly in position now projects anteriorly. Auricles now lie in front of the ventricle. The straight alimentary canal after torsion forms a loop. The long uncoiled pleurovisceral connective twisted to form a figure of '8'. The condition is called *chiastoneury*. The shell, which was dorsal, or exogastric now becomes ventral, or endogastric. There is a loss of symmetry due to displacement of anus on right side. In some cases, there is also loss of some paired parts. In *Patella,* the right kidney serves as gonoduct. In *Pila, Littorina*, the topographically placed right ctenidium, auricle, osphradium and kidney disappear.

### *Significance of torsion*

The following two contrasting views exist on this issue:

**Garstang's view:** Garstang advocated that torsion is an adaptive feature and is useful to larva for protection against enemies. The placement of mantle cavity at the anterior end in larva gives greater protection to head and associated structures. The mantle cavity provides a protected space for withdrawal of foot and head in the case of danger.

**Morton's view:** Morton emphasizes importance of anterior location of mantle cavity, both in larval and adult molluscs. Anteriorly placed mantle cavity housing the head with the sense organs, respiratory structure, etc. in adult add positive advantage to test water and to come in intimate contact for gaseous exchange with coming water respectively.

## 14.7 FOOT IN MOLLUSCA

The chief function of foot is locomotion. Being an important organ, it has undergone changes in shape and form for different habitats. Various shapes of foot are the basis of classification of Mollusca. Foot in different classes of Mollusca is described as follows:

**In Gastropoda:** Gastropoda (gaster, belly; podos, foot) means belly footed animals. The foot is a thick muscular fold, lying generally on ventral surface, beneath the head and digestive tract. The foot is generally divided into three parts. These are propodium, the anterior part of foot and are well developed in forms, which crawl on the wet sand. Example: *Sigaretus, Natica*. By foot the animal pushes sand away on both sides of its path. Mesopodium is the middle large portion of foot. Metapodium is the posterior part of the foot which bears the calcareous operculum. Gastropods creep upon the broad, flat, ventral foot by a process of body deformation. Small waves of contraction sweep along the length of the foot, one wave closely follows the wave. Area of foot in contracted region is lifted; at relaxation, it is replaced on the substratum, a little in front of point from which it was raised. During each wave of contraction, a small section of foot performs a little step. The summation of all these little steps results in gliding motion. Mucous produced by a large gland in the foot lubricates substratum. In some gastropods, waves sweep from the back to front (direct waves). In other species, waves move from front to back (retrograde waves), opposite to direction, in which animal is moving. Waves may extend all the way across the foot (monotaxic), or those on one side of foot may alternate with those on other side (ditaxic). Most common mode of muscular pedal creeping in gastropods is by ditaxic retrograde waves (Figure 14.15).

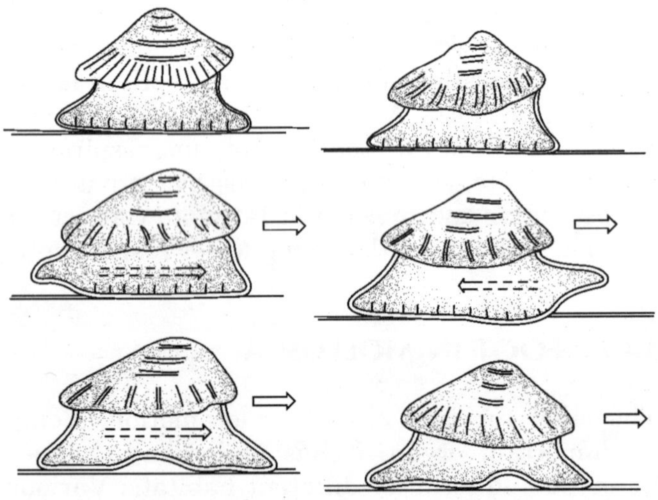

(a) (b)

**Figure 14.15** Creeping in molluscs. (a) Waves of contraction in same direction as animal, (b) Animal moves forward as the contraction waves pass in opposite direction.

### *Modification of foot in gastropods*

**Due to different condition of life:** In sedentary forms like *Vermetus,* the foot has been reduced to a simple small disc. In clinging forms like *Bathysciadium,* the ventral surface of foot functions as a sucker to cling to any object. In parasitic forms like *Stylifer*, the foot has atrophied and is represented only by a small ventral appendage. In crawling forms like *Bulia,* the foot has extended on all the sides of body and it slides over sand in order to progress forward.

**Structural changes in the form of appendages:** The epipodia is a distinct fold, extending from head to posterior end of foot as in *Haliotis.* The epipodia is well developed and gives out lateral tentacles as found in *Monodont*a. Two anterior angles of foot are prolonged as tentacles as found in *Valvata.* In *Atlanta and Rostellaria*, the posterior part of foot is separated from rest to form a lobe bearing the operculum.

Sometimes lateral broad, fine like lobes of sole of foot form a structure called the *parapodia*, which is generally used in swimming as found in *Aplysia*. In *Notarchus*, the extensions are united above the body forming a muscular sac. It is open in front and closed behind and at sides. By forcing water through the anterior aperture, the animal moves using sac as organ of locomotion.

### *In Scaphopoda*

Burrowing is accomplished by the muscular tongue. When fully extended, the foot is engorged with blood and in *Antails entails* (previously known as *Dentalium entails*) erectile side lobe are distended to help anchor foot in sand. Shell is then pulled down to foot.

### *In Cephalopoda*

Cephalopods crawl, or swim with the involvement of mantle. Water enters in the mantle cavity by the relaxation of circular muscle and simultaneous expansion of the mantle. Water is forced out through the funnel during contraction of the mantle giving the animal the power to move. Locomotion occurs by rapid undulation of the outer edges of fins in squids. *Octopus* is bottom crawler, swim rapidly by jet propulsion in disturbed situations. The finned octopods are provided with paddle-shaped fins that are used in hovering and slow swimming. *Nautilus* is a less active swimmer due to their inefficient funnel, which is made of two flaps instead of a single fused tube. *Nautilus* swims in the mid-water, or in the bottom by adjusting the gases of the chambered shell.

### *In Bivalvia*

Most bivalves burrow within the sediment. Burrowing involves foot, shell and siphons. Foot tends downwards in a probing motion and then expands to form anchor. Siphons close to prevent discharge of water. Adductor muscles soon close the valves ejecting water from ventral margin. This is immediately followed by the contraction of foot retractor muscles, pulling the animal downward towards the anchored foot. Lastly, the adductor muscles relax and the ligament opens the valves. Some bivalves bore into wood, or even solid rock.

*Lithophaga* secretes a weak acid from its mantle, erodes the weakened rock by rotating its shell. Ship worms burrow into wood using their shell as a file. Piddocks use mechanical abrasion to bore into soft rocks.

### In Monoplacophora

*Chitons* have a broad flat foot, which occupies most of the ventral surface. It serves both for locomotion and adhesion. Being very sedentary by nature, *Chiton*, especially the older individuals, stay in a very small area all their lives if an adequate food supply is available. For some species, this could be an area of as small as six square feet. The foot secretes a small amount of mucous and propulsion is accomplished entirely by muscular contraction. Both the foot and the girdle affect adhesion. Ordinarily, adhesion is accomplished just by means of the foot; however, when disturbed, the girdle clamps down on the hard substratum and the inner margin is raised. This creates a vacuum that enables the *Chiton* to grip the surface with great tenacity. This is also the reason that Chitons prefer smooth hard surfaces on which to live. The foot in its vestigial form is found in the primitive molluscs like *Neomenia* and *Chaetoderma* of order Aplacophora. In such cases, the foot is represented by a mid-ventral ciliated longitudinal groove. In the next higher order Polyplacophora (*Chiton*), the foot is represented by a broad muscular wave from the anterior to posterior end, which helps the animal in the creeping movement. It also helps the animal to adhere to any object firmly (Figures 14.16(a)–14.16(d).

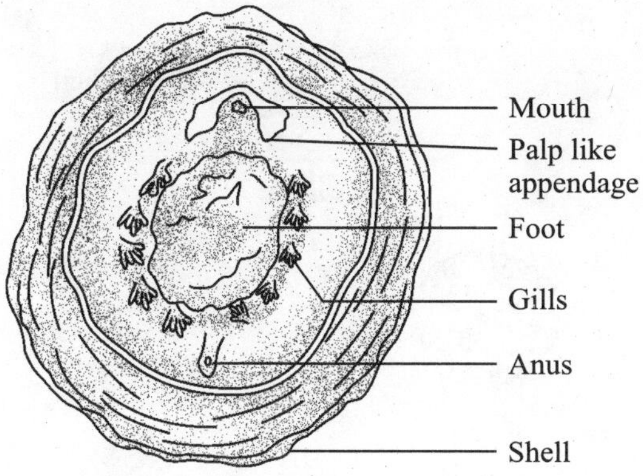

(a) in *Neopilina*

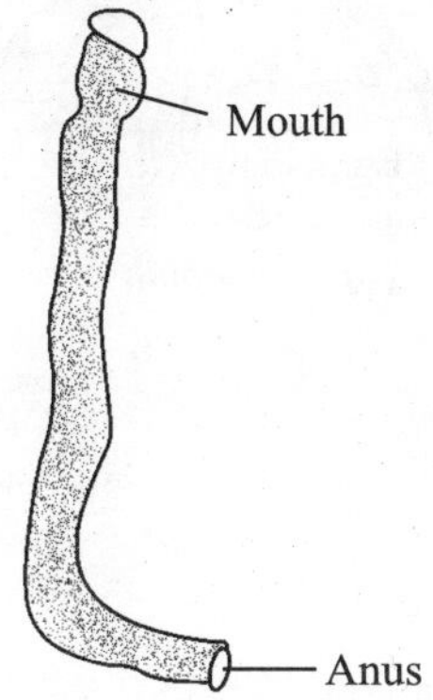

(b) in *Chaetoderma*

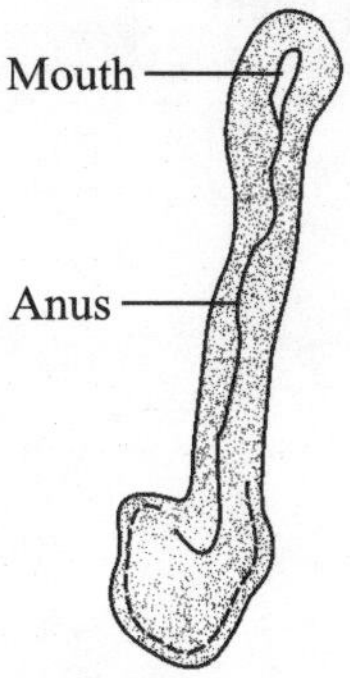

(c) in *Neomenia*

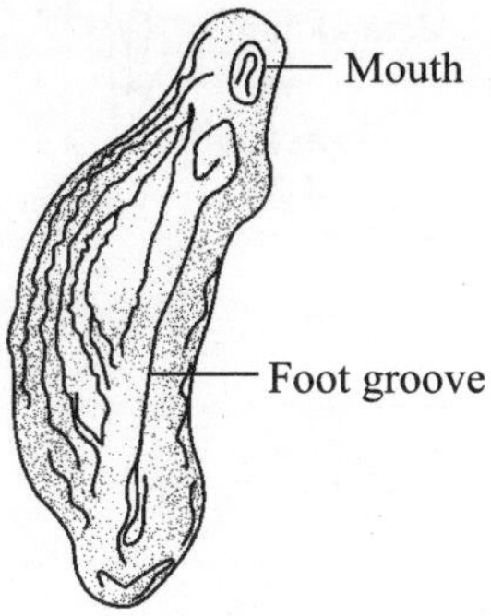

(d) *Proneomenia.*

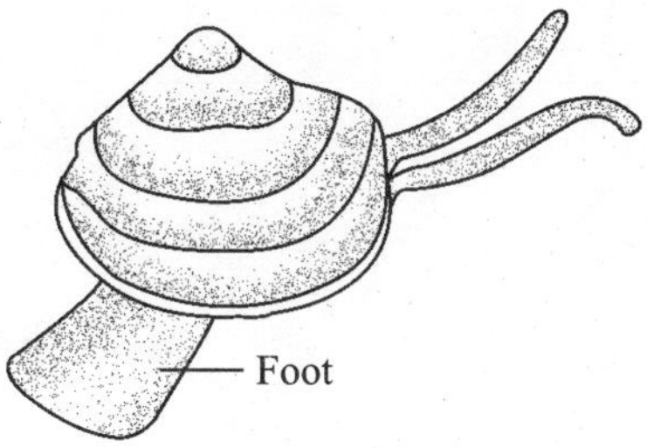

(*e*) *Tellina*

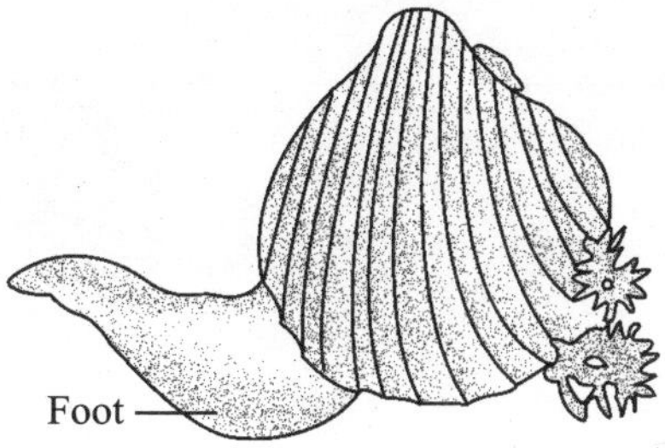

(*f*) *Cardium*

(g) *Ensis*

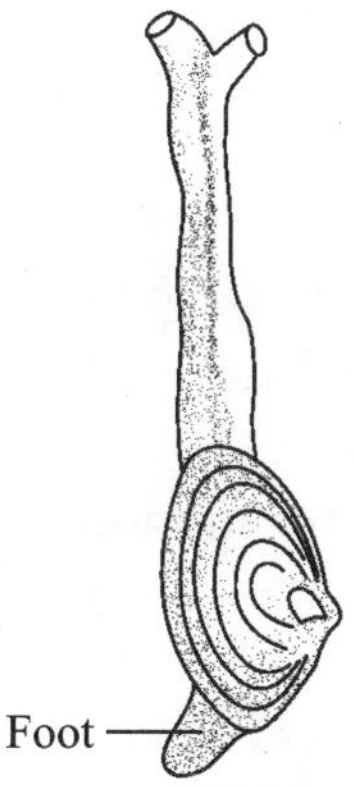

(*h*) *Mya*

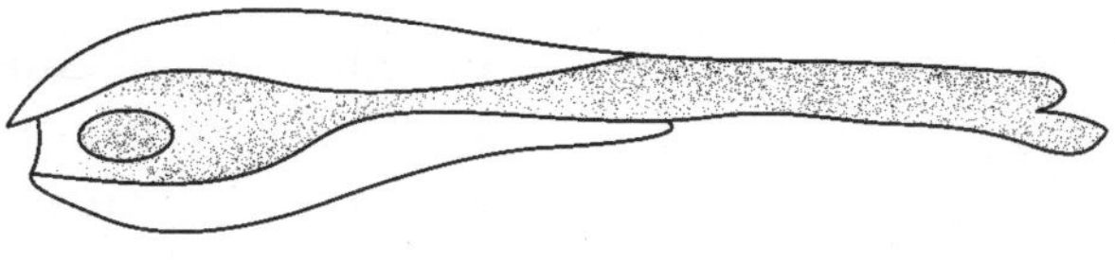
(*i*) *Pholas*

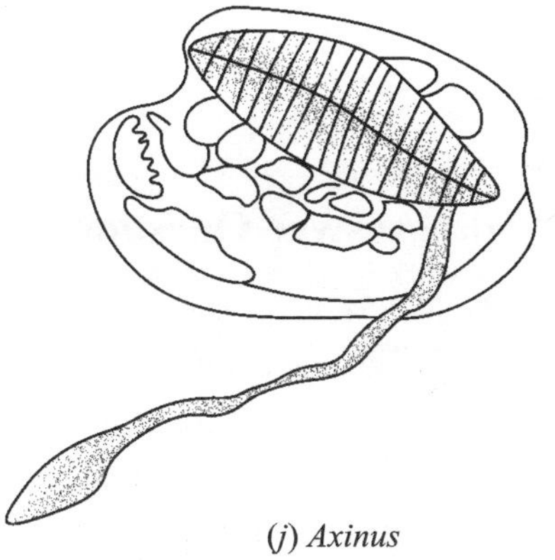
(*j*) *Axinus*

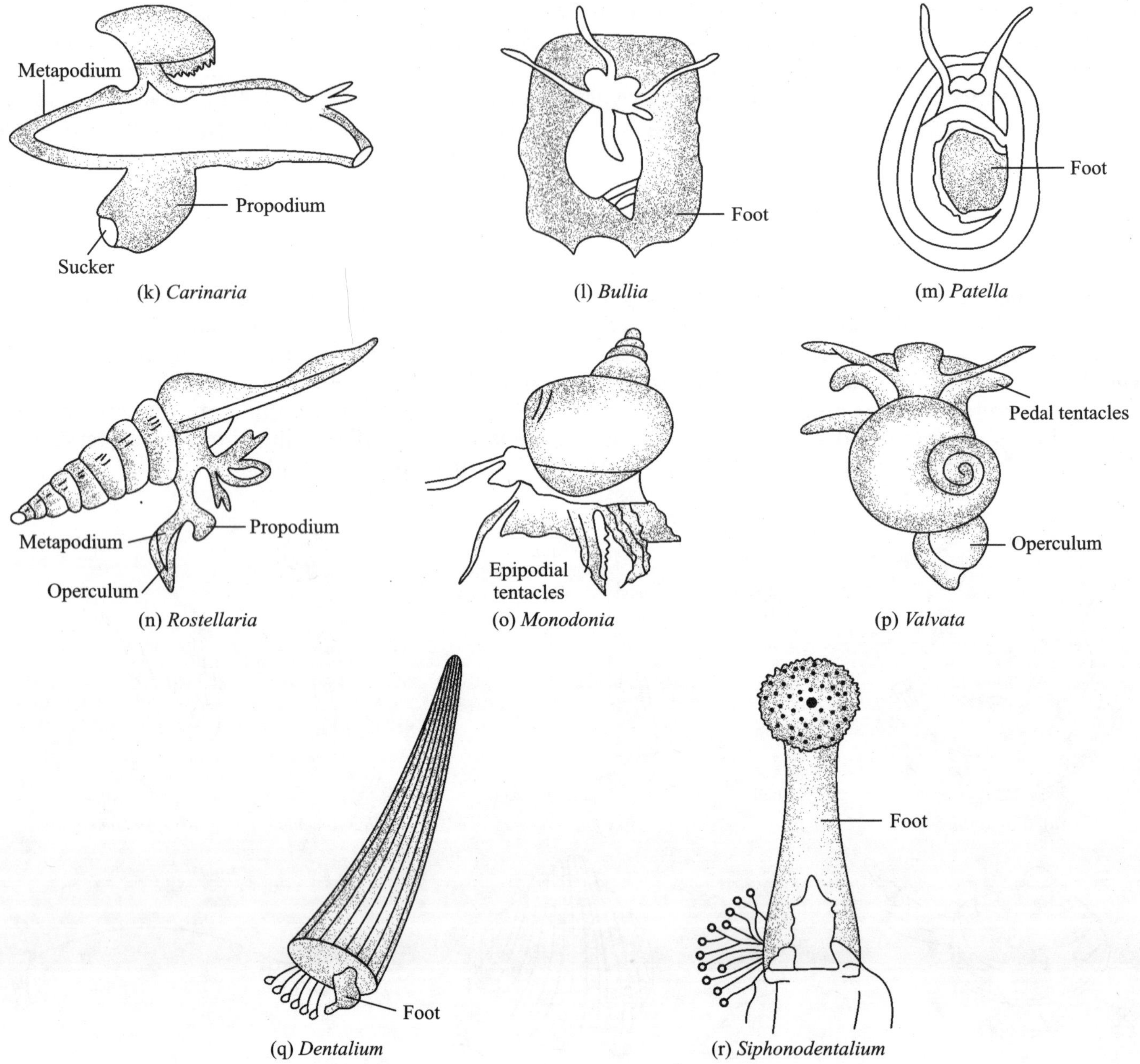


**Figure 14.16** (a)–(d) Foot in Mollusca, (e)–(j) Foot in bivalves and (k)–(r) Foot in scaphopods.

## REVIEW QUESTIONS

### Short Answer Questions

1. What is peristome?
2. What is umbilicus?
3. What is periostracum?
4. What do you understand by head-foot complex?
5. What is pallium?
6. What is the function of radula?
7. What are odontophores?
8. What do you understand by monopectinate gill?
9. What are nuchal lobes?
10. What is the difference between commissure and connectives?

11. What is the function of osphradium?
12. Mention a function of statocyst.
13. What is hypobranchial gland?
14. What is vesicular seminalis?
15. What are eupyrene sperms?
16. What are oligopyrene sperms?
17. What is receptaculum seminis?
18. Name a larval form of Mollusca?
19. What do you understand by torsion?
20. What the term " Mollusca" signifies?

## Long Answer Questions

1. Classify Phylum Mollusca up to order with examples.
2. Draw and describe the digestive system of *Pila*.
3. Give a brief account of respiration in *Pila with* suitable diagrams.
4. Draw and describe the nervous system of *Pila*.
5. Draw and describe the reproductive system of *Pila*.
6. Give a comparative account of respiration in *Pila* and *Unio*
7. Describe various modifications of foot in molluscs.
8. Give an account of torsion in molluscs with diagrams.

CHAPTER 15

# Phylum Echinodermata

Echinoderms are diverse with unique features like 5-fold symmetry and water vascular system. They constitute a monophyletic group for their 5-fold symmetry. Some may be polyphyletic. They are marine, benthic, radially symmetrical and coelomate. They are secondarily radially symmetrical deuterostomes, although their ancestors were bilaterally symmetrical. Their body wall includes a thick connective tissue, the dermis in which the calcareous ossicles are found. Ossicles form an endoskeleton of various forms in the various taxa. Calcareous spines of various sizes and shapes arise from the dermis and project from the body surface. Water vascular system mostly functions for supporting the locomotory tube feet, feeding, excretion and gas exchange.

Living echinoderms exhibit five types of organization. Their body parts occur in fives or multiples thereof. About 6,000 recent species and fossil of about 13,000 species are known. Echinoderms were included within the group Radiata along with coelenterates by Cuvier. The superficial resemblance in arrangement of parts of coelenterates and echinoderms is perhaps due to their sessile habit, which developed independently in two groups. Radial symmetry of coelenterates is the primitive, although in the echinoderms this is secondarily derived from a bilateral condition. The level of sensory and neural development is low.

## 15.1 ABORAL SURFACE

The calcareous, orange coloured madreporite is present on the aboral surface of the disc. Numerous microscopic pores found at the bottoms of the grooves open into the stone canal and axial canal of the water vascular system. A radial axis, passes from the disc's centre outward along the midline of arm, is called *radius*, or ambulacral axis. Such axes are five in number. Any axis bisecting the angle between any two adjacent arms is termed *interambulacral axis*, or *interradial axis*, which are also five in number. An interambulacral axis passes through the madreporite.

Aboral surface is covered by a single layer of ciliated epidermis. On the body wall of the aboral surface, numerous small fixed spines are found as the extensions of the calcareous endoskeleton. Each spine is surrounded by a circle of short-stemmed, white pedicellariae. Pedicellariae have endoskeleton of ossicles. An anus is located near the centre of the aboral surface. It is surrounded by tiny ossicles which are much smaller than the spines and stud the surface of the disc. This area of the disc is free of papulae.

Apical longitudinal muscle stretches the arm at median aboral side below the body wall. Each pair of ambulacral ossicle has an upper and a lower transverse adambulacral muscle. The upper or superior transverse ambulacral muscle widens adambulacral groove. The lower or inferior transverse adambulacral muscle makes adambulacral groove narrow. Between the adjacent adambulacral ossicles of each side, there is a longitudinal adambulacral muscle, which shortens arm and adambulacral groove.

## 15.2 ORAL SURFACE

The large mouth is present at the centre of the disc and remains surrounded by the thin peristomial membrane. The yellowish-orange, curtain-like folds of the cardiac stomach lies inside the mouth. Five deep ambulacral grooves radiate outward from the mouth, one along the midline of the oral surface of each arm. Each groove lies on an ambulacral axis. Numerous soft, tubular structures projecting into the groove from either side are termed *tube feet* or *podia*. Two rows of tube feet are present on each side of the groove. Tube feet bear suckers at their distal ends. Rows of long, flattened, movable spines on each side of the ambulacral groove are present.

The word ambulacrum is Latin for "covered way", an apt name as spines are used to cover the groove and to protect the tube feet. Sensory structures are arranged around the periphery, which in sea star are located at the tip of the arms. Several long, narrow sensory tube feet extend from the tip of each arm. They possess chemoreceptor and mechanoreceptor. Small circle of short, blunt movable spines are found at the tip of the arm. These spines surround a small, pale red, or yellow coloured eyespot and are not associated with the pedicellariae. Eyespot lies on the oral surface of the arm, almost at the tip. Small, lobed, rectal caecum surrounds the intestine and it obscures the junction with the body wall.

## 15.3 ENDOSKELETON

Definite skeleton provides rigidity to the body. Endoskeleton is unique as it is mesodermal in origin instead ectodermal origin of other invertebrates. It consists of numerous calcareous ossicles of various shapes and is bound together by the connective tissue. They form a reticulate skeleton leaving spaces for emergence of groups of papulae. Ossicles exhibit irregular arrangement on the aboral surface, but are regularly arranged around the mouth and in the ambulacral groove on the oral surface.

Five plates like ossicles called the *oral ossicles* remain arranged around the mouth. Each ambulacral groove is supported by the double rows of large, transversely placed, rod shaped ambulacral ossicles. Ossicles of two opposite rows are arranged like an inverted "V". Their aboral ends meet at apex of "V" like rafters supporting the roof of a shed and forming a conspicuous ambulacral ridges.

Ambulacral ossicles do not bear any spines, tubercles, or other external appendages. They are movably articulated in the ambulacral groove and permit opening, or closing of the latter. Each ambulacral ossicle has a notch on its outer margin and inner margin. Two notches of adjacent ossicle together form an oval aperture called the *ambulacral pore* for passage of the tube foot. Ambulacral pores are arranged to form two rows on each side of the ambulacral groove. At its outer end, each ambulacral groove articulates with one ambulacral ossicle forming the edge of groove and bear two, or three movable spines on the small tubercles. Two rows of ossicles called the supramarginal and inframarginal ossicles are found next to the ambulacral ossicle.

## 15.4 SECTION OF AN ARM

Arm is covered all around by the thin two-layered cuticles, a ciliated epidermis and an underlying thick dermis, which has many perihaemal spaces and ossicle. Epidermis and dermis are thin over the projecting spines, pedicellariae and papulae, but they wear off from the spines. Arm encloses a perivisceral coelom. The aboral wall has several irregular, fenestrated ossicles, some of which are calcareous. The dermal papulae project from the dermis and continue up to the coelom.

Many pedicellariae are found between the spines, some remain attached to the spines. Each lateral margin of arm bears two large spines called the *supramarginal spines* and below which there is an inframarginal spine. The mid dorsal arm has a large cardinal spine. On the oral surface, the ambulacral groove is supported by two elongated ambulacral ossicles meeting at the tip of the groove. At each end of the ambulacral groove, adambulacral ossicles and spines are present. Ambulacral ossicles produce two columns in the oral surface of each arm on each side. There is a single column of ambulacrals. The adambulacral spine can touch the substratum, or bend inward to protect the ambulacral groove. A radial canal runs above ambulacral groove, which is joined on each side by a podial branch to two ampullae and one tube foot. Below radial canal, a radial hypo neural sinus encloses a radial haemal channel (Figure 15.1).

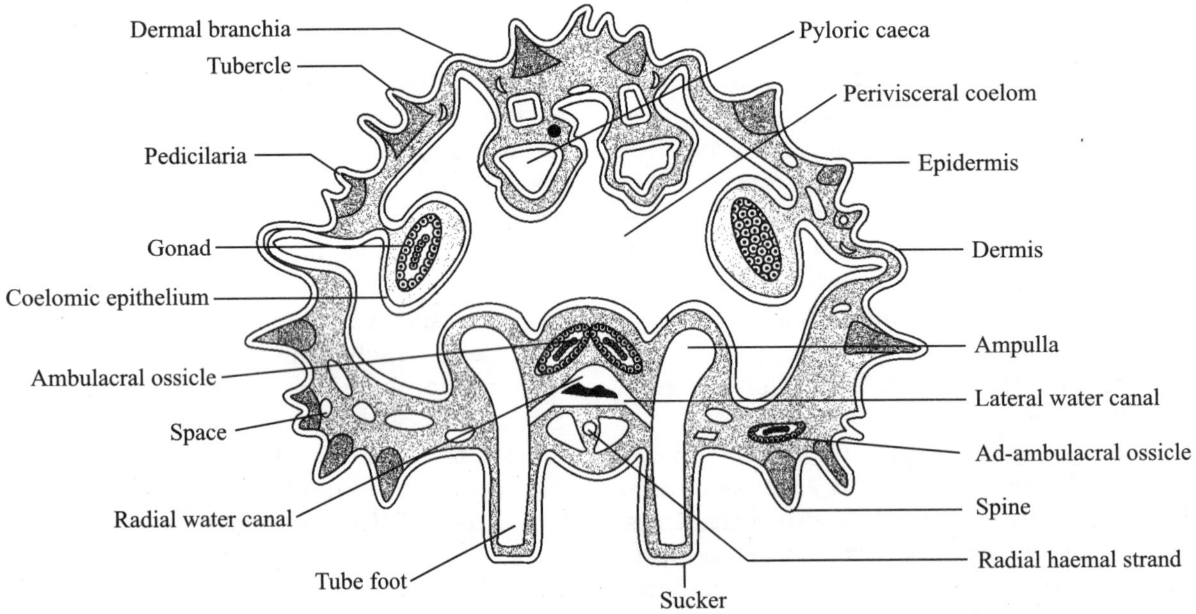


**Figure 15.1** Transverse section of an arm.

## 15.5 BODY WALL AND COELOM

Epidermis covers the body and overlies a dermis, which contains the ossicles. Internal to the dermis and ossicles, there are muscle layers and the peritoneum of the coelom. In sea urchins and sand dollars, ossicles are firmly attached to one another to form a rigid test. In sea cucumbers, ossicles are separate and lie scattered in the fleshy dermis. In the arms of sea stars and brittle stars, the body wall muscles are arranged in bands between the plates.

In some groups, skeletal plates are so developed that they nearly obliterate internal cavities. In brittle stars, each arm segment contains a central skeletal ossicle and the arm coelom becomes reduced to small channel. Arm coelom in crinoids is reduced by skeletal plates. Endoskeleton is calcareous, mostly made by $CaCO_3$ as calcite and small amounts of $MgCO_3$. Skeleton begins as numerous, separate, spicule-like elements, each element as a single calcite crystal. Additional material is deposited on the crystals. Each ossicle is porous with an internal meshwork of lattice-like, or labyrinth-like spaces and remains filled with dermal cells and fibres.

During skeleton formation, the plates may remain single, or they may fuse to form compound plates. They often cause bumps and knobs called the *tubercles*, granules, movable and fixed spines. In the asteroids and echinoids, the skeleton also produces pincer-like structures called the *pedicellariae* which respond to external stimuli independently of the main nervous system. They also possess neuromuscular reflex components.

Pedicellariae differ in their structural details, size and distribution on the body. Some are elevated on stalks; others lie nestled directly on the body surface, either singly, or in clusters. Some help to keep debris and others defend against larger organisms. *Toxopneustes* bears toxin-producing pedicellariae. In some sea urchins, the pincers grasp and hold objects for camouflage and protection. Few sea stars use their pedicellariae to capture prey. Movable spines and pedicellariae contain muscles and other tissues outside the main skeletal framework of the body wall.

Coelomic system usually develops as tripartite series, originating as paired protocoels, mesocoels and metacoels. The main body coeloms are derived from the embryonic metacoels and are well developed in most groups. Other coelomic derivatives include the water vascular system, gonadal linings and certain neural sinuses. The main body cavities or perivisceral coeloms are lined with ciliated peritoneum. Variety of coelomocytes are present in the body fluid and in the water vascular system. Many of these cells are phagocytic. Haemoglobin occurs in the coelomocytes of many holothurians and in few brittle stars.

## 15.6 SUPPORT AND LOCOMOTION

Except holothurians, the general body shape and structural support are maintained primarily by the skeletal elements. Podia and gills are supported mostly by hydrostatic pressure. In most sea cucumbers, body wall muscles form thick sheets, adding structural integrity to the body by working on coelomic spaces to provide a hydrostatic skeleton. Locomotion is determined by overall body configuration, the animals' habits and the nature of the skeletal-, muscular- and water vascular-systems.

Apart from the sessile sea lilies, most extant crinoids crawl and swim, by directing the oral side away from the substratum. Aboral cirri are used primarily for temporary attachment and for righting the position of animal, if overturned. Swimming is accomplished by up and down sweeps of the arms, which are divided into functional sets that move alternatively.

In ten-armed species, five arms move upward, while the other five arms move downward. Asteroids exemplify locomotion using podia. The action of a single podium involves power- and recovery strokes. Arms of the sea star are held about stationary relative to the central disc, even in species with a flexible skeletal framework and movement is accomplished by numerous podia on the oral surface. Control of podial action is not fully known. Most sea stars move very slowly. Some sedentary asteroids become extraordinarily rapid 'runners' on encountering a potential predator.

## 15.7 GENERAL CHARACTERS

1. Sessile, or slow-moving animals.
2. Marine.
3. A thin skin covers an endoskeleton.
4. Calcareous mesodermal endoskeleton is composed of separate plates, or ossicles.
5. Each plate originates as a single calcite crystal and develops as an open meshwork structure.
6. Adults are provided with basic pentamerous radial symmetry.
7. Coelomic water vascular system is composed of complex series of fluid-filled canals.
8. Gut is complete, may be secondarily incomplete or lost.
9. Excretory organ is absent.
10. Circulatory structures, when present, includes a haemal system derived from the coelomic cavities and sinuses.
11. Nervous system is diffuse, decentralized, usually consisting of a nerve net, nerve ring and radial nerves.
12. Mostly dioecious.
13. Development is direct, or indirect.
14. Unique feature is the presence of water vascular system, a network of hydraulic canals.
15. Tube feet function in the locomotion, feeding and gas exchange.
16. Excretion is carried out through the amoebocytes of the coelomic fluid.
17. Excretory products are liberated through gills.
18. Nervous system exhibits a low order of specialization.
19. Nervous system remains permanently associated with the skin and never migrates to an internal position.
20. Nerve ring and its branch are lowly organized.
21. Nerve cells do not form a brain and even ganglia.
22. Sense organs are mostly generalized.
23. Tactile organs are diversified.
24. Pedicellariae are sensory and serve the purpose of defense.
25. Eyes are commonly found in starfishes and sea urchins only.
26. Tubular digestive system is provided with both mouth and anus.
27. Both the sexes are separate.
28. Gonads are single in the sea cucumbers.
29. Generally there are 5 gonads, each with its duct, but with no other accessory organs.
30. Germ cells are liberated into the water for fertilization and subsequent development.
31. Sexual reproduction usually involves the release of gametes by the males and females into the sea water.

## 15.8 *Asterias*

*Asterias* is commonly known as the *starfish* which suggests an organism likely to be a star and fish. About 150 species of *Asterias* are known from the different geographical areas. *A. rubens* occurs on the English and North

European coasts, *A valugaris* on the North Atlantic coast of North Amarica, A forest on eastern sea shore to Gulf of Mexico, *A. amurensis* in Behring sea, Japan and Korea and *A. tenera* occurs on the sea shore from Nova Scotia to New Jersey. The common sea stars are *Pentaceros, Astropecten, Asterina, Heliaster, Solaster* and *Luidia*. The following account is a general idea about anatomical organization of genus *Asterias.*

Systematic Position

| | |
|---|---|
| Phylum | Echinodermata |
| Subphylum | Eleutheroza |
| Class | Asteroidea |
| Order | Forcipularia |
| Genus | *Asterias* |

### *Habit and habitat*

*Asterias* is exclusively marine, bottom dwelling or benthonic. They inhabit various types of bottom mainly the littoral zone, crawl mostly at a slow rate or may remain quiescent at times. *A. forbesi* is found abundantly on the hard, rocky, sandy or soft bottom. Other species prefer the rocky sea bottoms. Most species are generally solitary. To avoid direct sunlight or excessive drying, many individuals gather at some place and feed voraciously on the slow moving or sessile animals, chiefly on polychaetes, crustaceans, molluscs, echinoderms and even corpses. Many *Asterias* species exhibit parasitism and communalism . Sea stars in general exhibit remarkable power of autonomy and regeneration.

### *Shape, size and colouration*

Diameter generally varies between 10 and 20 cm, may be smaller, or longer. Colour is variable with shades of yellow, orange, brown and purple. Body has two surfaces. Upper convex and much darker side is called the *aboral* or *abactinal surface*. Less pigmented, flat lower surface is called the *oral* or *actinal surface*. Oral and aboral surfaces without ventral and dorsal surfaces correspond with the left and right sides of bilateral larva. The axes occupied by the arms are called *radii*. Regions of central disc between the arms are called the *interradii*. A well defined head is absent.

### *External features*

Body is divided into a central disc from which five arms radiate. The principal body axis and the axis of symmetry, the short oral-aboral axis passes vertically through the disc's centre. The pale lower side is the oral surface and the structures located in this side are the oral. Dark upper side is the aboral surface and structures on this side are the aboral. Structures remote from the axis are the peripheral. whereas those near the axis are the central (Figures 15.2–15.3).

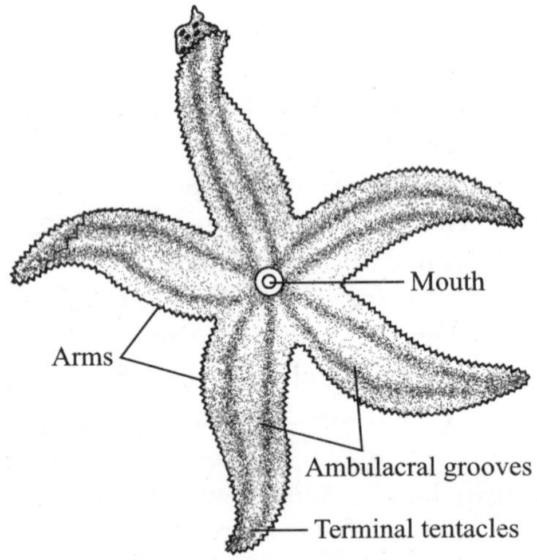


**Figure 15.2** Oral view of *Asterias*.

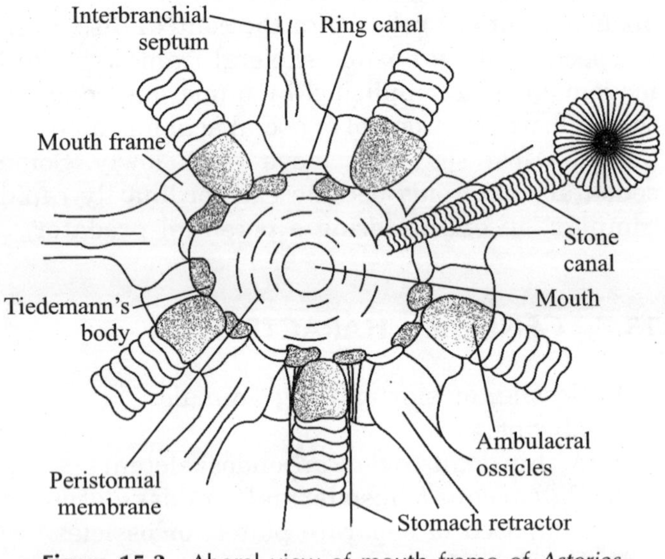


**Figure 15.3** Aboral view of mouth frame of *Asterias*.

**Mouth:** This is a pentagonal aperture with five angels, each directed towards an arm on oral surface in the centre of the pentagonal central disc. Mouth is surrounded by a soft and delicate membrane, the peristomial membrane or peristome and is guarded by five groups of oral spines or the mouth papillae.

**Ambulacral grooves:** From each angle of the mouth, a narrow groove radiates which is called the *ambulacral groove*, which runs all along the middle of the oral surface of each arm.

**Tube feet or podia:** Each ambulacral groove contains four rows of locomotory feet. Food capturing-, respiratory- and sensory-organ are called the *tube feet* or *podia*. Tube feet are soft, thin-walled, tubular retractile structures with the terminal discs or suckers, which function as the suction cups to afford a firm attachment on the surface to which they are applied.

**Ambulacral spines:** Each ambulacral groove is bordered and guarded laterally by two, or three rows of movable calcareous ambulacral spines, which are capable of closing over the groove. Spines often become larger, stouter and assemble in five groups one at each inter-radius of disc near the mouth and are called the *mouth papilla*. Outside the ambulacral spines, there are rows of stout immovable spines, beyond which another series of marginal spines along the borders of arms occurs, demarcating the oral surface from the aboral surface.

**Sense organs:** These include five unpaired, terminal tentacles and five unpaired eye spots. The tip of each arm bears a small, median, nonretractile and hollow projection

**Terminal tentacle:** It acts as a tactile and olfactory organ. A bright red photosensitive eye spot with several ocelli is found at the base of each tentacle.

**The aboral surface bears the following structures:**

**Anus:** This is a minute circular aperture situated close to the centre of the central disc of the aboral surface.

**Madreporite:** Aboral surface of central disc bears a flat, subcircular, asymmetrical and grooved plate called the *madreporite plate* or *madreporite* between the bases of two of five arms. Surface of madreporite is marked by several radiating, narrow, straight, or slightly wavy grooves with the pores in them. Madreporite is a sieve-like porous plate that leads to the stone canal of the water vascular system. The number of madreporite is generally one, may be more than one in some species, due to increase in number of the arms. Two arms with madreporite between their bases are collectively called the *bivium* and other three arms are called *trivium*. Symme-trical position of the madreporite converts radial symmetry of *Asterias* into the bilateral symmetry.

**Spines:** Aboral surface is covered with the numerous, short, stout, blunt, calcareous spines, or tubercles which are variable in size, arranged in irregular rows and run parallel to the long axes of arms. Irregularly shaped calcareous plates support them or ossicles that remain buried in the integument and from the endoskeleton.

**Papulae or gills:** Many minute dermal pores are present between the ossicles of the integument. Through each dermal pore, very small, delicate, tubular, or conical finger-like, thin-walled, membranous and retractile projection called the *dermal branchiates*, the gills, or the papulae emerge. Papulae are hollow evagination of the body wall and their lumen remains in continuation with the coelom. They are internally limited by the coelom and serve the respiratory- and excretory functions.

**Pedicellariae:** The whitish modified spines like tiny pinceros cover the entire aboral surface, or jaws called the *pedicellariae*. Each pedicellaria consists of a long, or short, stout, flexible stalk without the internal calcareous support. The stalk bears three calcareous ossicles or plates, a basilar piece or plate at extremity of the stalk and jaws or valves, which remain movable and articulated with basilar piece and secreted along their apposed edges. The pedicellariae having three calcareous pieces and a stalk are called the *forcipulate pedunculate pedicellariae*. Asterias possesses two types of forcipulate pedunculate pedicerllariae, the straight type and the crossed type.

- **The straight type pedunculate pedicellariae:** The simple pedicellariae with two almost straight jaws remain attached basally to the basal piece. When closed, they become parallel and meet throughout their length. Two jaws work against one another like the blades of a forcep, with the help of three pairs of muscles. There are two pairs of adductor muscles for closing the jaws and a pair of abductor muscles for opening them.

- **Crossed type pedunculate pendicellariae:** Here, basal ends of two jaws cross one another like the mandibles of a crossbill. Basal piece remains enclosed between their crossed portions. Two pairs of adductor muscles and one pair of abductor muscle operate jaws. Other pedicellariae without stalk are called the *sessile pedicellariae*, which serve the defensive purpose and keep the body surface free from debris. Organisms like sponges and coelenterates set on their body (Figures 15.4–15.5).

**Cuticle:** Cuticle consists of two layers—an outer thick homogeneous layer and inner delicate layer.

**Body wall:** Body wall consists of a thin, outer, ciliated epidermis and a thick, visible connective tissue, the dermis and the thin, inner peritoneum. Dermis contains the collagen fibres and many calcareous dermal ossicles or little bones. Some ossicles bear spines.

**Epidermis:** Just beneath the cuticle, a layer of ciliated epithelium extends over all the external appendages of body such as spines, pedicelariae, tube feed and gills. Epidermis is composed of a variety of cells like the ordinary flagellated, or ciliated columnar, neurosensory, mucous gland or goblet cells having finely granules. Uniform gland cells are filled with coarse spherules. Pigment granules provide external colouration to the animal.

**Nervous layer:** Beneath the epidermis, a nervous layer of varying thickness in various areas penetrates the epidermal cells.

**Basement membrane:** A delicate basement membrane separates the nervous layer and epidermis from the underlying dermis.

**Dermis:** This is composed of fibrous connective tissue developed from the mesoderm. It is the thickest layer of the body wall with outer and inner regions. The outer dermal region secretes and houses the ossicles of endoskeleton and binds them together, while the inner dermal region contains numerous blood containing space called the *periphaemal spaces*.

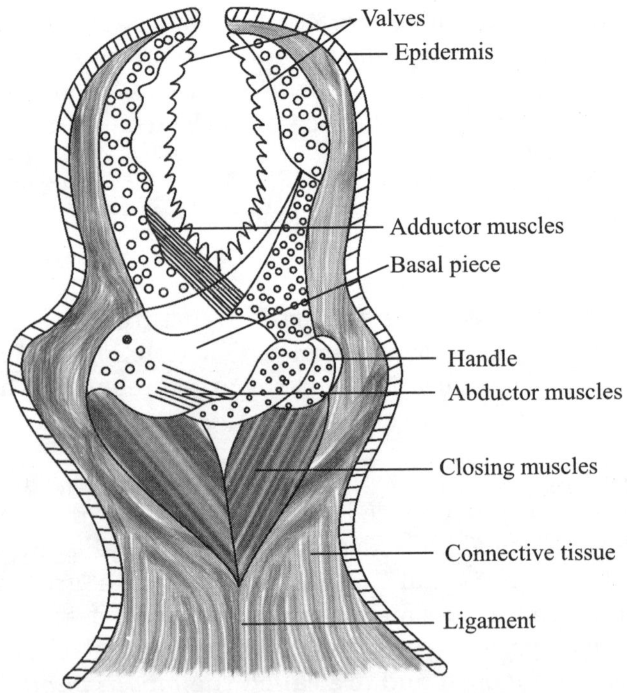


**Figure 15.4** Crossed type of pedicellariae.

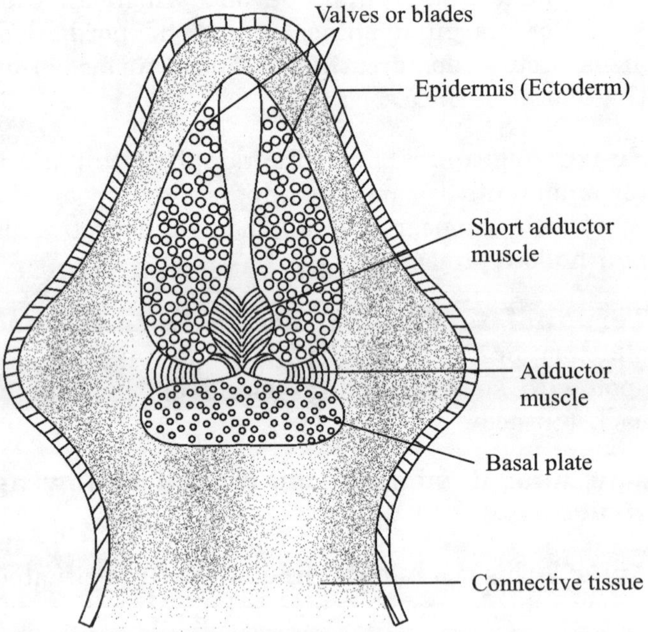


**Figure 15.5** Straight type of pedicellariae.

**Muscular layer:** This layer consists of the smooth muscle fibres and is differentiated into an outer circular muscle layer and an inner longitudinal muscle layer. Muscle layers are weakly developed except in the aboral wall, where strong longitudinal bundles radiate from the centre of disc along the mid-dorsal line of each arm to bend arms aborally.

**Coelomic epithelium:** This innermost layer of body wall lines the coelom and is composed of flagellated, cuboidal cells of mesodermal origin.

### *Coelom*

The coelom has many subdivisions but only the perivisceral coelom and the water vascular system will be dealt with. Perivisceral coelom, the largest of the coelomic compartments is the chief body cavity. Most of the space in the arms and disc is occupied by the perivisceral coelom and the viscera are located in it. This coelom is derived from the right and the left metacoel of the larva and is covered internally by a thin, transparent, ciliated epithelium, the peritoneum of the perivisceral coelom.

Cilia circulate the coelomic fluid to distribute food and oxygen to the surrounding tissues. Numerous small pores are found in the body wall and the peritoneum extends into them. These are openings of clusters of papulae. Many organs of perivisceral coelom like the pyloric caeca and gonads are suspended from the body wall by mesenteries. A true and spacious coelom is linked by a coelomic epithelium and consists of various compartments, viz.

(i) A perivisceral coelom which extends in the central disc surrounding visceral organs such as the digestive tract and gonads
(ii) Coelom of the water vascular system
(iii) Axial sinus
(iv) Perihaemal sinus and canals
(v) Genital sinuses, etc.

Coelom is filled with a colourless, alkaline, coelomic fluid, which contains various dissolved nutrient such as amino acids, fatty acids, glycerol and glucose. Coelomic fluid contains two main types of phagocytic amoeboid corpuscles—the amoebocytes with ordinary slender pseudopodia and the coelomocytes with petaloid pseudopodia. Coelomic fluid like haemolymph of arthropods bathes tissue and performs the circulatory function. It distributes nutrients to various cells and performs the respiratory—and the excretory functions also.

### *Digestive system*

Digestive system comprises the alimentary canal and the digestive glands. The canal is tubular, straight, short and extends vertically along the oral-aboral axis in the central disc. It comprises the following parts:

**Mouth:** The anterior most aperture of the alimentary canal. The mouth is situated in the centre of the peristomial membrane of the oral surface and is provided with a sphincter muscle and radial fibres. This is capable of great expansion and retraction. Mouth leads upwards into the oesophagus.

**Oesophagus:** This is a very short and vertical tube which opens aborally in stomach.

**Stomach:** A broad sac that fills interior of disc, typically divided by a horizontal constriction into a voluminous oral part—the cardiac stomach and a flattened aboral part, the pyloric stomach. Cardiac stomach with a muscular, highly folded wall bulges out to form five lobes, one opposite the arm.

Cardiac stomach is connected to the ambulacral ridge of each arm by the ligaments of muscles and connective tissues called the mesenteries or gastric ligaments, which anchor the cardiac stomach in place. During feeding, the stomach can be everted through the mouth by the contraction of muscles of body wall. Retraction of cardiac muscle is brought about by five pairs of retractor muscles arising from lateral sides of ambulacral ridge. Pyloric stomach is smaller, flat pentagonal sac and communicates with the intestine (Figure 15.6).

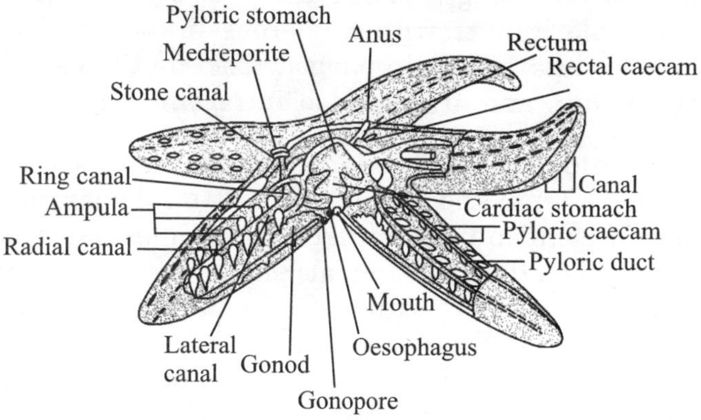


**Figure 15.6** Digestive system of starfish.

**Intestine:** This short, narrow tube runs straight upward to open out at anus. It gives off two, or three little, hollow, inter-radially placed diverticula called the *intestinal* or *rectal caeca*. Intestinal caeca are brown in colour and bears several short and irregularly shaped diverticula. The function of the intestinal caeca is disputed. However, they are considered as the excretory organ because they secrete a brown fluid. The short gut extends vertically from the mouth to the anus near the centre of the aboral disc. It consists, in order of, mouth, oesophagus, cardiac stomach, pyloric stomach, intestine and anus. It is lined internally with a ciliated epithelium and is surrounded by the perivisceral coelom.

Mouth opens into a short indistinct oesophagus. Oesophagus opens into the large, thin-walled, orange coloured cardiac stomach filling most of the perivisceral coelom of the disc. Digestion begins extracellularly in the cardiac stomach. Partially hydrolyzed materials are moved to the pyloric stomach by the ciliary currents, then to the hollow pyloric caeca, where both the extracellular and intracellular digestion takes place. Products of digestion can be stored in the cells of the pyloric caeca, or presumably diffused into the surrounding perivisceral coelom. Five pairs of stomach retractor muscles anchor the cardiac stomach to the sides of the ambulacral ridges.

Cardiac stomach opens at its aboral end into the smaller pyloric stomach. The pentagonal outline of the pyloric stomach makes it easy to recognize the cardiac stomach. Ten large pyloric caeca which are hollow diverticula emerge from the pyloric stomach. Two caeca of each arm share a common connection with the margin of the pyloric stomach. Each caecum extends through most of the length of its arm and consists of a long pyloric duct with numerous branches. The tiny food particles are phagocytized by the caecal epithelium and are digested intracellularly.

**Physiology of digestion:** *Asterias* is a carnivorous animal and feed voraciously on the warms, crustaceans and the bivalves. Sometimes *Asterias* also feed on the dead animals and under hazardous conditions may live without food for several months. Mode of feeding or ingestion in *Asterias* is most unusual. It swallows small sized animal directly through the mouth. Cardiac stomach is everted and wrapped round the food. Enzymatic secretions of the pyloric caeca poured out on prey. When digestion is completed, the stomach is withdrawn along with the digested food.

To feed shelled molluscs, *Asterias* creeps over clam and hold it with the tube feet, orients it to bring free merging of shell close to its mouth. It now arches its body with an umbrella like posture. Tube feet that are more proximal firmly grip both the valves of bivalve's shell, while ones that are more distal are attached to the substratum.

Cardiac stomach is now everted through the mouth by the concentration of body wall and pressure of coelomic fluid. Tube feet gripping valves of clams shell exert a steady pull as muscles in arms contract. As muscles of clam cannot maintain a continual state of contraction for a long time and sooner, or later adductor mussel become fatigued or exhausted and finally relax so that the shell opens.

*Asterias* inserts its already everted cardiac stomach into mantle cavity of clam, pours out its proteolytic enzymes for digesting the visceral organs of clam and convert the proteins of visceral organs into the peptones and amino acids. The polysaccharide and lipid contents transform into fatty acids and glycerol. When digestion is partially completed, the sea star withdraws its stomach along with digested food by its retractor muscles and moves on leaving behind the shell of prey.

Rest of the digestion occurs in cardiac stomach from which digested food diffuses to coelomic fluid for wide distribution and to pyloric caeca where it may be distributed to arm, or stored in storage cells of the epithelium of pyloric caeca. Undigested food is either egested out directly from the mouth, or egested out through the anus.

**Digestive glands:** Ten long brownish, or greenish glandular appendages attached to the pyloric stomach are called the *pyloric caeca*, *digestive glands*, *branchial caeca* and *hepatic caeca*. There are two pyloric caeca in each arm; each suspended from the aboral wall of arm by two longitudinal mesenteries that encloses between them the coelomic space, which is continuous at its central end with the general coelom of the disc. Each pyloric caecum consists of double series of hollow, lobulated sacs that open into a central tube duct.

Two ducts forming a pair of caeca of each arm unite to form a main pyloric duct that opens into pyloric stomach at one of its angles. Pyloric caeca are complex and are lined by the ciliated columnar epithelium and is composed of four types of epithelial cells. Pyloric caeca function like the pancreas of the vertebrates. Current producer cells bear the long flagella, maintain a steady circulation of fluids and digest the food in the cavity of caeca.

### *Water vascular (ambulacral) system*

Water vascular system is a modified part of the coelom. It consists of a system of sea water filled canal having certain corpuscles. It plays vital role in the locomotion and includes the madreporite, stone canal, ring canal, radial canal; Tiedemann's bodies, lateral canal and tube feet. A curved duct extends orally from the madreporite with calcareous skeletal rings for support. An obscure vertical partition, the interbrachial septum, is located between the bases of each pair of adjacent arms. *Asterias* possesses five interbrachial septa. Stone canal in the interbrachial septum lies between the arms I and II.

A soft *axial complex*, composed of axial gland of haemal system and axial canal of coelomic system, surrounds the stone canal. The stone canal extends to the inner surface of the mouth frame, where it joins the ring canal of the water vascular system. The thin, membranous walls of the ring canal are not calcified.

They adhere closely to the inner curve of the mouth frame and are not distinguished from it. The nine, small, soft, low, spongy Tiedemann's bodies on the inner margin of the frame mark its position. These organs are evaginations of the ring canal. Typically, out of ten such bodies, two associate with each interbrachial septum, but in *Asterias* one is missing, where the stone canal joins the ring canal and only nine are present. Lumina of these bodies are continuous with the ring canal. They were believed to remove the foreign particles by phagocytosis from the circulating fluid of the water vascular system. *Asterias* has no polian vesicles. Ten isolated ampullae of the buccal tube feet are found on aboral surface of the mouth frame. They pass through pores in the frame to connect with special enlarged buccal tube feet around the mouth.

Ring canals gives off five radial canals, one for each arm, which pass along the ambulacral groove outside the ambulacral ossicles of the skeleton. Single, unpaired lateral canals arise alternate to the right and left from radial canals and extend to the podial complices, one lateral canal to each. Each podial complex consists of tube foot and its ampulla.

Aboral end of each tube foot narrows and penetrates the overlying ambulacral ossicle and is continuous with an ampulla, which bulges into perivisceral coelom. Ampullae are covered by nonciliated peritoneum. Their lumina are also lined by peritoneum, since the space inside is a coelomic cavity. There is a very small, radial haemal vessel surrounding coelomic space. Radial water canal above the blood vessel lies between the superficial body wall, muscles and the ossicles and immediately oral to the ossicles.

**Madreporite:** A rounded calcareous plate occurs on aboral surface of the central disc in interradial position. Its surface bears several radiating, narrow, straight, or wavy groove, or furrows. Each furrow contains many minute pores at its bottom. Each pore leads into a very short fine tubular pore canal, which passes inward to the madreporite. There may be about 200 pores and pore canal. Pore canal unites to form the collecting canal, which opens into an ampulla beneath the madreporite.

**Stone canal:** Ampulla opens into S-shaped stone canal, which extends downwards (orally) and opens into a ring canal around the mouth. Walls of the stone canal are supported by a series of calcareous ring. Lumen of the stone canal is lined by tall flagellated cells. In embryonic stages and in the young, the stone canals remain as a simple tube but in the adult, the lumen of the stone canal possesses a prominent ridge with two spirally rolled lamellae, which by branching become more complicated in the structure. During its course, the thin walled tubular coelomic sac called the *axial sinus* unsheathes the stone canal.

**Ring canal:** Ring canal or water ring is located to the inner side of the peristomial ring of ossicles and directly above (aboral) the hyponeural ring sinus. It is wide and pentagonal. From its outer surface, the ring canal gives off a radial canal into the arm that runs through the length of arm and terminates as lumen of terminal tentacle. In the arm, radial water canal runs immediately to oral side of the ambulacral muscles. In each arm, radial canal gives out two series of short, narrow, transverse branches called the *lateral* or *prodigal canals*. Each lateral canal is attached to base of a tube foot and is provided with a valve to prevent backward flow of fluid into radial canal.

**Tiedemann's bodies:** Ring canal gives out inter-radially nine, small, yellowish, irregular, or rounded glandular bodies called the *racemose* or *Tiedemann's bodies* from its inner margins. These bodies rest on the peristomial ring of ossicles. The function of the bodies is still unknown but is supposed to be lymphatic glands for manufacturing amoebocytes of the water vascular system.

**Polian vesicles:** Ring canal gives off, on inner side in the radial position, one, two, or four little pear-shaped, thin walled, contractile bladders or reservoirs with the long necks called the *polian vesicles*, which regulate the pressure inside the ambulacral system and manufacture amoeboid cells.

**Radial canal:** There are four rows of tube feet in each ambulacral groove. A tube foot or podium is a hollow, elastic, thin walled, closed cylinder or sac like structure having an upper sac like ampulla, a middle tubular podium and a lower disc like sucker. Ampulla, lies within the arm, project into the coelom above the ambulacral pore, which is a gap between the adjacent ambulacral ossicles for passage of the podium. Tube feet are chief locomotory and respiratory organs.

### *Development of the water vascular system*

During the early larval development, an echinoderm is bilateral in form. Radial arrangement of the parts in the adult occurs due to the formation of the mesoderm. Mesothelial sacs are formed as lateral out pocketing of the entoderm just as in embryos of many other groups. Both right and left mesoderm pouches undergo a constriction, which separates each into an anterior and a posterior sac. The posterior sacs increase in size and ultimately form right and left coelomic cavities.

The anterior sacs, the rudiments of the water-vascular system are called the *hydrocoel sacs*. The left hydrocoel communicates with the body wall through a tubular outgrowth, which becomes the stone canal. The right hydrocoel fails to develop, remains vestigial for sometimes and finally disappears. The entire water vascular system is thus formed from the left hydrocoel.

As it increases in size, it encircles the oesophagus of the larva and becomes the ring canal. Five radial pouches extend outward from this ring canal and as they increase in length, they become recognizable as the radial vessels, one of which passes along the arm of the adult. The water vascular system thus originates from one of the mesodermal pouches of a bilaterally symmetrical larva, which in later development assumes a radial arrangement of its parts (Figure 15.7).

Skeletal system, the most characteristic features of the echinoderms varies in details arrangement and extent of development. Certain plates are constant in their fundamental relations in several

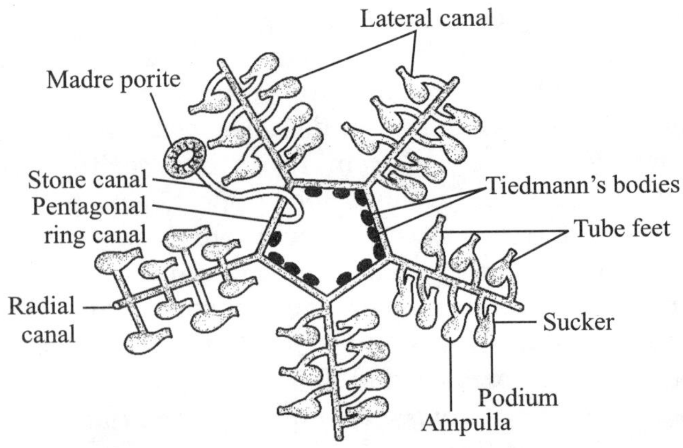


**Figure 15.7** Water vascular system of *Asterias*.

classes. Skeletal plates originate from the mesoderm and lie near the body surface directly beneath the outer body covering. Spines, often associated with these plates signify the name Echinodermata which is a Latin word for spiny skin.

In some cases, the plates are rigidly articulated to form a continuous shell or test which is unchangeable in form and within which most of the organs are located. In other classes (Asteroidea, Ophiuroidea and Crinoidea), some plates are movable and permit flexibility in parts which they cover. In the Holothuroidea, the skeletal plates are so poorly developed that their presence in the soft body wall is restricted to minute discs. However, the absence of polian vesicle in *Asterias* is noteworthy.

### *Locomotion*

Head or the anterior end is absent. Therefore, the organism is capable to move in the direction according to its desire. It can move on the horizontal as well as on the vertical surfaces by the help of the tube feet.

**Locomotion on horizontal surface:** When an *Asterias* desires to move on a horizontal surface in a given direction, the arms pointing in a direction is lifted. The raised arm contracts, valve in the lateral canal close and water of ampullae's is forced into the podia. Podia of the tube feet become extended, elongated and enlarged in general direction of the movement, due to hydrostatic pressure produced by the influx of water into them. Subsequently, terminal suckers of tube feet become attached to the substratum.

Central parts of terminal suckers are withdrawn to form suction cups. Due to vacuum so produced, the suckers firmly grip the substratum. Mucous secreted by tips of tubes aids in the attachment. Tube feet now pivot forward on their attached suckers, assuming the vertical position, force their fluid back into ampullae and release their sucker. Ampullae then contract again and whole sequence of event is repeated.

**Locomotion on vertical surface:** In climbing the vertical surface, the tube feet pull body forward. By alternate contraction and expansion of tube feet and by adherence of suckers of tube feet on surface, *Asterias* climbs on the vertical surface. *Asterias* employs its tube feet only when it moves on the hard rocky substratum. However, on the soft mud or sand, the suckers of the tube feet become useless. On such soft surface, the animals literally walk on its extended tube feet, which act like small legs. Beside locomotion, tube feet serve many functions such as clinging of animal's body to the substratum, tactile and respiratory functions (Figure 15.8).

### *Excretory system*

Specialized excretory organs are absent. Nitrogenous metabolic excretory waste usually contains ammonium compounds. The product of nitrogen metabolism is ammonia, which is eliminated by diffusion from papulae and tube feet. They pass from various tissues into coelomic fluid and from there, they diffuses through thin walls or from coelom. The coelomic and interstitial fluids are osmotically similar to sea water. Leakage of water vascular system fluid across the pressurized tube feet is countered by the slightly higher osmolarity of the fluid of water vascular system.

### *Circulatory system*

True blood vascular system is absent. The system responsible for the circulation of digested food to various body organs is often called the *circulatory system* which includes the following two systems—perihaemal system and haemal system.

### *Perihaemal system*

Perihaemal system like the water vascular system is derived from the coelom and is composed of many tubular coelomic sinuses such as the axial sinus, aboral ring sinus, genital sinuses, oral ring sinus, radial perihaemal sinuses, marginal sinuses and peribranchial sinuses.

**Axial sinus:** This is thin walled vertical, wide, tubular coelomic cavity enclosing the stone canal

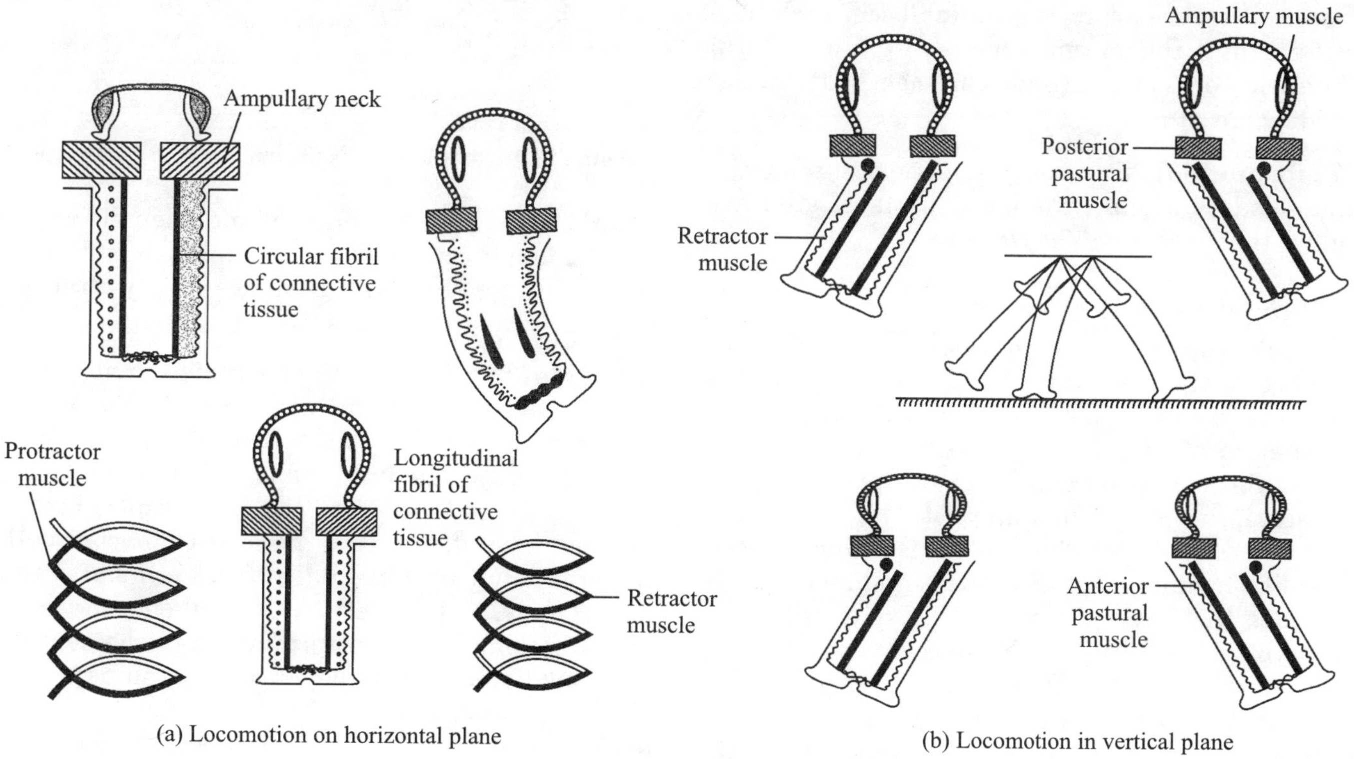


**Figure 15.8** Locomotion in *Asterias*.

and axial gland. The axial sinus, stone canal and the axial gland together form an axial complex.

**Aboral ring sinus:** This is a tubular, pentagonal channel or sinus around the intestine, lying just inside the aboral wall of the central disc. It communicates with the axial sinuses.

**Genital sinuses:** Aboral ring sinus gives off five pairs of genital branches, one pair in each arm. The genital sinuses surround the gonads.

**Oral ring sinus:** At its oral end, the axial sinus opens into the inner division of a circular channel, oral, peribuccal, perihaemal or hypo neural ring sinus, which runs around the mouth. This large tubular sinus is divisible into a narrow and an outer wide ring by an oblique circular septum called the *haemal strand*.

**Radial perihaemal sinuses:** Outer division of ring gives out five radial hyponeural or perihaemal sinuses, one of which extends through each arm between the radial nerve and radial water canal. Each radial sinus is divided longitudinally into two by a vertical partition which communicates with haemal strand. Radial perihaemal sinuses also give out fine channel into the tube feet.

**Marginal sinuses:** In each arm, two longitudinal marginal sinuses run longitudinally on each side just aboral to the marginal nerve cord. Fine lateral channels connect marginal channels with radial perihaemal sinuses.

**Peribranchial sinuses:** Sinuses occur as circular spaces around the basal parts of gills are called the *peribranchial sinuses*.

### *Haemal system*

Haemal system is a blood vascular system with a space in the connective tissue. It is atypical in that the blood vessels end blindly. Continuous circulation of the blood around a circuit does not occur. Blood vessels extend to the pyloric caeca, gonads, nervous system, water vascular system and perivisceral coelom. This system is reduced, open type like haemocoel of arthropods and molluscs. It includes the inter-communicating spaces derived embryologically from the blastocoel with coelomic epithelium. Coelomic fluid of the haemal system contains the coelomocytes and remains enclosed in coelomic spaces of perihaemal system. The main haemal sinuses are as follows:

**Oral haemal ring:** This is a circular haemal sinus, located around the mouth, just below the ring canal of the water vascular system. This ring, a fine channel or a ring of lacunar tissue, runs in septum dividing the hyponeural sinus and is connected with the aboral haemal ring through the axial gland.

**Radial haemal sinuses or strands:** These arise radially from the oral haemal ring and one extends into each arm along the floor of ambulacral groove just below the radial canal of a water vascular system (Figure 15.9).

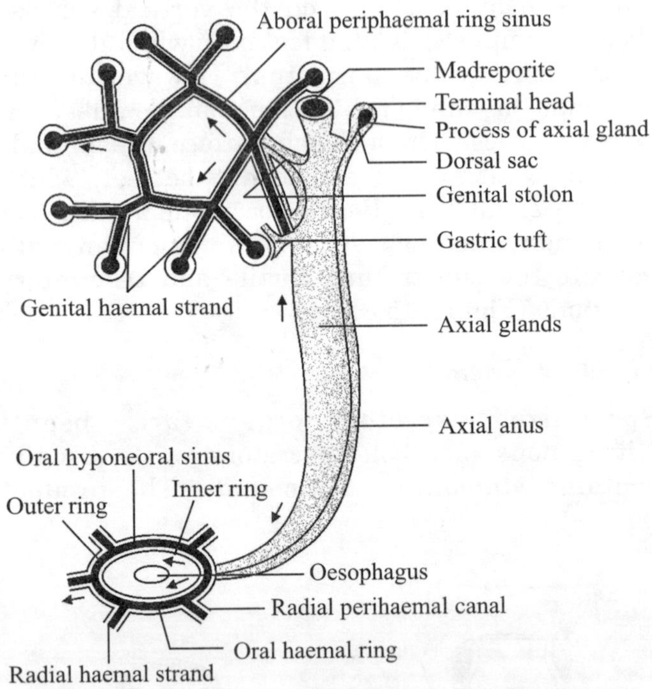


**Figure 15.9** Haemal and perihaemal system of *Asterias*.

**Axial complex:** Perihaemal and haemal system are intimately connected by a complicated structure called the *axial complex*. Both are closely attached with the walls of axial sinus by mesenteries.

**Axial gland:** This principal part of haemal system is an elongated, fusiform, brownish, or purple coloured, spongy body. It is covered externally by the coelomic epithelium and is called as the *heart*, or ovoid gland, or dorsal organ, or septal organ, or brown gland. An axial gland is connected with the oral and aboral haemal sinuses at its oral and aboral ends, respectively. At its oral end, the axial gland become thin and terminates in the septum that subdivides the hyponeural ring sinus.

At its aboral end, the axial gland has an aboral extension or terminal process, called the *head prose*, which is lodged in a separate closed contractile coelomic sac called the *dorsal sac*. Dorsal sac is situated below the madreporite and close to the ampulla of stone canal, but is not communicated with the ampulla of the stone canal.

A pair of gastric tufts arises from the haemal sinuses in wall of the cardiac stomach and passes into haemal circulation through the gastric tufts. Axial gland has an external lining of peritoneum and the connective tissue fill its interior numerous spaces containing the irregularly arranged coelomocytes. Axial gland has an intimate relation with the blood circulation in the perihaemal and haemal channels.

**Aboral haemal ring:** This is a pentagonal ring canal lying beneath the aboral surface of the central disc from which five pairs of genital haemal strands extend to gonads.

**Function:** Haemal system acts for the distribution of food substances carried by the coelomocytes. The flow of fluid is maintained by the contractile activity of the dorsal sac. The axial gland produces sex cells, which reach gonads through the aboral haemal ring and its branches. The haemal plexus of the axial gland is the heart of the haemal system.

### *Respiratory system*

Haemal system does not distribute oxygen to the tissues. Most coelomic spaces are associated indirectly with respiratory surfaces and the circulating coelomic fluid transports the gasses. Gills or papulae and tube feet are the respiratory structures for the perivisceral coelom. Tube feet serve this purpose for the water vascular system.

Papulae, the chief respiratory organs are simple contractile, transparent, hollow outgrowths of the body wall on the aboral surface having the ciliated epithelium at the outer and inner surfaces. They are derived from the coelom and their lumen remain in the direct communication with the coelom. An exchange of $O_2$ and $CO_2$ takes place between the sea water and body fluid. Cilia of epithelial cells play role in movement of the coelomic fluid and in increasing respiratory water currents in sea water. Thin walled richly vascular moist body parts also act as respiratory organs.

### *Nervous system*

Nervous system is intraepithelial and consists of a plexus extending through the body wall. Concentrations of plexus form a circumoral ring, from which arise the radial nerves. Five radial nerves are thickened bands of epidermis lying on the ambulacral axis of the oral surface of each arm. Each nerve is a thin, slightly raised, brown, or yellow line extending from the nerve ring to the eyespot at the tip of the arm. A radial nerve joins the circumoral nerve ring around the mouth. Nervous system is simple and primitive type and composed of nerve fibres and nerve nets which are closely associated with epidermis. The following four units are placed at the different levels in disc and arms:

**Oral or ectoneural or epidermal nervous system:** This is situated just beneath the epidermis. It is main part of the sensory nervous system, sensory in nature and comprises the following parts:

*Nerve ring:* This is pentagonal in shape and is circumoral, i.e. it occurs around the mouth in the peristomial membrane. It supplies nerve fibres to the peristomial membrane and oesophagus and at each radius gives off a radial nerve.

*Radial nerve:* Nerve ring gives off five radial nerves each of which runs throughout the length of arm in bottom of ambulacral groove. Each radial nerve terminates as a sensory cushion on the aboral side of terminal tentacle. This is continuous on its outer side with epidermis and separated on its inner side from the hyponeural sinus only by a thin dermis and coelomic epithelium. The radial nerve consists of fibrillae arranged in layers and interspersed with the bipolar and multipolar ganglion cells.

### *Subepidermal nerve complex*

This is an extensive network of nine cells and nerve fibres beneath the epidermis all over the body surface, including the gills and pedicellariae. It is connected with the radial nerve cords by the nerve fibres. The subepidermal nerve plexus is thickened into a cord, which forms two marginal nerves, each of which extends throughout the length of an arm on each side and gives off a longitudinal series of lateral motor nerves, which supply the ossicles muscles and coelomic epithelium and a nerve ring in the suckers of each tube foot. Nervous system act as the central nervous system. It has sensory and mother neurons.

### *Deep or hyponeural nervous system*

This occurs as a nervous layer in the lateral part

of oral wall of the hyponeural sinus beneath the coelomic epithelium lining the sinus. This nervous layer called the *Lange's nerve* is separated from the lateral part of radial nerve by a thin layer of the dermal connective tissue. The Lange's nerve gives off a series of nerve along arm into the adjacent lower transverse muscle extending between the ambulacral ossicles in roof of the hyponeural sinus. The Lange's nerve continues to the peristomial region, where it forms the five inter radial thickening in the floor of ring sinus that lies aboral to the main ring.

**Aboral or coelomic nerve system:** Situated just outside the parietal peritoneum on the aboral side, with connection to the marginal nerves by nerves fibres, it consists of a nerve ring in central disc and a nerve in each arm. It innervates body muscles of aboral side and is motor in function.

**Visceral nervous system:** Occurring in wall of the gut just outside the enteric epithelium, it innervates the muscles of the alimentary canal and is connected with the visceral receptors.

### *Sense organs*

**Eyes:** Eyes, the most significant sensory organs, are simple, pigmented and occur at the base of the terminal tentacles. On the oral surface, the base of each terminal tentacle occurs on optic cushion, which is composed of thick epidermis with many photoreceptors or pigmented cup ocelli. Each ocellus is a cup shaped, or funnel like pocket of the ectoderm covered externally by the cuticle beneath which a lens is formed by the epidermis in many species.

The wall of cup consists of the epidermal cells. These are shorter, stouter and filled distally with the red pigment granules and retinal cells disposed between the pigment cells. Retinal cells are elongated cells with a distal bulbous enlargement. A number of ocelli in one optic cushion or eyes range from 80 to 200 may even increase with age in different species. A transparent gelatinous tissue fills cavity of the ocellus, which are light perceiving organs and can detect changes in the light intensity.

**Terminal tentacles:** These are sensory cells, tactile and sensitive to food and other stimuli.

**Neurosensory cells:** Many neurosensory cells serve as tango receptor and chemo receptors.

### *Development*

Many developmental stages are spheres. The unfertilized female gamete, the egg or ovum is a large sphere recognized by the conspicuous female pronucleus containing a distinct nucleolus. Once fertilized, the egg becomes a zygote which is also a sphere but wrinkled. The transparent fertilization membrane is present around the cell and the pronucleus can no longer be seen.

The fertilization membrane produced immediately after the fertilization prevents penetration by the additional sperm. It persists until the blastula stage. The zygote, which is also a single cell, has the same size as the ovum. The blastula is ciliated. Several ages of blastulae may be present. The youngest would have about 32 large, recognizable cells. Interior spaces of blastula are the blastocoel.

As divisions continue, the cells in the blastula wall become smaller until they can no longer be distinguished from each other. Finally, one end of the blastula thickens for initiating the gastrulation. This end invaginates to form a double walled embryo called the *gastrula*. The early gastrula has only a short invagination, which rapidly increase in size. The blastocoel remains, but a second cavity, the archenteron is produced. The archenteron opens to the exterior by the blastopore.

The outer wall of the gastrula is ectoderm, whereas the wall of the archenteron is endoderm. Pouches of the archenteron form the coelom and mesoderm. The asteroid gastrula develops into a series of two larvae, the first is the bipinnaria larva followed by the brachiolaria larva. Larvae are ciliated, bilaterally symmetrical and feed on diatoms in the plankton. Bipinnaria has one or two locomotor ciliary bands. Young larvae have only one band, but older larvae have two.

The most distinguished feature of the larva is its C-shaped gut. The gut extends dorsally, then posteriorly and then ventrally to connect the mouth with anus. The locomotor band borders the embryo and passes immediately anterior to the mouth on ventral surface. On the left side of the embryo at about the level of the oesophagus-stomach junction, there is a small pouch connected to the dorsal surface by a canal and pore.

The pouch is the left protomesocoel, the canal is the pore canal and the pore is the hydropore, which is the larval nephridiopore. During metamorphosis, the left protomesocoel becomes the water vascular system. The pore canal becomes the stone canal and the hydropore is the presumptive madreporite. The right protocoel becomes the pericardial sac around the heart (Figure 15.10).

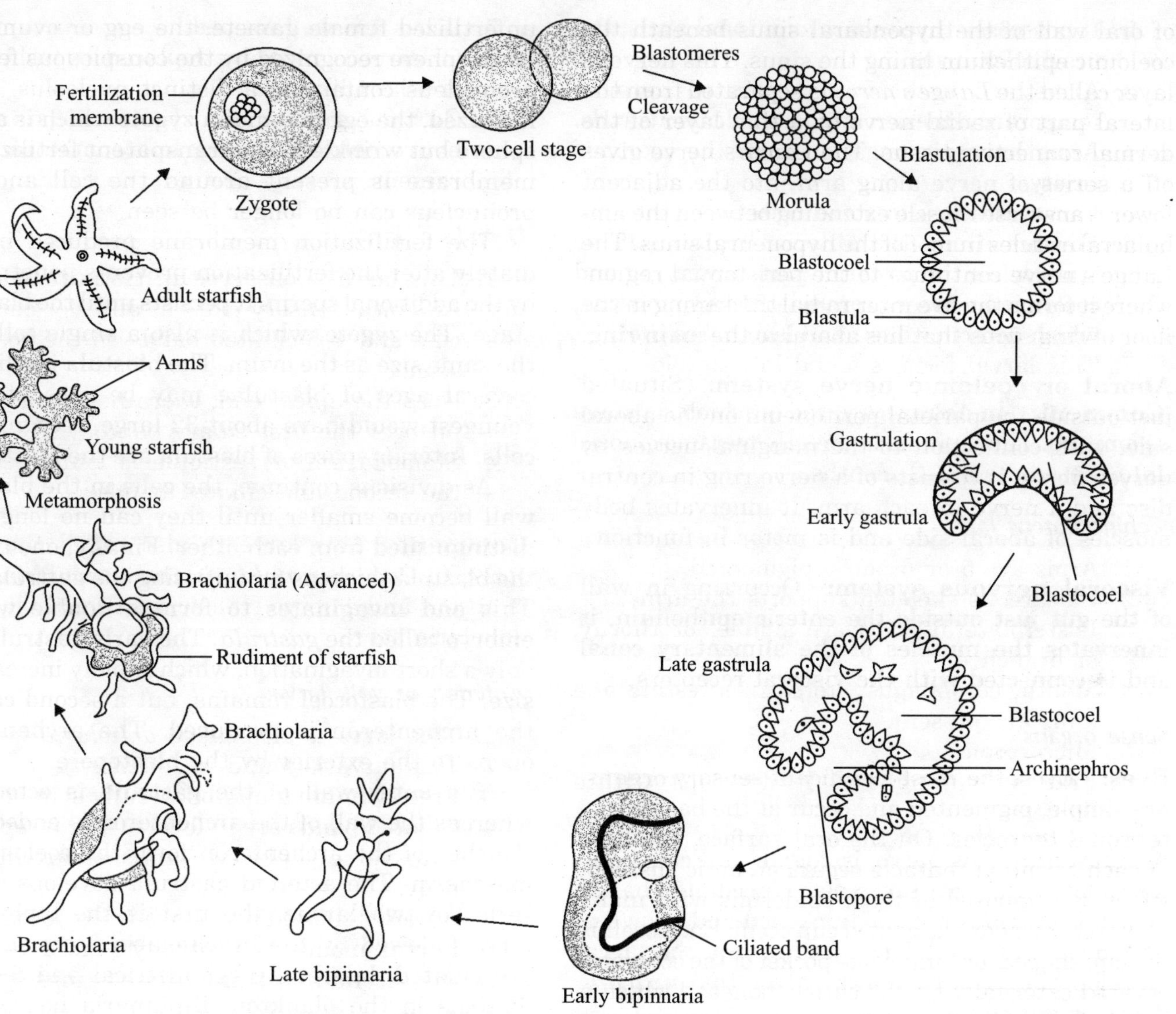


**Figure 15.10** Life history of starfish.

## 15.9 ECHINODERM LARVAE

Echinoderms mostly have free pelagic larvae. *Asterina* passes a considerable time in the egg membrane and larvae are not pelagic. A few species keep the young in brood pouches until they attain adult hood. Larval forms of echinoderms are as follows.

### *Bipinnaria larva*

1. Free swimming.
2. Bilaterally symmetrical.
3. Angular in shape.
4. Arms are dorsomedian, ventromedian, preoral, anterodorsal, posterodorsal, postoral and posterolateral.
5. Ciliated bands are preoral loop and postoral loop.
6. Arms are bordered by the ciliary bands.
7. Preoral loop encircles mouth.
8. Postoral loop surrounds the anus.
9. Presence of mouth, oesophagus, stomach, intestine and anus.

**Comment:** This larval form is found in the Assteroidea and bears a close resemblance with the auricularia. In *Luidia*, it has a slender anterior part, which ends into the two wide arms.

### *Brachiolaria larva*

1. Bipininnaria changes into the brachiolaria larva.

2. It possesses three short additional arms, the brachiolar arms.
3. Arms are beset with warts to help in temporary adhesion and lack calcareous rods.
4. Arms have prolongation from the coelomic cavity.
5. Larva attaches to some object.
6. For sometime, the anterior region acts as stalk, while the posterior part, having gut and coelomic chambers, converts into a young star, which detaches itself and lead a free life.
7. This larval form is found in Asteroidea.

**Comment:** In *Astropecten,* the brachiolaria stage is omitted. The bipinnaria directly metamorphoses into the adult in 2–3 months.

### *Echinopluteus larva*

1. Arms are 5 or 6 pairs, pigmented.
2. Calcareous skeleton supports the arms.
3. Posterolateral arms are simple, or thorny, or branched.
4. Mouth, oesophagus, stomach, intestine and anus are present.
5. Microscopic.
6. Free swimming and it develops within 30 days.

**Comment:** This larva is found in Echinoidea. Locomotion of larva takes place by ciliated bands, which sometimes become thickened and known as the *epaulettes*. In *Arbacia* and *Idaris,* the larvae develop the ciliated lobes, called the *auricular lobes* or *auricles*.

### *Ophiopluteus larva*

1. Arms are four pairs in number, slender.
2. Calcareous skeleton supports the arms.
3. Posterolateral arms are the largest and directed forward giving the larva a V-shaped appearance.
4. Ciliated bands are present on the edges of the arms.
5. Mouth, oesophagus, stomach, intestine and anus are present.
6. This is free-swimming larva of Ophiuroidea.

### *Auricularia larva*

1. Bilaterally symmetrical.
2. Present in Holothuroidea.
3. Pelagic, measures 0.5 to 1 mm in length.
4. Single longitudinal ciliated band is present.
5. Lobes are well formed.
6. Calcareous rods are replaced by spheroids, or wheel like bodies.
7. Curved gut, with sacciform stomach, hydrocoel, right and left somatocoels, is present.

### *Doliolaria larva*

1. This larva is observed in Hothuroidea.
2. An apical sensory plate at anterior end with cilia and 4–5 ciliated bands on the body are seen.
3. Near the apical end, there lies an adhesive pit over the first ciliary band.
4. Stomodaeum or vestibule is found between the second and ciliated bands.

**Comment:** This larva is also called the *pupa stage*. In *Cucumaria plancii*, the auricularia stage is absent and embryo transforms directly into the doliolaria stage.

### *Antedon or yolk larva*

1. This larva is also called *doliolaria larva*.
2. This larva is found in *Antedon*.
3. Free swimming and bilaterally symmetrical.
4. Ciliated bands are in form of 4 or 5 separate, transversely placed bands encircling the body.
5. Anterior ciliated ring is ventrally incomplete.
6. Larval mouth is ventrally placed between the second and third ciliated rings.

### *Cystidean or pentacrinoid larva*

1. This larva is found in the crinoids.
2. The larva possesses an elongated narrow stalk and becomes broader.
3. Closed ectodermal vesicle is present.
4. The larva resembles closely the adult *Pentacrinus* (Figures 15.11–15.18).

### *Significance of echinoderm larvae*

Except the larva of crinoids which becomes sedentary, larvae of the Holothuroidea, Arteroidea, Echinoidea and Ophiuridea are fundamentally similar in features such as the peroral and anal loops, V-shaped ciliated bands, gut and enterocoelic coelom. They exhibit many common characters. Their respective groups might have been originated from a common ancestor and the hypothetical common ancestor is named the *dipleurula*. The

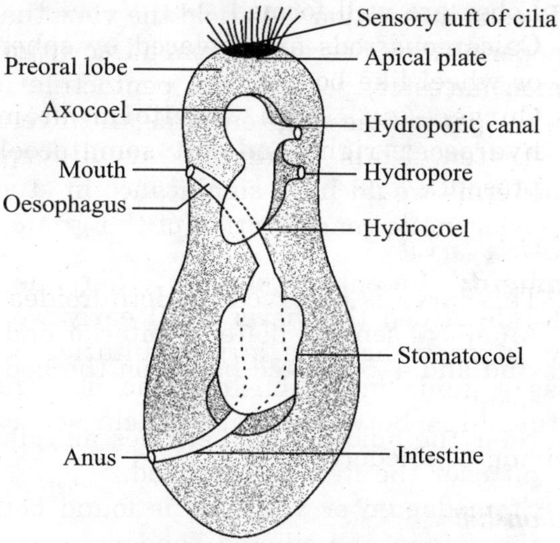


**Figure 15.11** *Dipleurula larva.*

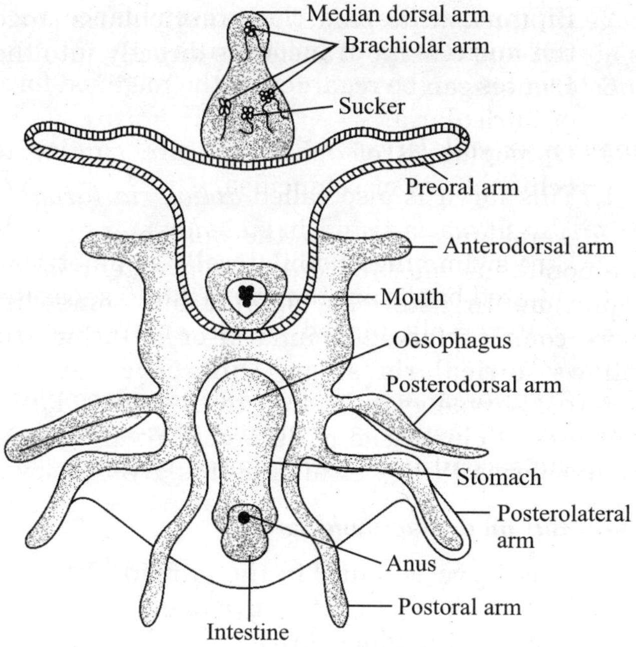


**Figure 15.12** *Brachiolaria larva.*

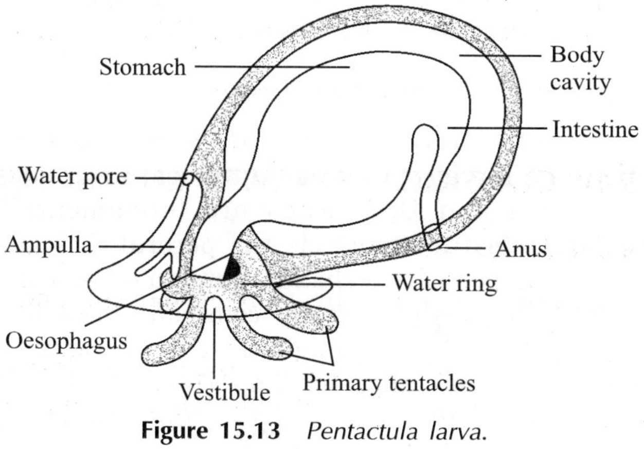


**Figure 15.13** *Pentactula larva.*

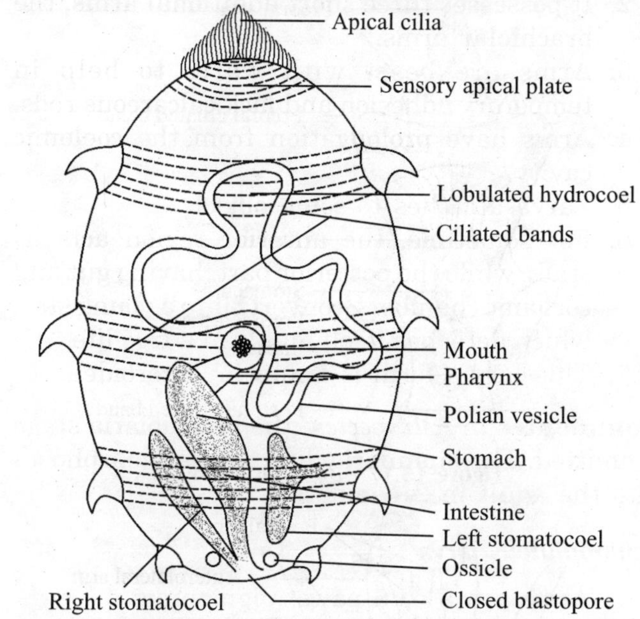


**Figure 15.14** Fully formed *doliolaria.*

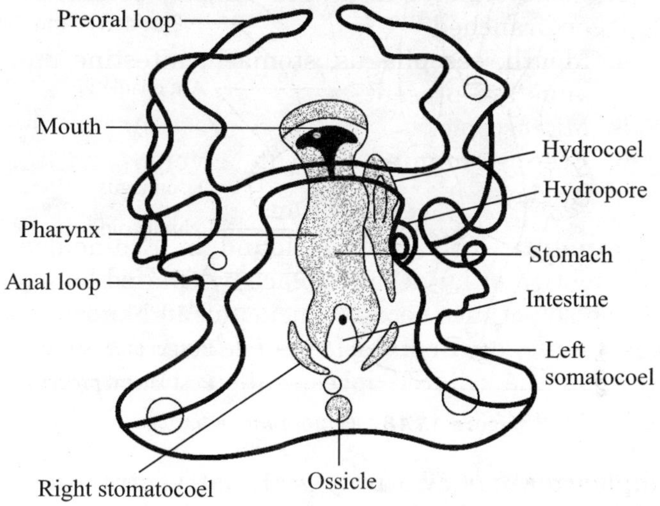


**Figure 15.15** *Auricularia larva.*

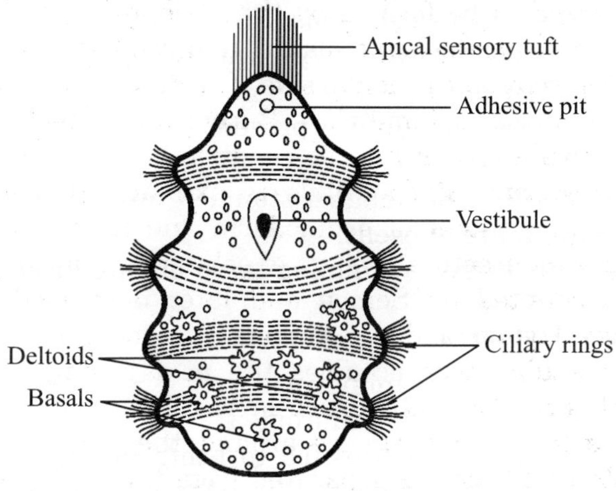


**Figure 15.16** *Doliolara larva.*

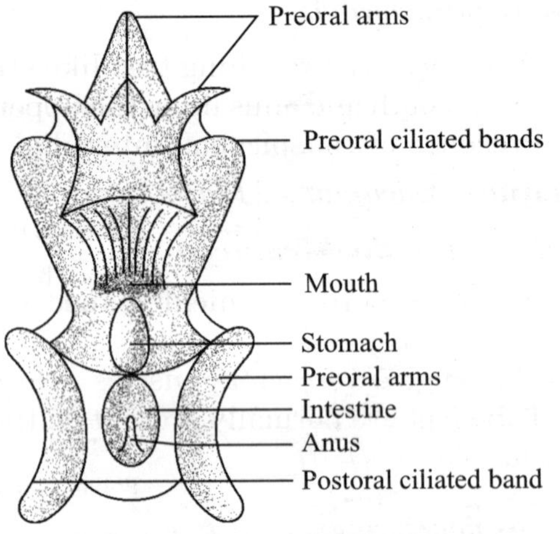


**Figure 15.17** *Bipinnaria larva.*

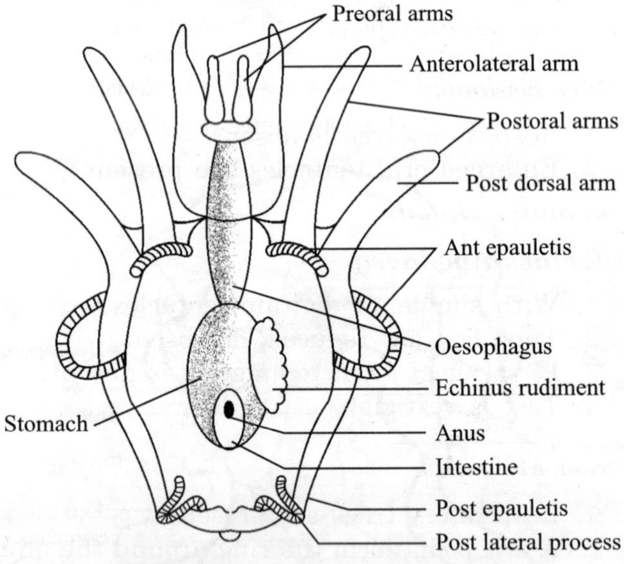


**Figure 15.18** *Echinopluteus larva.*

dipleurula has oval body with flat ventral sides. It has a mouth and a digestive tube having the stomodaeum, stomach and protodaeum. The anus supposed to be formed by the atriopre.

Three pericoelomic vesicles and two water pores are present on the dorsal side. The dipleurula stage is bilaterally symmetrical. Bather advocated that dipleurula represents the common ancestor of the echinoderms. Bather fails to explain the water vascular system of echinoderm and its coelomic origin. Pentactula concept of echinoderm ancestry was adopted by Semon and later developed by Berry, Hyman and many others. According to them, pentactula larva occupies the next evolutionary rank over the dipleurula larva. The pentactula larva has five tentacles around the mouth and hydrocoel becomes separated from the rest of the coelom to form the water vascular system.

Recent inclination is to hold the view that the echinoderms have descended from a free-swimming common ancestor, possibly the pentactula larva. From the pantactula stage onwards, the divergence has started. Thus, studying larval forms of echinoderms would have significance in studying the phylogenetic consideration of the group.

**Comment:** Recent workers on other line have totally discarded the views of all early workers. They suggest that the larval affinities cannot act as a guide in establishing the phylogenetic relationships between different classes of the surviving echinoderms.

***Relationship***

1. Brachiolaria is modified form of the bipinnaria.
2. Bipinnaria bears a close resemblance with the auricularia of holothurians.
3. Pleutes can be regarded as the modified form of auricularia.
4. Ophiopluteus of ophiuridea are similar to echinoplutes of echinoidea.

Resemblance between the ophiopluteus and echinopluteus must be due to convergent larval evolution. In asteroids, similar but somewhat more complex larva, bipinnaria or branchiolaria follows auricularia stage. Difference between Asteroid larvae may be due to divergent larval evolution. Comparison of different larval types of Echinoderms (Villee et al. 1984) is given below.

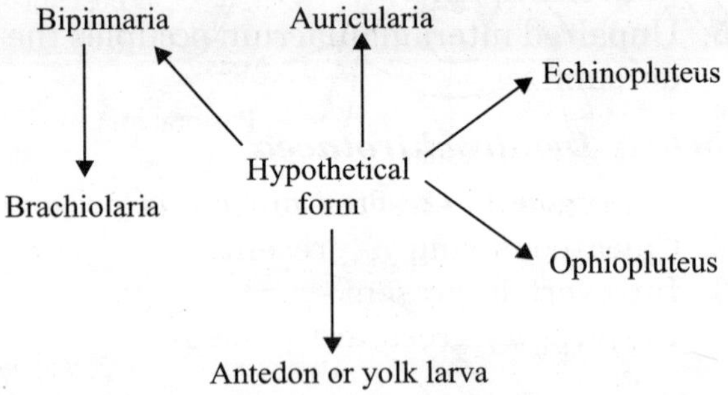


## 15.10 CLASSIFICATION WITH CHARACTERS

***Characters***

1. Skin is covered by thin ecotoderm and is often ciliated.
2. The complex skeleton is made of numerous plates or ossicles, each of which is composed of a single calcite crystal.

3. Body is radially symmetrical.
4. Alimentary canal extends from the mouth to the anus.
5. Presence of hydrocoel.
6. Water vessels of hydrocoel are carried by the ambulacra.
7. Excretory system is absent.
8. Mostly unisexual.
9. Reproductive organs comprise the interradial gonads with separate ducts.

Classification followed here is based on the classification scheme of Marshall and William (1972).

### *Subphylum Echinozoa*

1. Arms are absent.
2. Hydrocoel forms a ring around the mouth.
3. Water vascular system is meridional.
4. Mouth and anus are located at the opposite end.

### *Class Helicoplacoidea*

1. All are extinct.
2. Free living, fusiform, placoid echinozoans.

**Example:** *Helicoplacus*

### *Class Holothuroidea*

1. Bilateral symmetrical body.
2. Mouth and anus are situated at the opposite end.
3. Adoral axis is generally horizontal.
4. Unpaired ambulacrum is located in the midventral line.
5. Unpaired interambulacrum occupies the mid-dorsum.

### ***Subclass Dendrochirotacea***

1. Pharyngeal retractor muscle is present.
2. Calcareous ring is present.
3. Introvert is present
4. Respiratory trees are present.

### *Order Dactylochirotida*

1. Presence of digitate or feebly branched tentacles in some forms.
2. In some forms, the body is enclosed in a flexible test.
3. Mouth and anus are situated at the opposite poles.
4. Interradial valves sometimes protect the anus.

**Example:** *Thuroholia*

### *Order Dendrochiritida*

1. Tentacles are branching tree-like structures.
2. Both mouth and anus lie on the upper surface.
3. Presence of conspicuous interadial valves.

**Example:** *Cucumaria, Placothuria*

### ***Subclass Aspidochirotacea***

1. Presence of 10–30 shield-shaped tentacles.
2. Test is vestigial.
3. Pharyngeal retractor muscles are present.
4. Tube feet are normally present on the ventral side.

### *Order Aspidochirotida*

1. Presence of large branched respiratory trees.

**Example:** *Stichopus*

### *Order Elasipodida*

1. Respiratory tree is absent.
2. Enlarged oral tentacles are present.

**Example:** *Elpidia*

### ***Subclass Apodacea***

1. With simple, or digitate tentacles.
2. Tube feet are reduced, or absent.
3. Pharyngeal retractor muscle is absent.
4. Test is vestigial.

### *Order Molpadida*

1. Respiratory trees are present.
2. A few podia form the ring around the anus.
3. The plump body carries a tail like prolongation.

**Example:** *Molpadia*

### *Order Apodida*

1. Respiratory trees, podia, radial canals and tentacle ampullae are absent.
2. Body is elongated.
3. Spicules vary in shape.

**Example:** *Leptosynapta*

### *Class Edrioasteroidea*

1. All the members are extinct.
2. Apparently had the external water vessels with the tube feet resting in the small capsule.

**Example:** *Edriaster*

### *Class Echinoidea*

1. Mostly bear the radioles.

2. Test with moveable tentacles.
3. Periproct is present.
4. Many species are dull coloured.
5. Some species have brittle, hollow spines

### *Subclass Perischoechinoidea*

1. Usually with a flexible test.
2. All are fossils.

#### *Order Echinocystitoidea*

1. With a varying number of columns of plates.
2. Amb plates overlapped adorally.
3. Interamb plates overlapped upon adjacent amb.

**Example:** *Eothuria*

#### *Order Bothriocidaroida*

1. With a radially placed madreporite.

**Example:** *Bothriocidaris*

#### *Order Palaechinoida*

1. With a rigid test.
2. Ambs consisted of two or more columns.

**Example:** *Palaechinus, melonechinus*

#### *Order Cidaroida*

1. Ambs are found as the bicolumnar series of plates.
2. Each plate is provided with one zygopore.

**Example:** Lepidocidaris, Archaeocidaris

### *Subclass Euechinoidea*

1. Test is formed of two rows of ambulacral and interambulacral plates.
2. Anus is located in the posterior side.
3. Aristotle's lantern consists of jaws and plates.

#### *Superorder Diadematacea*

1. Presence of the lantern in the adult and juvenile stages.
2. Dentition is qulodont.

#### *Order Echinothuroida*

1. An endocyclic test is present.
2. Radioles are hollow.

#### *Order Diadematoida*

1. Amb plates do not extend across the peristome.

#### *Order Pygasteroida*

1. Anus is located in the posterior interamb.

#### *Superorder Echinacea*

1. Presence of the lantern in the adult and juvenile stages.
2. Stirodont, or camarodont dentition is present.
3. Test is endocyclic.

#### *Order Salenioida*

1. Dentition is striodont.
2. Each inteamb carries a large primary radiole.

#### *Order Hemicidaroida*

1. Absence of enlarged subanal plates.
2. Tubercles are perforated.

#### *Order Phymosomatoida*

1. Tubercles are imperforated.

#### *Order Arbacioida*

1. Usually 4, or 5 subanal plates are present.

#### *Order Temnopleuroida*

1. Lantern is camarodont.
2. Test surface is sculptured.

**Example:** *Mespilia*

#### *Order Echinoida*

1. Amb plates are invariably compounded in echinoid manner.
2. Test is sculptured.

**Example:** *Echinus*

#### *Superorder Gnathostomata*

1. Test is exocyclic.
2. Teeth are provided with lateral flanges.

#### *Order Holectypoida*

1. Test is inflated.
2. Teeth are provided with lateral flanges.

#### *Order Clypeasteroida*

1. Test is inflated.
2. No lateral flanges in the teeth.

**Example:** *Laganum*

***Superorder Atelostomata***

1. Lantern is absent in the adult stages.
2. Amb comb is simple.

***Order Cassiduloida***

1. Apical system is compact.

**Example:** *Nucleolites*

***Order Holasteroida***

1. Apical system is elongated in the anterior-posterior axis.

**Example:** *Echinosigra*

***Order Spatangoida***

1. Plates of posterior interamb form an amphisternous plastron.
2. Apical system is complex.

**Example:** *Echinocardium*

***Class Ophiocistoidea***

1. All the members are free living.
2. All the members are extinct.

**Example:** *Volchovia*

***Subphylum Homalozoa***

1. All are extinct.

**Example:** *Enoploura*

***Subphylum Crinozoa***

1. Spherical, placoid echinoderms.
2. With meridional symmetry.
3. Aboral plate is known as *calyx*.
4. Ambulacral may be arranged like spoke of a cycle wheel.

***Subclass Inadunata***

1. All are extinct.

**Example:** *Dendricrinus*

***Subclass Camerata***

1. All are extinct.

**Example:** *Cameaerinus*

***Subclass Flexibilia***

1. All are extinct.

**Example:** *Edriocrinus*

***Subclass Articulata***

1. All are living.
2. Calyx is dicyclic.
3. Uniserial arms bear pinnules.

**Example:** *Antedon*

***Subphylum Asterozoa***

1. Star shaped body.
2. Without stem.
3. Axial arm ossicles are biserial.
4. Gut sends caeca into the arm

***Class Stelleroidea***

1. The body is star shaped.
2. With the open ambulacral groove.

***Subclass Somasteroidea***

1. Skeleton is pinnate shaped.
2. Arms bear resemblance with the petals.
3. Intestine is closed.
4. Anus is absent.
5. Body is flat disc-like.
6. Having a central spherical plate.

**Example:** *Chinianster*

***Subclass Asteroidea***

1. Body is star shaped.
2. Oral and aboral arms are distinct.
3. Presence of ambulacral groove with the tube feet.

**Example:** *Asterias, Astropecten*

***Order Platyasterida***

1. Anus is absent.
2. Suckers are absent in the tube feet.

**Example:** *Luidia*

***Order Phanerozonida***

1. Suckers are absent in the tube feet.
2. Anus is sometimes present.

**Example:** *Plunaster, Pentagonaster*

***Order Spinulosida***

1. Anus is always present.
2. Suckers are present in the tube feet
3. Cpedicellariae are very simple.

**Example:** *Asterina*

***Order Euclasteroida***

1. Central part of body forms a small disc.
2. Tube feet are suctorial.

**Example:** *Brisingenes*

### *Order Forcipulatida*

1. Presence of forcipulate pedicellariae.
2. Anus is always present.

**Example:** *Sclerasterias*

### *Subclass Ophiuroidea*

1. Having a small central disc
2. Anus and pedicellariae are absent.
3. Arm is brittle

### *Order Stenurida*

1. Ambulacral groove is open.
2. ambulacral plates remain separate.

**Example:** *Pectimura*

### *Order Oesophiurida*

1. Ambulacral ossicles fuse to form vertebrae.

**Example:** *Ophicanops*

### *Order Phrynophiurida*

1. Lateral and ventral brachial plates are present.

**Example:** Astrothorax (Figures 15.19–15.22).

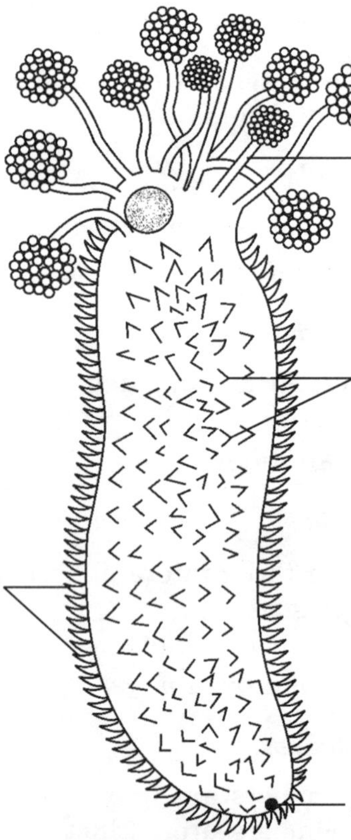


**Figure 15.20** *Holothuria.*

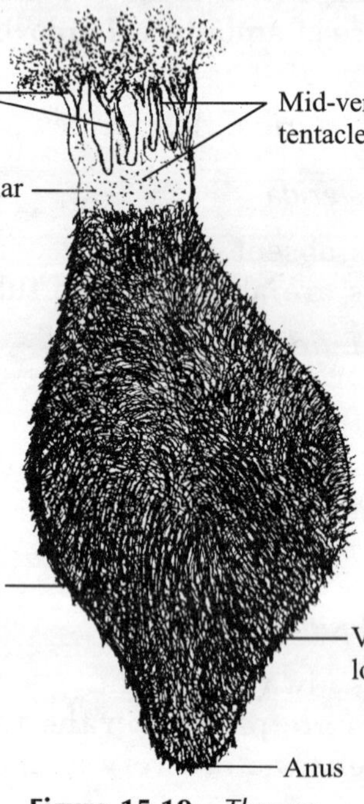


**Figure 15.19** *Thyone.*

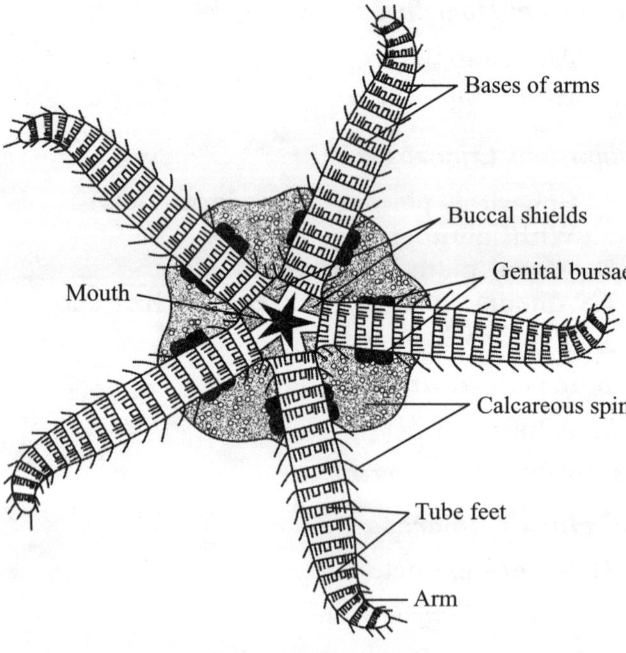


**Figure 15.21** *Ophiura.*

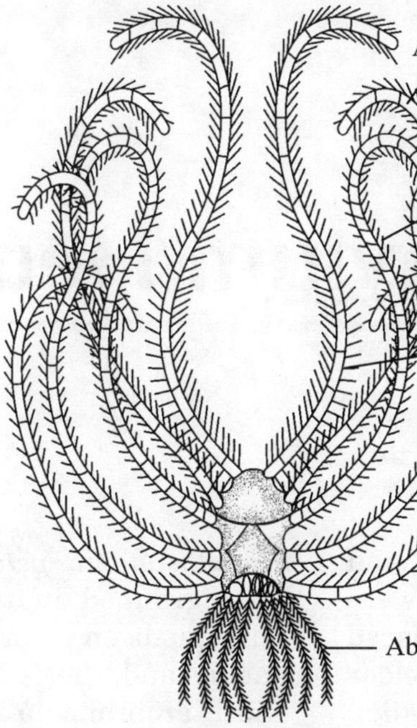


**Figure 15.22** *Antedon.*

## REVIEW QUESTIONS

### Short Answer Questions

1. What are radii?
2. What do you understand by oral surface?
3. What do you understand by aboral surface?
4. What are eye spots?
5. What is madreporite?
6. What are pedicellariae?
7. What are papulae?
8. What do you understand by bivium?
9. What do you understand by trivium?
10. What are adductor muscles?
11. What are abductor muscles
12. Comment on axial sinus.
13. Comment on perihaemal sinus.
14. Comment on genital sinus.
15. What is cardiac stomach?
16. What is pyloric stomach?
17. What is stone canal?
18. What is ring canal?
19. What are Tiedemann's bodies?
20. What are pollian vesicles?
21. What are lateral canals?
22. What is oral haemal ring?
23. What is the function of terminal tentacle?
24. What are bachiolar arms?
25. What is dipleurula?
26. Give two characters of the Phylum Echinodermata.
27. Give the systematic position of *Asterias*?
28. Give the systematic position of *Antedon*?
29. Give the systematic position of *Echinus*?
30. Give the systematic postion of *Echinocardium*?

### Long Answer Questions

1. Classify Phylum Echinodermata with examples.
2. Give an account of circulatory system of *Asterias*.
3. Draw and describe the water vascular system of *Asterias*?
4. Describe the locomotion of *Asterias* with suitable diagrams?
5. Draw and describe the digestive system of *Asterias*.
6. Give an account of larval forms of echinoderms.

CHAPTER 16

# Phylum Hemichordata

Hemichordates possess pharyngeal gill slits and most have a dorsal nerve cord. Once they were thought to possess a notochord and were placed in a subphylum of the Chordata. It is now clear that an evagination of the anterior gut called the *buccal diverticulum* is not homologous with the notochord. All adult hemichordates are benthic marine animals (except *Planctosphaera)*. Hemichordata contains the enteropneusts (acorn worms) and pterobranchs.

About 75 species of the class Enteropneusta are known. They are mostly intertidal, generally live buried in soft sediments and range from a few centimetres to over 2 m in length. Enteropneusts are solitary, elongate, vermiform, with bodies clearly divided into proboscis, collar and trunk which are homologous to the prosome, mesosome and metasome of deuterostomes. The proboscis is short and often conical. A short, thin proboscis stalk connects the proboscis to the collar. The anus terminates at the posterior end of the long trunk and bears mid-dorsal- and mid-ventral-external- and longitudinal ridges. The trunk is also differentiated regionally along its length. Most species bear a clear, anterior, branchial region of the trunk characterized by the presence of numerous gill (branchial) pores flanking the mid-dorsal ridge. *Balanoglossus* has a distinct genital region housing the gonads and external longitudinal genital wings. In *Balanoglossus*, the anterior portion of the intestine is thickened and visible through the body wall as a coloured area called the *hepatic region*.

Pterobranchs is largely colonial. Small zooids rarely exceed 1 cm in length, but colonies measure 10 cm, or more. *Saxipendium coronatum* is found in the deep-sea geothermal vent communities. *Planctosphaera pelagica* discovered in 1932 has been grouped in the class Planctosphaeroidea. This has been recorded in several localities in the Pacific and the Atlantic Oceans. Enteropneusta and Pterobranchia represent two divergent clades within the Phylum Hemichordata, each of which is adapted to different ways of living.

## 16.1 CHARACTERS OF HEMICHORDATA

1. Marine.
2. Solitary, or colonial.
3. Soft bodied.
4. Body is divided into proboscis, collar and trunk.
5. Gill slits, when present are variable in number.
6. The digestive tube is complete, straight, or U-shaped.
7. Presence of ciliated, pharyngeal gill slits.
8. Circulatory system is open and well developed.
9. Unique excretory structure, the glomerulus is found.
10. Preoral (buccal) gut diverticulum is found.
11. Deposit, or suspension feeders.
12. Notochord is absent.
13. Circular- and longitudinal muscles are present in the body wall of proboscis and collar of the enteropneusts.
14. Pterobranchs are provided with longitudinal muscle only.
15. Basement membrane of the enteropneusts in proboscis region is produced as rigid plates, the proboscis skeleton.
16. Short, dorsal, occasionally, hollow nerve cord is probably homologous with chordate nerve cord.

17. Dioecious, with external fertilization and indirect development
18. Asexual reproduction is common.
19. Cleavage radial, holoblastic and equal.
20. Strictly marine and except Planctosphaera, all are benthic.

The phylum includes two classes—Enteropneusta and Pterobranchia.

## 16.2 CHARACTERS OF ENTEROPNEUSTA

1. Solitary.
2. Body is elongated and vermiform.
3. Cuticle is absent and there is paucity of connective tissue in the body wall.
4. They inhabit soft marine sediments.
5. Numerous pharyngeal gill slits are found in the anterior gut.
6. Body is divided into the three regions.
7. Protosome is known as the *proboscis*, *mesosome* as the collar and metasome as the trunk.
8. Mouth is located between the proboscis and collar.
9. Trunk is located at the extreme posterior trunk end.
10. Anus is the terminal.
11. Both the sexes are separate.
12. They exhibit external fertilization.

**Example:** *Balanoglossus, Saccoglosssus.*

## 16.3 CHARACTERS OF PTEROBRANCHIA

1. Small, colonial and marine.
2. Sedentary.
3. Pyriform, or globular body exhibits the ancestral, tripartite regional division.
4. Protosome is reduced to a small plate called the *preoral disc*, which folds ventrally over the mouth.
5. Mesosome forms a collar that bears the mouth and forms two to several arms on which the tentacles are located.
6. Metasome is divided into a trunk and posterior stalk.
7. Dorsally arranged arms do not encircle the mouth.
8. Gut is U-shaped.
9. Anus lies antero-dorsally.
10. Colony consists of a number of zooids.
11. Zooids have proboscis, collar and trunk.
12. Gastrulation occurs through the invagination.

**Example:** *Cephalodiscus, Rhabdopleura.*

The three pterobranch genera are different from one another in habits and in anatomical details. Most pterobranchs live in colonies *(Rhabdopleura)*, or aggregations *(Cephalodiscus)* consisting of zooids. In *Rhabdopleura* colonies, the zooids are connected to one another by stolons, but no such inter-zooidal links occur in *Cephalodiscus*. The overall forms of the aggregations and colonies vary among species. The associated zooids are always the products of asexual reproduction initiated by a single sexually produced individual. The third genus is represented by a single species (*Atubariaheterolopha*).

## 16.4 BODY WALL, SUPPORT AND LOCOMOTION

Hemichordates possess a ciliated epidermis overlying a nerve plexus. The epidermis is supplied with gland cells, many of which produce mucous, particularly on the proboscis and collar of enteropneusts and on the tentacles of pterobranchs. In some enteropneusts, some epithelial cells produce mucopolysaccharide compounds that repel predators. Most of such secretions possess iodine-like odour. Both circular and longitudinal muscles are present in the proboscis wall and anterior collar of acorn worms, but only longitudinal fibres exist elsewhere.

Pterobranchs possess only longitudinal fibres in their body walls, some of which are produced by the peritoneum. A basement membrane lies between the epidermis and the musculature. In enteropneusts, it is produced as thickened, rigid plates, called the *proboscis skeleton* and as supportive structure of the gill slits. The peritoneum is reduced, or transformed into musculature in different parts of the body. However, the tripartite coelomic arrangement is clear in all forms. The usual three coelomic spaces are present as a single protocoel, followed by paired mesocoels and metacoels. These spaces occur, respectively, in the proboscis, collar and trunk of enteropneusts and in the cephalic shield, collar, arms, tentacles and trunk of pterobranchs.

Mesocoelic extensions are present in the arms and tentacles of pterobranchs. The mouth in pterobranchs lies outside the tentacular crown. In adult acorn worms, the buccal diverticulum functions as a skeleton. Hemichordates are sessile, or sedentary and capable of only limited movement. Enteropneusts crawl slowly, or burrow by peristaltic action of the proboscis. At least one species of enteropneust (*Glandiceps hacksii)* sometimes swarms at the surface in shallow water. The protraction and retraction of pterobranchs are due to the hydrostatic pressure and contraction of longitudinal muscles, respectively.

## 16.5 FEEDING AND DIGESTION

Enteropneusts that burrow through soft sediments are direct deposit feeders, ingest the substratum and digest organic material therein. Those live in permanent burrows are suspension feeders, selectively trap organic particulates from the water with the proboscis. Most species are perhaps capable of feeding by both suspension and deposit feeding. The suspension feeding has been observed in some acorn worms like *Protoglossus*. Food materials including detritus and live plankton are trapped in mucous of the proboscis and are moved posteriorly by ciliary currents.

The sorting of food occurs at the proximal end of the proboscis. The proboscis stalk passes most of the large particles over the lip of the collar. Such special rejection current removes particles. Most food, however, passes ventrally around the proboscis stalk over the preoral ciliary organ and condensed into a mucous cord that passes into the mouth. Swallowing appears to be facilitated by a combination of ciliary action, the flow of water in the mouth and removal of water through the gill pores. The gill pores facilitate water flow not as a filter-feeding device themselves. The cilia on the gill pores keep the pores clean and unclogged.

The digestive tract of enteropneusts is a straight, regionally specialized tube, extending from the mouth to the anus. Gut musculature is scant and the food is moved along largely by cilia. The mouth leads to a buccal tube, which is housed within the collar and gives rise anterodorsally to the forwardly projecting buccal diverticulum. Behind the buccal tube in the anterior part of the trunk, there is a pharynx. The pharynx bears a dorsal epibranchial ridge. The digestive portion of the pharynx is restricted to a thin ventral hypobranchial bridge, while the lateral and dorsal parts bear the gill slits, or pores.

The number of gill slits ranges from a few to over 100 pairs. Each one is a U-shaped opening in the wall of the pharynx that leads to a branchial sac and then to a dorso-lateral gill pore through which water exits to the outside. The septum between the adjacent gill slits and the partition between the arms of the U of each slit called a *tongue bar* is supported by skeletal elements derived from the basement membrane of the gut lining. Behind the pharynx, there is an oesophagus, which in some forms like *Saccoglossus* bears openings to the exterior through the dorsal body wall. The middle region of the oesophagus bears intrinsic muscles and moves the food into the intestine by peristalsis. The mechanical squeezing of the food material may expel excess water through the oesophageal pores. The intestine extends to a short rectum which terminates in the anus.

Digestion is probably largely extracellular in the intestine. The major feeding structures of pterobranchs are the arms and tentacles. *Rhabdopleura* bears one pair of arms, each with numerous tentacles, whereas *Cephalodiscus* bears five to nine pairs of arms. Pterobranchs are ciliary mucous suspension feeders. During feeding, they assume a position near an opening in their tubular cases and extend their arms and tentacles into the water. Tentacles of the adjacent arms interdigitate to produce a latticework across which a mucous-net is formed. Food is trapped in the mucous and is moved to the mouth by the action of cilia on the tentacles and arms.

In *Cephalodiscus,* the cilia over the body surface may move food to the mouth, in which the entire body surface functions as a feeding structure. In pterobranch, mouth is located under the antero-ventral edge of the collar. The gut is U-shaped, beginning with a buccal tube from which arises a buccal diverticulum. The pharynx bears one pair of gill slits in *Cephalodiscus* and *Atubaria,* but none in *Rhabdopleura.* The gill apparatus is much simple than that of enteropneusts as there is no supporting structures and branchial sacs are less well-defined. The slits open through pores on the exterior. An oesophagus connects the pharynx to a sacciform stomach at the bottom of the U and occupies most of the space within the trunk. The ascending portion of the gut, the intestine leads to the anus. Digestion probably occurs in the stomach and intestine.

## 16.6 CIRCULATION, GAS EXCHANGE AND EXCRETION

Enteropneusts have a well-developed, open circulatory system with blood vessels, sinuses and the heart vesicle. Two main longitudinal vessels lie in the dorsal- and ventral mesenteries along the length of the trunk and in the collar. Blood flows anteriorly in the dorsal vessel and posteriorly in the ventral vessel. The dorsal vessel expands in the collar as the venous sinus and receives blood anteriorly from a pair of lateral proboscis veins. The venous sinus leads to an elongated central sinus in the proboscis. The heart vesicle pulsates against the central sinus and aids in the movement of blood. From the central sinus, blood moves into the glomerulus within which metabolic wastes are presumably extracted. The glomerulus, buccal diverticulum, central sinus and heart vesicle altogether compose the proboscis complex of enteropneusts. Blood leaves the glomerulus and passes through the various vessels and sinuses supplying the anterior end and eventually reaching the ventral longitudinal vessel. Along the length of the trunk, blood leaves the ventral vessel to pass into the networks of sinuses supplying the gut and the body wall and from these sinuses, it then passes to the dorsal vessel.

Gas exchange occurs across the walls of the gill structures, especially the branchial sacs, which are richly supplied with blood from the ventral vessel. The gill septa and tongue bar bear cilia, which drive water into the mouth through the pharynx and out the gill slits. Enteropneust blood lacks pigments, contains few cells and gases are apparently carried in solution. There is a central sinus and a heart vesicle near the buccal diverticulum. Blood is carried in sinuses and lacunae that lack complete walls. Glomerulus is not well developed. The single pair of gill slits in *Cephalodiscus* and *Atubaria* may aid in the gas exchange, but the small diffusion distances throughout the body probably allow general cutaneous exchange, especially over the high surface area of the tentacles.

## 16.7 NERVOUS SYSTEM AND SENSE ORGANS

Nervous system of hemichordates consists of a netlike nerve plexus lying among the bases of the epithelial cells outside the basement membrane. A subepidermal neurochord is present in the collar of enteropneusts, but is reduced in pterobranchs. The plexus is thickened in enteropneusts as longitudinal tracts of neurons along the mid-dorsal and mid-ventral lines of the body. The neurochord is formed by an invagination of ectoderm and is hollow in some species. There are few types of sensory receptors in the hemichordates. Enteropneusts possess sensory cells over most of the body, probably serving as touch receptors. They also bear preoral ciliary organ, presumed to be a chemoreceptor which is used during feeding. Touch receptors are presumably present in the tentacles, perhaps on the cephalic shield and tip of the stalk in the noncolonial forms of pterobranchs.

## 16.8 REPRODUCTION AND DEVELOPMENT

Asexual reproduction occurs in some enteropneusts like *Balanoglossus* and in most pterobranchs. Hemichordates are dioecious, but possess no external evidence of sexual differences. Paired sacciform gonads lie in the trunk, outside the peritoneum. They are often elongated in the acorn worms. Among the pterobranchs, *Rhabdopleura* possesses a single gonad, which lies along the right side of the trunk coelom. Enteropneusts bear a pair of gonopores located dorso-laterally on the anterior trunk. The gonopores of *Cephalodiscus* are located at the base of the arms, whereas in *Rhabdopleura,* a single pore opens on the right side of the trunk.

Fertilization is always external. Spawning in enteropneusts involves the release of mucoid egg masses by the females, followed by shedding of sperm by neighbouring males. Once the eggs are fertilized, the mucous coat breaks down leaving the eggs into the sea water. Some species produce relatively yolky eggs and others produce eggs with very little yolk. In both cases, cleavage is holoblastic, radial and equal. A coeloblastula forms, which gastrulates by invagination. Species that produce yolky eggs develop directly to juvenile worms, without an intervening larval phase (*Saccoglossus*). In those that shed no yolky eggs (*Balanoglossus*), the hatching stage develops quickly to a tornaria larva with ciliary bands. This suspension-feeding larva soon elongates with three body regions which become externally prominent.

## 16.9 *BALANOGLOSSUS*

The name *Balanoglossus* has been originated from the Greek words, *balanos* meaning acron and *glossa* meaning tongue, which accords to the shape of the proboscis. This is a burrowing, exclusively marine, cosmopolitan animal and is found under stones, or aquatic weeds.

### *External features*

They are bilaterally symmetrical. The proboscis, or protosome, collar and trunk constitute the body. The body is divided into a long anterior proboscis which remains connected to the collar by a slender, dorsal, reinforced proboscis stalk. Body is soft, elongated and cylindrical. Worm-like, uniformly ciliated body has no exoskeleton. Body length varies between 2 cm and 2.5 m.

### *Proboscis*

The space inside the proboscis is called the *protocoel*, or *proboscis coelom*, which is largely occupied by the muscle fibres. Fibres give a spongy appearance to the protocoel. The heart-kidney complex lies at the posterior end of the protocoel. This large, white, or yellowish, heart-shaped organ consists of (i) a ventral glomerulus, (ii) a central heart and its surrounding pericardium with podocytes, (iii) the anterior tip of the stomochord and (iv) the proboscis skeleton. The haemal system, and excretory system functions together and constitutes the nutrient delivery system for the proboscis. The contraction of the heart forces an ultra filtrate of the blood across the walls into the protocoel to form the primary urine.

The proboscis muscles and other tissues absorb the nutrients such as the sugars and amino acids in the primary urine. Following this absorption, the ultrafiltrate become the final urine to be excreted through the proboscis pore. The proboscis narrows posteriorly as the proboscis stalk. Its cavity opens to the exterior through the proboscis pore. The proboscis sits in the collar like an acorn in its cup. The mouth is ventral and its lips are ventral edges of the collar region.

### *Collar*

The Y-shaped proboscis skeleton remains embedded in the dorsal wall of the collar just posterior to the heart-kidney complex. It is composed in part of a cartilage-like connective tissue known as the *chondroid*. The skeleton as the single rod continues anteriorly through the stalk into the heart-kidney complex. The proboscis skeleton attaches the proboscis to the collar. This also supports the heart-kidney complex.

The collar lies between the proboscis and the trunk. Its surface is marked by the elevations, depressions and circular grooves. The right and left coelomic cavities are separated from one another by the dorsal mesenteries and ventral mesenteries. The collar cavity is communicated to the exterior by the collar-pores. The ciliated tubes (canals) lead to the first gill-pouch.

**Trunk:** This is the posterior part of the body and is further divided into the branchiogenital region and abdominal region. The branchiogenital region has a double row of bronchial pores. Its lateral sides expand as the genital wings. The hepatic caecae are found in the trunk. The abdominal region has superficial annuli and anus.

### *Body wall*

The integument is ciliated. The animal is covered with a multiciliated, glandular, sensory, monolayered and columnar epithelium. Absence of cuticle, or exoskeleton is a characteristic feature. The integument is fragile and offers little mechanical protection against the infection, or damage. The epidermis secretes the mucous and 2, 5-di-bromo phenol, a bromine compound. This compound smells like iodine and is believed to be an antibiotic that protect the soft cuticle-free body of *Balanoglossus* from the infection. Some believe that the phenol prevents the bacterial growth on the walls of the burrow—thus reducing the competition for oxygen.

An iodine-like odour is easily detected during handling these worms. The reticulate cells, goblet gland cells and neurosensory cells are present in the epidermis. Below this epidermis, a network of nerve cells and fibres is present. Below this nerve net, there is a basement membrane followed by poorly formed muscles. In proboscis and collar, there is a circular and a longitudinal muscles layer. In trunk, only the layer of longitudinal muscles is present.

### *Supporting structures*

Definite endoskeleton is absent. The supporting structures are the buccal diverticulum, proboscis skeleton, branchial skeleton and pygochord. The buccal diverticulum's is a hollow preoral outgrowth, extending from the roof of buccal cavity to the proboscis. It consists of the endodermal, turgid vacuolated cells. The proboscis skeleton lies below the notochord in the form of an inverted 'Y'. It is a chitinoid structure with a plate-like main body and two horns. A midventral keel is present. In the branchial skeleton, the tongue bars and their septa are supported by skeletal rods. Each skeletal piece looks like a hairpin. One prong of skeleton is present in the septum and other lies in the tongue. The pygochord is longitudinal rod-like structure and extends from the ventral side of the intestine to the body wall. It may support the soft abdominal region.

### Coelom

The coelom is enterocoelous. The proboscis coelom opens to the exterior by the proboscis pore. Two cavities of the collar coelom are separated from each other by the mesenteries. Two cavities of the trunk coelom are separated by the dorsal mesenteries and ventral mesenteries. They are filled with coelom, which remains filled with the sea water.

### Digestive system

The digestive system consists of the mouth, buccal cavity and pharynx with gill pores, oesophagus and intestine. The intestine is divided into the hepatic region, post hepatic region and anus. The mouth and buccal cavity are located in the collar, whereas the remaining regions lie in the trunk.

**Buccal cavity:** The mouth opens into the buccal cavity in the interior of the collar. The stomochord is a dorsal evagination of the anterior end of the buccal cavity. It runs through the stalk into the proboscis. A tiny stomochord pore is present in the anterior dorsal wall of the buccal cavity, just posterior to the intersection of the two arms of the proboscis skeleton. At this point, the gut lumen evaginates into the hollow stomochord, which is homologous to the chordate notochord.

**Pharynx:** Posterior to the buccal cavity, there is a pharynx which is recognized by the U-shaped pharyngeal gill slits in its walls. The gill slits are long and narrow. The parallel openings occur in two rows dorso-laterally on the pharyngeal walls. Each slit opens into its own pouch, or atrium, which opens to the outside via the gill pore. The adjacent gill slits remain separated from each other by the gill arches. The ciliated epithelium of the gill slit produces the water current that enters the mouth and exits through the U-shaped gill slits, atria and the gill pores.

The centre of the 'U' is occupied by a vertical tongue bar that descends ventrally from the dorsal edge of the slit to the ventral edge. The tongue bars look like the gill arches that separate the adjacent slits. Unlike the arches, the tongue bars do not connect with the ventral border of the slit. The gill arches are attached at both ends, whereas tongue bars are only attached dorsally. The gill arches are partitions that separate the adjacent gill slits from each other.

The tongue bars simply subdivide a gill slit into the anterior halve and posterior halve. The tongue bar provides the surface area for cilia and contains most of the blood supply to the gill including a branchial plexus of capillaries. Each slit opens laterally into a water-filled internal pouch, called the *atrium*. The atria open to the exterior via the gill pores on the dorsal surface of the trunk. The outer openings of the gills are called the *gill pores*, whereas the inner openings in the gut wall are called the *gill slits*. This arrangement of gill slits, atrium and gill pores is similar to those found in cephalochordates. In the enteropneusts, each gill has its own atrium. The primary function of the gill slits is to allow the water from the feeding current to escape from the gut lumen, thus to concentrate the food. They also have a respiratory function and apparently also serve as filters to retain the fine particles as water exits the pharynx (Figures 16.1–16.9).

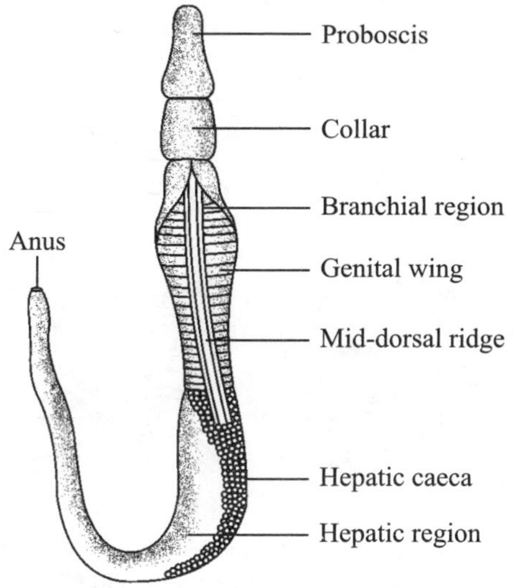


**Figure 16.1** *Balanoglossus:* External features in dorsal view.

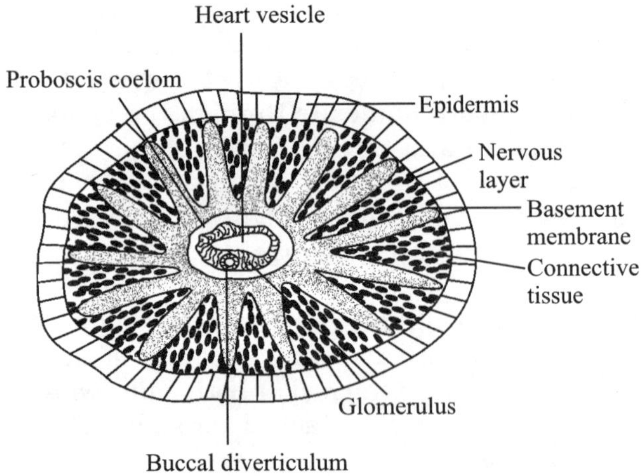


**Figure 16.2** *Balanoglossus:* T.S. of proboscis.

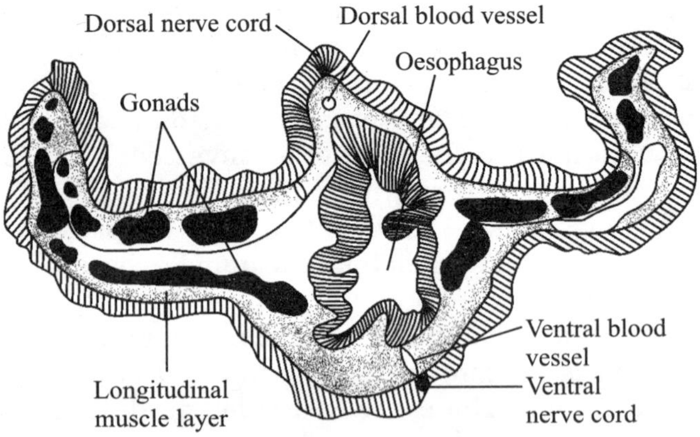


**Figure 16.5** *Balanoglossus:* C.S. through genital wings.

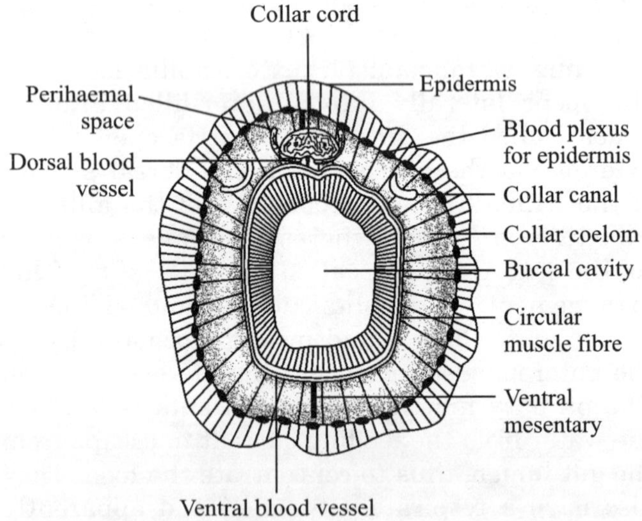


**Figure 16.3** *Balanoglossus:* T.S. of collar.

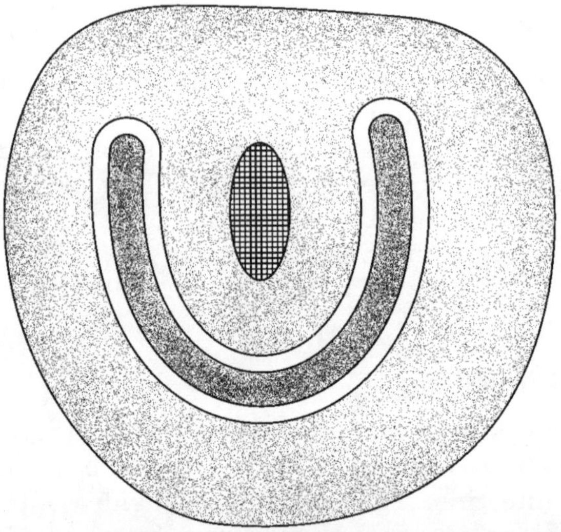

**Figure 16.6** *Balanoglossus:* Posterior view of proboscis showing the preoral ciliary organ.

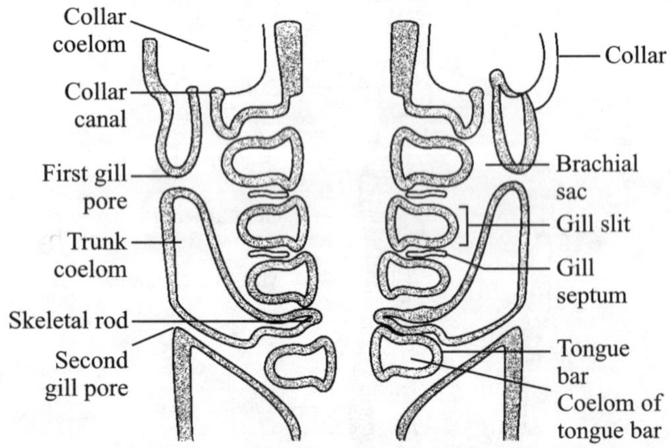


**Figure 16.4** *Balanoglossus:* H.L.S. of branchial portion of pharynx showing first four gill-slits opening by a common gill pore.

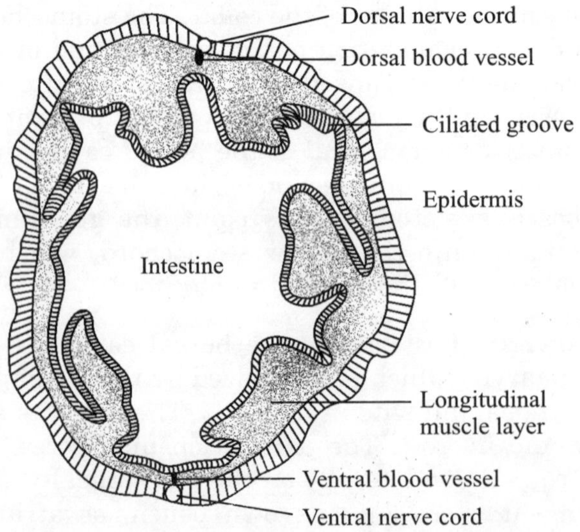


**Figure 16.7** *Balanoglossus:* T.S. through posthepatic intestine.

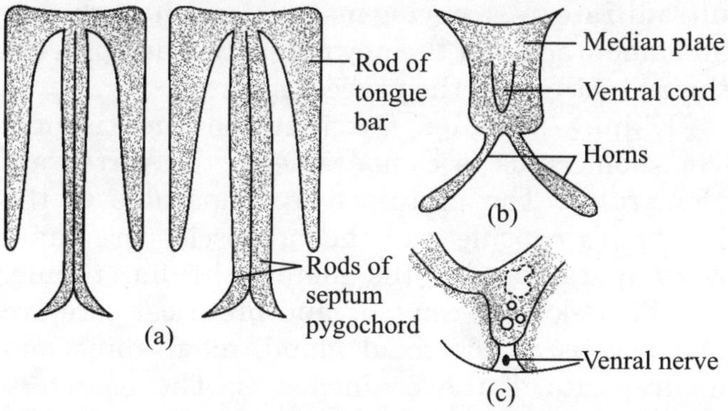


**Figure 16.8** *Balanoglossus:* Skeletal elements (a) Branchial skeleton (b) Proboscis skeleton (c) Pygochord.

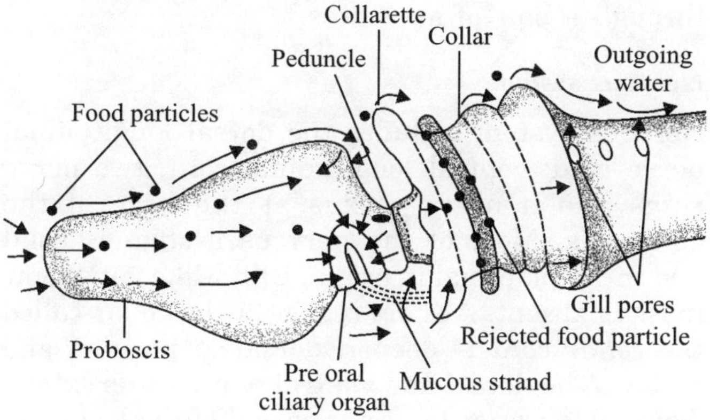


**Figure 16.9** *Balanoglossus:* Feeding current.

**Oesophagus:** Immediately posterior to the pharynx, there is an oesophagus. Its anterior walls are thin and its lumen is spacious. This is followed by a region which has thick folded walls and restricted lumen and then by a third region which has thin walls and the expanded lumen. Posterior to the pharynx, the gut has no respiratory function and it is assumed to be filled with the concentrated material from which much of the water is removed.

**Intestine:** The intestine runs from the posterior end of the oesophagus to the anus. The dorsal epithelium of the anterior end of intestine is modified to form liver-like hepatic tissue and is dark brown, or greenish in colour. This epithelium helps in storing the food following the intracellular digestion. In *Balanoglossus*, the gut wall forms the dorsal diverticula which are also called the *hepatic caeca*. The hepatic caeca appear at the body surface as papillae. They demarcate the hepatic region of the trunk. These caeca are lined by an epithelium. Main features of the digestive system are as follows:

1. Mouth leads into a straight alimentary canal.
2. Alimentary canal is divisible into the buccal cavity, pharynx and oesophagus, and intestine.
3. Wall of the roof of buccal cavity opens into the pharynx.
4. Pharynx has a dorsal branchial part having the gill clefts and a ventral gutter, which serve as the food channel.
5. Behind the last gill clefts, there is an oesophagus.
6. Intestine has hepatic region and intestinal region.
7. Hepatic caeca are present.
8. Posterior intestinal region runs straight to open through the anus.

### *Respiratory system*

1. Gill pouches (up to 700 pairs) open internally in the pharynx by the gill-pores and to the outside by the gill slits.
2. Gill slits are arranged in two rows.
3. Respiratory organs collectively from the branchial apparatus.
4. Branchial apparatus has a series of U-shaped gill clefts on each side, which perforates the pharyngeal wall.
5. Narrow dorsal wall of the pharynx between the two rows of gill clefts is known as the *epibranchial ridge*.
6. Pharyngeal wall between the two lines of a gill cleft is called the *secondary gill bar*, or *tongue bar*.
7. Between the successive gill clefts, the pharyngeal wall serves as a septum, or primary gill bar, which is solid without coelom.
8. Primary and secondary bars are joined by the synapticula.
9. Each gill cleft leads from the pharynx into a large bag known as the *gill pouch,* or *branchial sac*.
10. Each gill pouch opens to the exterior by a gill pore.

### *Blood vascular system*

Haemal system is well developed and includes the contractile dorsal and ventral longitudinal vessels and various connecting vessels. The heart is located in the protocoel on the dorsal surface of the stomochord. The heart itself is not contractile.

A pericardium partially surrounds the heart. A rich branchial blood supply to the pharyngeal gills is found. The dorsal and ventral blood vessels are located in the dorsal mesentery and ventral mesentery above and below the gut, respectively. The blood is colourless and acellular. Mesothelium covers the entire heart complex and stomochord. The dorsal surface of the heart remains surrounded by a contractile muscular fold of mesothelium which is known as the *pericardium*. The dorsal blood vessel carries blood anteriorly to the heart. The blood in the ventral blood vessel flows posteriorly away from the heart. The dorsal and ventral blood vessels connect to numerous vessels leading to and from the sinuses in the tissues.

The features of circulatory system can be summarized as follows:

1. Circulatory system is closed type and consists mainly of two longitudinal channels.
2. Dorsal blood vessel extends from the posterior end to the anterior end of collar.
3. Venous sinus receives a pair of lateral proboscis veins and extends forward as the central sinus, or heart.
4. Dorsal to the heart, there lies a pericardial, or proboscis sac.
5. From the anterior end of the heart, several vessels come out to from the glomerulus.
6. The efferent glomerular arteries from plexus in the collar region.
7. From each collar plexus comes out a vessel, which meet with the fellow of other side to form the ventral blood vessel.
8. The ventral vessel lies below the alimentary canal and extends up to the posterior end of body.

### Excretory system

Excretory organ is metanephric kidney which is intimately associated both functionally and spatially with the heart and pericardium. The anterior end of the heart forms numerous diverticula which are collectively known as the *glomerulus*. The peritoneum of the glomerulus possesses the podocytes. Blood flows anteriorly in the dorsal blood vessel and enters the posterior end of the heart. The contractions of the pericardium push blood through the inner wall of the glomerulus and force serum across the pericardium of the glomerulus to produce an ultrafiltrate in the lumen of the protocoel. Cells in and surrounding the protocoel modify the ultrafiltrate by removing materials such as glucose and amino acids for their own nutrition and prevent their loss through the proboscis pore.

Kidney provides nutrients to the tissues. The glomerulus is comparable to the vertebrate glomerulus. The protocoel is comparable to the Bowman's capsule and the proboscis pore canal is comparable with the metanephridial tubule. The heart-kidney complex and protocoel pore are homologous to the axial gland, axial sinus and madreporite of the echinoderms. The excretory organs are the two ciliated funnels present in the collar and the glomerulus present in proboscis. The ciliated funnels communicate with the exterior through a pair of apertures.

### Nervous system

Nervous system includes the dorsal longitudinal nerve cord, ventral longitudinal cord, two nerve rings and a nerve plexus at the base of the epidermis. Most of the nervous system is solid and intraepidermal in origin. Brain-like ganglionic mass is absent. A short region in the collar called the *collar cord* is deeper, dorsal in position and hollow. The dorsal and ventral nerve cords extend from the anterior to posterior end of the body and may be thickened in areas of the epidermis lying on their respective midlines. The dorsal nerve cords are usually tubular. The nerve cords are associated with a network of nervous tissue. At the junction of colour and trunk, two cords are connected with the help of lateral commisures.

### Sensory system

The preoral ciliary organ may be chemosensory. Isolated photoreceptive cells are found scattered through the epidermis. Striking light elicits the burrowing behaviour in some species and increase activity in others. Sense organs are absent in the adult. In larvae, there is a pair of eyespot.

### Reproductive system

The enteropneusts are gonochoric. They are provided with numerous pairs of gonads in low dorso-lateral genital ridges of the branchio-genital region. The genital ridges possess numerous separate pouch like gonads. Each gonad is a sac-like and bulges into the trunk coelom. It opens to the exterior by minute dorsal pore. The testes are pale orange in colour. Ovaries are gray, or lavender in colour. Genital ridges are swollen in reproductively active animal. Development in *Balanoglossus* and its relatives involves a long-lived planktotrophic tornaria larva.

The features of reproductive system are as follows:

1. Both the sexes are separate, no sexual dimorphism.
2. Gonads are present in the form of laterally arranged, coelomic, branched, or unbranched sacs.
3. Sperms and ova are discharged through the genital pores.
4. Fertilization is external.
5. Reproduces normally by the sexual process.
6. Asexual reproduction occurs rarely.

***Development***

1. Development (indirect) is followed by the metamorphosis of a well-developed larval form.

Prelarval development

1. Cleavage is holoblastic and occurs along the radial planes.
2. Blastomeres are approximately equal.
3. A blastocoel is produced which transforms into a gastrula by the invagination.
4. Developing embryo elongates along the anteroposterior axis and develops cilia.
5. Mouth and anus are formed a new.
6. Embryo ultimately escapes as a larva.

The larva was first observed by Johannes Auller who gave the name tornaria, because of its habit of rotating in circles.

Larval development

1. Presence of a preoral lobe.
2. Preoral lobe has an apical plate which bears the nerve fibres, ganglionic nerve cells, a tuft of immobile cilia and a pair of pigmented eyespots.
3. Preoral and postoral ciliated bands unite for a short distance at the apical plate.
4. Ciliated band present around the anus is called the *circumanal ciliated band*, or *telotroch*.
5. Possesses a pair of gill slits.
6. Five coelomic cavities are present.
7. Tornaria sinks to the bottom and metamorphoses into an adult.

Figures 16.10–16.13 illustrate various aspects of *Balanoglossus*.

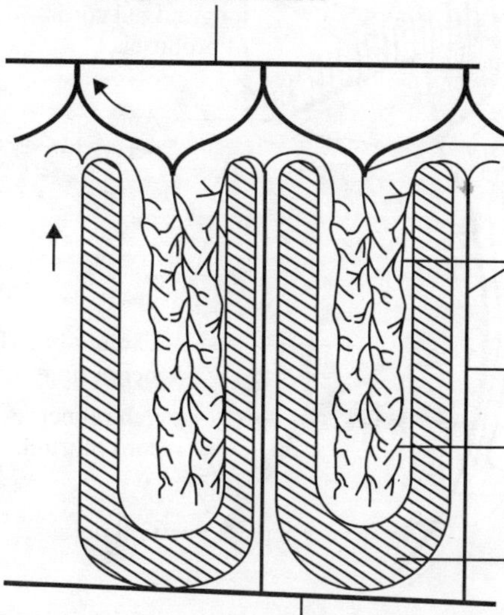


**Figure 16.10** Branchial vessels in portion of the branchial wall.

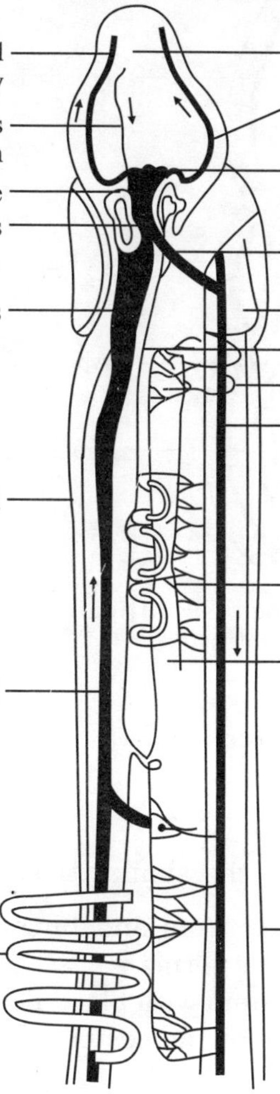


**Figure 16.11** *Balanoglossus:* Blood vascular system in anterior end in lateral view.

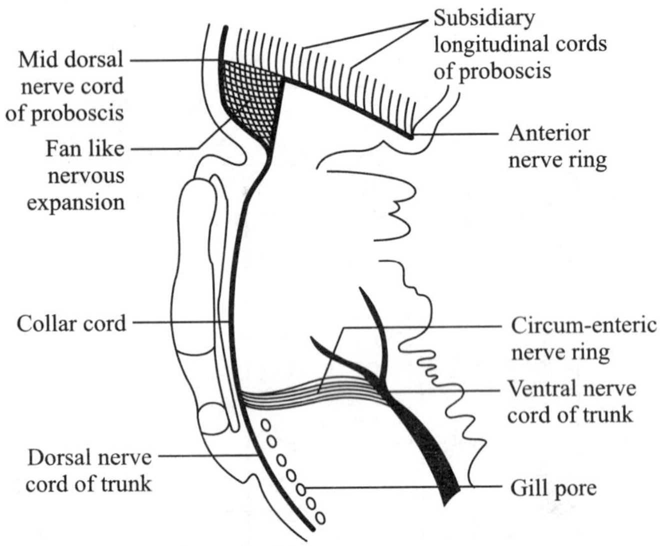


**Figure 16.12** *Balanoglossus:* Nerve cords in the anterior region of body.

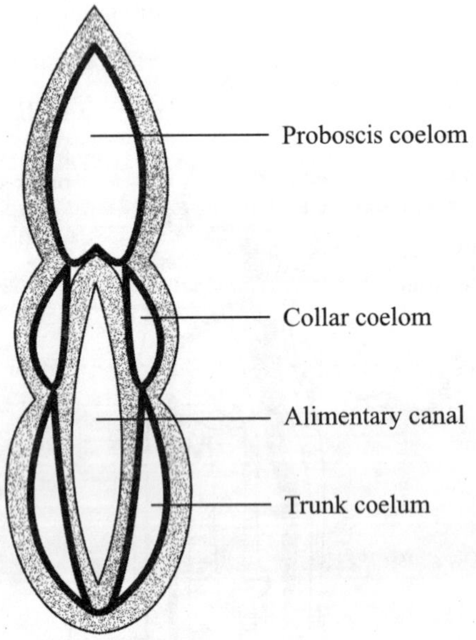


**Figure 16.13** *Balanoglossus:* Tripartite embryo showing coelomic cavities.

## 16.10 AFFINITIES

### *With Annelida*

The following are the similarities:

1. General body from and burrowing habit of the tubicolous forms.
2. Vascular system with blood flowing interiorly in dorsal vessel and posteriorly in the ventral vessels.
3. Tornaria appears like a modified trochophore larvae.

The following are the dissimilarities:

1. Gill slits are absent in annelids.
2. Paired nerve cord is present in the annelids.
3. Trochophore and tornaria differ in the following ways:
   - Nephridia are absent in the tornaria larva.
   - Preoral coelom is absent in the trochophore larva.
   - In the trochophore, blastopore becomes mouth, while in tornaria it becomes anus.

**Comments:** The differences between the two groups are so marked that no phylogenetic relationship between them could be established.

### *With Phoronida*

Actinotrocha larva of *Phoronis* and Tornaria larva show the following similarities:

1. Similar disposition of the coelom.
2. Anus is surrounded by a ring of cilia.
3. Proboscis pore is comparable to the water pore of Actinotrocha.

**Comments:** Masterman supported the placing of phoronids under a separate order of Hemichordata.

### *With Pogonophora*

The following are the similarities:

1. Enterocoelous coelom with similar division.
2. Presence of intraepidermal nervous system.
3. Gonads are present at the trunk region.
4. The septum is present between the mesosome and metasome.

The following are the dissimilarities:

1. Main nervous system in pogonophores is concentrated at protosome, but concentrated at mesosome in *Balanoglossus*.
2. Pogonophore does not have any alimentary canal.

**Comment:** The proposed affinities are superficial and may be due to common phylogenetic connection.

### *With Echinodermata*

The following are the similarities:

1. Ciliated band is identical and follows the same course in tornaria, auricularia and bipinnaria.

2. Shape of the alimentary canal and its divisions are similar.
3. In both groups, blastopore becomes larval anus.
4. Cleavage and gastrulation follow the same pattern.
5. Coelom is enterocoelous in the origin.

The following are the dissimilarities:

1. Eyespot is absent in the bipinnaria.
2. Apical plate and telotroch are absent in bipinnaria.
3. Protocoel is paired in echinoderms, while unpaired in tornaria larva.

**Comment:** Two groups are closely related and probably they arose from a common ancestor. Echinoderms have deviated greatly from the ancestral type, while hemichordates are closer to it.

***With chordates***

The following are the similarities:

1. Position, formation of dorsal nerve cord from dorsal epidermis is similar.
2. Hollow collar cord is comparable with the neural cord of vertebrates.
3. Branchial apparatus having the tongue bars, M-shaped skeletal rods and synapticula are exactly like those of *Amphioxous*.

The following are the dissimilarities:

1. Chordates do not have the body and coelomic regions corresponding to those of hemichordates.
2. Circulatory system of hemichordates is not similar with chordates.
3. Post anal tail is absent in the hemichordates.

**Comment:** Similarities between the hemichordates and chordates can be only due to the common ancestry. Gregenbaur established the group Enteropneusta to which *Balanoglossus* belongs. Metschnikoff and Spangel showed its relationship with the echinoderms and annelids respectively. Sedgwick and Huxley suggested their affinities with the vertebrates. Bateson included them in the Phylum Chordata. In the present state of knowledge, it is safe to give it and its allies a separate status of a phylum Hemichordata by ignoring doubtful homologies and by emphasizing the presence of gill slits. Clearly *Balanoglossus* does not stand on the direct line of chordate ancestry, but a divergent off shoot from main stock of chordate evolution.

## REVIEW QUESTIONS

### Short Answer Questions

1. Why the hemichordates are called invertebrate chordates?
2. What the name *Balanoglossus* signifies?
3. Name the larval form of *Balancglossus*.
4. What is mesosome?
5. What is collar?
6. What is buccal diverticulum?
7. What is collar-trunk septum?
8. What is proboscis coelom?
9. What is tongue bar?
10. What is branchial sac?
11. What is pygochord?
12. What are genital wings?
13. What are para-branchial ridges?
14. Comment on branchial skeleton.
15. Comment on collar coelom.

### Long Answer Questions

1. Describe the endoskeleton of *Balanoglossus* with suitable diagrams.
2. Describe the mechanism of ciliary feeding in *Balanoglossus*.
3. Draw and describe the respiratory system of *Balanoglossus*.
4. Draw and describe the blood vascular system of *Balanoglossus*.
5. Discuss the affinities of *Balanoglossus* with other animal groups and comment on its systematic position.

CHAPTER 17

# Need for Studying and Conserving Invertebrates

Invertebrates comprise 1.4 million described species and have been estimated to contain 2–10 million unknown species. About three quarters of all living species are invertebrates, of which about 1 million are insect species. Invertebrates of the tropical forests and deep seas are poorly known but supposed to be mainly insects and arachnids. Species conservation aims to maintain the ecological processes and life support systems of the earth, to maintain genetic diversity for improvement of cultivated plants and domesticated animals and to ensure the sustainable utilization of species and ecosystems.

As components of food chains, invertebrates play role in ecological processes, in nutrient's cycling and in maintenance of soil structure and fertility. Insects pollinate plants. Some invertebrates are vectors of diseases of human, plants and animals, some are causative organism of diseases and some are pests of crops. Invertebrate fisheries supply food to people in various parts of the world. Some crustaceans and molluscs supply protein requirements of the coastal people. Shrimps are valuable for export trade. Marine invertebrates produce antiviral, antimicrobial, tumour-inhibitory, anticoagulant substances, or with substances of cardioactive or neurophysiological properties.

Coral reefs contain novel biochemical compounds. Arthropods are used extensively in biological and integrated pest control programmes. Some invertebrates are indicators of pollution and other environmental changes. Products such as silk, dyes, lime and building materials are derived from the invertebrates. Some insects such as Drosophila are also used in research. Some invertebrates are valued for aesthetic reasons. Corals and shells are used for personal decoration. Most invertebrates are short-lived and produce many offsprings. However, they can recover through recolonization. The highest diversity of them occurs in areas under severe anthropogenic threat. At least 50% of the world entomofauna live in the tropical forest.

Some invertebrates with poor dispersal abilities survive in small ranges like single springs, caves and patches of vegetation, some adjust ranges through dispersal by flight, or by the agency of man and other animals, or by air and sea currents as planktonic larvae. Many non-arthropod invertebrates are bisexual in which self-fertilization occurs. About 70% of marine invertebrates are dispersed as freely-drifting planktonic larvae. The highest marine species diversity is found in the Indo-Pacific region. A few cases of endemism with unique forms are found in the Red Sea. High genetic variation is found in the animals of coral reefs and certain deep sea areas.

### *Threat to invertebrates*

Population pressure and activities of human have modified wilderness areas into man-made landscapes that preclude coexistence of many wildlife. Invertebrates are not exception to this rule. Pollution, commercial exploitation and introduction of exotic species are the major causes of decline of invertebrate species. Terrestrial invertebrates are mostly affected by deforestation, habitat fragmentation, agricultural conversion and intensification, alteration of grasslands, industrialization and urbanization. Deforestation in tropical equatorial regions poses a serious threat to insects, other arthropods and pulmonates. Terrestrial molluscs are the second most seriously affected group by forest loss. Forest loss on the Pacific islands has led to extinction rate of 50–66% of terrestrial molluscs in the Hawaiian Islands. Pollution directly affects the water quality. Alteration and loss of wetlands and water bodies through drainage, channelization, impoundment and many other processes affects leeches, molluscs, spiders and caddis flies.

Tropical reefs and the continental slope are susceptible to environmental changes. Commercial important invertebrate fisheries depend on coastal habitats. Young prawns and shrimps depend on protected estuarine waters adjacent to marshes, mudflats and mangroves. Loss of these habitats reduce production from offshore fisheries. In most maritime countries, alteration of coastal habitat is taking place rapidly. Atmospheric pollution and acid rain affect aquatic molluscs, crustaceans and insects.

Synthetic fertilizers used in modern agriculture reduce invertebrate diversity. Some species of protozoans, annelids, crustaceans and molluscs may apparently benefit from organic pollution. Tubificid worms and some leeches occur in large numbers in polluted sites and at the same time reduce the number of less tolerant species. Mangroves and coral reefs are most vulnerable to oil spills. Industrial and domestic effluents, siltation from soil run-off and mining activities adversely affect filter-feeding and benthic organisms. Introduced species pose threats through competition for food resources or habitat and through hybridization.

## 17.1 PROTOZOA

Protozoa form the basis of most food chains and along with other microbes constitutes a vast portion of the biomass of aquatic systems. Planktonic flagellates and ciliates are important element in food chains. Other species as decomposers play a role in soil breakdown and humus formation. Ciliates occur in the rumen of cattle and sheep. Some protozoa cause disease.

Many protozoa feed on bacteria and ciliates purify water by removing pathogenic organisms from polluted water. Ciliates by their predatory action on bacteria clarify the effluent of the sewage. Protozoan are used in environmental, ecological and biomedical research. Ciliates are often used for water quality testing. *Terahymena pyriformis* species complex are used as micro assay organisms in the detection of amino acids, vitamins and proteins. Disruption of protozoan communities could affect the food webs.

Although some species are able to survive adverse conditions by encysting, many are readily destroyed. Some marine ciliates are sensitive to sudden increases in temperature as occur at power plants and other industrial discharge sites. One study has shown that *foraminiferan* species diversity is depressed close to a sewage outfall although at some distance offshore there is an enormously high diversity compared to unpolluted sites. *Stentor* is potentially important in biomedical research, since a wider range of microsurgery and cell growth experiments can be performed on these organisms. Timber management, highway and residential development and other uses of the bay margins could threaten *Stentor* population.

## 17.2 PORIFERA

Sponges serve as model in tissue culture and cell reaggregation. At the same time, they ensure that the whole cell mass pumps sufficient water to effect all essential exchanges. The endemic sponge faunas have been termed thalassoid sponges, since they occur in lakes isolated from the sea in relatively recent times. Fresh water sponges could be used in the palaeolimnological studies of ancient lacustrine habitats. Sponges were first studied for new biochemical compounds. Many novel compounds have been found including nucleotides, nucleosides, amino acids and glycoproteins, which are important in medicine.

The only drug being developed so far is an important synthetic antiviral agent. D-arabinosyl cytosine has been produced from the discovery of spongouridine isolated from a Jamaican sponge, *Tethya crypta*; three derivatives of this compound have been patented as antiviral and anticancer drugs. *Spongia* and *Hippospongia* are used for personal and household purposes. During the first half of the 19th century, sponge fishing was started on a commercial basis in the Mediterranean. Sponges of the eastern Gulf of Mexico and West Indies are regarded as being of second best quality. The sponge industry has tended to decline in many areas as a result of the development of artificial sponges. Natural sponges now fetch high prices. In some regions, sponges could be a major natural resource if the industry was developed and managed efficiently. Leather Sponge *Chondrosia reniformis* is collected and eaten locally. Finger Sponge *Axinella polycapella* is used for liquid shoe polish.

The Egyptians may have used fresh water sponges as a pottery temper and some Amazonian Indians use these still now, crushed and burnt, for the same purpose. Sponges may be used to purify water for their filtration activities. Removal of bacterium *Escherichia coli* and fungus *Candida albicans* from waste effluents by the Red Beard Sponge, *Microciona prolifera* are noteworthy. The commercial sponge fishery has been urged to implement conservation measures for many years. Minimum size limits were imposed in the early days of the industry but these were generally recognized as insufficient. Further efforts should be made to develop commercially viable methods of sponge culture including research and surveys for suitable habitats.

In Tunisia, sponges are still collected in shallow water and dredging at depths less than 50 m has been recommended. Tunisia interested in culturing sponges lacks the necessary expertise for the same. In Tunisia, usable species are common far beyond the established fishing areas. Areas with diverse sponge faunas should be protected. Deep shelf (200–300 m) faunas of the southern oceans are at risk from trawling activities. Protecting the endemic fresh water 'thalassoid' sponges of Lakes Malawi, Tanganyika, Baikal and Tiberias should be ensured. The only known conservation measures has been taken for a fresh water sponge in the USA.

## 17.3 CNIDARIA

Cnidarians play unique role in the marine ecosystem, provide habitat and food for many other invertebrates including fish. Reef-building corals form the basis of the productive marine ecosystem. Coral reefs, the nursery grounds for many species, prevent coast erosion, support fisheries and attract the tourists. Black and red corals are used in medicines. Jellyfishes are used in China for treatment of blood pressure and bronchitis. Antimicrobial and antitumour compounds are found in gorgonian corals and Hawaiian zoantharians respectively. Aequorin, the bioluminescent protein of *Aequorea aequorea*, enables detection of calcium concentrations in biological fluids.

Aequorin may be used in cardiac irregularities. Prostaglandins have been isolated from *Plexaura homomalla*. The extract of *Rhodactis howesi* has an anticoagulant factor. Stony coral *Porites* is in use as a template for artificial bone. *Porites, Favia, Favites, Leptastrea* and *Platygyra* are used as building material and for preparation of calcium carbide and cement. Seriatopora and Pocillopora have ornamental value. Jellyfishes are commercially exploited for food in China, Japan and South Korea. Gorgonacean and antipatharian corals are used in jewellery. Gold- and bamboo-corals are commercially important. Sea fans have ornamental value. The hydroids *Sertularia* and *Hydrallmania* are fished in Germany that are used for decoration. Decline of *Adamsia carciniopados* and *Aiptasia mutabilis, Aurelia aurita,* some hydroids, gorgonian sea fans and cup corals in UK waters are reported recently.

*Amphianthus dohrnii* is now seems to be very scarce. In British waters, it occurred on the sea fan *Eunicella verrucosa* which has declined. Over-exploitation of precious and semiprecious corals led to their local extinction. The scouring of the sea bed by trawlers, and dredging probably affects sea fans, sea pens, alcyonarians and sea anaemons. Fresh water cnidarians may be vulnerable to thermal and chemical pollution with drastic effects on *Cordylophora lacustris*. Destruction of coral reefs is the greatest threat to cnidarians.

Reefs can survive in a state of considerable natural damage. However, man made changes have long-term effects on the reconstructional ability of damaged reefs. Silt would have a major impact on stony corals, preventing the settlement of planula larvae and hindering the photosynthesis in symbiotic algae within the corals. Siltation rates are increasing due to soil run-off from the land. Certain sewage treatments and effluent concentrations enhance algal growth at the expense of corals. Coral mortality may not be directly related to effluent toxicity, it may be the result of competition with algae for space and light. Gorgonians are the most tolerant to siltation and dredging-induced turbidity. Although scleractinians are tolerant to siltation to some extent, prolonged exposure caused loss of zooxanthellae. Sedimentation damages reefs in Thailand and in Hawaii.

By-products of oil drilling, in the form of mud with toxic affect *Montastrea annularis*. Studies in the Red Sea show that oil affects the reproductive potential of corals. Upper reef corals in shallow water suffer severe damage, especially oil slicks coat the corals during low tides. Heated effluents affect coral growth and fast-growing species are less tolerant to thermal pollution. Corals are easily damaged by boats and anchors, trampling and fishing. Habitat destruction could have a serious effect on coral species.

Some stony corals have limited ranges. Collection of stony corals for building and industrial purposes has caused their loss to a significant extent at Philippines, Sri Lanka, Malaysia and Indonesia. Ornamental coral trade is also serious factors in Philippines and Hawaii. The collection of coral colonies by tourists may have a local impact. *Fungia scutaria* was declined as a result of the combined effects of over-collection and pollution. The coral reef reserves including the Great Barrier Reef Marine Park are now established in many countries and survey work is being carried out to locate potential sites. The Australian Government has a management regime for multiple use of the reef. *Eunicella verrucosa* is potentially vulnerable, since recovery of exploited populations is slow. They are now threatened locally by over-collecting.

Temperature has an important effect on sexual reproduction of *C. rubrum*. Precious coral was an effective antidote to poison in the middle Ages. Coral jewellery was fashionable in the middle of the 19th century. Slow-growing, long-lived precious coral is vulnerable to over-exploitation. Over collection of *C. rubrum* has made the commercial grade scarce in most accessible areas, although small colonies are still abundant, particularly in caves. Coral stocks off the western coast of Greece are reported to have been exhausted by Italian coral collectors at the beginning of the century.

Coral divers are now travelling further to find fresh stocks. In Tunisia, foreign divers may keep 60% of their catch but the rest is retained by the host country. Japanese and Taiwanese coral fishery is characterized by a cyclical pattern of discovery of new grounds, exploitation and depletion of stocks. The fishery has gradually changed from one based on many small boats to one in which most of the coral is collected by fewer, larger boats capable of remaining at sea for periods of up to six months and going much further field. Attempts have been made to prohibit coral fishing in several places but invariably these have been unsuccessful. When recovery of the beds has occurred, it has been accidental as a result of interruption of fishing by war.

In the Middle and Far East, Black Coral has long been accredited with medicinal and magical properties. Synthetic Black Coral is often sold to unsuspecting tourists. It has recently become a popular material for souvenir jewellery and is sold to tourists in many tropical countries, providing income to local people. An unknown but probably fairly large amount is involved in international trade. Black Coral is exploited commercially in Jamaica at Montego Bay and Ocho Rios. Over-exploitation has occurred in Mexican waters. In the other areas in Mexico where commercial collection occurs, stocks have been severely depleted. In the Bay of Cabo San Lucas, South Baja California, attempts are being made to manage the fishery. In areas like the Chinchorro Bank, Black Coral stocks are under-exploited. In New Zealand, *A. aperta* is under threat from collection for the jewellery market. Black Coral in waters around the Chagos Archipelago could be vulnerable.

In the Galapagos, the shallow depths at which Black Corals occur make them especially vulnerable to over-collection.. Black Corals may also be affected by habitat disturbance. The order Antipatharia is listed on Appendix II of the Convention on International Trade in Endangered Species of Wild Fauna and Flora (CITES). Trade in Black Coral between parties is allowed only if a valid export permit has been issued by the country of origin. Black Coral is protected under national legislation in several countries. Antipatharians are listed under the Netherlands legislation for protected non-native species.

## 17.4 PLATYHELMINTHES

Planarians are used in research work on the molecular mechanism of memory. Fresh water planarians occurring in clean waters and they may be important in water pollution study. They can withstand prolonged starvation and tend to disperse. The Pollution appears to be the main threat to many groundwater cave dwelling Planarians. Lake Baikal and Lake Ohio have a large numbers of endemic species which could be affected by pollution.

## 17.5 ANNELIDA

Earthworms are studied as experimental animals and for their effect on soil fertility. Bait market is a major consumer of earthworms. Huge quantities of introduced earthworm species are exported from Canada to USA per year for bait. Vermiculture is a big business in some countries. Earthworms may be the source of new drugs. Folk medicine prescribed earthworms for jaundice, smallpox, haemorrhoids and rheumatism. Earthworm extracts have an antipyretic activity. Some earthworms have a bioluminescence reaction which may be used in the clinical test for this substance. Use of leech in medicine is well known. *Haementeria ghilianii*, the largest leech in the world, is used in research for its simple nervous system and its large eggs permit embryological studies.

*Lumbrineris brevicirra* produces a toxin which is used in the development of an insecticide, effective against pests such as Colorado Beetle. It is nontoxic to warm-blooded animals and decomposes rapidly in biological tissues and natural environment. Aquatic oligochaetes play role in the retrieval of organic matter from lakes and rivers sediments. Leeches are potential indicators of pollution. In South Africa, earthworms of the families Acanthodrilidae and Microchaetidaeare are probably already extinct in Transvaal and Natal due to deforestation and soil erosion. Lumbricids can colonize disturbed habitats. In South America, native Acanthodrilidae are disappearing from Chilean-Patagonia subregion and Peregrine Lumbricidae are becoming more common.

In Brazil, a same condition exists with the native Glossoscolecidae and peregrine Megascolecidae and Octochaetidae. Changes in agricultural practices also affect earthworm populations. Large scale arable farming causes a lowering of the organic content of soil and reduces the food of earthworms. Several insecticides affect numbers and biomass of earthworms. In the long term, this could have a serious effect on earthworm populations and on soil aeration. The periodical burning of stubble leads to calcinations of the superficial layers of the soil affected earthworm population. Earthworms with restricted ranges are vulnerable to human activities. Island species like *Microscolex* and two lumbricid species could also be at risk.

Less information is available on the status of aquatic annelids but several species have very limited ranges. Family Lumbriculidae is known only from Lake Baikal, where 90% oligochaetes of open waters are endemic and is under considerable threat from pollution. Endemic tubificids are found in Yugoslavia and Albania in the lakes Ochrid, Prespa, Dojranand Skadar. Endemic species in the Haplotaxidae, found in groundwater springs and caves in karst areas in Pyrennees and Roumania, could be threatened by pollution. *Haplotaxis ornamentus*, may be vulnerable to changes in the lake level. Both lakes have had their levels raised for hydroelectric purposes but this species has only been found in original levels of lakes. A number of leech species are threatened. The data sheet on the Medicinal Leech *Hirudo medicinalis* is an example.

In South Carolina, one marine leech parasitic on skates and rays might now be locally extinct. *Ozobranchus margoi* a marine leech is found exclusively on *Chelonia mydas* and *Caretta caretta*. Marine polychaetes are less likely to be seriously threatened. There have been local depletions of populations of lugworms *Arenicola* and *Nereis*, due to destruction of habitat. Oil pollution may affect some species. Worms living in sandy or muddy beaches may be killed due to toxic oil pollution. Different species vary in sensitivity to oil. Tolerant, hardy species, which first recolonize oil damaged sediment, dominate the early stages of restoration of community. Lugworms eat contaminated sediments, bring oily residues to surface and hasten their degradation.

The Palolo Worm of the Pacific is becoming rarer. It has been noticed that the Palolo has apparently completely disappeared from almost the entire northern coast of Upolu in Samoa. A number of endemic earthworms of New Zealand are now not under any major threat, but action may be required for future. Appropriate management techniques are needed for annelid reserves. Protection may be necessary for the breeding areas of the Palolo Worm in the Pacific. Knowledge about the ecology, distribution and population biology of many threatened species is essential. Threats to Arachnida are poorly documented. The major threat is undoubtedly loss of habitat, several categories of habitat destruction may be recognized.

## 17.6 MOLLUSCA

Molluscs play an important role in food chains. Bivalves provide food for water fowl and fish. Benthic fish often depend on bivalves. Whales and fish consume pelagic opisthobranchs and cephalopods. Sperm whales consume mainly squid. Gastropods like limpets and top shells control intertidal growth of algae, sponge and some encrusting organisms. Molluscs are source of protein for people, especially in coastal areas. Oysters contribute 25% of world mollusc production. Twelve species of gastropods, 50 bivalves and 13 cephalopods are now exploited in the Mediterranean. About 38 bivalve species, 4 of chitons and 4 of gastropods are used as food in the Caribbean and western central Atlantic. *Strombus gigas* is important in the Caribbean.

Cephalopods are the most important fishery export item after shrimp and are highly valued in southern Europe, South East Asia and South America. The world catch varies between 0.8 and 1.2 million tonnes, 70% of which is due to squid, 17% due to octopus and 10% due to cuttlefish. Molluscs like *Aplysia californica* is used in studies on the memory because of its large nerve cells. Antiviral and antibacterial compounds have been isolated from abalone, oyster, Queen Conch and squid. Mercenene, an extract of *Mercenaria mercenaria* contains an antitumour factor. Muricid, Urosalpinx, feeds on bivalves and damage to oyster beds. Ship worm bores into timber, damage piers, boats and other wooden structures but reduce the amount of sea-borne wood.

Molluscan shells are used for ornamental objects, currency, road construction and buildings and endowed with magical attributes. Ground shells produce the lime for pottery glazes, poultry food additives and for chewing with betel nuts in South Asia. Mother-of-pearl is used for decorative inlay work, buttons, jewellery and ornamental items. The most valuable mother-of-pearl is found in *Pinctada maxima, P. margaritifera, Trochus niloticus* and *Turbo marmoratus.* An annual demand of about 6,000 tonnes of unprocessed *Trochus* shell is estimated. The Philippines and Indonesia export *Pinctada* spp. and Indonesia, the Solomon Islands and other Pacific islands export *Trochus niloticus* and *Turbo marmoratus.* In the South Pacific, the export of mother-of-pearl is the only source of cash income for people on some of the remote islands. Japan is the main producer of pearls and techniques for culturing these molluscan having been developed in this country first. Colourful gastropods from the Indo-Pacific and Caribbean are Queen Conch, Tiger Cowry, Pearly or Chambered Nautilus and helmet shells, cones and volutes. Small cowries, dove shells and other shells are used in large quantities to make necklaces, lampshades, decorated boxes and figurines. The Philippines is the main supplier and is the centre of capiz shell industry. The Giant Clams are possibly threatened with extinction. Many commercially important species are locally over-exploited.

Abalone has been over fished in Japan, China, South Africa, New Zealand, South Australia and Mexico. *Haliotis tuberculatus* has almost disappeared from the English Channel. Oysters are now used as a luxury in most countries. Natural oyster grounds get destroyed through coastal development, domestic and industrial pollution, poor management and commercial-exploitation. Scallops were fished to virtual commercial extinction in some country. In Chile, *Choromytilus chorus* and *Concholepas concholepas* have been over fished. Cephalopods have been over fished in some areas. Cuttlefish catches in the eastern central Atlantic have declined. In the Andaman and Nicobar Islands, new diving techniques have caused rapid depletion of *T. niloticus*. Intensive exploitation has led to the disappearance of *Pinctada margaritifera* from Polynesian atolls.

The ornamental shell trade is causing depletion of local populations of some popular shells. Shells heavily collected on Kenyan coast by amateur shell collectors, tourists and commercial operators has resulted in the scarcity of large colourful species on the accessible parts of the coastal reef. In the Philippines, species such as *Tridacna* and *Charonia tritonis* are becoming rare. Easter Island and Mauritius have endemic cowries such as *Cypraea cribellum* and *C. esontropius.* With the growth of tourism, these species could be over-collected. *Cymbiola rossiniana* is highly sought after fetching high prices on the international market.

*Harpa costata* is a very valuable, apparently rare species, endemic to Mauritius. Oil on the shore and in the intertidal zone can kill molluscs and filter feeders and deposit feeders may ingest and accumulate dispersed or sedimented oil. Moderate and chronic exposure of bivalves to low concentrations can have an important commercial impact. Bivalves filter a large volume of water and ingest hydrocarbons which accumulate in the fatty tissues. Molluscs may be more vulnerable to toxic dispersants than to the oil spill itself. Pollution could be serious for brackish water and estuarine species. *Doridella batava*, a brackish water species has been affected by habitat modification. The *Panopea bitruncata*, the largest western Atlantic bivalve, has been designated as rare.

Most countries have legislation for their edible molluscs, such as closed seasons, minimum size limits for collecting and commercial quotas. In most cases, controls become effective force after evidence of over-collection. In the Channel Islands, fishing for the *Haliotis tuberculatus* is controlled. In Jersey, skin-diving and SCUBA gear are prohibited and there is a closed season during the period from May to August. Strict legislation governs the abalone industry in California, USA and includes size limits, closed seasons and prohibition on abalone export. *Choromytilus chorus* is now protected in Chile and is listed on Appendix II of CITES. In Papua New Guinea, there is a minimum size limit for the collection of *Trochus niloticus*. In New Caledonia, there is a minimum size limit and fishing is only permitted under licence.

In Palau, the collecting season is limited to one month and collection of shells smaller than 76 mm is prohibited. In addition, collection is prohibited from 16 sanctuaries. Similar legislation exists for Guam and Vanuatu. Ornamental species are protected under national wildlife legislation in few countries. In Bermuda, *Strombus gigas, S. costatus, Conus bermudensis, Oliva reticularis, Pecten ziczac, Argopecten gibbus, Pinctada imbricata* are protected by law. In Australia, *Charonia tritonis, Cassis cornuta, Tridacna gigas* and *Cypraea friendi* are protected. *Harpa costata* collection is prohibited in Mauritius. Aquaculture is being intensively developed in many areas for a number of molluscs which may answer some of the problems of over-exploitation. As filter feeders, bivalves are well suited to cultivation, can be grown at high densities and are economic plant protein transformers. Mussels are easy to cultivate. Spain, France and Holland are the main producers using various substrates including rafts, suspended ropes and the sea bed for spat settlement.

Bivalves, including mussels, oysters and Venus shells are cultured in the Mediterranean, coastal lagoons and brackish water lakes. About 20 molluscs are cultured in the Indo-Pacific region. In Indonesia, *Pinctada, Anadara* and *Mytilus* are cultured at low intensity in marine and brackish water. *Crassostrea gigas*, the Pacific oyster, is cultured extensively in Japan in fully controlled hatcheries. In Japan, large populations of abalone are cultivated in semi-enclosed areas with sea urchins. Large scale aquaculture of cephalopods still poses problems, although some species have been commercially cultivated in Japan. Marine monocultures increase the risk of disease as are terrestrial monocultures that may affect the natural fauna in surrounding areas and in the case of mussels, it can cause fouling problems. Many marine molluscs are valuable resources if properly managed and for such species, conservation efforts should be directed for exploiting them sustainably. For small populations with restricted ranges, research is required to determine the conservation method.

At least a dozen naiad species are presumed to have become extinction and about 20% of those remaining are seriously endangered in the USA. In Ohio, the Ohio River, Lake Erie and their respective tributaries had a total of 78 different species and subspecies, of which almost 25% can no longer be found and nearly half of the remaining 59 species are uncommon or rare.

The Mississippi River, considered one of the least damaged North American rivers, has lost 27% of its original naiad species. In the Coosa and Tennessee Rivers, Tulatoma in the family Viviparidae and Apella in the Pleuroceridae and some 30 other endemic pleurocerid species have become extinct. In Italy, the most threatened mollusc species are thought to be *Unio elongatulus, Microcondylaea compressa, Sphaerium corneum* and *Theodoxus fluviatilis*. Species vulnerable in the UK are *Theodoxus fluviatilis, Myxas glutinosa, Segmentina nitida* and *Margaritifera margaritifera*. *S. nitida* is particularly rare and its remaining localities in ponds and ditches are threatened by factors such as pollution, infilling and drainage.

Fresh water molluscs decline due to pollution and alteration of watercourses, although in the past over-exploitation of valuable species has occurred. The North American fresh water mollusc fauna has been affected to a great extent by the alteration of watercourses. Light penetration may be reduced as a result of siltation and production of food mate.

Channelization and dredging cause the habitat unsuitable for many molluscs to live. Modern agriculture also leads to heavy siltation, creating substrates for larval settlement and clogging the gills of mussels preventing respiration. Sand grains may have an abrasive effect on the shell. Species endemic to springs or similarly restricted localities are generally vulnerable to drainage and alteration of groundwater level.

Twelve fresh water hydrobiid snails, apparently endemic to springs in Ash Meadows, a valley in USA, are threatened by ground water pumping. Acid rain causes a lowering of pH values in lakes and rivers. In Sweden alone about 20,000 lakes are believed to have been affected, as a result of pollutants originating in Central Europe and the UK. pH of 5.5 are deleterious to most benthic organisms. Sphaerid mussels from lakes with low pH possess very thin and quite soft shells. When pH drops to about 6.0, five sphaerid species disappear and snails are similarly affected. Gastropods exhibit sensitivity to zinc, copper, mercury and silver pollution.

Introduced species in many cases affect native fresh water molluscs. Native fresh water molluscs in North America may be threatened from competition with the introduced Asian clam, *Corbicula manilensis*. Dredging destroys stream bed and adults died as a result of being unable to re-bury them once they had been uprooted. USA is one of the few countries to have taken major steps towards the conservation of its fresh water mollusc fauna. Some naiads are listed under the US Endangered Species Act. The American Malacological Union has taken an active role in conservation efforts. Critical habitat has not been designated for any species of mollusc and that molluscs have the lowest priority for conservation effort and research funds within the US Department of the Interior.

## 17.7 ARTHROPODA

In the Mojave Desert, California, the endemic whip scorpion *Trithyreus shoshonensis* survives the cold winters in the upper Shoshone Cave, which is heated by thermal ground water. Malheur Cave with 1,100 m long lava tube contains a 400 m lake which supports a diverse fauna including endemic *Apochthonius malheuri*. Spiders are often endemic to isolated caves systems, like *Nesticus sp.* in Bat Cave and *Ivesia carolinensis* in Linville Cave. *Dolomedes plantarius* is restricted to Red grave and Lopham Fens Nature Reserve in Suffolk, UK.

*Acanthophyma gowerensis* is found only in salt marshes in Glamorgan, Wales and south-west Ireland, while *Glyphesis cottonae* occurs only in Sphagnum bogs in the New Forest and Cumbria, UK. Felling of tropical rainforests is threatening many arachnid species. At least six of the 13 arachnid orders are confined to the tropics and subtropics and many others have their greatest diversity there. *Dendrochernes cyrneus* and *Lepthyphantes midas* live in ancient trees in UK woodlands. *Thyreostenius parasiticus* the common spider in UK is also found in artificial habitats.

*Cyclocosmia torreya* has a special concern in Florida. Similarly, in Florida *Eustala eleuthra* may be threatened by destruction of tropical hardwood hammocks. *Sphodros abboti* and *S. rufipes* both build tubular webs on the sides of trees and bushes in woodlands in south-east USA. *Sosippus placidus* is possibly extinct because of the development of citrus plantations in its type locality. The same threat applies to *Geolycosa xera*. *Lycosa ericeticola* occurs only on a small area next to an expanding open-cast mine in Interlachen, Florida. In Florida, population of *Paraphrynus raptator* was formerly common in outside lavatories and cisterns the population may now been endangered. *Cesonia irvingi* is known only from restricted ranges. Two mite species, *Eremobates* and *Erythraeustuberculatu,* are endemic to the dune remnants, which are threatened by development and introduced plants.

*L. Polyphemus* preys on clams and clam diggers and when they moved into clam beds they were collected. The collection of *L. polyphemus* for eel baiting in USA water is significant source of mortality. Huge quantities of *L. polyphemus* and *T. tridentatus* are taken for *Limulus* lysate test. *T. tridentatus* population in Japan has declined as a result of land reclamation. *Limulus* exposed to oil show increased respiratory rates and delayed moulting. The Red-knee Tarantula is known from Mexico west and from a single record from Costa Rica.

*Brachypelma smithi* is a burrowing species of semi-deserts. They burrow in sand or soil banks protected from rain. Their food is insects and small vertebrates. Tarantulas are favoured research animals for physiologists and are used in studies of blood pressure, nerve action potentials and metabolic investigations. No conservation measures have been taken specifically for the Red-knee Tarantula. There is no apparent restriction on import of the Tarantula into the USA. Many Crustaceans are scavengers, filter-feeders and predators and are most vulnerable in the period before the new skeleton hardens.

Copepoda constitutes a link between phytoplankton and fish. Some cladocerans are the basic food for commercial fish, particularly juveniles and the isopod *Asellus aquaticus* is a major food for fresh water fish in Europe. Krill is an important food for most vertebrates including 5 whale species, 4 seal species, many fish and birds in addition to several squid in the southern oceans. Disappearance of fresh water crayfish can cause premature ageing of lakes. The scavenging habits of crayfish contribute to the recycling of organic matter because of their high fecundity and wide range in sizes. They serve as prey for a number of secondary consumers, including trout, perch, reptiles, amphibians, birds and mammals.

Other scavenging crustaceans are the hermit crabs which recycle dead organic matter on beaches. Many fresh water and terrestrial crustaceans are interesting to biologists due to their relict distributions and interesting evolutionary history. Both fresh water and terrestrial isopods have evolved directly from marine ancestors and their current distributions provide useful data on early geological events. The main value of crustaceans to man is as a food resource and they are a major element in both commercial and subsistence fisheries. Greatest economic importance are given to the prawn and shrimp fisheries. The lobster is probably the most valuable and popular crustacean. *Homarus americanus* in the West Atlantic and *H. gammarus* and *Nephrops norvegicus* from the eastern Atlantic and Mediterranean have been exploited intensively. Other marine crustaceans including barnacles are used as food. *Artemia* is used as fish food. Hermit crabs have become popular as pets in the USA.

Fresh water and terrestrial crustaceans with restricted ranges are subject to habitat loss and pollution. Although this has the greatest impact on fresh water and terrestrial species, marine species with limited coastal ranges may be vulnerable. In California, *Balanus aquila* appears to be endemic to the area between San Francisco and San Diego and is one of the last surviving members of a group which is known mainly from fossils. It is vulnerable to alteration of its coastal habitat. *Parapinnixa affinis* lives symbiotically with polychaete worms and has only been recorded from a few bays in southern California. Fresh water crayfish are vulnerable to impoundments which significantly reduce faunal composition and variety and cause siltation.

Channelization and dredging reduce habitat diversity. Dredging removes sheltered portions and shallow areas that could support plant growth and collect detritus material, thereby reducing food resources and/or habitat. Rock cover may be removed or embedded into the substrate by heavy earth-moving equipment, while the remaining rock habitat is silted in as result of disturbances to the substrate or stream banks. Silt, mining, deforestation, farming and changes of land contours reduce crayfish populations.

Silt in suspension is not in itself detrimental but heavy deposits on stream beds smother burrows and obliterate retreats under rocks and debris. In east central Kentucky, crayfish in Leatherwood Creek was reduced by 90% when the stream bed became covered with clay silt due to strip mining. Sawmill operations that use streams for discarding sawdust have a particularly serious effect as the sawdust fills the interstices between rocks and affects the flow of water through them, so that the oxygen becomes depleted and in such areas the crayfish tends to be nonexistent.

Ash wastes smothering the substrate are reported to be detrimental. Some species have extremely localized distributions. The crayfish *Cherax crassimanus* and *Engaewa subcoerulea* are known only from single localities in the Karri Forest in south-west Australia. Over fishing of marine crustacean stocks has taken place in many areas and in others stocks tend to be fully exploited.

Over fishing of lobster stocks are the most serious problems. The fisheries for *Nephrops norvegicus, Homarus gammarus* and *H. americanus* are serious. The *N. norvegicus* fishery has undergone a major expansion and catches of *H. gammarus* in European waters are falling, although their prices continue to rise. The spiny lobster fisheries are not under immediate threat, but in many areas they have declined. The Jasus edwardsii fishery in the Chatham Islands, New Zealand, had declined dramatically by the early 1970s. The continual decline in stocks of *Palinurus interruptus* in southern California is said to be due to the taking of undersized lobsters.

Reductions in spiny lobster *Panulirus argus-landings* in Florida are probably associated with reduced postlarval recruitment and fishing activity in the nursery areas and the recreational fishery in

Biscayne Bay has significantly altered the density and age structure of population. The spiny lobster fishery is reported to have declined in Thailand and in Ceylon. Data available from most of the major spiny lobster fisheries show that annual total production is likely to fall very dramatically unless controls are placed on fishing effort. There are many other examples of valuable crustacean stocks that are being over exploited as a result of poor management of the fishery or lack of controls.

The barnacle *Pollicipes pollicipes* is thought to be declining in northern Spain where it is collected as a luxury food. The sand crab *Emerita emeritus* (Hippidae) is classed as threatened in Thailand where it has disappeared from heavily collected beaches in Phuket. Blue Crab fisheries of Chesapeake Bay and Delaware Bay in the USA have come under threat from over exploitation and pollution. Large scale krill harvesting has become a major issue as it could become a threat to the recovery of whale stocks.

Commercial crustacean culture is in progress. Pond culture of fresh water crayfish was carried out in Europe for centuries but was then largely ignored until more recent times. Crayfish are popular for aquaculture since their ability to convert vegetation, dead fish and surplus fish feed into edible protein allows them to be combined with other animals such as fish. Furthermore the percentage survival to harvestable stages is relatively high since there is no free-living planktonic larval stage and hatcheries are not needed. Several species are currently farmed. The largest commercial enterprise is the farming of Procambarus clarkii in the southern USA.

The yabby, *Cherax destructor*, is cultured in Australia and *Pacifastacus leniusculus* in Europe in at least 17 countries. No commercially viable lobster farming operations exist on account of the difficulties of rearing lobsters to marketable size. *Homarus americanus* appears to have particular potential. Lobster hatcheries aimed at restocking fishing grounds with hatchery-reared juveniles or extending fishing grounds by stocking new areas that have suitable habitat.

Success of such operations has never been satisfactorily monitored and the cost of rearing individuals tends to make the process uneconomic. In France, fishermen are now supporting a scheme to restock over-collected grounds with hatchery-reared juveniles. Although no marine crustaceans are cultured on a commercial basis in Europe or North America, certain species have been cultured for many years in the Indo-Pacific, where warm water and cheap labour abounds. A number of species of prawn have been cultured experimentally in nearly 50 countries, some of which have moved into commercial production. Prominent among these is Japan which produces over 1000 tonnes of cultured prawns a year. The crab *Scylla serratahas* long was an incidental product of brackish water pond culture.

Over fishing of this species resulted in the development of culture techniques and crabs are now raised as a subsidiary crop to milkfish. The ponds are seeded with small crabs which reach marketable size in six months. Crab culture in general is not a commercial proposition on account of the slow growth rates of the larger species. Culture techniques have been developed for the Blue Crab of the east coast of the USA and juveniles are being reared to restock depleted bays.

Very little interest has been shown in the conservation of noncommercial crustaceans although some countries include vulnerable species in their lists of protected species. Crustaceans are included in the national Red Data Books for Switzerland and Germany. In the USA, threatened fresh water crustaceans have been identified in many states. For fresh water crustaceans, the main requirements are increased pollution controls and protection of critical habitat.

The causes of decline and extinction of insects may be broadly divided into two categories, natural threats and man-induced threats. Natural changes in populations of insect species, evolution of new species and extinction of unsuccessful species through natural selection, immigration and emigration are important to mention. Man-made changes are currently by far the most important factors affecting insects worldwide. Through technological innovation, man is able to create or destroys landscapes and biotopes within a geological instant, permitting no opportunity for natural selection of adaptable genotypes.

Man-induced changes that affect insects are divided into five categories—changes to land, changes to water, atmospheric pollution, changes to closely associated fauna and flora and specific threats to the species in question. The first four categories deal with the insect's habitat or biotope, the components which make up the integrated network of physical and biological parameters of which the species is a part and which are essential to the species survival. Impact in these categories is generally created as a direct result of or enhanced by increasing human population. The fifth category includes threats peculiar to the species itself, but not necessarily to its environment.

Agriculture is one of the most extensive methods of land use and has probably resulted in great loss of native insect populations, particularly when it follows the forest clearance.

Little extinction has so far resulted from agriculture in temperate latitudes because of a low incidence of restricted endemicity in the recent alluvial soils best suited to farming. Many such areas have only developed since the last Ice Age ended 10,000 years ago, a short evolutionary time compared to, for example, the 60 million years of relatively stable conditions over Africa from 20°N to the Cape. In most of lowland UK, butterflies are now unable to survive the intensification of farming and populations have been reduced. Intensification includes a range of factors including use of insecticides and herbicides, destruction of hedgerows, drainage, short fallow periods and use of fertilizers.

In North America, the conversion of prairie to agriculture has had disturbing effects. Two sphinx moths, *Euproserpinus euterpe* and *E. wiesti*, have been reduced by planting of cereal crops and recently by insecticidal sprays. In Nepal, agricultural encroachment in the Kathmandu Valley is threatening the Relict Himalayan Dragonfly. The greatest threats are in tropics where reductions in populations and in species diversity may be occurring daily as a result of conversion of tropical forests and woodlands to agriculture. The conversion of lowland and montane forest in Hawaii to pineapple and sugar cane plantations has resulted in loss of several hundred species of native insects. In Jamaica, timber operations and coffee plantations have divided and reduced the populations of Homerus *Swallowtail*. Introductions of exotic animals and plants invariably upset the balance of arthropod species, sometimes irreparably.

Introduced plants may dominate native species, indirectly causing loss of native insects. This has occurred in Hawaiian and Californian coastal dunes after the introduction of *Ammophila arenaria* and *Mesembryanthemum spp.* Introduction of *Lepus capensis* to Laysan in leeward Hawaiian Islands caused extinction of several noctuid moths. Accidental introduction of *Rattus rattus* to Lord Howe Island caused probable extinction of *Dryococelus australis*. In New Zealand, *Deinacrida spp.* have been severely threatened by rodent introductions throughout their range. The once numerous *Megalagrion acificum* has probably been exterminated by *Gambusia affinis* which was introduced into Hawaii to control mosquitoes. Widespread use of artificial chemicals is often cited as a major factor in loss of insect's populations. Decline of insects in rural areas of Europe is due to intensification of agriculture, only one aspect of which is the use of insecticides and herbicides.

Pesticide use in native habitats, particularly on islands, should be viewed with great caution. Severely reduced populations may be susceptible to aerial application of insecticides. The final extinction of the English population of *Aporiacrataegi* has been attributed to orchard spraying. International agreement is contained in the Convention on International Trade in Endangered Species of Wild Fauna and Flora (CITES), which controls and monitors import and export of listed species and to which 79 countries have now acceded. Appendix I, which contains no insects, is a list of species in which trade is subject to strict regulation and is virtually prohibited except to fulfill scientific needs. Appendix II lists species in which trade is regulated or monitored.

## 17.8 ECHINODERMATA

The eggs of sea urchins are used in fundamental research on cell structure, fertilization and experimental embryology. Sea-urchin eggs have short development period which allows researchers to observe the teratological effects of drugs and potential for use of sea urchin larvae in bioassay techniques is being explored. The eggs have even been shot into space in rockets to know whether cosmic rays have an effect on living organisms. A steroid saponin isolated from *Actinopyga agassizi* suppresses the growth of tumours in mice. Crude extracts of *Asterias forbesi, Acanthaster planci* and *Asterina pectinifera* are effective against influenza B virus in embryonated chicks.

*Stichopus japonicus* is the source of two extracts, one of which has antitumour activity, the other has antifungal activity. Sea urchins are marketed as a luxury food. In Japan, *S. intermedius, Heliocidaris* and *Hemicentrotus* are fished commercially. Dried sea cucumbers were the first products of South

Pacific to enter international commerce as an aphrodisiac, tonic and a cure for high blood pressure.

In the UK, sea urchins are collected for curio trade. Dried starfish and sand dollars are used for ornamental purposes. Sea urchins play an important role in controlling algal growth. *Acanthaster planci* of the Indo-west Pacific, cause damage to coral reefs. Many echinoderms are indicators of clean water but they may be killed by toxic oil. *Holothuria tubulosa* and *H. forskali* disappears in areas of high pollution. A large concentration of *Lytechinus variegatus* urchins found in Cartagena Bay, Colombia show skeletal deformations probably due to oil and waste water pollution. The starfish *Solaster papposus, Marthasterias glacialis, Luidia sp.* and *Henricia sanguinolenta* decline around Milford Haven, UK, perhaps due to over fishing by divers. *Loxechinus albus* has been over-exploited in Chile.

Heavy exploitation of sea cucumbers for trepang trade is known. Industrial land-fill, pollution and over-exploitation are cause of concern for urchins. Commercial fisheries clearly need to be managed on a sustainable yield basis. Fukui Prefectural Fisheries Research Station in Japan has developed and tested an artificial substrate for growing juvenile sea urchins. Nylon fibres are set into concrete blocks to form nests on which juvenile urchins settle. The highest mortality in populations occurs at this stage, after metamorphosis, when juveniles settle into algal beds. Efforts are continuing to improve the sea cucumber fishery.

## REVIEW QUESTIONS

### Short Answer Questions

1. Mention two reasons behind conservation of coral reef.
2. What are the aims of species conservation?
3. Mention two useful roles of insects.
4. Why tropical forest conservation is very important?
5. Mention five reasons for loss of invertebrates?
6. What do you understand by anthropogenic threat?
7. Give scientific name of two edible arthropod species.
8. Mention two beneficial roles of protozoans.
9. Mention two beneficial roles of poriferans.
10. Comment on the role of cnidarians in future research.
11. Comment on the role of earthworms in ecosystem.
12. Give two examples of edible molluscs.
13. What are the causes of decline in insects?
14. Why echinoderm conservation is required?

### Long Answer Questions

1. Why invertebrate study is required. Illustrate your answer with examples.
2. Discuss the conservation measures of invertebrates in brief.
3. Describe the importance of study of arthropods.
4. Describe the importance of study of molluscs.
5. Describe the importance of conservation of cnidarians

# Glossary

*Abductor muscle*: A muscle that moves a structure away from the middle of the body.

*Aboral side*: Opposite side to the side in which mouth is located.

*Acanthocephala*: A phylum consisting of parasitic worm-like pseudo coelomates.

*Aciculum*: A chitinous stiff basal seta.

*Acid gland*: Glands secreting acid in many invertebrates, may be the poison gland, the HCN-secreting glands or gland of stinging.

*Acoelomate*: Animals which lack a body cavity and a solid type of body construction.

*Acorn*: The unsegmented head of annelids.

*Actinal*: Star-shaped.

*Actine*: The single ray of a star-shaped spicule.

*Actinotrocha*: A free-swimming ciliated larva of the Phoronida bearing tentacles.

*Actinula*: A larval stage that looks like a polyp or medusa, depending upon whether the mouth is turned upward or downward.

*Adoral zone*: Region within the buccal cavity of certain ciliates.

*Aerial*: Living or occurring in the air.

*Afferent nerve*: Nerve that conducts impulses from the periphery towards a nerve centre.

*Afferent neuron*: Sensory neuron that conveys inward impulses perceived by a sense organ from external sources.

*Aflagellar*: Without flagella.

*Ambulacral area*: Five sections of the oral area containing podia of echinoderms.

*Ambulacral groove*: The furrow extending from the mouth of echinoderms.

*Ambulacrum*: Groove, ridge, or double band of the tube feet.

*Ametabolous*: Development in which the young insects are identical to adults except for size and sexual maturity. No instar has wings and there is no metamorphosis.

*Amictic egg*: A thin-shelled diploid egg which cannot be fertilized and develops by parthogenesis into an amictic female as fond in rotifer.

*Ammonotelic*: Animals producing ammonia as the end product of nitrogen metabolism.

*Amphiblastula*: Sponge larva which is hollow with one hemisphere having flagellated cells and the other of large nonflagellated macromeres.

*Amphid*: Paired, anterior chemosensory and mechanosensory organs of nematodes.

*Amphimixis*: The union of male and female gametes in sexual reproduction, as opposed to automixis.

*Ampulla*: The bulb at the end of lateral canal in Astroida.

*Anastomose*: Branching and rejoining in a complex way.

*Annulus*: A spiraled transverse groove of dinoflagellates.

*Apical field*: The anterior cilia-free area surrounded by the circumapical band as found in rotifers.

*Apopyle*: Outlet from a flagellated chamber to an excurrent canal in sponges.

*Arborescent:* Branching in a tree- or bush-like way.

*Archenteron*: The embryonic gut formed during gastrulation.

*Archeocytes*: Phagocytic cells found in the mesohyl of sponge.

*Archinephridium*: An excretory organ of some invertebrate larvae; a solenocyte.

*Arenicolous*: Burrowing in sand.

*Aristotle's lantern*: A complicated masticating apparatus consisting of several ossicles surrounding the mouth of sea urchins; called Aristotle's lantern as it resembles an early Greek lantern.

*Articular membrane*: A thin, chitinous membrane joining segments of the arthropod body.

*Ascon*: A simple sponge with unfolded pinacoderm and choanoderm.

*Asconoid*: A sponge body that is a small, simple cylinder.

*Asexual reproduction*: Reproduction not involving fertilization including fission, fragmentation, spore production, budding, vegetative reproduction, and gemmule formation.

*Asplanchnic*: Pertaining to the absence of an alimentary canal.

*Asteroid*: Refers to star-like.

*Asteroidea*: A subclass of the phylum Echinodermata characterized by having five-radiate to multiarmed rays, usually not sharply offset from the central disk.

*Asymmetrical*: Not alike on both sides of an axis.

*Asymmetry*: Absence of symmetry.

*Asymptomatic*: Lacking the subjective evidence of a disease.

*Asynchronous*: Not occurring at the same time.

*Atoll*: A circular coral reef surrounding a central lagoon.

*Atrium*: External pore of sponges.

*Attachment disc*: Series of zig-zag lines that serve to anchor the draglines of spiders.

*Auditory organ*: Any specialized structure capable of being stimulated by sound vibrations like tympanal organs and auditory hairs.

*Auditory*: Pertaining to the organs or sense of hearing.

*Auricularia larva*: A free-swimming bilaterally symmetrical larva characterized by the main ciliated band being increased in length due to sinuosity.

*Auto gamy*: Nuclear reorganization without conjugation or exchange of micronuclear material between two protozoans.

*Auto zooid*: Typical feeding zooids of bryozoans and some anthozoans.

*Auxotroph*: Nutritionally dependent cells, or individuals whose growth depends on a specific nutrient in addition to their basic food medium.

*Avicularium*: An enlarged pedunculate operculum of polymorphs, resembling a bird's beak and more intricately reinforced than those of ordinary feeding autozooids.

*Axenic*: Free from associated organisms.

*Axial cells*: Central cells comprising a nucleus and germinal cells.

*Axial filament*: A protein core around which siliceous spicules are organized.

*Axial gland*: A dark elongated mass of spongy tissue extending along the length of the stone canal, forming part of the haemal system.

*Axial rod*: Tough, collagenous endoskeleton of gorgonians.

*Axial*: Pertaining to an axis; relative to the central axis of a cylindrical body or organ.

*Axilla*: A triangular or rounded sclerite on each side of the scutellum of hymenopterans.

*Axillary lobe*: One of the sclerites covering the base of a wing.

*Axillary membrane*: The membrane of the wing base extending from the tegula at the base of the coastal margin to the axillary cord at the base of the anal area.

*Axillary plate*: In Odonata, a large plate hinged to the tergum and supported by an arm from the pleural wing process.

*Axillary region*: Area of a wing base containing the axillary sclerites.

*Axillary vein*: A vein in the anal area of the hind wing.

*Axis*: A line of reference around, along or across where symmetry is established.

*Axon*: A long-unbranched, nerve fibre which conveys impulses away from the cell-body of its neuron.

*Axoneme*: Microtubules and other proteins composing the core of flagella and cilia.

*Axoplasm*: Cytoplasm or neuroplasm of an axis cylinder.

*Axopodium*: Fine, needle-like pseudopodium that contains a bundle of microtubules.

*Barrier reef*: Reef whose platform is separated from the adjacent land mass by a lagoon.

*Basal body*: An organelle located at the base of flagellum or a cilium.

*Basal lamina*: Thin, collagenous, fibrous sheet secreted by epithelial cells and on which they rest.

*Basement membrane*: The layer of fibrous connective tissue under the epidermis consisting of the basal lamina plus additional connective tissue.

*Basopodite*: A crustacean appendage process attached to the coxapodite and which bears the endo- and exo-podites.

*Bdelloid*: A class of fresh water rotifers.

*Benthos*: Community of organisms that lives on or in the bottom of a water body.

*Bilateral symmetry*: Body plan in which there is a single plane of symmetry.

*Binary fission*: Asexual division that produces two similar individuals.

*Bipectinate*: A gill in which the filaments arise on both sides of the axis.

*Bipinnaria*: Primary free-swimming larval stage of asteroids.

*Biradial symmetry*: Body plan with two planes of symmetry.

*Biramous*: Two branched; sometimes extended to cover more than two.

*Blastocoel*: The fluid or gel-filled embryonic cavity beneath the germ layers.

*Blastomere*: A cell produced during cleavage of an ovum; a blastula cell.

*Blastopore*: Primary opening of the archenteron to the exterior of the embryo.

*Blastostyle*: A reduced, finger-like gonozooid that bears gonophores.

*Blastozoid*: A reproductive bud of tunicates.

*Blastula*: A sphere of blastomeres created by repeated cleavage divisions of the zygote.

*Blood-vascular system*: Circulatory system that develops within the connective tissue.

*Body ciliature*: Cilia distributed over the general body surface of ciliates.

*Body whorl*: The last and largest whorl of the gastropod shell.

*Book gills*: Leaf-like lamellae of Limulus which serve as organs of gaseous exchange, i.e. gills.

*Brachiolaria*: Second asteroid larva, following the bipinnaria, marked by the appearance of three adhesive arms at the anterior end.

*Brachiole*: Slender, pinnule-like projection of fossil echinoderms.

*Branchium:* A gill.

*Brood*: To care for developing eggs outside the body.

*Brown body*: A dark sphere of waste-containing cells that remains lodged in the coelom following regression by bryozoan polypides.

*Buccal cavity*: A cavity just inside the opening of mouth.

*Buccal field*: A large ventral ciliated area which surrounds the mouth of some rotifers.

*Buccal*: A cavity just inside the opening of mouth.

*Bulbous pharynx*: Platyhelminth pharynx characterized by a sucking muscular bulb.

*Bursa*: A female reproductive chamber for the reception and temporary storage of sperm received at copulation.

*Byssus*: A bundle of secreted protein threads used to attach some bivalves to the substratum.

*Caecum*: A pouch or sac within the body which has only one opening.

*Calcareous*: Composed of calcium carbonate.

*Calyx*: The body and tentacles of an entoproct.

*Campanulate*: Bell-shaped.

*Capitulum*: Body of stalked barnacles exclusive of the stalk. In stalked Cirripedia, main body, excluding the stalk.

*Carapace*: A chitinous exterior structure covering the crustacean thorax.

*Carina:* Posterior median plate of the exoskeleton of barnacle.

*Caudal gland*: A posterior spinneret as found in many free-living nematodes.

*Cephalic gland*: Slime secreting gland of nemerteans.

*Cephalization*: Tendency to coalesce the segmental ganglia into a large neural centre.

*Cephalon*: The interior body of trilobites.

*Cephalothorax*: Head and thorax together.

*Cercaria*: Free-swimming developmental stage of digenean trematode.

*Cerebral organ*: One of a pair of ciliated sensory canals associated with the brain.

*Chelate*: Refers to pincer-like appendages consisting of movable and immovable fingers.

*Chelicera*: Claw-like feeding appendages of chelicerata.

*Cheliped*: A chelate thoracic appendage found in decapods crustaceans.

*Chloragogen*: A substance which serves as a centre for glycogen and fat synthesis in oligochaetes.

*Choanocyte*: Flagellated and collared cells which create water currents and extract food as found in sponge.

*Chromatophore*: A cell that expands or contracts to change the colour of the organism.

*Cilia*: A hair-like projection of the cell wall used for locomotion.

*Circumapical band*: A ribbon of cilia encircling the anterior end of the rotifer's head.

*Cirri*: A sensory appendage or an eversible copulatory organ.

*Cirrus sac*: Sac containing the internal seminal vesicle, prostate glands, and cirrus of some platyhelminths.

*Clitellum*: A glandular segment of oligochaetes which forms a reproductive cocoon.

*Cnida*: An eversible cnidarian organelle that occurs in a cnidocyte.

*Cnidocil*: A short, stiff, bristle-like cilium that is borne on a cnidocyte.

*Cnidocyte*: Cnidarian cell containing an eversible cnida.

*Coelenteron*: The body cavity and gut of cnidarians and ctenophores.

*Coeloblastula*: Blastula having a well-developed blastocoel.

*Coelom*: Body cavity lined by a mesodermally derived epithelium.

*Coelomate*: An animal having a coelom.

*Coelomocyte*: Circulating coelomic cell which may or may not contain a respiratory protein.

*Coelomoduct*: Mesodermally derived duct leading from a coelom to the exterior.

*Coenosarc*: Tissue underlying the cuticular perisarc of hydroids.

*Collagen*: Common animal fibrous protein that forms extra cellular skeletal materials.

*Colloblast*: In ctenophores, cells with adhesive properties and which are analogus to nematocysts.

*Colony*: Structurally joined zooids that share resources.

*Columella*: Central axis of asymmetrical shells around which whorls are coiled.

*Columnar epithelium*: Epithelium of elongated cells.

*Comb row*: One of eight ciliary bands of ctenophores each composed of a series of combs.

*Comb*: A flat paddle of fused cilia in ctenophores.

*Commensalism*: Symbiotic relationship in which one species benefits from the relationship and the other species (host) is neither benefitted nor harmed.

*Commissure*: A band of fibres joining symmetrical parts or similar ganglion.

*Compound eye*: An eye composed of multiple ommatidia.

*Compressed*: Flattened laterally, from side to side.

*Conchiolin*: Molluscan protein which makes the periostracum, byssus, and operculum.

*Conjugant*: One of a pair of fused ciliates during exchanging genetic material.

*Connective tissue*: Body layer between epithelia, composed of a fluid with or without cells.

*Connective*: Nerve that connects two ganglia of different pairs.

*Conspecific*: Belonging to the same species.

*Contractile vacuole*: Large spherical vesicle responsible for osmoregulation.

*Copraphagy*: Ingestion of faeces.

*Coracidium*: A ciliated free-swimming developmental stage of cestodes.

*Corona*: Ciliated organ at the anterior end of rotifers.

*Cortex*: An outer ectoplasmic layer.

*Cosmopolitan*: Worldwide distribution.

*Coxa*: Basal section of the leg of chilicerates and insects.

*Coxal gland*: Thin-walled sacs which serve as excretory organs of chilicerates.

*Coxapodite*: Crustacean appendage process to which the basopodite is attached.

*Crown*: Pentamerous body of a crinoid.

*Cryptobiosis*: Metabolically inactive, resistant condition.

*Ctenidium*: A molluscan gill.

*Cuboidal epithelium*: Epithelium in which the cells are roughly cubical in shape.

*Cuticle*: Protective or supportive, nonliving, external, layer secreted by the epidermis.

*Cyclomorphosis*: A seasonal change in the body shape.

*Cyclosis*: Streaming protoplasm in Amoeba.

*Cyphonautes*: Planktotrophic larva of some species of nonbrooding gymnolaemate bryozoans.

*Cypris*: An ostracod-like, settling larval stage of barnacles.

*Cysticercus*: The second stage of development of cestodes.

*Cystid*: Exoskeleton and body wall of bryozoans.

*Cytopharynx*: Permanent oral canal of ciliates that is separated from the cytoplasm by the cell membrane.

*Cytoproct*: Permanent cellular anus of some ciliates.

*Cytopyge*: Anal opening in ciliates.

*Cytostome*: Mouth or oral chamber of ciliates.

*Dactylozooid*: A finger-shaped, defensive, hydrozoan polyp.

*Definitive host*: Host for the adult stage of a parasite.

*Demibranch*: One of a pair of lamelli branch gills.

*Dendritic:* Structure like tree.

*Deposit feeding*: Feeding upon detritus that has settled to the bottom of aquatic environments.

*Depressed*: Flattened dorsoventrally.

*Dermis*: The inner layer of skin.

*Detritus*: Fragments of dead plants or animals.

*Deuterostome*: Member of a major branch of the animal kingdom in which the site of the blastopore is posterior—far from the mouth, which forms as a new opening at the anterior end.

*Deutocerebrum*: Median region of the brain of arthropods.

*Diagnostic character*: A characteristic that is unique to a certain group.

*Diapause* A period of arrested metabolism to survive adverse environmental conditions.

*Digenea*: A parasitic organism whose life cycle involves two or more hosts.

*Dimorphism*: Exhibiting two shapes or appearances.

*Dioecious*: Having separate sexes; gonochoric.

*Diploblastic*: With only two embryonic germ layers.

*Direct development*: Lacking a larval or nymphal stage.

*Distal*: *Distant* from the centre, origin, or midline.

*Ditaxic*: A fine muscular contraction which sweeps across half of the foot of some Mollusks.

*Diurnal*: Active during the day.

*Diverticulum*: A small passage diverging from a larger channel.

*Doliolaria*: Barrel-shaped larval stage following the auricularia of holothuroids.

*Dormant egg*: An egg capable of withstand adverse conditions for long periods before hatching.

*Ecdisone*: Hormone which causes ecdysis.

*Ecdysis*: Shedding of arthropod skeleton.

*Echinoplutens*: Second and planktonic larval stage of echinoderms.

*Eclosion*: Emergence of the imago from the pupa or last nymphal cuticle but sometimes, confusingly used as a synonym for hatching from the egg.

*Ectoderm*: Embryonic germ layer composing the outer wall of the gastrula.

*Ectoparasite*: Parasite that lives on the outside of its host.

*Ectoplasm*: Outer portion of cytoplasm of cells.

*Ectosymbiont*: A symbiont living outside its host.

*Elytra*: Plate-like scale modified from a dorsal cirrus that is borne on a short stalk of polychaetes.

*Embryonated egg*: An embryo enclosed in an egg shell.

*Encystment*: Forming resistant cysts in response to unfavourable conditions.

*Endemic*: Species found in a restricted geographic area.

*Endoderm*: Embryonic germ layer composing the archenteron wall.

*Endogastric coiling*: When shell coils posteriorly over the foot.

*Endoparasite*: Parasite that lives inside its host.

*Endoplasm*: The inner portion of cytoplasm of cells.

*Endopodite*: Appendage process which is attached to basopodite of crustaceans.

*Endosymbiont*: A symbiont living inside its host.

*Enterocoel*: Coelomic cavity formed from an out pocketing of the embryonic archenteron.

*Enteronephric*: Typical or modified nephridia that open into various parts of the digestive tract of earthworms.

*Entoproct*: Sessile bryozoan-like animals with the mouth and anus within the tentacular crown.

*Ephemeral*: Short-lived, brief.

*Ephyra*: An immature scyphomedusa.

*Epicuticle*: Exterior portion of the chitinous arthropod skeleton.

*Epidermis*: The outer layer of skin.

*Epigastric coiling*: When the shell coils anteriorly over the head.

*Epimorphic*: Development in which the young hatch with the full complement of segments.

*Epipodite*: Appendange process which is attached to, or is apart of the coxapodite of crustaceans.

*Epistome*: A narrow crescent fold which overhangs the mouth of phoronids.

*Epitheliomuscular cell*: A cnidarian contractile cell that has characteristics of both epithelial and muscular cells.

*Epitoke*: In annelid, reproductive individual adapted for bottom dwelling or pelagic existence.

*Epitoky*: Reproductive phenomenon in some polychaetes: the production of a reproductive individual (epitoke).

*Epizoic*: An external parasitic form.

*Equilateral*: Anterior and posterior ends of a bivalve valve are of similar in shape and size.

*Equivalve*: Two valves of a clam being the same size and shape.

*Eukaryote*: A cell whose nucleus is enclosed within a membrane.

*Eulamellibranch gill*: Bivalve gill with filaments joined together by continuous sheets of tissue.

*Euryhaline*: Tolerant of a wide range of osmotic conditions.

*Exocytosis*: Process in which indigestible material is released from a cell to the exterior by fusion of the residual vesicle with the cell membrane.

*Exopodite*: An appendage process which is attached externally to the basopodite of crustaceans.

*Extrinsic*: A muscle extending from one structure to another.

*Exumbrella*: Exterior upper surface of a medusa.

*Filibranch gill*: Bivalve gill in which filaments are held together by tufts of cilia.

*Filiform*: Having the shape of a filament or thread.

*Filopodia*: Pseudopodium that is slender, clear, and sometimes branched.
*Filter feeding*: Suspension feeding in which organic particles are removed from a water current by a filter.
*Fission*: Formation of daughter cells by division.
*Flabellum*: Spatulate process on last leg of Limulus used to clean gills.
*Flagella*: A long slender motile appendage of cells. It is typically long and its motion is a complex whip-like undulation.
*Flame cell*: A protonephridial terminal cell that has many flagella, which beat synchronously and resemble a minute flickering flame.
*Foliaceous*: Erect, leaf-like.
*Food vacuole*: Cellular vesicle containing ingested food.
*Fringing reef*: Reef that extends seaward directly from the shore.
*Frontal gland*: Anterior aggregation of secretory cells in platyhelminths.
*Funiculus*: A mesothelial cord extending across the bryozoan coelom.
*Fusiform*: Spindle- or cigar-shaped, i.e. thick in the middle and tapered bluntly at both ends.
*Gamete*: Reproductive cell.
*Gametogonia*: Primary gametocytes.
*Gamogony*: Multiple fission that forms gametes.
*Ganglion*: An aggregation of neuronal cell bodies.
*Gastric filament*: One of several cnidocyte-bearing threads that extend into the scyphozoan stomach from the septa between gastric pockets.
*Gastric pouch*: One of four pockets in the wall of the scyphozoan stomach.
*Gastro dermis*: Nutritive muscle cells of cnidarians and ctenophores.
*Gastrolith*: A calcareous concretion in the stomach of some crustaceans for calcium storage.
*Gastrozoid*: Nutritive or feeding polyp of cnidarians which is similar to a short hydra.
*Gastrula*: A cup-like early stage of germ layer formation characterized by two layers of cells.
*Gastrulation*: *Developmental* establishment of germ layers.
*Gemmules*: A spore-like aggregate of sponge material which winters over and grows into a larva in the spring.
*Geniculate*: Bent at a sharp angle like an elbow.
*Genital atrium*: Small chamber in parasitic platyhelminths that receives the openings of both male and female reproductive systems.
*Gill*: An outward expansion of the body surface for the purpose of gas exchange in water.
*Gnathobase*: Heavily armed processes at base of appendages for grinding food in Limulus.
*Gnathochilarium*: A broad flattened plate formed of a fused pair of maxillae in millipedes.
*Gnathopod*: Each of the second and third thoracic appendages of amphipods.
*Gonangium*: Type of gonozooid that consists of a central blastostyle bearing gonophores and is surrounded by an extension of the perisarc.
*Gonochoric*: Having Separate sexes.
*Gonoduct*: Principal duct providing for the transport of sperms or eggs in any reproductive system.
*Gonophore*: A hydroid reproductive bud that bears the germ cells and may become a free-swimming medusa or a variously modified sessile medusa.
*Gonopore*: External opening of any reproductive system.
*Gonotheca*: An extension of the perisarc around a gonozooid.
*Gonozooid*: A hydrozoan reproductive polyp which is often reduced, lacking mouth and tentacles, and bears gonophores.
*Gravid*: Bearing developmental stages, such as eggs or embryos, internally.
*Gross*: Large scale, not fine or delicate, i.e. not microscopic or ultrastructural.
*Halteres*: Reduced second pair of dipteran (fly) wings, functions as a gyroscope to maintain stability in flight.
*Haptor*: Attachment organ that bears hooks and suckers.
*Hatschek's groove*: A shallow ciliated invagination of the dorsal wall of the vestibule of cephalochordates.
*Hectocotylus*: A cephalopod arm modified as a reproductive organ.
*Haemal system*: Blood-vascular system.
*Hemimetabolous*: Insect development characterized by nymphs that do not closely resemble adults but that do not undergo a radical metamorphosis.
*Haemocoel*: Blood-filled cavity, occupying much of the body.
*Haemocyanin*: Oxygen carrying component of the blood of arthropods, cephalopods, crustaceans and molluscs.
*Hermaphroditic*: Having both male and female reproductive systems in the same individual.

*Heterogony*: Alternating sexual and asexual phases in a life cycle.

*Heterotrophic*: Nutrition in which organic compounds are obtained by consuming other organisms.

*Heterozoid*: A modified bryozoan individual which serves many different functions.

*Hinge ligament*: A noncalcified, elastic, proteinaceous band joining the two valves of a bivalve.

*Holoblastic cleavage*: Cleavage furrows extend completely cut through the egg mass.

*Holometabolous*: Insect development in which larvae and adults are distinctly different and a major change is required to transform the juvenile into the adult.

*Holonephridium*: A typical, segmental metanephridial duct of an oligochaete.

*Holothurin*: Toxic substance released in the Cuvierian tubules of certain holothuroids.

*Homolecithal egg*: Egg in which the yolk is uniformly distributed.

*Homology*: Similarity of structure attributable to common ancestry in two or more species.

*Hyaline*: Translucent or transparent, clear.

*Hydranth*: Oral end of a hydroid polyp bearing the mouth and tentacles.

*Hydro theca*: A cuticle that encloses the hydranth.

*Hydrocaulus*: Stalk of a hydroid polyp.

*Hydrocoral*: Colonial, calcified polypoid hydrozoan with either an encrusting or an upright growth form.

*Hydroid colony*: A collection of polyps in which each polyp is connected to the others.

*Hydromedusa*: Hydrozoan medusa.

*Hydrorhiza*: Root-like structure at the base of a polyp.

*Hyperparasitism*: A parasite parasitized by another parasite.

*Hypobranchial gland*: Mucous-secreting epithelium on the molluscan mantle roof.

*Hypodermis*: In arthropods, the underlying layer of integumentary epithelial cells.

*Hypognathus*: Insect head orientation that causes the mouthparts to directed downward.

*Hypostome*: The mouth of a hydra.

*Imago*: The final instar or adult stage of an insect life cycle.

*Incomplete cleavage*: Cleavage furrows do not completely cut through the egg mass.

*Incurrent canal*: Tubular invagination of the sponge pinacoderm that leads into the flagellated chambers.

*Incurrent pore*: Small opening on the surface of sponges that leads into an incurrent canal.

*Indeterminate cleavage*: Fate of the blastomeres is fixed relatively late in development.

*Indirect development*: Having a larval stage(s) between egg and adult.

*Inequilateral*: Anterior and posterior ends of a bivalve valve are dissimilar.

*Inequivalve*: Two valves of a clam of different sizes.

*Infraciliary system*: Entire assemblage of ciliary basal bodies or kinetosomes, and the fibres that link them together in ciliates.

*Instar*: Stages between growth periods in arthropods.

*Integument*: Outer layer of the body wall.

*Intercellular junction*: Membrane specialization that binds cells together which helps to regulate transport across an epithelium.

*Intermediate host*: Host for larval stages of a parasite.

*Interstitial*: Situated between the epidermis and epithelio-muscular cells of hydras.

*Interstitial cell*: Small, rounded totipotent cnidarian cell, located between cells of the epidermis and gastrodermis.

*Interstitial fauna*: Organism living in the spaces between sand and grains.

*Intrinsic*: Confined within a structure.

*Introvert*: Head and anterior part of the body which can be retracte into the body.

*Isogamy*: Fusion of two gametes with the identical genetic material.

*Isolecithal egg*: Egg in which the yolk is uniformly distributed.

*Ivagination*: A type of morphogenetic movement in which the cells of the vegetal hemisphere fold into the interior to form the archenteron.

*Jellyfish*: A cnidarian medusa.

*Kinetodesma*: *A* fine striated fibre that connects kinetosomes of ciliates.

*Kinetoplast*: Conspicuous mass of DNA that is situated within the single, large, elongated mitochondrion of kinetoplastid protozoans.

*Kinetosome*: Organelle to which the base of the cilium or flagellum is attached.

*Kinety*: One row of cilia, kinetosomes, and kinetodesmata of ciliates.

*Labial palp*: Flap-like folds associated with proboscides which act as food sorting structures.

*Labium*: The lower lip of arthropods.

*Labrum*: Lip-like upper prominence of mouth parts of chewing insects.

*Lamella*: In bivalves, each of the gill surfaces.

*Lappet*: Lobe formed by the margin of the scyphozoan bell.

*Larva*: An independent, motile, developmental stage that does not resemble the adult.

*Larviparous*: Eggs brooded internally within the female and are later released as larvae.

*Lateral canal*: Part of the water-vascular system that joins the radial canal and tube feet.

*Laurer's canal*: Short, inconspicuous canal that extends from the seminal receptacle of trematodes to dorsal surface, where it may open at a minute pore.

*Leuconoid*: A type of sponge body organization built around flagellated chambers and an extensive system of canals.

*Ligament sac*: A connective-tissue sac in the acanthocephalan haemocoel containing the gonads.

*Lobopodia*: A blunt pseudopodium.

*Longitudinal cord*: Ridge that extends the length of the body created by the inward expansion of the epidermis as found in nematodes and some gastrotrichs.

*Lophophoral organ*: An area on the phoronid lophophore where spermatophores are formed.

*Lophophore*: A circular or horseshoe-shaped fold of the mesosomal body wall encircling the mouth, bearing hollow ciliated tentacles, and excluding the anus.

*Lorica*: Shell, test or thickened encasement of bdelloid rotifers.

*Luciferase*: Enzyme that catalyzes the bioluminescence reaction.

*Luciferin*: Substrate of luciferin capable of bioluminescence.

*Macromere*: One of several large blastomeres located in the yolky vegetal hemisphere of early embryos.

*Macronucleus*: Large, ciliate nucleus concerned with the synthesis of RNA and DNA.

*Macrophagous*: Collecting and ingesting large food particles.

*Madreporite*: A modified pore in the aboral surface of echinoderms.

*Malpighian tubule*: Excretory system consisting of a blind, tubular, contractile tube as found in insects.

*Mandible*: Short, heavy mouth parts of crustaceans used for biting and grinding.

*Mantle cavity*: Protective chamber created by the overhang of a mantle.

*Mantle*: A body wall fold that secretes a shell, as in molluscs.

*Mantle-palium*: In bivalves, a sheet of muscular, sensory and secreting tissue lying beneath the valves.

*Manubrium*: Tube like extension, bearing the mouth that hangs down from the centre of the cnidarian subumbrella.

*Mastax*: Cuticular pharyngeal jaw apparatus of a rotifer.

*Mastigoneme*: One of the many fine, lateral branches of some flagella.

*Mastigont system*: Complex formed by groups of flagella and microtubular and fibrillar organelles.

*Mastigophora*: Zoo or phytoflagellales which have flagellae.

*Maxillae*: Pairs of feeding appendages of crustaceans.

*Maxillipeds*: First three appendages on the thorax of malacostrans which are modified to manipulate food.

*Medulla*: Central part of the heliozoan cell that is composed of dense endoplasm, one to many nuclei and the bases of the axial rods.

*Medusa*: Form of cnidarian with a well developed gelatinous mesoglea and is generally free-swimming.

*Megalops*: Crab post larva with a large abdomen and full complement of appendages.

*Megasclere*: A large spicule forming one of the chief supporting elements in the sponge's skeleton.

*Mehlis's gland*: Conspicuous unicellular gland cells associated with the reproductive system of trematodes.

*Meroblastic*: Cleavage furrows do not completely cut through the egg mass.

*Merozoites*: Individuals produced by fission of sporozoan schizont.

*Mesenchyme*: Cells which develop into connective tissue, blood and blood vessels.

*Mesenteron*: Central part of anthozoan gastrovascular cavity.

*Mesentery*: A longitudinal sheet of tissue that divides the body cavity of bilaterally-symmetrical animals.

*Mesoderm*: Embryonic germ layer that forms the tissues situated between ectoderm and endoderm.

*Mesoglea*: Thick gelatinous material making up the major part of the inner body of medusae.

*Mesohyl*: Sponge connective tissue.

*Mesothelium*: Single, nonstratified epithelium lining the coelom.

*Metacercaria*: Encysted final stage of digenean development.

*Metachronal*: Wave-like beating of fields of cilia or flagella.

*Metachrony*: Wave pattern that results from the sequential coordinated action of cilia or flagella.

*Metamere*: A body segment.

*Metamerism*: Division of body into segments along an amteropostorior axis.

*Metamerism*: Division of the body into a linear series of similar modules.

*Metamorphosis*: Transformation from a larva into an adult.

*Metanauplius*: One of several instars following the crustacean nauplius.

*Metanephridial system*: Excretory system composed of a vascular ultrafiltration site, a coelomic space, and a metanephridium tubule.

*Metanephridium*: The nephridium in which the preseptal end is an open ciliated funnel.

*Metatroch*: A second girdle of cilia that develops posterior to the prototroch of a trochophore.

*Metazoans*: Asexually-produced progeny that either remains attached to the parent as a colonial zooid or undergoes differentiation before being released as a separate individual.

*Micromere*: One of many small blastomeres located in the animal hemisphere of the cleaving zygote.

*Micronucleus*: Small, ciliate nucleus concerned primarily with the synthesis of DNA.

*Microphagous*: Specialized for feeding on small food particles.

*Micropyle*: An opening in the eggshell or resting stage from which the primordium eventually emerges.

*Microsclere*: A tiny sponge spicule.

*Microtrich*: Type of microvillus found on the tegument of tapeworms.

*Mictic egg*: Type of fertilized rotifer egg that is thin-shelled, haploid and can be fertilized.

*Miracidium*: Ciliated, free-swimming, first larva of digenean trematodes.

*Molt*: To shed the old cuticle as a new cuticle is being secreted.

*Monoecious*: An individual which produces male and female gametes.

*Monogenea*: A parasitic organism having only one form or generation in the life cycle.

*Monopectinate*: Refers to a gill in which the filaments occur on only one side of the axis.

*Monophyletic group*: All species descended from a common ancestor.

*Monoplacophora*: A class of mollusc having a single symmetrical shell.

*Monospecific*: A taxon consisting of a single species.

*Monotaxic*: A fine muscular contraction which sweeps synchronously across the whole surface of the foot of gastropods.

*Monotypic*: A taxon consisting of a single species.

*Mosaic development*: In which cell fate is determined early in development and is the result of the action unevenly distributed factors, like pieces of a mosaic, in the cytoplasm of the uncleaved egg.

*Mucocysts*: Mucigenic bodies that are arranged in rows and discharge a mucoid material.

*Mucous*: Secretion utilized in a variety of ways as an adhesive, protective cover, or lubricant.

*Mutualism*: A symbiotic relationship in which both species benefit.

*Myonemes*: Long contractile fibrils in ciliates.

*Naupliar eye*: Median crustacean eye composed of three or four ocelli.

*Nauplius*: Earliest hatching stage and basic crustacean larva with three pairs of appendages.

*Nectophore*: Mouthless, pulsating swimming bell of siphonophores.

*Nematocyst*: Toxic stinging structures carried by cnidocysts in hydrozoans and scyphozoans.

*Nematodesma*: One of several microtubular rods that line and support the wall of the ciliate cytopharynx and assist in the inward transport of food vacuoles.

*Nephridium*: An excretory tubule usually opening to the exterior via a nephridiopore.

*Nephrocyte*: A large phagocytic cell in the haemocoel of many arthropods.

*Nephrostome*: Fan-shaped open and ciliated end of metanephridia in polychaetes.

*Neurogenic*: Originating with a neuron.

*Neuropodium*: Lower part of the parapodium.

*Nocturnal*: Circadian behaviour characterized by activity at night.

*Notopodium*: The upper part of the parapodium.

*Nuchal organs*: Ciliated sensory pits in the head region of polychaets.

*Nudibranchs*: Gastropods characterized by a secondarily bilateral symmetry and loss of shell, mantle and original gill.

*Nutritive-muscle cell*: A muscle cell in the cnidarian gastrodermis that usually bears a cilium and is responsible for intracellular digestion.

*Ocelli*: Light sensitive organs.

*Odontophore*: The cartilaginous base for the radula of molluscs. A muscular and cartilaginous mass in the buccal cavity of many Mollusks, it supports the radula.

*Oligomery*: Division of the body into three linear regions, characteristic of many deuterostome animals.

*Ommatidium*: A long cylindrical structure possessing the ability of light reception in the arthropod eye.

*Onchosphere*: The ciliated free-swimming larva of cestodes.

*Oncomiracidium*: Ciliated larva of monogeneans.

*Ontogeny*: Development of the individual from fertilization to death.

*Ootype*: Small, centrally-positioned sac within the female reproductive system of most parasitic platyhelminths.

*Oozooid*: Zooid developing from the fertilized egg of urochordates.

*Operculum*: A lid or covering of an opening or chamber.

*Ophiopluteus*: Planktotrophic larva of many species of ophiuroids.

*Opisthosoma*: The posterior portion of pogonophores.

*Oral arm*: One of the four, often frilly extensions of the scyphozoan manubrium.

*Oral ciliature*: Cilia that are associated with the mouth region of ciliates.

*Oral disc*: Area around the mouth of an anthozoan polyp which bears eight to several hundred hollow tentacles.

*Oral shield*: One of a series of large plates that frame the ophiuroid mouth and also form a chewing apparatus with five triangular, interradial jaws at the centre.

*Oral sucker*: Organ that surrounds the trematode mouth, prevents dislodgement and aids in feeding.

*Osculum*: Excurrent opening of the water circulation system of the sponge, the large opening at the top of atrium of asconid sponges.

*Osmoconformation*: Internal osmolarity is allowed to vary with external osmolarity.

*Osmoregulation*: Maintenance of an internal osmolarity unlike the external.

*Osphradium*: Sensory areas on the poster margin of efferent gill membranes which have a chemoreceptor function.

*Ossicle*: Calcareous rods, crosses or plates which are the skeleton of asteroids. An internal skeletal piece, commonly calcareous as in echinoderms.

*Ostium*: A small incurrent opening or pore on the surface of a body, gill, or heart. Lateral openings in the heart of arthropods. Also the incurrent pore of sponges.

*Ovigerous leg*: The third appendage of pycnogonids, used by the male to brood the fertilized eggs. Bearing eggs externally.

*Oviparous*: Egg-laying.

*Paedogenesis*: Achievement of sexual maturity as a larva, without attaining adult morphology.

*Palium*: The epidermis underlying the shell of molluscs.

*Papula*: Finger-like, respiratory evagination of the aboral body wall of some asteroids.

*Paraphyletic*: A taxon containing some, but not all, of the descendants of an ancestor.

*Parapodium*: Lateral, fleshy, paddle-like appendage on polychaete annelids. Lateral appendages extending from the segments of annelids and from the body or foot of some molluscs.

*Parasitism*: A symbiotic relationship in which one species (parasite) benefits from the relationship and the other species (host) is harmed but usually not killed.

*Parasitoidism*: A prolonged intimate symbiosis in which one member eventually kills the other.

*Paratomy*: The phenomenon of linear budding in some turbellarians and annelids.

*Parenchyma*: A spongy mass of cells. Connective tissue compartment between the body wall musculature and gut of platyhelminths.

*Parthenogenesis*: Reproduction by the development of an unfertilized egg. Unisexual reproduction with unfertilized eggs and no contribution by males.

*Patch reef*: A small circular or irregular reef that rises from the floor of a lagoon behind a barrier reef or within an atoll.

*Pecten*: Comb-like chitinous plates which project from the sides of scorpions and which are sensory.

*Pectinate*: Having teeth or side branches arranged like a comb.

*Pectine*: A comb-like sensory appendages unique to scorpions.

*Pedal disc*: In some sea anaemons, a flattened disc at the aboral end of the column for attachment.

*Pedal laceration*: Method of asexual reproduction in some anaemons in which parts of the pedal disc are left behind as the animal moves.

*Pedicellaria*: A small, specialized jawlike appendage of asteroids and echinoids which is used for protection and feeding.
*Pedicle*: A cylindrical extension of the ventral wall of orachiopods which attaches to substrate.
*Pedipalp*: A sensory appendage arising out of prosoma and modified to perform various functions. The second chelicerate appendage, it is modified for a variety of functions.
*Peduncle*: Muscular, flexible attachment stalks of goose barnacles.
*Pelecypod*: A class of molluscs which have two shell valves and a large foot, both laterally compressed, and gills which have assumed a food collecting function.
*Pellicle*: Body covering of ciliates.
*Pelmatazoa*: Animals which travel on soles of foot (i.e. gastropods).
*Penetration anchor*: An anchor that holds one part of a burrowing animal's body in place as another part penetrates and advances into sediment.
*Pentactula*: Metamorphosing stage of holothuroid development that bears five primary tentacles.
*Pentamerous*: Divided into five parts, characteristic of the body of echinoderms.
*Pericardial*: The region nearest the heart.
*Perihemal space*: Separate extensions of the coelam; part of the bloodvascular system of asteroides.
*Periostracum*: Outer proteinaceous layer of a molluscan shell, composed conchiolin.
*Periproct*: Membranous area, often bearing ossicles, around the anus of echinoids.
*Perisarc*: A supporting, nonliving chitinous cuticle secreted by the epidermis surrounding most hydroids.
*Peristalsis*: A wave of muscular contraction moving along a body or internal tube or vessel. *Peristome*: Pre-oral chamber of ciliate.
*Peristomium*: First true segment, immediately posterior to the prostomium, of an annelid.
*Peritoneum*: Innermost, noncontractile layer of a stratified coelomic epithelium; separates the coelomic fluid from the musculature.
*Perivisceral*: Region nearest the intestines.
*Perrostracum*: Outer layer of the mollusk shell.
*Phagocyte*: Cells which engulf food particles or foreign bodies.
*Phagocytosis*: Engulfment of large particles, such as bacteria and protozoans, by evagination of the cell surface.
*Pharynx*: Area between the mouth and nasal passages and the larynx.
*Phasmid*: Unicellular glands in the tail of nematodes.
*Photocyte*: Specialized cell within which light is produced.
*Photophore*: A light-producing organ.
*Phyllobranch*: Having flat, leaf-like gills.
*Phyllopod*: Flattened, leaf-like appendage.
*Phytoflagellate*: A photosynthetic flagellate.
*Phytophagous*: Plant eating.
*Pinacocyte*: Flattened cells which together make up the sponge pinacoderm.
*Pinnate*: Having side branches, such as feather.
*Pinnule*: Jointed appendages on arms of crinoids giving a feather-like appearance.
*Pinocytosis*: Uptake of liquid by cellular engulfment.
*Planula*: A cnidarian larva that is elongated and radially symmetrical but with anterior and posterior ends.
*Plasmodium*: Amoeboid syncytial mass.
*Pleopod*: The anterior abdominal appendages of malacostracans.
*Plerocercoid*: The final stage in the life cycle of certain tapeworms.
*Pleuron*: The two lateral cuticular sections of the skeleton of arthropods.
*Plicate gill*: A molluscan gill which the surface of the lamellae has been increased by folding along the long axis.
*Pneumostome*: Pore-like opening in pulmonate snails leading into lung or mantle cavity.
*Podium*: Tubular projections on the ambulacural groove of asteroids which serve as locomotor organs.
*Podocyte*: Cell with branching interdigitating toe-like processes over the surface of a blood vessel.
*Polian Vessicles*: Elongated muscular sacs suspended in the coelom of some asteroids which serve as expansion chambers of the water vascular system.
*Polychaete*: A class of annelids with paired lateral parapodia.
*Polyembryony*: Development of multiple embryos from a single cell mass.
*Polymorphism*: Two or more individuals of a species modified for different functions.
*Polyp*: Form of sessile cnidarian that has a thin layer of mesoglea and is generally sessile.
*Polyphyletic*: A taxon that includes the descendants of more than one ancestor.

*Polypide*: Innermost parts of a bryozoan zooid, including the introvert, lophophore and viscera but not the body wall or zooecium.

*Polyplacophore*: Chitons which have eight overlapping transverse plates as shells.

*Porocyte*: A sponge cell that surrounds an opening which extends from the external surface to the spongocoel.

*Preoral pit*: Developmental precursor of the wheel organ and Hatschek's groove that opens on the left side of the head of larval cephalochordates.

*Pretrochal region*: Apical plate, prototroch, and mouth region of a trochophore larva.

*Proboscis*: Any tubular process of the anterior part of the gut, usually used in feeding and often extensible.

*Procercoid*: Developmental stage of certain tapeworms between oncosphere and plerocercoid.

*Proctodeum*: Invaginated embryonic ectoderm joining the anus with the endodermal midgut.

*Proglottid*: Linearly arranged individual reproductive segments of the strobila of cestodes.

*Prokaryote*: A one-celled organism whose genetic material is distributed throughout the cell including bacteria and blue-green algae.

*Propodium*: The front of a gastropod foot which acts like a plough and anchor. The fore part of a podium.

*Prosobranch*: Primitive molluscs of the sub order Archeogastropods.

*Prosoplyle:* Pores opening into the incurrent canals of syconoid sponges.

*Protandry*: Type of hermaphroditism in which the male and female attains maturity at different time.

*Protocereberum*: Anterior portion of the brain of arthropods.

*Protoconch*: Shell of the veliger which remain at the apex of the adult shell.

*Protonephridium*: Primitive excretory and osmoregulatory organ.

*Protoplasm*: All cellular material.

*Protopod*: Basal part of a crustacean appendage, consisting of combined coxa and basis.

*Protopodite*: Two most proximate elements of the crustacean biramous appendage.

*Protostome*: A major branch of the Animal Kingdom in which the blastopore contributes to the formation of the mouth.

*Prototroch*: Free-swimming larvae of molluscs characterized by a girdle of ciliated cells.

*Protozoan*: A single-celled animal.

*Protozoea*: Third larval stage of a decapod (shrimp) after the metanauplius and before the zoea.

*Proximal*: Close to the origin, centre, or midline.

*Pseudo branch*: In aquatic pulmonates, a gill formed secondarily from folds of mantle.

*Pseudocoel*: Fluid-filled body cavity that occupies the connective tissue compartment.

*Pseudofeces*: Material too large to be digested and egested by bivalves.

*Pseudolamellibranch gill*: Bivalve gill with filaments bound together with small tissue junctions.

*Pseudopodia*: A flowing extension of the body of sarcodinians.

*Pteropods*: An order of swimming pelagic opisthobranchs.

*Ptychocyst*: *A* cnida that discharges a thread used to weave a tube.

*Pupa*: In holometabolous insects, the stage between the last larval instar and the adult.

*Pygidium*: The terminal, nonsegmental part of the body of a segmented animal typically bears the anus.

*Pyriform*: Pear-shaped.

*Racemose*: Formed of a number of coalescing ducts draining to a central cavity.

*Radial canal*: One of five fluid-filled channels of the echinoderm water-vascular system that join the ring canal to the lateral canals.

*Radial cleavage*: Cleavage pattern in which the cleavage spindles or cleavage planes are at right angles or parallel to the polar axis of the egg.

*Radial symmetry*: Arrangement of similar parts around a central axis.

*Radiole*: Each of the several pinnate tentacles on the head of a sabellid, serpulid, or spirorbid polychaete.

*Radula*: A belt of transverse rows of teeth supported by the odontophore.

*Ramus*: A branch.

*Raptorial*: Animals that capture prey.

*Recent*: The current epoch of the Quaternary Period.

*Redia*: Third developmental stage in the trematode life cycle.

*Regulated compartment*: A space, such as an organelle, gut region, or body cavity, where the chemical environment can be controlled.

*Regulative development*: In which cell fates are determined by a network of cellular communication in the embryo.

*Respiratory tree*: One of two respiratory organs of most holothuroid echinoderms and consisting of a network of thin-walled tubules in the perivisceral coelom that originates from the cloacal wall.

*Reticulopodium*: A pseudopodium that forms a thread-like branched mesh and contains axial microtubules.

*Retinula*: A network of rhabdomes which line the omatidium.

*Retractor muscle*: Muscle that withdraws an eversible body part.

*Retrograde*: *Passing* in a direction opposite the direction of motion of the animal.

*Retroperitoneal*: Outside the coelom but typically bulging into the coelom and covered by peritoneum.

*Rhabdite*: Platyhelminth epidermal secretion droplets which are characterized microscopically by a specific, layered ultrastructure.

*Rhabdome*: An enclosed central space, lined with rhabdomeres, in the arthropod's eye.

*Rhabdomere*: Area of retinular cells oriented at right angles to axis of omatidium of arthropods.

*Rhagon*: Developmental stage immediately following the metamorphosis of a demosponge larva.

*Rhinophore*: One of the second pair of sensory tentacles.

*Rhmophore*: Modified opisthobranch tentacle containing tactile and chemoreceptor cells.

*Rhopalium*: Concentrations of neurons in the margins of scyphozoans.

*Rhynchocoel*: A fluid-filled coelomic cavity that houses the retracted nemertean proboscis.

*Rhynchodeum*: In nemerteans, the short anterior canal that joins the proboscis pore to the proboscis.

*Ring canal*: Part of the echinoderm water-vascular system that joins the stone canal to the radial canals.

*Rostrum*: Middorsal projection in some rotifers that bears cilia and sensory bristles at its tip and is also adhesive.

*Sanguivorous*: Feeding on blood.

*Saprophyte*: A plant which obtains its nutrients from dead organic material.

*Saprozoic*: An organism which obtains its nutrients from dead organic material.

*Sarcodinas*: Protozoans having flowing extensions of the body.

*Scaphognathite*: Paddle-like projection of the second maxilla that produces a ventilating current.

*Schizocoel*: Coelomic cavity derived from the separation of a solid mass of mesodermal cells.

*Schizogony*: Multiple fission that produces merozoites.

*Sclerite*: *Thickened* area of cuticle in the exoskeleton of arthropods.

*Scleroseptum*: One of the many radiating calcareous partitions in the skeletal cup of stony corals.

*Sclerotized*: Highly darkened, and thickened arthropod exoskeleton.

*Scolex*: Anterior head region of tapeworms that is adapted for adhering to the host.

*Scutum*: One of the calcareous plates forming the barnacle operculum.

*Scyphistoma*: A scyphozoan polyp.

*Scyphozoan*: The cnidarians in which the medusa is the dominant and conspicious individual in the life cycle.

*Segmentation*: Body composed of a linear series of repeating units, or segments.

*Seminal receptacle*: Chamber in the female gonoduct for the reception and storage of allosperm.

*Seminal vesicle*: A coelomic pouch which provides for maturation of gametes of oligochaetes.

*Sensillum*: Arthropod sense organ involving a specialized part of exoskeleton.

*Sensu lato*: In the broad sense.

*Sensu stricto*: In the strict sense.

*Septal filament*: Free edge of an anthozoan septum that is trilobed.

*Septibranch*: A subclass of molluscs having degenerate gills modified to muscular septa.

*Septum*: A double-walled tissue partition in the cross-sectional plane of a bilaterian or a radial plane of a cnidarian.

*Sessile*: Attached directly to the body surface and not stalked.

*Seta*: *An* exoskeletal bristle composed of a chitin.

*Setiger*: A segment with setae.

*Setose*: Bearing setae.

*Sexual dimorphism*: Male and female of a species with different shapes or appearances.

*Shell*: A rigid skeleton on the outside of an organism.

*Shield*: A small calcareous plate in certain echinoderms, especially ophiuroids.

*Sieve tracheae*: Arthropod tracheal system in which the spiracle opens into an atrial chamber from which a great bundle of tracheae arises.

*Siliceous*: Composed of silica.

*Simple epithelium*: Composed of a single layer of cells.

*Sinus*: Any cavity, recess, or passage in tissue or bone.

*Siphon*: A tubular fold of the molluscan mantle used to direct water to or from the mantle cavity.

*Siphonoglyph*: Ciliated groove in the pharyngeal wall of some anthozoans that moves water into coelenteron.

*Siphonozooid*: A highly-modified pennatulacean polyp that pumps water into, or allows it to escape from, the interconnected gastrovascular cavities of the colony.

*Slug*: Opisthobranch or pulmonate in which the shell is absent or reduced and buried in the mantle.

*Solenocyte*: A protonephridial terminal cell with one flagellum and a long microvillar collar.

*Solenogaster*: A class of worm-shaped molluscs.

*Somite*: A body segment.

*Spermatheca*: Chambers for storage of sperm.

*Spermatophore*: Sperm contained within a membrane.

*Spicule*: A small needle-like or rodlike skeletal piece.

*Spinneret*: Spinning organ of spiders.

*Spiracle*: Slitlike external opening of arthropod tracheal system.

*Spiral cleavage*: Type of cleavage pattern in which cleavage spindles are oblique to the polar axis of the egg.

*Spire*: All the whorls of a gastropod shell above the body whorl.

*Spirocyst*: Cnida with a long adhesive thread that functions in capture of prey and in attachment to a substratum.

*Spongin*: A large, collagenous, connective tissue fibre of sponges.

*Spongocoel*: Interior cavity of asconoid sponges.

*Sporocyst*:. Nonciliated second stage in the life cycle of digeneans. Arises from a miracidium and gives rise to rediae.

*Sporogony*: Cycle in which the spores are formed.

*Sporosac*: Incomplete gonophore that remains attached to the polypoid colony.

*Sporozoite*: Stage in life cycle of some protozoans, the development of which follows sporogony.

*Spring tides*: Tides occurring on new and full moons characterized by large tidal amplitude.

*Squamous epithelium*: Epithelium of flattened tile-like cells.

*Statoblast*: Resistant reproductive bodies of bryozoans which germinate under favourable conditions.

*Statocyst*: A sense organ that can provide orientation to the pull of gravity.

*Stenohaline*: Restricted to a narrow range of salinities.

*Stenopod*: A narrow, cylindrical, leg-like appendage.

*Stereo blastula*: A solid blastula, lacking an internal cavity or blastocoel.

*Stereo gastrula*: A solid gastrula, lacking an archenteron cavity.

*Sternite*: Ventral plate of cuticle of each segment of an arthropod.

*Sternum*: The ventral cuticular skeleton of arthropods.

*Sterogastrula*: A blastula changed to a solid gastrula by ingression, a process by which cells move into the blastocoel.

*Stolon*: Root-like extension of the body that interconnects colonial zooids.

*Stomodeum*: Invaginated embryonic ectoderm joining the mouth with endodermal midgut.

*Stratified epithelium*: Composed of two or more layers of cells, only one of which rests on the basal lamina.

*Streptoneury*: Gastropod nervous system twisted by torsion into an asymmetrical figure 8.

*Stridulate*: To produce sound by rubbing one body part against another.

*Strobila*: A scyphozoan polyp that buds medusae, or the posterior part of a tapeworm that consists of proglottids.

*Strobilation*: Process by which scyphomedusae arise as buds that are released by transverse fission of the oral end of the scyphistoma.

*Style sac*: Posterior conical section of the stomach of Mollusks.

*Stylet*: Dagger-like structure associated with various systems of different animal groups.

*Subchelate*: A pincer in which the movable finger closes against a flat palm.

*Subradula organ*: Cushion-shaped chemosensory structure.

*Subterminal:* Located some distance from the end.

*Subumbrella*: Lower exterior surface of medusae.

*Sulcus*: A longitudinal groove of dinoflagellates that bears the posteriorly directed flagellum.

*Suspension feeding*: Feeding on organic particles suspended in water.

*Suture*: Junction between the septum and the wall of a cephalopod shell.

*Syconoid sponge*: A radially symmetrical sponge that has a body wall folded into radially oriented canals.

*Symbiosis*: An intimate, long-term, physical interraction between two species in which at least one of the species is dependent to various degrees upon the other.

*Symmetrogenic*: Producing mirror-image daughter cells as a result of fission.

*Synanthropic*: Living with humans.

*Synapse*: Connections between neurons.

*Syncytium*: Tissue in which nuclei are not separated by cell membranes.

*Synkaryon*: Zygotic nucleus of ciliates.

*Systole*: Contraction phase of a heart beat.

*Tagma*: An arthropod body region of arthropods.

*Tarsal organ*: Cup-like spider chemoreceptor for detecting pheromones.

*Taxodont*: Hinge dentition and consisting of uniform alternating teeth and sockets in a row.

*Taxon*: A group of organisms with a common ancestor.

*Tegument*: Nonciliated outer syncytial layer of body wall of parasitic platyhelminths and acanthocephalans.

*Telolecithal*: Type of egg in which the yolk material is concentrated to one side of the egg.

*Telopodite*: Movable part of an appendage extending outward from an immovable protopod.

*Telotroch*: A ring of cilia encircling the anus at the posterior end of a trochophore larva.

*Tentacle*: Evagination of body wall surrounding the mouth which aids in capture and ingestion of food.

*Tentacle sheath*: In Bryozoa, the part of the withdrawn body wall that encloses the withdrawn tentacles of the lophophore.

*Tergite*: Dorsal, sclerite of each arthropod segment.

*Tergum*: Combined tergites.

*Terminal cell*: Tubular flagellated cell attached to the inner end of the protonephridial tubule.

*Terminal*: At the end.

*Test*: An encasing skeleton, typically covered externally by cytoplasm.

*Tetramerous*: Radial symmetry in which a basic pattern is repeated in multiples of four.

*Theca*: Nonliving cuticle around the hydranths of thecate hydroids.

*Thecate*: Refers to hydroids with a hydrotheca surrounding polyp proper.

*Thoracopod*: Any thoracic appendage of an arthropod.

*Tiedemann's body*: One of the interradial outpockets of ring canal of many echinoderms.

*Tongue bar*: A downgrowth of pharyngeal tissue that divides a developing gill opening into two side-by-side slits.

*Tornaria*: Transparent, long-lived, planktotrophic larva of enteropneusts.

*Torsion*: In Mollusks, the visceral mass, mantle and mantle cavity being twisted 180° counterclockwise.

*Trachea*: Respiratory tubes of insects and arachnids.

*Trichobranchiate*: Having filamentous gills.

*Trichocyst*: Organelles containing fluid hardens into rod-like structures for defense or attachment.

*Trilobite larva*: Horseshoe crab larva that superficially resembles trilobites.

*Triploblastic*: Embryos possessing ectoderm, endoderm and mesoderm.

*Tritocerebrum*: Posterior region of the arthropod brain.

*Trivium*: Three ventral ambulacral areas of holothuroids.

*Trochophore*: Type of larva found in Mollusks, annelids, and other groups in which the larval body is ringed by a girdle of cilia, the prototroch.

*Trochus*: Anterior band of cilia of divided corona of some rotifers.

*Trophi*: Cuticular hard parts of rotifer mastax.

*Trophosome*: Central mass of tissue in trunk of pogonophore that is packed with symbiotic bacteria.

*Trophozoite*: Feeding stage that occurs when the sporozoite invades the host.

*Trophozooid*: A feeding or nutritive polyp.

*Tropic hormone*: A hormone whose target is an endocrine cell.

*Tube tracheae*: Simple branched or unbranched trachea.

*Tubicolous*: Tube-dwelling.

*Tubules of Cuvier*: Eversible toxic tubules associated with the bases of respiratory trees of some holothuroid echinoderms.

*Typhlosole*: A ridge projecting internally from intestinal wall to increase its surface area.

*Ultra filtration*: Passage of fluid across a fine-mesh filter to retain proteins and larger particles.

*Umbo*: A dorsal protuberance of a bivalve valve rising above hinge.

*Uncinus*: A minute seta modified into a hook.

*Undulating membrane*: Type of ciliary organelle that is a row of adhering cilia forming a sheet.

*Uniramous*: Having one branch.

*Ureotelic*: Producing urea and uric acid as the end product of nitrogen metabolism.

*Uropods*: Sixth abdominal appendages of most malacostracans but the 4th, 5th, and 6th of amphipods.

*Vacuole*: A liquid-filled cavity in a cell.

*Vascular plug*: Specialized nemertean exchange site across which an ultrafiltrate passes from blood to rhynchocoel.

*Vegetative nucleus*: Macronucleus.

*Velarium*: Velum-like structure of cubozoans.

*Veliger larva*: Molluscan larva in which the foot and shell make their appearance that follows the trochophore.

*Velum*: Marginal flap attached to subumbrella at margin of bell of medusae.

*Vessel*: A small tubular blood channel.

*Vestigial*: Reduced to a nonfunctional remnant.

*Vestimentum*: Collar-like body region of a vestimentiferan that helps to secrete the animal's tube.

*Visceral mass*: One of three primary parts of molluscan body.

*Vitellarium*: A structure in flatworm ovary which produces yolk eggs.

*Viviparous*: Embryos gestated internally within female where supplemental nutrition is supplied.

*Whorl*: *Any* complete turn (360°) of a coiled molluscan shell.

*Xylophagous*: Feeding on wood.

*Zoarium*: Form of a bryozoan colony.

*Zoea*: Second stage of crustacean larvae. Penultimate larval stage of many decapod crustaceans, preceding the post larva.

*Zoochlorella*: Unicellular green algal symbiont of certain animals, especially fresh water sponges.

*Zooecium*: Cuticle, or exoskeleton, of a bryozoan zooid.

*Zooflagellate*: A flagellate that has one to many flagella, lacks chloroplasts, and is heterotrophic.

*Zooplankton*: Microscopic animals suspended in the water of oceans and fresh water lakes.

*Zooxanthella*: A golden-brown alga, usually a dinoflagellate, symbiotic with various marine animals, especially cnidarians.

*Zygote*: A cell created by union of two gametes.

# Suggested Reading


Anderson, O.R., 1987, *Comparative Protozoology: Ecology, Physiology, Life History*, Springer-Verlag, Berlin, p. 482.

Anderson, D.T., 2001, *Invertebrate Zoology*, 2nd ed., Oxford University Press, Oxford.

Audesirk, T. and Audesirk, G., 1999, *Biology: Life on Earth*, Prentice-Hall International, Inc.

Barrington, E.J.W., 1981, *Invertebrate Structure and Function*, 2nd ed., ELBS and Nelson.

Brusca, R.C. and Brusca, G.J., 2002, *Invertebrates*, 4th ed., Sinauer Associates.

Chapman, R.F., 1998, *The Insects: Structure and Function*, 4th ed., Cambridge University Press.

Cox, F.E.G., 1982, *Modern Parasitology*, Blackwell Scientific Publications, Oxford, London, p. 346.

Hyman, L.H., 1951, *The Invertebrates: Platyhelminthes and Rhynchocoela, the Acoelomate Bilateria*, Vol. 2, McGraw-Hill, New York, pp. 1–51.

Levine, N.D. et al., 1980, A newly revised classification of the Protozoa, *J. Protozool.*, 27(1): 37–58.

Mandal, F.B., 2011, *Human Parasitology*, PHI Learning, New Delhi.

Meglitsch, P.A. and Schram, F.R., 1991, *Invertebrate Zoology*, Oxford University Press, Oxford.

Morton, J.E., 1967, *Molluscs*, 4th ed., Hutchinson University Library, London.

Parkar, T.J. and Haswell, W., 1972, *Textbook of Zoology*, Vol. I, ELBS and Macmillan.

Ruppert, E.E., Fox, R., and Barnes, R.D., 2003, *Invertebrate Zoology: A Functional Evoloutionary Approach*, 7th ed., Brooks Cole.

Sinha, K.S., Adhikari, S., and Ganguly, B.B., 1977, *Biology of Animals*, Vol. I, 3rd ed., New Central Book Agency, Kolkata.

Tembhare, D.B., 2006, *Modern Entomology*, Himalaya Publishing House.

Wilmer, P., 1990, *Invertebrate Relationships: Patterns in Animal Evolution*, Cambridge University Press, Cambridge, p. 400.

[illegible]

Morton, [illegible] Library, London.

Pfeiffer, Richard [illegible] Macmillan.

Plummer, [illegible] Zondervan [illegible] Brooks [illegible]

[illegible]

Tennant, D.F., 2009 [illegible] Publishing House.

Williams, [illegible] 1980 [illegible] University Press [illegible]

# Index


Abactinal surface, 296
Abdominal ganglia, 242
Aboral
  haemal ring, 305
  organ, 116
  ring sinus, 304
  surface, 292
*Acanthaster planci*, 339
Acanthostrongyles, 70
Acanthostyle, 69, 70
Acanthoxeas, 70
Accessory sex organs, 166
Acetylcholine, 164
*Achatina*, 175
Acicula, 192
Aciculum, 211
Acidic fauna, 119
Acoela, 273
Acone, 263
Acronematic, 54
Actin, 48, 212
Actinomyocin, 48
Actinotrocha, 133, 326
  larva, 131
Acute
  amoebic dysentery, 43
  form, 38
Adambulacral
  groove, 292
  muscle, 292
ADP, 212
Adult, 250
*Aedes*, 251, 253
*Aedes aegypti*, 182, 254
Afferent
  ctenidial vein, 280
  renal vein, 280
Agriculture, 338
Alary muscles, 239
Alima larva, 260
Alkaline fauna, 119
Allopatric, 13
Alpha taxonomy, 15
Alternation of generations, 97
Alveolus, 277
Amastigotes, 40
Ambulacral
  ossicle, 292
  pore, 293
  ridges, 293
Amebocytes, 90
Amictic ova, 126
Ammonia, 214
Ammonites, 267
Ammonotelic, 206
*Amoeba*, 54, 55
Amoebic
  abscesses, 42
  granuloma, 44
  ulcers, 43, 44
Amoebocytes, 67, 78, 198
Amoeboid
  cells, 48
  form, 32
  movement, 43
  pigment cells, 263
Amoebulae, 43
Amphiblastula, 76
Amphidisc, 69
Amphids, 165
Amphineura, 271
Amphinucleus, 58
Ampulla, 57, 130
Anaerobic, 6
Anal, 230
Ancestrula, 139
Anchoras, 70
*Ancylostoma*, 188
  *duodenale*, 170, 173, 189
Ancylostome dermatitis, 174
Ancylostomiasis, 170
Anematic, 54
*Angiostrongylus cantonensis*, 175
*A. costaricensis*, 175
Aniosochelas, 69
Anisakiasis, 175
*Anisakis simplex*, 175, 176
*Anopheles*, 17, 32, 253, 255
*A. barbirostris*, 181
*A. gambiae*, 254
*Antails entails*, 288
*Antedon*, 308
Antennae, 119
Antennal
  hairs, 256
  sockets, 234
Antennary gland, 249
Antennules, 223
Anterior loops, 203
Anthropogenic threat, 328
Antifungal, 338
Antimalarial therapy, 35
Antitumour factor, 333
*Aortic ampulla*, 278
Apex, 275
Aphrodite, 192
Apical
  field, 121
  gland, 149
Aplacophora, 272
*Aplysia*, 333
Apodacea, 311
Apodida, 311
Apolysis, 233
Apopyle, 72
Apposition image, 263
*Apus*, 223
Aquaculture, 334
Arachnid species, 335
*Arca*, 273
Archaea, 2
Archaeocytes, 66, 67
Archaeogastropoda, 272
Archenteron, 306
*Argulus*, 220
Arolium, 237
Arthrobranch, 257
Arthropods, 215
Articular membranes, 216
Artificial
  chemicals, 338
  classification, 20
Ascariasis, 177
*Ascaris*, 163, 177, 188
  *lumbricoides*, 177

Ascaron, 177
Asconoid type, 73
Ascopore, 138
Aspidochirotacea, 311
*Asplanchnopus*, 123
*Asterias*, 295, 296, 300
*Asterina*, 296
*Astropecten*, 296
Asulcal side, 88
Asymptomatic, 37, 180
Atlanta, 288
Atolls, 105
ATP, 48, 212
ATPase, 50
ATP sensitive proteins, 48
Atrial
  chamber, 259
  muscles, 259
Atrium, 321
*Aurelia*, 97
Auricles, 6
Auricularia, 326
  larva, 308
Autogamy, 59
Autoheteroxenous, 167
Autoinfection, 153
Autonomy, 296
Autoskeleton, 68
Autotrophic, 24
Autozooid, 135, 139
Autumn cycles, 126
Avicularia, 134
Axial
  complex, 304
  gland, 301
  sinus, 303
Axoneme, 49
Axopodia, 46

*Balanoglossus*, 320
*Balanus*, 220
Bancroftian filariasis, 255
Basal, 230
Basement membrane, 263, 298
Basipinacocytes, 67
Basipodite, 221
Basitrichous isorhizas, 94
Basommatophora, 273
Bathysciadium, 288
Beef tapeworm, 151
Behavioural approach, 18
Behavioural isolation, 18
Bell margin, 99
*Belostoma*, 222
Benthic, 114, 316
Benzene hexachloride, 257
Beta taxonomy, 15
Binary fission, 25
Biodiversity, 15
Biogenetic law, 261
Biological
  control, 34
  evolution, 11
  species concept, 16
Biomedical research, 329
Biotopes, 337
Bipinnaria, 326
  larva, 307
Biramous, 211
Biting and chewing type, 234
Bivalvia, 273
Bivium, 297
Black coral, 331
Bladder worm, 152
Blastea, 9
Blastocoel, 126, 306, 325
Blastoderm, 232
Blastomeres, 172
Blastopore, 79
Blastula, 232
*Blatta orientalis*, 233
Blood glands, 204
Book
  gills, 226
  lung, 6
Boring sponges, 81
Bowden's (1943) spiral wave action theory, 51
Brachiolaria larva, 306, 307
Brachiopoda, 142
Brackish water, 334
Bradybaena, 175
Branchial
  caeca, 300
  filaments, 285
  sac, 323
  skeleton, 320
Branchiogenital region, 320
Branchiostegites, 257
Bristles, 163
*Brugia*, 180
  *malayi*, 183
  *timori*, 185
Brugian filariasis, 183
Buccal
  commissure, 281
  diverticulum, 318
  field, 121
  ganglia, 281, 282
  receptors, 208
Budding, 25
Buds, 77
*Bugula*, 136
*Bulia*, 288
*Buthus*, 219

Calcareous
  cells, 277
  spines, 292
Calciferous glands, 199
Calcoblasts, 67
Calthrops, 69
Calycophorans, 89
Campaniform sensilla, 218
Canal
  of laurer, 147
  system, 77
Carapace, 221
Carcinoma, 176
Cardia, 138, 251
Cardiac stomach, 299
Cardo, 235
Carnivores, 71, 114
Cell rosettes, 112
Cephalization, 143
Cephalothorax, 223
Cercaria, 150
Cerci, 237
Cerebral commissure, 281
Cerebrobuccal connective, 281
Cerebropleural connective, 281
*C. fatigans*, 254
Chaetae, 190
*Chaetoderma*, 272, 289
Chaetodermomorpha, 272
*Chaetogaster annandalei*, 190
Chagas disease, 37
Chain saw movements, 276
Change of viscosity theory, 48
Cheilostomate ectoprocts, 137
Chelas, 69, 70
Chemical evolution, 11
Chemoprophylaxis, 35
Chemoreceptor, 125, 319
Chemosensory, 125
Chiastoneury, 287
*Chiton*, 268, 272, 289
Chloragocytes, 201
Chloragogen cells, 197
Chloroquine, 35
Choanocytes, 63, 66
Choanoflagellata, 9
Choanoflagellate protozoan, 63
Choanosome, 73
Chondroid, 320
*Choromytilus chorus*, 334
Chromidia, 23
Chromosomal aberration, 14
Chronic' amoebiasis, 44
Chronic exposure, 333
*Chrysops*, 185
Cibarium, 236
Cilia, 7
Cingulum, 121
Cipangopaludina, 175
Circumapical band, 121
Circumenteric connectives, 246
Circumfluence, 55
Circumoesaphageal
  commissure, 165
  connectives, 246
Circumoral ring, 305
Circumpharyngeal connective, 207
Circumvallation, 55
Cirrus, 8
Cites, 338
Clasping organs, 8

Classification, 15
*Cliona*, 76
Clitellar region, 195
Clitellum, 192, 210
Closed type of circulatory system, 6
Clypeus, 228, 234
Cnida, 82
Cnidocil, 93
Coagulocytes, 240
Coastal marshlands, 4
Cocoon, 210
Coelenteron, 82, 99, 115
Coeloblastula, 131
Coelom, 9, 191
Coelomate, 128
  ancestor, 9
Coelomic
  fluid, 197
  nerve system, 306
Coelomocytes, 294
Coelomoduct, 281
Coelosepta, 195
Coenosarc, 88
Coenosteum, 91
Cold stage, 34
Collagen, 68
Collar, 77
  cord, 324
Collencytes, 67
Colleterial glands, 241
Colloblast, 111
Collotheca, 121
Colon, 251
Colonial theory, 8
Colony, 24
Columella, 275
  lip, 275
Combined therapy, 174
Comb
  plates, 111
  rows, 115
Commercial extinction, 333
Commissure, 242, 281
Compensation sac, 138
Compound eyes, 218
Conchin, 268
Condyles, 216
Cone cells, 263
Congenital infection, 37
Congo trypanosomiasis, 37
Conjugation, 26, 58
  nucleus, 58
Connective, 281
  tissue, 5
Constriction furrow, 57
Contaminative route, 35
Contracaecum, 175
Contractile
  hypothesis, 93
  vacuole, 23
Coolie's itch, 174
Copepodid, 224
Copepods, 158
Copulation, 210
Copulatory
  bursa, 166
  spicules, 171
Coquillettidia crassipes, 181
Coracidia, 154
Coral fishing, 331
Corallines, 134
Corallite, 91
Corallium, 90
Corallum, 91
Coral reefs, 104, 328, 330
Coral sea, 105
Cornea, 263
Corneagen cells, 263
Coronal muscles, 92
Costa, 230
Coxa, 225
Coxal glands, 227
Coxopodite, 221
Crab culture, 337
Craspedote, 90
Crayfish, 337
Crepis, 69
Critical habitat, 335
Critically endangered, 190
Cross-fertilization, 209
Crustaceans, 221, 336
Crylaster, 69
Cryptocyst, 138
Ctenes, 111
Ctenidia, 268, 285
Ctenocephalides, 155
Ctenostomata, 134
Cubitoanal, 230
Cubitus, 230
*Cucumaria*, 308
*Culex*, 228
*C. fatigans*, 254
*C. quinquefasciatus*, 180, 254
Cutaneous leishmaniasis, 39
Cuticulin, 216
Cyclical transmission, 38
*Cyclops*, 158, 176, 188, 220, 224
Cyphonautes, 141
*Cypris*, 220
Cypris larva, 260
Cyst, 23, 152, 170
Cysticerci, 151
Cysticercoids, 155
Cysticercosis, 151, 153
*Cysticercus bovis*, 153
Cytocytes, 240
Cytogamy, 59
Cytological approach, 18
Cytopharynx, 49
Cytoproct, 25
Cytostome, 25

Dactylochirotida, 311
Dactylostyles, 91
Dactylozooids, 88, 101
Dall's code, 19
DDD, 257
DDT, 252
Decomposers, 329
Decorticated egg, 178
Definitive host, 151, 170
Deirids, 165
Dendrochiritida, 311
Dendrochirotacea, 311
Dengue haemorrhagic, 253
*Dentaliun*, 272
Denticles, 163
Deposit feeders, 333
Dermal branchiates, 297
Dermal ostia, 77
Dermis, 298
Desma, 69, 70
Deutocerebrum, 217
Dextral spiral, 275
Diaphragms, 239
Diastole, 239
Dictyonina, 63
Didelphic, 166
*Dientamoeba fragilis*, 180
Dimorphic, 45
  anaemia, 174
Dimorph stage, 77
Dinein, 50
Dioecious, 8, 95, 131, 162, 191
Dioptical region, 263
*Diphyllobothrium latum*, 151
Dipleurula, 308
Diploblastic, 64
  animals, 6
Diptera, 232
Discontinuity layers, 111
Ditaxic retrograde waves, 288
Diurnal periodicity, 185
Doliolaria larva, 308
Domains, 3
Domestic species, 255
Donax, 273
*Doris*, 273
Dorsal cirrus, 211
*Dracunculus*, 163
*Dracunculus medinensis*, 187
Dragonfly, 338
  nymph, 252
*Drosophila*, 17, 328
*Drosophila melanogaster*, 254

*Echinococcus*, 146
*E. granulosus*, 151, 156
Echinopluteus larva, 308
*Echinus*, 312
Ecological
  approach, 17
  isolation, 18
Ectoderm, 64
Ectodermal, 213
Ectoneural, 305
Ectosome, 73

Effective stroke, 52, 211
Efferent
  ctenidial vein, 280
  renal vein, 280
Egestion, 201
Egg, 250
Eiffelia, 63
Ejaculatory duct, 240
Electropotential theory, 53
Elephantiasis, 181, 183
Elytra, 229
Embryological approach, 18
Embryonation, 168
Embryonic nephridia, 213
Emigration, 337
Endemic, 38, 39, 329
Endite, 236
  projecting, 223
Endocuticle, 216
Endocytosis, 32
Endoderm, 64
Endogastric, 287
Endogenous, 170
Endogenous phase, 32
Endomixis, 26
Endoplasm, 43
Endopod, 225
Endoskeleton, 77, 293
*Enterobius vermicularis*, 179
Enterocoelous, 321
Enterocoel theory, 9
Enterocoely, 10
Enteronephric, 205
  nephridia, 205
Enteropneusts, 316
Entodermal cells, 66
Entolecithal, 144
Entomofauna, 328
Enzyme, 50
Eosinophilia, 174
Ephyra, 100
Epibranchial ridge, 323
Epicaridian larva, 260
Epicranium, 229, 234
Epidemiology, 181
Epidermal receptors, 208
Epifaunal, 128
Epimastigote, 35, 36
Epipharynx, 236
Epipodites, 257
Epistome, 134
Epitaenia, 280
Epochs, 12
Ericthus larva, 260
Erythrocytic schizogony, 32
Estuaries, 4
Euchlanis, 121
Eucone, 262
*Euglena*, 49
Euglenoid locomotion, 51
Eukaryotes, 2
Eulamellibranch, 286
*Eupagurus*, 223
*Euplectella*, 68
Eupyrene-sperms, 284
*Eurypterus*, 219
*Euspongia*, 65
Eutelic, 118
Evolution, 11
Evolutionary species, 16
Ex-conjugant, 58
Excretion, 78
Excretory duct, 249
Excurrent canals, 74
Excystment, 150
Exocone, 263
Exogastric, 287
Exogenous, 170
Exonephric, 204
Exopod, 225
Exoskeleton, 236
Exotic species, 328
Extinct, 14
Extinction, 334
Extracellular digestion, 270
Exumbrella, 94
Exuviae, 233
Eyes, 306

Facultative pathogen, 44
Feathery antennae, 229
Febrile paroxysm, 34
Feeder canals, 57
Feeding current, 130
Femur, 225
Fertilization, 210
Fertilized eggs, 179
Filariasis, 180
Filariform larva, 172
Filopodia, 46
Filter feeders, 71, 334
Fisheries, 328
Flagella, 49
Flagellar
  locomotion, 49
  roots, 53
Flagellated chambers, 64
Flame
  bulb, 125
  cell, 160, 165
Flatworms, 143
Flora, 337
Folk
  concept of species, 16
  medicine, 332
Food chains, 332
Freshwater habitat, 4
Frill, 256
Fringing reefs, 105
Fronto-lateral cilia, 285
Fulcrum, 122
Fungi, 2
Funiculus, 136
*Fusulina*, 45
Gaits, 217
*Gambusia affinis*, 257
Gametocytes, 33
Gamma taxonomy, 15
Gamonts, 25
Ganglia, 7
Garstang's view, 287
Gastrea, 9
Gastropoda, 272
Gastrovascular system, 6
Gastrozooid, 88, 101
Gastrula, 79, 306
Gastrulation, 79, 306
Gemmulation, 76
Gemmule, 76
  formation, 64
Genetical control, 34
Genetic differences, 13
Genital
  alae, 166
  atrium, 148
  operculum, 226
  papillae, 177, 209
  sinuses, 304
Geographical speciation, 14
Geological formations, 12
Geological time scale, 12
Geotaxis, 139
Germarium, 126
Germovitellaria, 126
Giant squid, 267
Gill, 297
  filaments, 285
  lamellae, 277
  roots, 258
Gills, 297
Gizzard, 138, 199
*Globigerina*, 45
*Globotruncana*, 45
Glossae, 236
Gnathobases, 225
*Gnathostoma spinigerum*, 176
Goblet cells, 196
Golgi cycle, 31
Gonad, 276
Gonangium, 89, 96
Gonochoric, 142, 324
Gonocoel theory, 9
Gonophores, 89
Gonopophyses, 231
Gonozooids, 101, 141
Gorgonin, 90
*Grantia*, 73
Granulocytes, 197
Gravid segments, 152
Gray's (1928) wave action theory, 51
Green glands, 249
Ground itch, 174
Guinea worm, 187
Gustatoreceptors, 208
Gymnolaemata, 134

Habitat
    destruction, 331
    management, 34
    modification, 334
Haemal channel, 293
*Haementeria ghilianii*, 191, 332
Haemocoel, 215, 217
Haemocyanin, 270
Haemocytes, 239
Haemoglobin, 174, 207, 270, 294
Haemolymph, 270
Haemopoietic system, 174
*Haliotis*, 272, 286, 334
*Haplochilus*, 257
Hatcheries, 337
Hatching, 178
Head appendages, 257
Head-foot, 268
Health promotion, 44
Heart-kidney complex, 320
Heavy exploitation, 339
Heliaster, 296
Helicoplacoidea, 311
Hematophagous, 252
*Hentkenina*, 45
Hepatic caeca, 300
Hepatopancreas, 277
Herbicides, 338
Herbivory, 268
Hereditary variation, 59
Hermaphroditic, 8
Hermatypic, 104
Heteractinellida, 63
Heterogonic, 170
Heteroinfection, 153
Heteropia, 73
Heterotrophic, 25
Heteroxenous, 167
Heterozooids, 135
Hexactinellida, 63
*Hippa*, 222
Hirudinaria, 193
Holoblastic, 126
Holometabola, 229
Holonephridia, 205
Holotrichus, 51
Holozoic, 54, 78
Homogonic, 170
Homonym, 19
Hookworm disease, 171
Horseshoe crabs, 224
Hot stage, 34
Hydatid disease, 156
Hydatidosis, 156
Hydrocaulus, 88
Hydrorhiza, 94
Hydrostatic
    pressure, 198
    skeleton, 245
Hygiene, 179
Hymenolepiasis, 154
Hymenoptera, 232
Hypnozoites, 32
Hypobranchial glands, 283
Hyponeural
    nervous system, 305
    sinus, 304
Hypopharynx, 228, 236

Ileum, 251
Imaginal tissues, 233
Immigration, 337
Immune
    response, 174
    system, 183
Import, 55
Incidence, 40
Incubation period, 32, 41, 175
Incus, 122
Indirect metamorphosis, 256
Infaunal, 128
Infective, 170
    larvae, 185
Infrabranchial chamber, 285
Inframarginal ossicles, 293
Ingestion, 55
Inhalant chamber, 285
Insecticides, 338
Insects, 328
    vector, 37
Integumentary nephridia, 204
Interambulacral axis, 292
Interbreeding, 12
Intermediate hosts, 152
Intermoult period, 233
International code, 19
International Commission of Zoological
        Nomenclature, 19
International Congress of Zoology, 19
Interradial axis, 292
Interradii, 296
Interstitial
    cells, 87
    fluids, 303
Intestinal amoebiasis, 43
Intracellular digestion, 5, 55, 277
Intra-cellular parasites, 186
Introduced plants, 338
Introvert, 136
Invagination, 55
Inversion, 79
Invertebrata, 3
Invertebrate phase, 32
Ircinia, 70
Iris pigment, 264
    sheath, 264
Isochelas, 69
Iucn red list, 190

Japanese encephalitis, 255
Jellyfish, 115
Jugal, 230
Juveniles, 337

Kala-azar, 41
Karyosome, 22, 25
Kenozooids, 134
Kentrogen larva, 260
Kidney, 240, 270, 276
Kinetodesmal region, 52
Kinetodesmata, 52, 212
Kinetoplast, 23
Kinetosomes, 52
Knapsack sprayer, 255
Krill, 336
K/T boundary, 15

Labella, 251
Labial palpi, 228
Labium, 228, 251
Labyrinth, 250
Lacinia, 235
Lacunae, 238
Lagoon, 105
Landscapes, 337
Land use, 338
Lange's nerve, 306
Larva, 76, 250
    migrans, 176
Lateral
    galea, 235
    hearts, 203
    line, 164
Latero-oesophageal hearts, 203
Law of priority, 19
*L. braziliensis*, 39
Leeches, 332
Leishman-donovan bodies, 40
*Leishmania*, 39, 40
Leishmaniasis, 40
*Lepas*, 220
*Lepisma*, 221, 222
Leucilla, 72, 74
Leucocytes, 198
Leuconoid type, 73
Leucon type, 72
*Leucosolenia*, 72
Limax, 46
*Limulus*, 219, 224, 225, 227, 335
*L. polyphemus*, 224
Lingua, 236
Lithistida, 63
Littoral zone, 119
*Loa loa*, 185
Lobopodia, 46
Lobopods, 243
Lobster fisheries, 336
Local depletions, 332
Localized contraction theory, 53
Locks, 216
*Loligo*, 274
Long-standing disease, 185
Lophocytes, 67
Lophophoral
    cavity, 131
    ring, 138

Lophophore, 128
Lophopus, 140
Lorica, 119
*L. tropica*, 41
*Lumbricus*, 197
*Lumbrineris brevicirra*, 332
*Lutzomyia*, 39
*Lymnaea bulimoide*, 149
Lymphatic
  filariasis, 181
  tissues, 38
Lymph glands, 198

Macrogamete, 34
Macrogametocytes, 23
Macrophages, 40
Macrotaxonomy, 15
Madreporite, 297
  plate, 297
Malaria, 31
Malayan filariasis, 185
Male
  gametes, 34
  gonopore, 240
Mallei, 123
Malleus, 122
Malpighian tubules, 7, 217, 240
Mandibles, 228
Mangroves, 329
*Mansonella*, 186
*A. perstans*, 187
*Mansonia*, 180
*A. uniformis*, 182
*Mantis*, 222
Mantle cavity, 267, 268
Manubrium, 90
Marginal, 230
  sinuses, 304
Marine
  eco-system, 330
  habitat, 4
Mass extinction, 14
Mastax, 118, 122
Mastigoneme, 49
Mastigophora, 49
Maxillae, 228, 236
Maxillary, 228
  palp, 235, 236
Mechanical transmission, 38
Mechanoreceptors, 165
Media, 230
Mediocubital, 230
Medusae, 90
Megalopa larva, 260
Megalospheric phase, 45
Megascleres, 68
*Megascolides australis*, 191
Memory, 331
Merozoites, 32
Mesenchyme, 74
Mesenterial filament, 88
Mesenteries, 83
Mesenteron, 237, 251
Mesocoel, 130
Mesoderm, 162
Mesoglea, 84
Mesonotum, 236
Mesosome, 128, 316
Mesosternum, 236
Mesothorax, 229
Mesozoan, 9
Metacercaria, 150
Metachronal, 217
  rhythmic movement, 52
  waves, 52
Metacoel, 136
Metacyclic trypanosomes, 35
Metagenesis, 97
Metameres, 195
Metamorphosis, 116, 232, 306, 325
Metanauplius, 224
  larva, 260
Metanephridia, 128, 131, 204, 213, 271
Metanephridial system, 7
Metanotum, 236
Metasome, 128, 316
Metasternum, 236
Metathorax, 229
Metazoans, 3
Metazoea larva, 260
*Michrochaetus*
  *michrochaetus*, 190
  *rappi*, 190
Microfilaraemia, 181
Microfilariae, 181
Microgamete, 34
Microgametocyte, 23
Micronucleus, 58
Microscolex, 332
Microspheric phase, 45
Microstrongyles, 69
Microtaxonomy, 15
Microtubules, 38, 53
Migration, 167
Mineral skeleton, 68
Miners anaemia, 171
Miracidium, 149
Mixocoel, 244
Mixonephrium, 213
Molecular clock, 3
Molpadida, 311
Molting, 215
Monaxon, 69
Monaxon spicules, 69
Monocondylic, 237
Monocrepid, 69
Monocultures, 334
Monodelphic, 166
Monodonta, 288
Monoecious, 8
Monopectinate gill, 277
Monophyletic, 292
Monoplacophora, 271
Monotaxic, 288
Monoxenous, 167
Morphological approach, 17
Morton's view, 287
Mosaic visions, 263
Mosquitoes, 250
Motorium, 52
Moulting, 172, 178
Moulting fluid, 233
Mouth papilla, 297
Mouthparts, 250
Mucocutaneous leishmaniasis, 39
Mucocytes, 198
Multiple fission, 25
*Musca*, 221
Myocytes, 67
Myofilaments, 212
Myonemes, 86
Myosin, 48, 212
Mysis, 220
  larva, 260
*Mytilus*, 273

*Natica*, 287
Natural
  classification, 20
  predators, 253
  selection, 11
  threats, 337
Nauplius
  eye, 224
  larva, 259
Nautilus, 270, 288
*Necator americanus*, 170
Nectonema, 163
Nematocyst batteries, 93
Nematocysts, 82
Nematoda, 163
*Neomenia*, 272, 289
Neomeniomorpha, 272
Nephridia, 193, 270
Nephridiopore, 125, 213, 306
Nephridiostome, 205
Nephridium, 217
Nephroblast, 213
Nephrocoel theory, 9
Nephromyxia, 213
Nephromyxium, 213
Nephrostome, 7, 204, 213
*Nereis*, 192, 210, 211
Neurochord, 319
Neurocysticercosis, 152
Neuroid fibre theory, 53
Neuropodium, 211
Neurosensory cells, 125, 306
Nexin, 53
Nocturnal periodicity, 182
Nodulus, 195
Nomenclature, 15
Nominalistic species concept, 16
Noncommercial crustaceans, 337
Noninvasive infection, 44
Non-pathogenic, 42
Notarchus, 288

Notoneuron, 10
Notopodium, 211
Nuclear
 column, 182
 reorganization, 59
*Nucula*, 273
Nummulites, 46
Nutrient delivery system, 320
Nymphs, 232

*Obelia*, 95
Obligatory asexual populations, 16
Oceanic temperatures, 4
Ocelli, 218, 306
Ocellus, 99
*Octopus*, 274
Odontoblasts, 276
Odontophore, 276
Oesophageal gland, 171
Olfactory setae, 262
Oligopyrene-sperms, 284
Olynthus, 72
Olynthus stage, 79
Ommatidia, 218
Omnivorous, 201
*Onchocerca volvulus*, 186
Onchocerciasis, 186
Oncosphere, 151, 152
*Oniscus*, 220, 222
Ontogeny, 261
Onychophora, 248
Oocysts, 34
Ooecium, 141
Ookinete, 34
Ootheca formation, 210
Ootype, 148
Ophiopluteus larva, 308
Opisthorchiasis, 157
*Opisthorchis*, 146
Oral
 ciliature, 51
 haemal ring, 304
 ring sinus, 304
 surface, 292
Organ
 of bojanus, 280
 for respiration, 6
 system, 5
Ornamental value, 330
Orthoptera, 232
*Oscarella*, 72, 74
Oscula, 71
Osculum, 77
Osmoconformers, 165
Osmolarity, 303
Osmoregulation, 55, 78, 160
Osmoregulatory, 207
Osmotic hypothesis, 93
Osphradium, 282
Ossicles, 292
Ostia, 217, 246
Otolith, 8
Ovaries, 241
Over-exploitation, 331
Overgrown, 232
Ovicell, 141
Oviducts, 241
Oviparous, 126, 166
Oviposition, 168, 254
Ovoviviparous, 166
Oxeas, 68, 69
Oxyhaemoglobin, 207

*Palaemon*, 257, 263
Palaeolimnological, 329
Pallial
 cavity, 268
 complex, 275
Pallium, 268
Palmate hairs, 256
*P. americana*, 233
Pantonematic, 54
Papillae, 243
Papulae, 297, 305
Parabasal body, 53
Paraglossae, 236
Parallel speciation, 17
Paramoecium, 26, 51, 53
Parapodia, 191
Parapodium, 210
Parasite, 43
Parasitic, 163
 forms, 23
 nematode, 165
 rotifers, 119
Parasitism, 167
Parastrongylus, 175
Paratenic hosts, 158
Parietal ganglia, 271
Paromomycin, 42
Parthenogenesis, 126
Passive dispersal, 255
*Patella*, 225, 272, 286
Pathogenesis, 183
Pathogenic, 42
Pathogens, 177
*P. aurelia*, 26
*P. caudatum*, 26
Pedal ganglia, 271, 281
Pedicel, 234
Pedicellariae, 292
Pedipalp, 225
Pelagic, 111, 333
Penis sheath, 283
Pentaceros, 296
Pentacrinoid larva, 308
Pentacrinus, 308
Pentacronematic, 54
Pentactula concept, 310
Pentamerous, 295
Penultimate whorl, 275
Perforate, 275
Peribranchial sinuses, 304
Pericardial
 cells, 246
 sinus, 239
Pericardium, 276
Pericoelomic vesicles, 310
Perihaemal system, 303
Perineurium, 207
Peripatidae, 243
Peripatopsidae, 243
*Peripatus*, 219, 243, 248
Peripheral blood, 182
*Periplaneta americana*, 233
Peristaltic activity, 44
Peristome, 275
Peristomium, 190
Peritrophic membrane, 251
Perivisceral
 cavity, 197
 coelom, 299
Permanent nephridia, 213
Pervisceral coelom, 293
Petaloid pseudopodia, 299
pH, 4
Phagocytic cells, 270
Phaosome, 208
Pharyngeal nephridia, 204
Phasmids, 163, 165
Phenetic, 20
Phenetic (morphological) species concept, 17
*Pheretima*, 193
Pheromones, 244
*Phlebotomus*, 39
Phokanema, 175
*Pholas*, 273
*Phoronis*, 131
Photoreceptors, 208
Phototactic, 139
*Phyllodoce*, 206
Phyllosoma larva, 260
Phylogenetic
 relationship, 19
 species concepts, 17
Phylogeny, 20, 261
*Physalia*, 89
Physonectans, 89
Pigment granules, 33
*Pila globosa*, 283
Pilose, 250
Pinacocytes, 66, 67
Pinacoderm, 66
Pinocytosis, 25
Pistia, 274
Pistillum, 102
Placids, 162
Placozoa, 9
Plakula, 9
*Planaria*, 146
Planktonic
 larvae, 328
 rotifer, 119

Planorbis, 273
Planula, 96, 97
Plasma
  gel, 48
  sol, 48
Plasmatocytes, 240
Plastids, 23
Platoon spicules, 69
Pleura, 216, 229, 237
Pleural ganglia, 271
Pleurites, 236
*Pleurobrachia*, 115
Pleurobranch, 258
Pleuropedal, 281
Plumose, 250
Pneumatophore, 89, 101
Pneumonitis, 179
Podia, 293, 297
Podites, 216
Podobranch, 257
Podura, 221
*Poecilia reticulata*, 257
Poikilothermal, 194
Polian vesicles, 302
Pollution, 329
Polychaeta, 190
Polyembryony, 139, 215
Polymorphic, 38
Polymorphism, 101
*Polynoe*, 192
Polyphyletic, 292
  group, 12
Polypide, 134
Polyplacophora, 272
Polyploids, 14
Polyploidy, 14
Polypoid symmetry, 88
Polyps, 87
Polyspermy, 113
Pond culture, 337
Porocytes, 67
Post-clitellar region, 195
Postmating mechanisms, 18
Post-tarsus, 227
Post-typhlosolar region, 200
Pouch of receptaculum, 284
Precambrian, 12, 267
Pre-clitellar region, 195
Precyst, 43
Predation, 268
Premating mechanisms, 18
Premunition, 23
Preparatory stroke, 211
Pretarsus, 237
Pre-typhlosolar region, 200
Prevalence, 42
Primaquine, 35
Primary
  exo-erythrocytic schizont, 32
  urine, 217, 247
Principal reservoirs, 43
Principle of nomenclature, 18
Proboscis
  coelom, 320
  skeleton, 317
*Procambarus clarkii*, 337
Procercoid, 154
Proctodeal rectum, 138
Procuticle, 216
Prodigal canals, 301
Proglottids, 151
Pro-haemocytes, 239
Prokaryotes, 2
Promastigote, 36
Pronotum, 236
Prophylaxis, 35
Prosobranchia, 272
Prosome, 316
Prosopyles, 72
Prostate glands, 209
Prosternum, 236
Prostomium, 190
Prothorax, 229
Protista, 2, 12
Protobranchia, 273
Protocerebrum, 217
Protofilaments, 49
Protonephridia, 125, 163
Protonephridial system, 7
Protonephridium, 160
Protonephromyxia, 213
Protonephromyxium, 213
Protopodite, 216, 221
Protospongia, 63
Prototroch, 285
Protozoan shell, 22
Protozoea larva, 260
Proventriculus, 251
Pseudocoel, 122, 163
Pseudocoelomates, 162
Pseudocoelomocytes, 165
Pseudo cone, 262
Pseudofaeces, 270
Pseudopodia, 46, 275
Pseudoterranova, 176
*Pseudoterranova osculates*, 175
Pseudotracheae, 217
Pterobranchs, 316
Ptychocysts, 93
Pulmonary
  abscesses, 43
  vein, 280
Pupa, 250, 253
Pupation, 256
Pygidium, 190
Pygochord, 321
Pyloric
  caeca, 299, 300
  stomach, 300
Pyrethrum, 255

Quadrinucleate amoeba, 42
*Quinqueloculina*, 45

Radial
  canal, 73, 83
  haemal sinuses, 304
  nerve, 305
  perihaemal sinuses, 304
Radially symmetrical, 292
Radiata, 292
Radio-medial, 230
Radius, 230, 292
Radula, 268
Ramate mastax, 124
Raptorial feeding, 124
Rat nematode, 175
Recapitulation theory, 261
Receptaculum seminis, 8, 284
Recovery, 44
  stroke, 52, 217, 295
Rectal caeca, 299
Rectum, 251
Redia, 150
Reef edge, 105
Regeneration, 77, 296
Rejuvenation, 59
Remigium, 230
Renal chamber, 279
Renette cells, 164
Replacing cells, 196
Replicating molecule, 11
Reptant, 216
Respiration, 6
Respiratory trumpets, 253
Reticulate skeleton, 293
Reticulopodia, 46
Retinal pigment sheath, 264
Retinella, 208
Retinula cells, 263
Retrocerebral organ, 125
Rhabdites, 144
Rhabditiform larvae, 170, 172
Rhabdome, 263
Rhabdomeres, 218, 264
Rhopalia, 98
Rhopalial
  canal, 99
  ganglion, 99
Rhythmic behaviour, 182
Right handed, 275
Ring canals, 301
River blindness, 186
RNA world hypothesis, 11
Romana's sign, 37
Rostrum, 121
Rotalidae, 45

*Saccamina*, 45
*Sacculina*, 220, 259
Salivarian, 37
Salivarium, 236
Salivary reservoir, 246
Saprophytic, 24
Scalids, 162

Scape, 234
Scaphognathite, 258
Scaphopoda, 272
Scavengers, 335
*Schistosoma*, 146
Schistosomes, 159
Schizocoelous, 191
Schizocoely, 9
Schizogony, 23
Schizonticidal drugs, 35
Schizonts, 32
Schwargerina, 45
Sclerites, 68, 216, 228
Scleroblasts, 68, 90
Sclerocytes, 67
Scolex, 145
*Scypha*, 63, 72, 76
Sea
  anaemones, 92
  cucumbers, 338
Seaward flat, 105
Secondary
  amoebiasis, 43
  gill bar, 323
Second maxilla, 236
Secretory cells, 277
Sectorial, 230
Segmental ganglia, 207
Seminal
  receptacle, 210, 240
  vesicles, 209, 240
Sensilla, 218
Sensory
  cells, 246
  papillae, 246
*Sepia*, 274
Septal
  nephridia, 204
  pouches, 209
Sessile pedicellariae, 298
Seta, 197
Sex pheromones, 165
Sexual dimorphism, 181
Sheep liver fluke, 146
Shell, 45
Shrimp fisheries, 336
Sibling species, 17
Sigaretus, 287
Sigmas, 70
Sigmaspire, 69
Signet ring, 32
Silicoblasts, 67
*Simulium*, 186
Sinuses, 238
Siphonoglyphs, 88
Siphonophores, 101
Skeleton, 294
Sleeping sickness, 37
Sliding filament model, 53
Slime glands, 244
Slit sensilla, 218
Slow creeping locomotion, 211
Soil fertility, 332
Soil-pollution, 179
Soil-transmitted nematodes, 167
Solaster, 296
Solenia, 88
Sol gel, 48
Somatic
  cells, 181
  ciliature, 51
  layer, 197
  type, 179
South American trypanosomiasis, 35
Spadix, 97
Sparganum, 154
Speciation, 12
Species, 15
  conservation, 328
  diversity, 329
Spermatheca, 192, 210
Spermatocyte, 78
Spermatogonia, 209
Spermatogonium, 78
Spermatophore, 240
Spermatozoa, 78, 209
Spheres, 69
Spherulous, 68
Spiculin, 68
Spiders, 329
Spiracles, 227
Spires, 69
Spirocysts, 93
Splanchnic layer, 197
Sponge fishery, 330
Sponges, 62, 329
*Spongilla*, 65, 74
Spongin, 62
Spongocoel, 64
Spongocytes, 67
Sporocyst, 23, 149, 159
Sporogony, 23, 34
Sporonts, 25
Sporozoites, 32, 34
*Squilla*, 220
Stadium, 233
Starfish, 295
Statoconia, 283
Statocysts, 94, 283
Statolith, 99, 112
Stellate cells, 165
Stenolaemata, 134
*Stentor*, 329
Stercocarian, 35
Stereoblastula, 126
Stereogastrula, 97
Sternite, 237
Sternum, 229
Sterrasters, 69
Stichonematic, 49, 54
Stigma, 259
Stipes, 235
Stolons, 317
Stomoblastula, 79
Stomochord, 321
Stomodaeum, 223
Stone canal, 301
Stotocysts, 262
Streptasters, 69
Stride, 212
Strobila, 100
Strobilation, 100
Strobilization, 145
Strongylaster, 69
*Strongyloides stercoralis*, 170
Styles, 69
Stylifer, 288
Styongyles, 69
Subcosta, 230
Subepidermal nerve complex, 305
Subfibril
  A, 49
  B, 49
Submarginal, 230
Submolar region, 235
Subperiodic strains, 180
Sub-pharyngeal ganglia, 207
Subpopulations, 14
Subradular organ, 276
Subumbrella, 94
Suctoria, 24
Suctorial apparatus, 257
Summer cycles, 126
Superposition image, 263
Suprabranchial chamber, 285, 286
Supraintestinal ganglion, 281
Supramarginal spines, 293
Suspension
  feeders, 122, 318
  feeding, 268
Sutures, 275
Sweating stage, 34
Syconoid type, 73
Symbiotic relationship, 81
Sympatric speciation, 14
Synanthropic, 233
Syncytial theory, 8
Syncytium, 118
Syngamy, 25
Synkaryon, 58
Synthetic fertilizers, 329
Systolic condition, 57

Tabulae, 91
Tactile organs, 262
Taeniasis, 153
Tagmata, 224, 243
Tango-receptors, 208
Tapeworms, 151
Tarantula, 335
Tarranova, 175
Tarsus, 225
Taxodonta, 273
Taxonomic
  categories, 20
  hierarchy, 20
Taxonomists, 17
Taxonomy, 15

*T. cruzi*, 37
Tegmina, 229
Telopod, 216
Telson, 224
Temporal isolation, 18
Tension hypothesis, 93
Tentacles, 114, 115, 271 , 283
Tentaculocysts, 84
Tentillae, 111
Teratological effects, 338
Terga, 239
Tergite, 237
Terminal ganglion, 242
Terrestrial
  habitat, 4
  invertebrates, 328
Testis sacs, 209
Tests, 45
*Tethya*, 63, 76
Tetracrepid, 69
Tetracts, 68
Tetraxons, 68
Tetraxon spicules, 69
Thermal pollution, 331
Thoracic ganglia, 242
Threadworms, 163
Threatened, 332
Tibia, 225
Tiedemann's bodies, 301
Tissues, 5
Tongue, 236
  bar, 318, 321
Tornaria, 133, 325
Tornotes, 69
Torsion, 285, 286
Touch receptors, 319
Toxas, 70
Toxopneustes, 294
*Toxorhynchites*, 251, 253
Tracheae, 6, 227
Tracheal type, 179
Tracheoles, 231
Traditional epidemiology, 35
Transcursion fauna, 119
Transitional haemocytes, 239
Transplacental route, 172
Transverse binary fission, 57
Triacnes, 69
Triacts, 68
*Trichinella spiralis*, 168
Trichinelosis, 168
Trichinosis, 168
*Trichomonas*, 49
Trichoplax, 9
Trichoptera, 232
*Trichuris trichiura*, 167
Tricrepid, 69
Trilobite larva, 227
Triploblastic, 191
Triploid, 14
Trivium, 297
Trochaea, 10
Trochanter, 225
Trochophore, 326
Trochosphere, 285
Trochus, 121, 333
Trophocyte, 78
Trophozoite, 23, 32
Tropical reefs, 329
Trunk coelom, 128
Trypomastigotes, 35
Tsetse flies, 37
Tube feet, 293, 300
Tubercles, 294
Tuberculatus, 334
*Tubifex*, 190
Tylasters, 69
Tylostyle, 69, 70
Tylotes, 69
Typhlosolar region, 200
Typhlosole, 200
Typological species, 15

Umbellula, 89
Undulating membrane, 23
Unicinate mastax, 123
Unilocular, 275
*Unio*, 273, 285
Uniramous, 227
Ureotelic, 206
Ureter, 249
Uterus, 166

Vaccine candidates, 40
*Vallisneria*, 274
*Vampyroteuthis*, 274
Vasa deferentia, 209
Vas deferens, 283
Vectors, 328
Vegetative macronucleus, 58
Velarium, 98
*Velella*, 89
Veliger, 285
Velum, 94, 285
Ventral
  gland, 165
  phallomere, 240
Ventricles, 6
Vesicular seminalis, 148, 247
Visceral
  leishmaniasis, 39
  lump, 268
  nervous system, 306
  segments, 237
Vitellaria, 8
Vitellarium, 126
Vittellaria, 157
V-spot, 182
Vulnerable, 330
Vulvar
  opening, 171
  waist, 177

Water-current system, 62
Water
  tubes, 285
  vascular (ambulacral) system, 300
  vascular system, 292
Wheel animalcules, 118
Winterbottoms sign, 38
*Wolbachia*, 182
Wriggler emerges, 256
*Wuchereria*, 180
*Wuchereria bancrofti*, 181

*Xiphophorus conchianus*, 14
*X. variatus*, 14

Yolk
  cells, 232
  larva, 308
  reservoir, 148

Zoea, 224
  larva, 260
Zoecia, 137
Zoecium, 134, 136
Zonites, 162
Zooids, 134, 316
Zoological nomenclature, 18
Zoology, 1
Zoonosis, 169, 175
Zoonotic
  cycle, 181
  species, 37
Zoophilism, 256
Zoophyte, 62, 134
Zooplankton, 270
Zootrophic, 54
Zygoneury, 281
Zygote, 34, 78